9th International Conference on Fibre Reinforced Concretes, Textile Reinforced Concretes and Ultra-High Performance Concretes (FIBRE CONCRETE 2017)

IOP Conference Series: Materials Science and Engineering
Volume 246

Prague, Czech Republic
13 - 16 September 2017

Editors:

Iva Broukalova
Michaela Frantova
Alena Kohoutkova
Ippei Maruyama
Petr Stemberk
Michaela Vaitova

ISBN: 978-1-5108-4994-5
ISSN: 1757-8981

Printed from e-media with permission by:

Curran Associates, Inc.
57 Morehouse Lane
Red Hook, NY 12571

Some format issues inherent in the e-media version may also appear in this print version.

Printed by Curran Associates, Inc. (2017)

For permission requests, please contact the Institute of Physics
at the address below.

Institute of Physics
Dirac House, Temple Back
Bristol BS1 6BE UK

Phone: 44 1 17 929 7481
Fax: 44 1 17 920 0979

techtracking@iop.org

Additional copies of this publication are available from:

Curran Associates, Inc.
57 Morehouse Lane
Red Hook, NY 12571 USA
Phone: 845-758-0400
Fax: 845-758-2633
Email: curran@proceedings.com
Web: www.proceedings.com

Table of contents

Volume 246

FIBRE CONCRETE 2017
13–16 September 2017, Prague, Czech Republic

Accepted papers received: 22 September 2017
Published online: 16 October 2017

Preface

011001
OPEN ACCESS
FIBRE CONCRETE 2017

011002
OPEN ACCESS
Committees

011003
OPEN ACCESS
Conference participants

011004
OPEN ACCESS
List of Conference Partners

011005
OPEN ACCESS
Peer review statement

Papers

012001
OPEN ACCESS
Structural Applications of Fibre Reinforced Concrete in the Czech Republic

A Kohoutková and I Broukalová.....1

012002
OPEN ACCESS
Glassfibre Reinforced Concrete: a Review

P J M Bartos.....10

012003
OPEN ACCESS
Replacement of minimum steel bar reinforcement with steel fibres in structural concrete members

F Junker, K Holschemacher, T Müller and H Kieslich.....32

012004
OPEN ACCESS
Probabilistic design of fibre concrete structures

R Pukl, D Novák, T Sajdlová, D Lehký, J Červenka and V Červenka.....41

012005
OPEN ACCESS
Comparison of experimental and theoretical data for steel fiber reinforced concrete slabs on ground under a point load according to design codes of the Republic of Belarus and the Russian Federation

A Valodzin.....51

012006
OPEN ACCESS
Dramix® Steel fibres in residential foundation slabs in Czech Republic: design approach

S Pouillon.....56

012007
OPEN ACCESS
On the residual properties of damaged FRC

R Zerbino, M C Torrijos and G Giaccio.....66

012008
OPEN ACCESS
Crack control in concrete members reinforced by conventional rebars and steel fibers

G Tiberti, I Trabucchi, M AlHamaydeh, F Minelli and G Plizzari.....76

012009
OPEN ACCESS
Production of footbridge with double curvature made of UHPC

J Kolísko, D Čítek, P Tej and M Rydval.....86

012010
OPEN ACCESS
Application of BCN test for controlling fiber reinforced shotcrete in tunnelling works in Chile

S Carmona and C Molins.....94

012011
OPEN ACCESS
Influence of steel fibres on the blast response of normal-strength and high-strength reinforced concrete columns

A Hammoud and H Aoude.....102

012012
OPEN ACCESS
Influence of mixing procedure on mechanical properties of high-performance concrete

J Fládr and P Bílý.....109

012013
OPEN ACCESS
Utilisation of Fibre Reinforced Sprayed Concrete for Primary Lining of the Považský Chlmec Tunnel

M Baláž, M Uhrin, P Růžička and M Hilar.....114

012014
OPEN ACCESS
Recent development in blast performance of fiber-reinforced concrete

R Hajek, M Foglar and A Kohoutkova.....123

012015
OPEN ACCESS
Effects of aggregate grading on the properties of steel fibre-reinforced concrete

M Acikgens Ulas, K E Alyamac and Z C Ulucan.....134

012016
OPEN ACCESS
Non-Destructive Testing of Steel Fibre Reinforced Concrete

D Corbett and G Tronca.....144

012017
OPEN ACCESS
Fibre reinforced concrete in flexure and single fibre pull-out test: a correlation

M Manca, D Ciancio and P Dight.....152

012018
OPEN ACCESS
Micromechanical performance of interfacial transition zone in fiber-reinforced cement matrix

V Zacharda, J Němeček and P Štemberk.....160

012019
OPEN ACCESS
Vibrated and self-compacting fibre reinforced concrete: experimental investigation on the fibre orientation

A Conforti, G A Plizzari and R Zerbino.....170

012020
OPEN ACCESS
Reactive powder concrete reinforced with steel fibres exposed to high temperatures

T Kh Alrekabi, V M C F Cunha and J A O Barros.....180

012021
OPEN ACCESS
Flexural stiffness of the composite steel and fibre-reinforced concrete circular hollow section column

A Tretyakov, I Tkalenko, F Wald, J Novak, R Stefan and A Kohoutková.....190

012022
OPEN ACCESS
Multi-Scale Pull-Out Resistance of Steel Reinforcing Bar Embedded in Hybrid Fiber Reinforced Concrete (HYFRC)

A Lin and C P Ostertag.....197

012023
OPEN ACCESS
Application of UHPC thin-walled elements for multi-purpose use table

O Slabý, J Vašková and V Veselý.....207

012024
OPEN ACCESS
Quantification of cracking localization in fibre-reinforced concrete beams

S Werzbrger, Y S Karinski and A N Dancygier.....213

012025
OPEN ACCESS
Punching Shear in Steel Fibre Reinforced Concrete Slabs Without Traditional Reinforcement

K H Tan and A Venkateshwaran.....219

012026
OPEN ACCESS
Seismic Behaviour of Synthetic-Frc Columns

P Paultre and R Eid.....228

012027
OPEN ACCESS
The effect of the length of macro synthetic fibres on their performance in concrete

K P Juhász and V Kis.....236

012028
OPEN ACCESS
Creep behaviour of macro glass fibre reinforced concrete beams

P Löber, B Heiden and K Holschemacher.....244

012029
OPEN ACCESS
Application of trilinear softening functions based on a cohesive crack approach to the simulation of the fracture behaviour of fibre reinforced cementitious materials.

A Enfedaque, M G Alberti and J C Gálvez.....254

012030
OPEN ACCESS
How to predict the orientation factor of non-rigid macro-synthetic fibre reinforced concrete

M G Alberti, A Enfedaque and J C Gálvez.....262

012031
OPEN ACCESS
Effect of Different Matrix Compositions and Micro Steel Fibers on Tensile Behavior of Textile Reinforced Concrete

J Esmaeili, I Sharifi, K Andalibi and J Kasaei.....272

012032
OPEN ACCESS
Influence of steel fibers on the shear and flexural performance of high-strength concrete beams tested under blast loads

O Algassem, Y Li and H Aoude.....280

012033
OPEN ACCESS
Characterization of polymer concrete with natural fibers

M Barbuta, A A Serbanoiu, R Teodorescu, B Rosca, R Mitroi and G Bejan.....288

012034
OPEN ACCESS
Feasibility of using fibrous waste in cement-based material

M Beddar, A Meddah and L Belagraa.....294

012035
OPEN ACCESS
Probabilistic Analysis of Structural Member from Recycled Aggregate Concrete

I Broukalová and K Šeps.....306

012036
OPEN ACCESS
Studies for understanding effects of additions on the strength of cement concrete

R D Bucur, M Barbuta, P Konvalina, A A Serbanoiu and J Bernas.....314

012037
OPEN ACCESS
The influence of concrete mixture's rheological properties on the quality of formed concrete surfaces

M Daukšys, A Klovas and L Venčkauskas.....320

012038
OPEN ACCESS
Sustainability and durability analysis of reinforced concrete structures

A Horáková, I Broukalová, A Kohoutková and J Vašková.....329

012039
OPEN ACCESS
Evaluation of Blast Resistance of Fiber Reinforced Composite Specimens under Contact Blast Load

O Janota and M Foglar.....339

012040
OPEN ACCESS
Effect of the loading rate on fibre reinforced concrete beams

K P Juhász, P Schaul and L Nagy.....349

012041
OPEN ACCESS
Rigid-body-spring model numerical analysis of joint performance of engineered cementitious composites and concrete

Y Khmurovska, P Štemberk and V Křístek.....356

012042
OPEN ACCESS
Design of footbridge with double curvature made of UHPC

P Kněž, P Tej, D Čítek and J Kolísko.....362

012043
OPEN ACCESS
UHPC panels utilized as permanent formwork of in-situ cast reinforced concrete deck bridges

J Kolísko, J Čech, P Tej and P Kněž.....367

012044
OPEN ACCESS
On the effect of addition of carbon nanotubes on the electric conductivity of alkali-activated slag mortars

I Kusak and M Lunak.....372

012045
OPEN ACCESS
Fibre reinforced concrete exposed to elevated temperature

J Novák and A Kohoutková.....380

012046
OPEN ACCESS
Two innovative solutions based on fibre concrete blocks designed for building substructure

J Pazderka and P Hájek.....387

012047
OPEN ACCESS
Composite Material from By-products and Its Properties

K Šeps, I Broukalová and J Vodička.....394

012048
OPEN ACCESS
Research of UHPC properties prepared with industrial mixer

E Šerelis, V Vaitkevičius and V Kerševičius.....400

012049
OPEN ACCESS
Realization of FRC interior and exterior furniture

Š Šonka, M Frantová, P Štemberk, J Havrda and P Janouch.....409

012050
OPEN ACCESS
Heat transfer in hybrid fibre reinforced concrete-steel composite column exposed to a gas-fired radiant heater

R Štefan, J Procházka, J Novák, J Fládr, F Wald, A Kohoutková, L Scheinherrová and M Čáchová.....414

012051
OPEN ACCESS
The Overall Research Results of Prestressed I-beams Made of Ultra-high Performance Concrete

P Tej, J Kolísko, P Kněž and J Čech.....424

012052
OPEN ACCESS
Tensile properties of structural fibre reinforced concrete

M Tipka and J Vašková.....433

012053
OPEN ACCESS
Influence of Basalt Mesh Induced Increase of Heterogeneity of Cement Composites with Dispersed Fibers on Its Resistance under Near-Field Blast

J Zíma and M Foglar.....441

012054
OPEN ACCESS
Production of a textile reinforced concrete protective layers with non-woven polypropylene fabric

J Žák, P Štemberk and J Vodička.....450

IOP Conf. Series: Materials Science and Engineering **246** (2017) 011001 doi:10.1088/1757-899X/246/1/011001

FIBRE CONCRETE 2017

9th international conference on fibre reinforced concretes (FRC), textile reinforced concretes (TRC) and ultra-high performance concretes (UHPC)

PREFACE

The Fibre Concrete Conference series is held biennially to provide a platform to share knowledge on fibre reinforced concretes, textile concretes and ultra-high performance concretes regarding material properties and behaviour, technology procedures, topics of long-term behaviour, creep, durability; sustainable aspects of concrete including utilisation of waste materials in concrete production and recycling of concrete. The tradition of Fibre Concrete Conferences started in eighties of the last century. Nowadays the conference is organized by the Department of Concrete and Masonry Structures of the Czech Technical University in Prague, Faculty of Civil Engineering.

The 9th International Conference Fibre Concrete 2017 had 109 participants from 27 countries all over the world. 55 papers were presented including keynote lectures of Professor Bažant, Professor Bartoš and Dr. Broukalová.

The conference program covered wide range of topics from scientific research to practical applications. The presented contributions related to performance and behaviour of cement based composites, their long-term behaviour and durability, sustainable aspects, advanced analyses of structures from these composites and successful applications.

This conference was organized also to honour Professor Zdeněk P. Bažant on the occasion of his jubilee and to appreciate his merits and discoveries in the field of fibre reinforced composites, structural mechanics and engineering.

IOP Conf. Series: Materials Science and Engineering **246** (2017) 011002 doi:10.1088/1757-899X/246/1/011002

Scientific committee:

Jiří Adámek	(Czech Republic)
György L. Balázs	(Hungary)
Joaquim A. O. de Barros	(Portugal)
Eugen Brühwiler	(Switzerland)
Frank Dehn	(Germany)
Antonio D. de Figueiredo	(Brazil)
Takashi Hara	(Japan)
Alena Kohoutková	(Czech Republic)
Konstantin Kovler	(Israel)
Rathinam Kumutha	(India)
Yuriy P. Linchenko	(Ukraine - Russia)
Ippei Maruyama	(Japan)
Ludovít Naď	(Slovakia)
Kyungsoo Park	(South Korea)
Kypros Pilakoutas	(United Kingdom)
Giovanni Plizzari	(Italy)
Pierre Rossi	(France)
Luc R. Taerwe	(Belgium)
Tatsuya Tsubaki	(Japan)
Jan L. Vítek	(Czech Republic)
Vladimir D. Yakimovich	(Belarus)

Editorial committee:

Iva Broukalová	(Czech Republic)
Michaela Frantová	(Czech Republic)
Alena Kohoutková	(Czech Republic)
Ippei Maruyama	(Japan)
Petr Štemberk	(Czech Republic)
Michaela Vaitová	(Czech Republic)

IOP Conf. Series: Materials Science and Engineering **246** (2017) 011002 doi:10.1088/1757-899X/246/1/011002

Organizing committee:

Iva Broukalová	(Czech Republic)
Tomáš Dvorský	(Czech Republic)
Josef Fládr	(Czech Republic)
Michaela Frantová	(Czech Republic)
Yuliia Khmurovska	(Czech Republic)
Terézia Němcová	(Czech Republic)
Martin Petřík	(Czech Republic)
Karel Šeps	(Czech Republic)
Petr Štemberk	(Czech Republic)
Michal Števula	(Czech Republic)
Michaela Vaitová	(Czech Republic)
Helena Včelová	(Czech Republic)
Jan Vodička	(Czech Republic)

FC
2017

Vladimír KŘÍSTEK
Zdeněk P. BAŽANT
FC
2017

IOP Conf. Series: Materials Science and Engineering **246** (2017) 011004 doi:10.1088/1757-899X/246/1/011004

CONFERENCE PARTNERS

BEKAERT
better together

EUROVIA
VINCI

STEEL WIRE FIBRE

krampe harex

proceq

beto tech

ČERVENKA CONSULTING
Concrete Engineering & Software Solutions

ZÚS

BETON
TECHNOLOGIE • KONSTRUKCE • SANACE

ČBS

ČESKOMORAVSKÝ BETON
HEIDELBERGCEMENT Group

IOP Conf. Series: Materials Science and Engineering **246** (2017) 011005 doi:10.1088/1757-899X/246/1/011005

Peer review statement

All papers published in this volume of *IOP Conference Series: Materials Science and Engineering* have been peer reviewed through processes administered by the proceedings Editors. Reviews were conducted by expert referees to the professional and scientific standards expected of a proceedings journal published by IOP Publishing.

IOP Conf. Series: Materials Science and Engineering **246** (2017) 012001 doi:10.1088/1757-899X/246/1/012001

Structural Applications of Fibre Reinforced Concrete in the Czech Republic

A Kohoutková[1] and I Broukalová[1]

[1] Department of Concrete and Masonry Structures, Faculty of Civil Engineering, Czech Technical University in Prague

E-mail: iva.broukalova@fsv.cvut.cz

Abstract. The paper presents improvement of function and performance of the precast structural members by using fibre reinforced concrete (FRC) instead of ordinary reinforced concrete and attempts to transfer innovative technologies from laboratory in academic sphere into real industrial production which is cost-effective and brings about savings of labour and material. Three examples of successful technology transfer are shown – application of FRC in an element without common rebar reinforcement, in the element with steel rebar reinforcement and SFRC pre-tensioned structural element. Benefits of FRC utilization are discussed.

1. Introduction

For centuries, concrete has changed minimally. In recent decades, rapid development has taken place - new materials are invented, non-traditional materials and technologies are used in the production of concrete, or known materials are used in an unconventional way.

Various desired properties of the composite may be achieved by careful design of the concrete components and the structure of the composite material, so that the resulting component has completely different properties and potential applications compared to conventional concrete. The use and composition of concrete is currently being modified and adapted for different needs - from construction, architecture, concrete precast structural elements and products to everyday objects (e.g. furniture, applied arts).

Trends in the development of cement based composites can be seen in several major areas:

- Development and consolidation of the position of fibre reinforced concrete and other fibre-reinforced cement composites, use of special types of reinforcing – in addition to the fibres dispersed in the cement matrix, fibres arranged in unwoven or woven fabrics (TRC – Textile Reinforced Concrete) are used.
- A significant trend is the development of High Performance and Ultra High Performance cement composites (HPC, UHPC).
- One of the trends is the use of waste materials in the production of cement composites and the development of new technologies that will minimize the environmental burden.

The focus across all areas is on increasing durability and service life, and design and production according to the principles of sustainable construction.

IOP Conf. Series: Materials Science and Engineering **246** (2017) 012001 doi:10.1088/1757-899X/246/1/012001

Utilisation of fibre reinforced concretes is no longer exceptional. Fibres are used as a partial or full substitution of conventional concrete reinforcement. The fibres can be combined in such a way that each of the fibre types has its specific function - for example, short fibres eliminate shrinkage cracks and longer fibres have a structural function.

For industrial floors, the steel fibre reinforced concrete (SFRC) is now used more often than conventional reinforcement with weld mesh sheets. The steel fibre reinforced concrete was also used for slab structure. Experimental structures have confirmed that steel fibre reinforced concrete (SFRC) can be used as a ready-mix concrete, transported to the site and achieve the required quality and uniform distribution of the fibres in the mixture. The use of steel fibre reinforced concrete (SFRC) for tunnel lining is also a common application.

2. Advantageous application in precast elements

Very advantageous is application of fibre reinforced concrete (FRC) in precast elements. The technology of fibre reinforced concrete production is a bit more demanding than the production of traditional ordinary concrete; and a good control of production on the precast plant provides a high quality fibre reinforced concrete mixture with evenly dispersed fibres. Fibre reinforced concrete (FRC) in precast elements prevents shrinkage cracks and other types of cracking and thus helps to achieve better appearance and higher durability of the product.

Application of fibre reinforced concrete implies increased toughness and impact resistance, prevents spalling of the element's edges during transport and handling. An advantage is improving the cohesion and anchoring of lifting and handling anchors. Another benefit is increased bond and anchoring of rebar reinforcement and lifting anchors. Fibre reinforced concrete, namely steel fibre reinforced concrete is used as a substitution of conventional reinforcement, which may result in reduction of thickness of the element, decrease of the total weight and consequent savings in transport and manipulation. In prestressed elements the steel fibre reinforced concrete substitutes shear reinforcement; the dispersed reinforcement enables reduction or omitting of conventional rebar shear reinforcement. Likewise, in compressed elements the conventional rebar reinforcement may be replaced by dispersed fibre reinforcement.

Several successful applications of fibre reinforced concrete elements have been developed in cooperation of the Department of Concrete and Masonry Structures of CTU with construction companies. In designing and preparing production of FRC element the following procedure was verified:

1. Choice of an element where the application of fibre reinforced concrete is profitable.
2. Preliminary structural design.
3. Selection, design and examining of fibre reinforced concrete parameters.
4. Detailed check of resistance and analysis of the structural member behaviour.
5. Full scale tests of the element in case of need.

Application of fibre reinforced concrete shall not be the aim itself. The application of FRC in the element should lead to saving and/or enhancement of the element properties and behaviour. FRC is beneficial in elements with complex shape and thin elements subjected to tensile stresses.

The main features and advantages of FRC application in precast elements can be summarized as follows:

- FRC can become a partial substitution of steel rebar reinforcement or FRC can replace the rebar reinforcement.
- Thanks to FRC application the element may be thinner.
- The secondary cost reduction is due to decrease of the weight of the element – precast elements with lower weight are not so demanding as far as transport, manipulation and assembling concerns.
- Fibre reinforced concrete decreases labour demandingness.
- FRC reduces cracking, which implies high quality of the element.

- The reliability and durability of the element are enhanced.
- The damage of precast member during transport, manipulation and assembling is reduced or eliminated.
- The fire resistance is higher where FRC with synthetic fibres is applied.
- Utilization of FRC may reduce the cost of the product. The economic analysis is usually one of the decisive criteria for the acceptance and practical use of fibre concrete in the production of the precast element.

3. Transfer of technology

The team of the Department of Concrete and Masonry Structures of the CTU has conducted long-term investigations of fibre reinforced concrete (FRC) since the seventies; advantageous properties were observed, beneficial features were referred to. Knowledge of mix design and technology of FRC was gained and effect of various types of fibres on resulting properties of a particular fibre reinforced concrete.

A common problem of technology transfer of scientific and technology developments from laboratory to real applications retarded wider disseminating of FRC in structures and building products. Implementation of technological advances and new processes faces many obstacles – aversion of employees to new technologies, doubts of the firm management about profit from FRC use, fear of delays in production, needs to change the production process, demand to test a new technology on a company's existing production facilities and related interruption of the production.

Thanks to the open positive attitude of visionary management and their understanding of the potential benefits of FRC application we succeeded in tying cooperation with several firms, which led to practical application of FRC in construction industry. The following text presents examples of technology transfer and points to the benefits that use of FRC has brought.

3.1. Concrete sewage pipes

This application is a result of collaboration of the Department of Concrete and Masonry Structures of the Czech Technical University, firm Betotech, Ltd. and producers of fibres [1]. The firm produces sewage pipes of various sizes. Traditionally, the pipes are made from reinforced concrete. The reinforcement cage is formed from spiral wires welded to longitudinal rebars. The cage is made on spot welding machine and it is quite laborious.

Regarding required properties and behaviour of the underground pipe, the rebar reinforcement is not necessary because there are no tensile stresses in the element. The reinforcement is inserted to prevent damage of the pipes during transport and manipulation and because of the production technology; the pipes are made in vertical position and they are demoulded immediately after concreting and vibration.

The development of the fibre reinforcement pipe accounted principle requirements on the pipes: The basic demand is water tightness of the pipes; so that the cracking must be avoided. The early demoulding was considered, too.

From several variants of tested fibre reinforced concretes, SFRC has proven the most efficient. The dosage was 40 kg/m^3 of steel fibres with length 50 mm, diameter 1 mm and strength ca 1 000 MPa. The composite mix is in the table 1.

Tests of material properties showed compressive strength 54 MPa (mean value) and mean tensile strength 5.2 MPa.

Making of pipe with actual sizes proved manufacturability of the elements. The resistance of pipes was measured in full scale tests of pipes with diameter 600 mm and 1 400 mm. The pipe was supported by two linear restraints at the bottom and loaded by linear load at the apex (figure 1).

Table 1. SFRC mixture composition

Component	Units	Dosage
Aggregate 0-16	kg/m^3	1800
Cement CEM I 42,5 R	kg/m^3	390
Plasticizer	kg/m^3	1,5
Water	kg/m^3	140
Steel fibres	kg/m^3	40

Figure 1. Full scale test of the sewage pipe

The application of fibre reinforced concrete in the production of sewer pipes was motivated by efforts to streamline, simplify and reduce the cost of production. The investigations focused in substitution of classical rebar reinforcement by dispersed fibre reinforcement and determine suitable type and dosage of fibres. The modification of the technology brings profit; the production is more simple, less time consuming and cheaper. In general, it can be concluded that certain types of fibre reinforced concretes may substitute conventional rebar reinforcement; however, it is necessary to choose the appropriate type of fibre and to design the appropriate composition of the mixture for the given application.

3.2. Bridge cornice

The application of fibre reinforced concrete in precast bridge cornice was designed in cooperation of the Department and firm SMP CZ from Vinci Group [2]. The primary aim of the investigations was verification of possibilities of production of the fibre reinforced element with an existing production facility without the need for demanding changes of production technology or needs for additional equipment of production machines. The economic indicators were naturally carefully monitored; analysis from a range of angles showed that the introduction of fibre reinforced precast member into the production, in addition to the increased lifetime of the element, also generates immediate profits.

The precast bridge cornice is a common part of equipment of some bridges. The cornice is not a load-bearing part of the bridge; nevertheless it can be subjected to substantial strain, salts treatment and atmospheric effects. Because of aesthetic requirements on cornice, to provide bond of anchors and prevent corrosion and pull-out of anchoring elements it is desirable to avoid excessive cracking. The main reasons of cracking of cornice are principally volume changes. Fibre reinforced concrete with synthetic fibres is usually an effective solution to this kind of cracks. Knowing the stress state and considering demands on durability of the cornice, fibre reinforced concrete mixture composition was proposed. After testing the workability of the FRC mixture in the precast-plant devices the mixture was modified a bit. The final composition shows the table 2.

Table 2. Composition of the fibre reinforced concrete for the bridge cornice

Aggregate	extracted aggregate fraction 0-4 and 4-8
	crushed aggregate fraction 0-16
Cement	CEM I 52,5 R
Filler	ground limestone
Plasticizer	two types of carboxylate based plasticizers
Water	
Additives	air entrainment additive
	stabilizing additive
	microsilica
Synthetic fibres	polypropylene fibre of length 54 mm; dosage 3,5 kg/m^3

Measured in standard tests, the mean compressive strength was determined 52 MPa with a standard deviation of 2,99 MPa.

The manufacturability of the cornice has been verified in a pilot plant (figure 2) and the elaborated cornice was used in a bridge construction.

The innovation of the element consists of substitution of part of the steel rebar reinforcement of the original element with polypropylene fibres dispersed in concrete matrix. The changed properties and behaviour of the FRC element enabled reduction of conventional rebar reinforcement by 42% and decrease of thickness of the element. The consumption of concrete was reduced by 27%. The total weight of the cornice is 25% lower. The lower material consumption led to decrease of cost; the price is 10% lower.

Fibre reinforced concrete shall be a contribution to better service, behaviour and durability of the element. Other problem that may be solved by use of FRC is cracking and failure during transport and resulting fall of the element and injure of pedestrians and damage of passing vehicles. Use of polypropylene fibres changed behaviour of the element, which is more tough and more impact resistant and has higher deformation capacity. Indirect savings consist in reduced transport costs, factory handling and assembly. Small cargo is transported, thus more elements are transported at once and handling and manipulation is easier, and storage space requirements are lower. Fibres tribute to higher durability and longer life cycle, what decreases maintenance cost. Tougher material better resist impact load and so there is no risk of spalling of corners during transport and manipulation, what decreased number of complaints.

Figure 2. Bridge cornice manufactured in the pilot plant

Research and development, laboratory and field tests of fibre reinforced concrete with synthetic fibres proved possibility of industrial production of precast fibre reinforced elements without the need to modify the existing production equipment even without necessity to change the production quality management system. At the same time, some limiting features of fibre reinforced concrete, such as the influence of fibres on fresh concrete rheology and relatively higher costs per cubic meter have been successfully eliminated.

3.3. Prestressed precast girder

The last example of practical application of the fibre reinforced member presents utilization of the steel fibre reinforced concrete in the prestressed girder. The girder was designed as a load-bearing element for the prestressed foot bridge with the span 13,6 m. It has a solid cross-section in the shape of irregular hexagon with dimensions ca 400 mm x 1000 mm and embodies only pre-stressed reinforcement without any mild rebar reinforcement [3].

The dispersed reinforcement are steel fibres of diameter 0,8 mm and 60 mm long, with strength 1 200 MPa. Composition of the SFRC is in the table 3.

Table 3. SFRC mixture composition

Component	Units	Dosage
Fine aggregate 0-4	kg/m^3	700
Aggregate fraction 4-8	kg/m^3	290
Coarse aggregate 8-16	kg/m^3	750
Cement CEM I 52,5	kg/m^3	400
Plasticizer	kg/m^3	1,5
Water	kg/m^3	167
Steel fibres	kg/m^3	40
Ground limestone		
Microsilica		
Plasticizer		

The tests of SFRC proved parameters required for the concrete class C 55/67. The concreting procedure was recommended in the direction from end of girders to the centre. The SFRC flew in below the strands and the fibres are evenly distributed.

The load-bearing capacity of the beam was controlled by full-scale tests, where the deflections were measured and cracking was monitored. Naked eye observing before, during and after test did not discover any cracking, damage and other abnormalities in the girder behaviour.

Figure 3. Load test of the prestressed SFRC girder

Additional tests were conducted for verification of the railing anchors. The pull-out tests showed that SFRC has much better anchoring capacity than ordinary concrete without dispersed fibre reinforcement. The failures occurred at loads 7,5 times higher than guaranteed load-bearing capacity of the anchors. The confinement of fibres prevents cracking in concrete and changes the failure mode from pull-out to rupture of the steel anchor.

The application of SFRC in pre-tensioned girder demonstrated ability of dispersed steel fibres to substitute shear reinforcement and enhance the anchoring. Minimizing of conventional rebar reinforcement diminishes workability and time consumption in the manufacturing of the element and thus the manufacturing becomes cheaper.

4. Conclusions

The potential of fibre reinforced concrete is undeniable. FRC is used as partial or total substitution of conventional reinforcement, what is economically profitable since higher cost of FRC material is compensated by reduction in labour cost of manipulation and storage of conventional reinforcement. Moreover, there are a lot of small or thin concrete elements that are not subjected to significant loads, where rebar reinforcement is required to prevent a brittle failure due to accidental actions and shrinkage cracking. In these cases fibres are very suitable, because production process of elements without rebars is easier and quicker and placing of rebar reinforcement in such tiny element is very demanding and labour-consuming. The dispersed reinforcement is able to enhance parameters of concrete in a number of ways; FRC have higher mechanical properties, higher impact resistance, better control of cracking, favourable behaviour in fire, etc. Utilization of fibre reinforced concrete increases durability and life cycle of the structure or structural element, thus it is favourable also regarding sustainable building and environmental assessment.

The FRC is usable either in a form of plain FRC, or with rebar reinforcement, or in a prestressed member. The most efficient is utilisation of FRC in precast members, thanks to better control of manufacturing the composite mixture. The paper introduces three examples of application of fibre reinforced composites in precast structural members, presents composites employed in particular application, mentions benefits of each application.

Each application of fibre reinforced concrete must be carefully planned and prepared. The realization of the fibre reinforced concrete element should be successful not only because of the immediate success and profit of the producer, but also because the eventual failure of the fibre reinforced concrete can cause distrust and inhibit or stop the development of the use of fibre reinforced concrete as a structural material.

Acknowledgement

This outcome has been achieved with the support of the Czech Science Foundation, project: GA15-19073S and project „The durability of concrete with recycled composites materials", SGS17/123/OHK1/2T/11 supported by the Ministry of Education, Youth and Sports.

References

[1] Vodička, J.; Veselý, V.; Kolář, K.; Krátký, J.: Practical Application of Fibre Concrete, In: Fibre Concrete 2007. Praha: Vydavatelství ČVUT, 2007, pp. 181-184. ISBN 978-80-01-03740-9.

[2] Kohoutková, A.; Broukalová, I.: Optimization of Fibre Reinforced Concrete Structural Members, In: Concrete and Concrete Structures 2013. Žilina: University of Žilina, 2013, pp. 100-106. ISSN 1877-7058.

[3] Kohoutková, A.; Broukalová, I.: Concrete Behaviour and Crack Control of Prestressed Concrete Beams with Fibres, In: Fibre Reinforced Concrete: challenges and opportunities. Bagneux: Rilem Publications s.a.r.l., 2012, pp. 299-300. ISBN 978-2-35158-132-2.er reference

[4] Kohoutková, A.; Broukalová, I.: Structural Performance and Crack Control of Fibre Concrete Beams with Conventional Reinforcement, In: Brittle Matrix Composites 8. Oxford: Elsevier Ltd, 2006. pp. 343-350. ISBN 9781845690311.

[5] Bílý, P.; Fládr, J.; Kohoutková, A.: Behavior of anchorage areas in the steel-concrete composite structure loaded by longitudinal shear forces, In: Procedia Engineering. Lausanne: Elsevier,

2017. pp. 104-110. Modern Building Materials, Structures and Techniques. vol. 172. ISSN 1877-7058.

[6] Šeps, K.; Fládr, J.: Deformation properties and testing of synthetic fibres, In: Fibre Concrete 2013 - Technology, Design, Application. Praha: České vysoké učení technické v Praze, Fakulta stavební, 2013, pp. 431-438. ISSN 2336-338X. ISBN 978-80-01-05240-2.

IOP Conf. Series: Materials Science and Engineering **246** (2017) 012002 doi:10.1088/1757-899X/246/1/012002

Glassfibre Reinforced Concrete: a Review

P J M Bartos[1]

[1] Em. Professor, UWS, Scotland, UK

E-mail: p.bartos@ntlworld.com

Abstract. Introduced to construction about 40 years ago, GRC has come of age. It is now widely used all over the world and in quantities very likely greater than most of the other types of fibre reinforced concrete, although it remains less known. A brief history of GRC is followed by review of the basic make-up of this complex composite. Methods of production are identified, properties reviewed and modes of fracture which are unique to GRC are explained. Benefits which are already available and exploited by its users are summarised and the wide spectrum of current applications of GRC is outlined.

1. Brief history and introduction

Reinforcement of brittle matrices by fibres in order to improve construction materials dates back millenia; there is nothing new in the current exploitation of this principle. Beginning of the 20th century brought about advances in production of glass fibres mainly of the borosilicate type (E-glass). The first production of Glassfibre Reinforced Cement dates from late 1950s, when the E-glass fibres were combined with non-alkaline matrices [1]. It was already understood, that the alkaline environment of hardened ordinary portland cement caused corrosion and loss of tensile strength of the E-glass fibres.The earliest GRC production therefore required unusual cements and it remained very limited.

Special glass formulations ,which generated alkali resistance had existed but it was not until late 1960's that such fibres became commercially available. The new fibres and their applications were covered by patents and production of the GRC in early years was subject to licensing.

Production methods for GRC were relatively simple but the material itself was sophisticated and a much greater quality control was required than that used for ordinary precast concrete. Inadequate quality control of early GRC production had led to problems with performance of the composite and its rise in acceptance was temporarily curtailed. Early versions of GRC also showed a loss of the initial pseudo-ductility with age (embrittlement) when exposed to wet environment. Manufacturers of the AR glass fibres responded in early 1980s by developing a new generation of fibres, with different surface treatments. EN standards were also developed, the quality control was tightened and the best practice in production and design became strongly promoted by the International Glassfibre Concrete Association (iGRCA). The new fibres improved significantly the long-term performance of hardened GRC. Beginning of the new Millenium saw a very rapid increase in production of GRC, fuelled by the construction boom world-wide. The post-2008 global economic crisis slowed the pace of the rise for a time, however the GRC industry has now recovered and GRC is firmly established as a high-tech material of choice for leading architects and other end-users world-wide.

This paper is aimed at both construction practitioners and potential researchers with a limited knowledge of GRC. It provides the basic information and highlights a few points of special interest. More detailed information ,supported by additional sources, can be found in the newly published book [2] and extensive technical guidance and data are available from the International Glassfibre Reinforced Concrete Association [3] .

Concrete reinforced principally with a *glassfibre fabric or textile (*TRC) differs significantly from GRC. Principal reinforcement of TRC are glass fibre strands acting as very long, continuous reinforcing elements interwoven as a fabric/textile. *Position of reinforcement in a textile-reinforced concrete is therefore pre-determined by the nature of its 'textile' pattern.* Both the production processes and fracture mechanisms in TRC differ from those of GRC. Existing GRC manufacturers are rarely involved in production of pure TRC and practical production of TRC is currently limited to a few, namely German, specialist companies. Textile reinforced concrete is outside of the scope of this paper. Wider information can be obtained from publications focused specifically on the TRC [e.g. 4].

2. Basic constituent materials

2.1. Binders

Suitability of a hydraulic cement for production of matrices for GRC depends on how compatible it is with the type of glass fibre used as reinforcement. All common varieties of Portland cement produce matrices which remain strongly alkaline throughout their service life. Such matrices can be only used in combination with alkali-resistant (AR) glass fibres. White Portland cement is used where brighter colouring of GRC surfaces is required.

Alkalinity of portland cement based GRC matrices can be reduced by using pozzolanic additions such as *microsilica* (silica fume), *pulverised fly ash (pfa), ground granulated blastfurnace slag* (GGBS) and *metakaolin*. Such additions reduce alkalinity and modify the microstructure and properties of the hardened matrix but do not replace the basic alkaline binders. An adequately performing GRC can be made using alternative binders which produce non-alkaline matrices. These include cements of the aluminium-silicate and sulpho-aluminate type, such as the *High Alumina cemen*t (HAC), *Calcium Aluminate cements* (CAC) and *Supersulphated cements* (SSC). However, higher costs of such very low-alkali matrix materials tend to outweigh the saving in the cost of fibre reinforcement and other negative side-effects occur. Such alternative binders are currently little used, although applications have been reported from China.

2.2. Aggregates

The aggregate content of a typical mix has increased over time to the currently prevailing cement : aggregate ratio of 1:1 or more. The preferred shape of the particles is round and the maximum size is generally limited to 1.2 mm (spray-up process) and 2.4 mm (premix process). The content of very fine particles (passing 150 μm sieve) is limited to 10% of the total weight of the sample. Silica sands with SiO_2 content > 96% are widely used. Moisture content of the aggregate must be controlled. Mix designs are usually based on moisture content < 2% and an allowance for the free water in aggregate is made in the mix design.

2.3. Fibres

Tensile strength of glass in the form of very thin individual fibres /filaments (dia. < 20 μm) is very high (> 3GPa) compared with that of the glass in bulk. It relies on the glass fibres retaining their pristine, undamaged surfaces with very few, if any, defects. Production of the fibre itself and that of the composite is designed to maintain the fibres undamaged as much as possible throughout the GRC production process. The freshly drawn filaments are immediately coated - a '*size*' usually tens of nanometres (10^{-9} m) thick is applied for protection and keeping the fibres in a bundle. *It is important*

IOP Conf. Series: Materials Science and Engineering **246** (2017) 012002 doi:10.1088/1757-899X/246/1/012002

to note, that the content of glass fibres in GRC is traditionally given in % by total weight (mass) of the composite. Spraying or sprayed-premix processes permit a higher packing density of the alkali-resistant fibre strands and an optimum content of approx. 5% b.w. The premix process has the optimum alkali-resistant fibre content of about 3.0 %. Higher than the optimum contents of fibres for each of the production processes reduce workability of the fresh mix which becomes difficult to compact adequately.

Tensile strength of fibres influences the ultimate direct tensile and flexural strength of the composite. The high strength of the fibres can be exploited only if bond between the fibre and the matrix (and between the fibres within the strand) were high enough and the fibre was long enough (and bonded adequately to the matrix) to enable the high strength to be fully utilised. Tensile load-bearing capacity of the fibres decreases when their effective cross-sections are reduced by corrosion caused by a reaction between the glass and the surrounding highly alkaline cementitious environment. The alkali-attack also creates localised weak spots on the fibres which also substantially reduce the very high strength of the initially pristine fibres. These reductions in load-bearing capacity are minimised by using alkali resistant (AR) fibres. Level of performance of fibres, which makes them suitable for GRC is specified [5] and relative performance of fibres in different alkaline environments is assessed by the standardised Strand-in-Cement (SIC) test [6].

Tensile strength is also reduced when products of hydration of the cementitious matrix precipitate on the surfaces of fibres inside of a strand These cause local damage and act as stress concentrators, reducing the tensile strength of the fibre, namely when a displacement occurs. Both the individual fibres and bundles/strands of fibres are flexible (they can bend), however, when they bridge a crack in the matrix at an angle other than at or near to 90 deg , the 'edge' of the crack begins to act as a severe stress-concentrator and 'tensile' load-bearing capacity of the strand is sharply reduced.

Modulus of elasticity of fibres E_f also has a strong influence on properties of the composite.*Fibres with* $E_f > E_{matrix}$ can increase both toughness and strength of the composite. Alkali resistant glass fibres with E_f around 72-74 GPa therefore can reinforce the typical hardened cement paste paste with E_{matrix} = approx. 35 GPa.

3. Production methods

3.1. Batching and mixing

All the basic ingredients of the matrix are batched by weight. Special dispensers are used for adding precise amounts of small quantities of admixtures and additives. The weighing equipment requires an accuracy within ± 2% of the stated target batch weight. Input of glass fibres is usually controlled by the number of strands/rovings and the speed at which they are fed into the chopping mechanism when the spray-up production method is used. Performance is checked continuously by a fibre output rate monitor or the machine needs to be regularly tested for the accuracy of its delivery rates and ratio of the fibres and the slurry.

Most of the mixers used are designed specifically for GRC production. They are usually of a forced-action type, with a batch size of approximately 0.060 m3 or 125 kg of a fresh mix. Larger batch mixers and continuous action mixers are used in integrated batching & mixing plants capable of a high output of GRC in a variety of mix-designs. A typical mixer for the spray-up process is capable of generating a high shearing action by mixing tools rotating at a high speed. Such mixers are usually operated in two modes/stages: first, a high speed (high shear) action produces a highly workable matrix mix (slurry) which is the pumped into a spray gun where it combines with the fibres. The first stage is followed by a period of low-speed (low shear) action for the blending-in of pre-cut fibre strands when the Premix process is used. It is possible to have separate batching, mixing and placing/spraying equipment, however the highest productivity and often the highest uniformity/quality of the GRC are achieved when an integrated system is used.

3.2. Production of GRC elements

3.2.1. Compaction and orientation of the fibres

Production is of fundamental importance as it governs the formation of the internal structure of the composite. GRC is an inherently porous material and the *degree of compaction* of the fresh mix affects its properties when hardened. A common production method is by spraying, which generates compaction by the impact of the mix hitting the mould. However, an additional compaction is routinely applied, usually a manual one using hand-held rollers. Special care has to be taken regarding corners in elements of complex shapes. Compaction of the premix GRC is usually achieved by the mix being hand-packed or trowelled into a freshly filled mould, which is then vibrated externally. The requirement for compaction is eliminated when a very highly workable, flowing self-compacting mix is used.

It is important to understand that the type of the *production process has an effect on the orientation of the fibres. It therefore determines the degree of anisotropy of the hardened GRC* produced. Spraying tends to arrange most of the fibres in a 2-dimensional random distribution, parallel to the flat surface of the mould and 'in-plane' of the sheet/panel produced. The length of the fibres usually exceeds significantly the thickness of GRC elements and a random 2-D rather than a 3-D distribution of fibres is achieved. Such a distribution can have a positive effect by improving the flexural strength of a hardened flat, sheet-like GRC. Earlier production methods used twin-head spray guns, which, incorrectly used, could produce non-uniform distribution of fibres 'in-plane' of the sheet, making the 2-D distribution non-uniform. Modern concentric spray heads have eliminated this type of anisotropy.

3.2.2. Simultaneous spray-up

A freshly made cement/sand slurry is pumped to the nozzle the concentric spray gun where it is mixed with chopped strands of fibres and projected out of the nozzle by compressed air. The fibres are cut to a pre-set length, usually between 25 to 37 mm. The chopping rate of the fibres and the pumping rate of the cement:sand slurry are controlled and typically set to produce GRC with 4.0-5.0 % of fibres b.w of total mix.

A '*mist coat*' is applied by spraying the first layer as thin as possible, using matrix without fibres. Subsequent layers of fresh matrix with fibres are applied without delay, to ensure integrity. In case of a '*facing mix*', the mix is allowed to stiffen before the bulk of the fresh GRC is built-up on it in layers. The facing layer is usually 3-5 mm thick, depending on the type of surface treatment. Each pass of the spray gun deposits a layer approx. 4 to 6 mm thick. Care has to be taken to ensure that an adequate thickness of the GRC is produced in corners and edges of moulds with complex shapes. Thickness is checked using a depth gauge or a template. The 'back face' is then finished as required by either the compaction roller or by a float. *Thickness stated in a specification for GRC is the minimum thickness.*

3.2.3. Premix and casting process

Fresh cementitious matrix is produced first and between 2% - 4 % b.w. of pre-cut AR glass fibres are then blended in. The length of the pre-cut fibre is usually 6 or 12mm but other lengths are available. The matrix is produced in a high-shear mixer, to obtain maximum workability and the chopped fibre strands are added in the second stage, often in a low-speed mixing regime. This facilitates their dispersion at the highest practical volume content with a minimum of damage to the fibres. The resulting mix can be hand-packed or trowelled into a mould or cast - in. Casting is either accompanied by vibration of the whole element or without vibration, if a self-compacting mix is used.

The fresh premix-GRC can be also sprayed in a process known as *sprayed premix* using a specially developed spray gun. The method is particularly suited to manufacture of large numbers of standard products. Current production plant is capable of production of up to about 2 tonnes of such a fresh mix per hour.

IOP Conf. Series: Materials Science and Engineering **246** (2017) 012002 doi:10.1088/1757-899X/246/1/012002

3.2.4 Other production methods

Filament-winding

Filament winding consists of wrapping continuous AR glassfibre reinforcement, impregnated with fresh matrix material, around a mandrel which defines the final shape of the finished product. It produces GRC with up to 18% volume fraction of the fibres. The process is very suitable for making hollow tubular products. It has been used in practice for GRC transmission poles and for pipes in the U.S.A and Australia.

Extrusion

Panels and similar elements in which one or both faces were to be smooth and which were required in large numbers with a very high repetition, can be made by this method. Trials were carried out but the process has not been widely adopted into GRC production practice, mainly due to the high cost of the equipment and the uncertainty about its adequate exploitation.

Manual lay - up

Simple flat and 3-D shapes can be produced by this process. The mould is coated with a plain matrix and then lined with a chopped strand mat, fabric or a mesh. Another layer of fresh matrix is applied, penetrating the fabric, and the next layer of the mat, or of a glass fibre fabric is placed. The process is repeated until the required thickness is achieved, making it close to that for TRC. A similar technique produces *light-transmiting concrete,* where a bulk element made of layers of parallel glass strands is made first. It is then cut precisely into 'slices' with fibres perpendicular to the cut. Special light-transmitting fibres are used.

3-D printing

Production of GRC using *3-D printing* has been developed very recently in China, e.g. by WinSun Co. [7]. Thin 3-D elements are 'printed' at this stage and then joined up by stacking to make a structure or a large structural element [2]. 'Hollow' printed elements can be also stacked and used as permanent formwork, with the voids in-filled with concrete (Fig.1). Complete buildings have been assembled in this manner.

3.3. *Curing*

GRC has a high cement content and it is mostly used to make elements with thin cross-sections. The *curing regime for GRC has to be controlled much more accurately than in the case of ordinary concrete.* It must ensure that the composite attains an adequate strength by the time it is to be removed from moulds/formwork and handled and that an early drying shrinkage and any subsequent distortion are minimised.

Early production methods relied on curing GRC elements by sealing them in polythene sheets or keeping them in warm fog-rooms. Current production processes avoid it by an adjustment of the mix design. Small amounts of acrylic polymers in the fresh mix keep the internal moisture in and prevent its loss by evaporation. Curing environment still has to be ideally at approx. 20 °C and 95% R.H. Sudden and rapid drying-out or large temperature changes must be avoided to ensure that the GRC reaches adequate strength adequate strength for the element to be safely removed from a mould.

Figure 1. 3-D printed elements stacked to form a wall unit. Photo courtesy of Y. Che.

3.4. Moulds and formwork

The exceptional capability of GRC to manufacture products of very complex shapes requires very high quality moulds/formwork. Performance requirements for moulds and formwork for GRC are greater than in ordinary concrete technology. The mass of the material placed is less and formwork pressures are lower in case of GRC and the moulds can be much larger, more complex and re-used more often. However, greater accuracy and tighter dimensional tolerances are also required. The moulds have to be designed with an appropriate allowance for the expected volume changes of the GRC element (shrinkage or changes in temperature).

Two-component moulds, in which a rigid base is lined with another material, such as timber lined with a polymer resin or steel lined with polyurethane rubber are now common. Such moulds or mould-liners are usually based on silicone rubber and polymeric elastomers [2] and produce elements even with very deep textures and under-cuts (Fig.2). Mould-release agents are also used for GRC. The mould must be shaped is such a manner that it can be lifted or stripped easily and without damage the finished GRC product. The recent trend to use curved elements for flowing, free-form shapes of buildings generated new challenges for the mould-making. The latest developments integrate the digital design of facade elements with new 'adaptive-mould' forming technology [2,8] and reduce significantly the amount of manual work required.

3.5. Surface finishes and treatments

The initial surface finish of the GRC product reflects the surface of the mould used. It is essential that there are no defects and visible joints in the surfaces of the moulds. The original surface of the GRC can be subsequently treated to produce an even greater range of finishes.

Surface finishes vary from very smooth, glossy ones to a flat matt one and a wide array of textured and patterned surfaces with or without pigments. GRC can be made with:

► *An ex-mould finish, either smooth or textured, according to the mould surface.*

IOP Conf. Series: Materials Science and Engineering **246** (2017) 012002 doi:10.1088/1757-899X/246/1/012002

Texturing makes minor imperfections less visible.

► *An exposed, fine aggregate finish (with or without a texture)*

Water-washing away a freshly set, but not hardened, cement paste is sometimes assisted by an application of a coating by a retarder onto the mould surface prior to the facing mix being applied. In a typical treatment, the thin top layer is removed by a water jet to reveal a specially selected decorative fine aggregate (the 'exposed aggregate' process) on the external surface of a panel. *Strong acids* can also be applied to the surface to expose the decorative sand or fine aggregate. This is usually done immediately after demoulding as the surface of GRC rapidly becomes impervious and difficult to penetrate. *Mechanical treatments* aim to produce a surface with a texture simulating natural materials such as either a rough or polished stone. This can be achieved by wire-brushing and grit-blasting or grinding / polishing at a suitably early age of the composite.

Surfaces of GRC can be painted. The paint is usually applied onto a pre-pigmented GRC of a similar base colour. There are a number of suitable proprietary paints and penetrating stains now on the market which can be used to finish the appearance of a GRC product.

Hydrophobic coatings and surface sealants can be applied and must be adequately permeable to water vapour. They can reduce lime-based efflorescence on the external surfaces and a reduce the effects of a polluted environment.

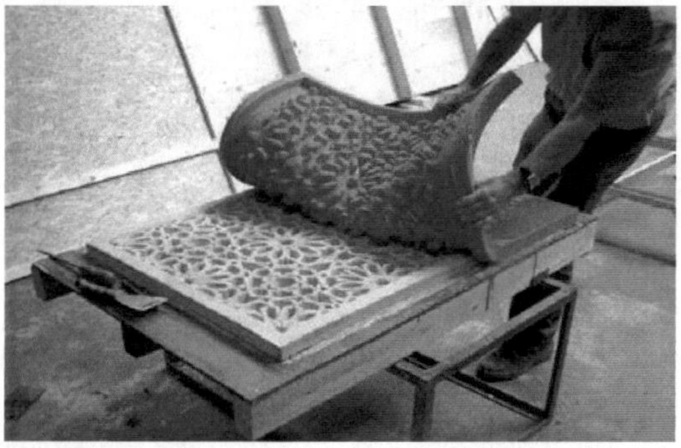

Figure 2. Polyurethane rubber mould being stripped-off [2]

3.6. Handling, transport, storage and repairs

Thin-walled GRC, particularly very large panels are vulnerable to permanent damage if not handled, transported and stored in an appropriate manner. Handling of a GRC panel, including the effect of an asymmetrical lift should be considered as part of its structural design. It is essential to use only lifting points identified in design and special lifting frames are used for handling of large panels. *Embedded fixings can be used for lifting of GRC elements only if they were also designed for this purpose.* Care must be taken to *minimise any vibration and to restrain their movement during transport.* All the points of support must be vertically aligned and in correct locations if the units are stacked. *Incorrect storage may lead to permanent distortions and unacceptable deflections/deformations and lead to a rejection of otherwise perfect products.* Any damage to load-bearing GRC elements must be assessed. Minor damage, if not structurally significant, can be repaired using fresh GRC mix similar to that used for the production of the element.

4. Fracture mechanisms

Microstructure of GRC is determined largely by the composition of the mix and the effects of mixing and production processes. It continues to change and develop with age in a non-linear manner, highly dependent on service environment. A cross-section of a typical GRC in Fig.2 shows the primary reinforcement: bundles /strands of glass fibres embedded in a matrix.

Internal stresses, including tension, develop when an external load is applied to a GRC element . The tensile stress becomes critical when it exceeds the tensile strength of the matrix and causes it to crack. Mode of fracture (failure mechanism), which follows the first cracking of the matrix, depends on properties of the matrix and fibres, and on the bond between them [9]. As these change with age, so does the mode of fracture. Interaction between a fibre strand and the matrix which it is embedded in, is measured by the strength of their bond. The mode of fracture also reflects *bond between individual filaments within a strand.* Ageing and conditions of exposure affect both the strand/matrix bond and bond between fibres within a strand to a different extent. Bond between fibres within a strand determines the degree to which the strand acts as a single 'solid' element or as a 'composite' fibre bundle.

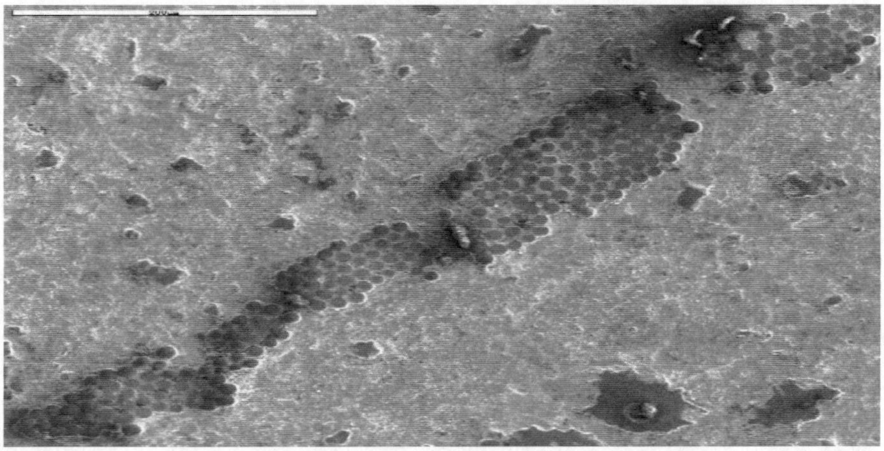

Figure 3. Cross-section of GRC showing strands of glass fibres. Note that most of the fibres remain in bundles of different sizes.

The first crack in the matrix occurs when the internal stress approaches the level of stress corresponding to the LOP and the tensile strength (or strain capacity) of the matrix is exceeded. Three basic categories of failure mechanism which follow the first crack were identified [9,10,11]. At this point the stress, which until then was taken by the matrix, is instantly added to that already carried by strands of fibres crossing the crack. How much of the additional stress an individual strand crossing the new crack can take depends on many factors, such as its embedment length, angle in relation to the direction of the stress and, importantly, its peripheral and internal bond. A further increase in load (stress) leads to an eventual failure of the strand.

The three principal modes of fracture (Fig.4) are [11]:

a. ***Complete pullout of the strands.***

Bond between the fibres at the perimeter of the strand and the surrounding matrix is low but bond between fibres within a strand is high. It enables the whole strand to act as a single, integral reinforcing element. Tensile strength of the inner fibres, which make up most of the strand and their reinforcing capability is therefore poorly utilised. The strand tends to pull out as one single element. The work of fracture is low. *This mode of fracture is usually associated with GRC at an early age.*

b. ***'Telescopic' or 'combined' mode of failure [11].***

Peripheral fibres of the strand, in contact and bonded well to the matrix, take up most of the additional stress transferred from the matrix, and eventually fail in tension. At this point all the stress previously carried out by the peripheral fibres transfers by bond onto the next layer of inner fibres within the strand. The inner fibres then begin to fail in stages, and the process continues until all the fibres are broken. The process is usually accompanied by partial internal pullouts of the inner fibres as they break at different points and the frictional part of the bond continues to provide additional means for fibre-fibre transfer of some of the internal stress. *The highest amount of work of fracture (the highest pseudo-ductility or toughness) are recorded,* while the strength of the composite is also enhanced. However, this mode of fracture tends to be a transient phenomenon, associated with current types of GRC at a lower age, when magnitudes of external bond (with the matrix) and bond within a strand are in an appropriate proportion. When and for how long this mode of failure exists depends very much on factors such as external service conditions, mix design and age. The failure of the GRC in this mode is gradual and a complete fracture occurs only after a substantial deformation/extension of the composite. It is the mode of failure most desirable for GRC in practice.

Pullout of the whole strand Mode (a) Telescopic fracture Mode (b) Tensile fracture of the whole strand Mode (c)

Figure 4. A composite of microphotographs of the three typical modes of tensile fracture of strands of glass fibres embedded in a cementitious matrix and crossing a crack. The fractures were observed during tensile tests carried out inside an SEM.

c. Complete tensile fracture of the strand.

This mode occurs when *bond between both the fibres within the strand and between the strand and matrix is very high.* As in the mode (a), *the strand then acts as a single reinforcing element and fails in a complete tensile fracture with a minimum extent of pulling out, or with no pullout at all.*

Strength of the composite is enhanced, but work of fracture is lower than in (b). The composite shows a typically brittle fracture with a sudden failure at a much lower ultimate strain/deformation then in (b). Fractures in the mode (c) are typical of a highly aged GRC, particularly in a humid environment.

A pseudo-ductile composite with failure modes (a) or (b) is much preferred to a brittle one (c) in structural applications. There are no sharp boundaries between the three basic modes of fracture and ore than one type can be present in a given GRC at any age. Moreover, the fracture mechanism is also influenced by factors other than bond, such as the *position of a reinforcing strand in relation to the applied stress* and *the line (or path) of the crack in the matrix.*

Fractures shown in Fig.3 are typical of a strand bridging a crack in a position perpendicular to the fractured surface of the matrix and the modes of fracture can be identified. However, in practice, only a very small percentage (5-8%) od strands would be in such a position.

Fracture processes of strands crossing a crack at a more acute angle is even more complicated. Limited studies have indicated [12] that peripheral fibres of the strand fracture first as the edge of the matrix along a crack concentrates the stress. The broken fibres then provide a 'cushion' over which the inner fibres can still pull out at an angle (Fig. 5).. There are indications [12], that strands at a slight angle to the principal stress provide more 'toughness' than those perfectly perpendicular. Strands lying perpendicularly to the direction of the applied stress and weakly bonded to the matrix may reduce the strength of the composite. Contribution of the strands at higher angles is extremely difficult to examine directly. Numerical modelling of the fracture processes in GRC has been tried. However without an access to values for bond obtained independently (experimentally) for all its different aspects and considering its inherently variable nature, *no reliable and genuinely predictive numerical model for GRC* has been developed to date.

The fracture process depends very strongly on *fibre-matrix bond,* which, in turn, depends on the micro/nano – properties of the contact zone (often identified as the interfacial transition zone – ITZ) between either a fibre and the matrix or between fibres within a strand. Two types of bond which control the 'composite' action between matrices and reinforcement can be broadly identified [10], namely:

- *Adhesive bond* between glass and hardened cement. It tends to be low and keeps deformations of both of the fibre and of the matrix proportional / compatible in early stages of loading. When the interfacial stress exceeds that of the adhesive bond, de-bonding begins and proceeds rapidly along the remaining embedded length of the strand.
- *Frictional bond* resists further displacement (pullout) of the strand (or fibres within a strand) during multiple cracking of the matrix, up to and beyond the ultimate tensile or flexural strength (MOR) of the composite. It governs any remaining resistance against pullout of a fibre crossing a crack after the failure of the adhesive bond. Frictional bond is neither uniform nor constant, its magnitude changes with the speed of the loading and displacement and with the fracture process.

Glass fibres in GRC are not provided with any 'anchorage' at their ends, unlike many types of fibres in steel fibre reinforced concrete (SFRC). This type of 'bond' therefore has a negligible effect on the GRC fracture processes.

Figure 5. Fracture of a strand bridging a crack at an angle.

An accurate quantitative assessment of bond in GRC and its control (design) remain a big challenge on the way to a greater exploitation of the potentially higher performance of GRC. Direct measurements by 'pullout' tests such as are used for ordinary concrete and other fibre reinforced concretes are not practicable in GRC research, where the basic reinforcement is by bundles of fibres. A breakthrough came with first applications of nanotechnology in mid-1990s [13]. A nano-indenter was developed, which permitted precise loads (mN) to be applied onto very small targets (10^{-6} m) and enabled displacements to be continuously measured at nano-scale (10^{-9} m). Such extremely high precision testing can be only carried out in specialist laboratory premises with very closely controlled temperature and in a vibration-free environment. The nano indentation technique permitted 'micro-mechanical' properties of the matrix to be assessed both near the fibre-matrix interfaces and within the bulk of the matrix or the fibre strand and a single-fibre push-out test was developed. In this test, a very thin (< 0.5 mm) slice was cut perpendicular to the axial direction of a strand of glass fibres embedded in the matrix material. The push-through test specimen was placed into the nano-indenter and the tip (approx. 5-7 µm in diameter) was positioned above the centre of the selected fibre. The original sharp tip (diamond cube) of the indenter was later modified by re-shaping it to provide a flat contact area with the cross-section of the fibre tested [14]. Such shaping was only possible by using focused ion beam technology. The cost of production of the flat tips was very high and it curtailed further progress of a systematical investigation of bond. Direct information about bond between fibres within a strand and between peripheral fibres and the surrounding matrix was obtained but much more work is still required to build up an adequate base of the data on all aspects of bond. Such data on bond within GRC are a pre-requisite to a substantial advance towards full exploitation of the structural potential of GRC.

5. Basic properties

5.1. Mix design and fresh GRC

Properties of both fresh and hardened GRC depend on many factors which include constituent materials such as:

- ► matrix (type of cement, w/c, cement content, additives and admixtures)
- ► filaments (size/diameter, strength, modulus of elasticity, surface treatment)
- ► strands (number of filaments, shape, length)

Their proportions are established in the *process of mix-design, which aims to produce GRC with the required performance characteristics when fresh and hardened.*

The mix design takes into consideration *additional factors*, such as :

- ► production process - which includes mixing, compaction, curing, distribution and orientation of fibres/strands
- ► application of the product - which determines the length of service life (age) and expected exposure / service environment (levels/variations of temperature and humidity incl. freeze/thaw cycles), chemical and biological attack and chemical/physical fibre-matrix interactions
- ► development / changes in fibre-matrix interaction over the period of service life and conditions for a given GRC mix design.

Mix-design of common GRC matrices includes superplasticisers and admixtures developed for self-compacting concrete. A highly workable fresh mixes are produced without a reduction of strength when hardened. Workability is fundamental for an effective production and it is monitored and maintained within narrow limits. A standardised mini-slump test [2] is used, serving both as an indicator of optimum workability and as an indirect measure of uniformity of the GRC produced. The content of glass fibres in an uncured, green GRC can be obtained by using a 'wash-out' test [2], to check compliance with specifications.

Freshly made GRC in the form of "*green*" sheets can be re-shaped by folding and pressing to produce 3-D elements. Care must be taken to ensure that fibres were not damaged in the process and no cracking or another discontinuity is generated in corners and bends.

Guides and aids, which help in the selection of grades (strength) of GRC for a range of applications are available [e.g. 15]. The choice of the grades has to be confirmed by the specifier, a the GRC manufacturer and a competent engineer. Mix proportions for a typical grade 18 GRC, both with and without the addition of a polymer for curing are given in Table 1. Designs with proportions falling

IOP Conf. Series: Materials Science and Engineering **246** (2017) 012002 doi:10.1088/1757-899X/246/1/012002

outside of the guidelines may be acceptable but performance of such GRC has to be verified by trials before use.

5.2. Assessment of performance

Assessment of properties of GRC requires an understanding that factors which influence them are both numerous and inter-related to a very large and variable extent. Magnitude of the effect of one single factor on a given property usually depends on the value(s) of other factor(s). Such multi-factorial interactions also tend to be non-linear and of a varied degree of significance. *Reliable, general conclusions therefore cannot be drawn from a test-series or from an experiment, in which only two factors/variables are correlated, without regard to any of the other ones.*

Table 1. Mix proportions of Sprayed Grades 18 & 18P [3]

Sprayed GRC Grade	Grade 18	Grade 18P
Aggregate/cement ratio	0.5 -1.5	0.5 -1.5
Water/cement ratio	0.30 - 0.38	0.30 - 0.38
Glassfibre content (% by weight of total mix)	4.0 - 5.5%	4.0 - 5.5%
Polymer solids content (% by weight of cement)	Nil	4 -7%

Properties of a cement-based matrix change with age. It is impractical to assess the effect of age effect directly and *accelerated ageing tests* are therefore used. The long natural exposure is replaced by a shorter one in more aggressive warm/hot solutions simulating pore solutions in a specific GRC matrix material. *Accelerated ageing tests are useful for comparative purposes, however, performance in real service life may depend on additional factors, such as the history and effects of applied stresses and micro or macro-cracking already present in the composite.*

As the hydration of cement continues, at a decreasing rate with time, changes in the fibre-matrix interaction occur simultaneously. It is essential to quote the age and curing/exposure history whenever properties of hardened GRC are mentioned or comparisons made.

Dispersion of fibres within GRC is never completely random or uniform. The composite therefore always exhibits a *variable degree of anisotropy, which means that its performance parameters also depend on the direction of the applied load.* This is particularly significant for all types of strength-related properties, where the value of a property depends on the angle between the direction of the applied load to that of the prevailing number of fibres.

5.3. Mechanical properties

Typical values at the age of 28 days for current GRC are shown in Table 2

Table 2. Range of properties of a typical hardened premix or sprayed (italics) GRC [17]

Dry density (kg/m3)	Ultimate strength (MPa) at 28days in bending (MOR)	uniaxial tension	transverse tension
1900-2100 ; *1800-2000*	5-14 ; *18-30*	3-6 ; *8-12*	7-11 ; *5-8*
Modulus of Elasticity GPa	Tensile strain at MOR %		Impact resistance kJ/m 2
10-20 (both)	0.1- *0.8*		7-12 ; *15-25*
Thermal movement	Drying shrinkage strain		Poisson's ratio
10-20 x10 -6 per 1 °C (both)	300-1200 x10 -6 (both)		0.24-0.25 (both)

5.3.1. Flexural / bending strength

Bending (flexural) strength is the principal characteristic of hardened GRC. It is routinely assessed by a standard 4-point test [2,17] (Fig. 6) which provides information for safe structural design and indicates fracture mechanisms involved. Practical experience has shown that a correctly performed flexural test provides a good indicator of the strength and variability of GRC, as required for design [e.g. 16]. Performance in bending is characterised by two parameters which survived from the early days of GRC and which are sometimes difficult to fit into a modern context of behaviour of fibre reinforced composites. The first is the *Modulus of Rupture* (MOR, MPa), which indicates the maximum flexural stress which a test specimen was able to resist. The other one is the *Limit of Proportionality* (LOP, MPa) indicating the stress at which the stress-strain (load-deflection) curve begins to deviate from linearity.

The MOR is equivalent to the *bending strength of the material* tested. It is not a 'modulus', such as the modulus of elasticity. Neither is it associated with a 'rupture'. As is typical for a pseudo-ductile composite, the GRC normally does not fail entirely at the MOR level of stress. Instead, it continues to resist gradually reducing flexural stresses until the ultimate tensile strain ε_{ult} (%) (deflection, extension) is reached. Only then the composite loses its integrity and its load-carrying capacity is reduced to zero. With the ε_{ult} being normally much greater than the ε_{MOR}, the failure of GRC is usually a gradual process providing a useful safety margin in its structural applications.

5.3.2. Tensile strength and Poisson's ratio

Behaviour of a GRC composite is best characterised by its stress/strain diagram obtained from a direct uniaxial tensile test [2] which provides a genuine *tensile strength (UTS)* of the GRC. The test produces the value of the ultimate tensile strength as the stress at the peak of the stress/strain curve. As in the bending test, the ultimate tensile strain does not normally coincide with the ultimate tensile strain and occurs only after an additional strain develops and the tensile stress has already decreased. Uniaxial tensile tests are difficult to carry out reliably by simple test equipment (difficulties with gripping of specimens, size-effect, etc.). Such tests are expensive and limited to specialist laboratories.

Instead, tensile strength of GRC is obtained *indirectly,* derived from an assumed elastic stress distribution occurring during from the 4-point standard GRC bending test. Poisson's ratio describes ratio between strains parallel and perpendicular to the applied tensile stress, measured in direct tensile tests. The Poisson's ratio of a typical GRC is approx. 0.24.

5.3.3. Modulus of elasticity.
Young's modulus of GRC reflects primarily that of the matrix, with fibres making a contribution. Elasticity of the composite can be adjusted by changing the modulus of elasticity of the matrix. The adjustment is usually downward, to make the hardened composite more 'ductile'. In such case the ultimate tensile strain and the toughness of the composite are increased. A very high-modulus (and usually also a very high-strength) of the matrix tends to make the composite more brittle. Modulus of the fibres has a lesser effect, which depends on the degree of internal stress transfer available, controlled by bond.

5.3.4. Compressive strength
A typical GRC load-bearing product is made of thin sheets of the composite, where the compressive strength alone tends to be only a secondary parameter. Products where 'bulk' GRC is used tend to be largely non-load-bearing. There is no standard procedure for assessment of and little data exists on compressive strength of GRC. Tests are more likely to be carried out on full-scale prototype elements, subject to loads expected in service, instead of on individual test specimens. For approximate guidance, compressive strength of GRC can be estimated as 40-60 MPa or 50-80 MPa for cast premix and spray-up process respectively [2].

Figure 6. Typical test jig for the standard bending test.

5.3.5. Transverse tensile and inter-laminar shear strength
Transverse and inter-laminar shear strength are closely related, both depending largely on the tensile strength of an unreinforced matrix.

Transverse strength resists a tensile load applied perpendicular to the surface of a cross-section of a sheet-like GRC element. Transverse strength relies greatly on adhesive bond between the matrix and the reinforcement. There is no standard test, any measurement is strongly dependent on the size/shape of the test element and on the manner of its support.

Interlaminar shear strength resists loads applied 'in-plane' of a flat thin-walled GRC element. This is of significance during fracture of the GRC in bending. The interlaminar shear strength ensures that the element will not fail by an internal displacement, a 'delamination', when subjected to bending. The value of this strength is estimated as 3-5 MPa, lower than the tensile strength of the matrix [2]. There is no standard test and few data are available.

Shear strength and punching shear strength perpendicular to the plane of a typical flat sheet of GRC are inter-related. The shear strength is difficult to assess as a material's property as it is strongly

IOP Conf. Series: Materials Science and Engineering **246** (2017) 012002 doi:10.1088/1757-899X/246/1/012002

dependent on shape/size and support conditions of any test specimen. Punching resistance improves with a higher fibre content and then it follows any changes in the ultimate bending strength (MOR) with age. It is estimated as 25-35 MPa (sprayed) and 4-6 MPa (premix) in a typical GRC. Full-scale load trials simulating conditions in service are required for verification of performance of any structural GRC elements in which punching resistance was considered significant.

5.3.6. Toughness and impact resistance

GRC is unusual amongst fibre reinforced composites in use because the fibre reinforcement is by bundles (strands). Depending on the bond between the matrix and fibres and between fibres within a strand, an applied load generates complex stresses and displacements which provide the composite with a degree of an inelastic behaviour, a *pseudo-ductility is observed.* GRC is therefore inherently more impact resistant than plain concrete and some other fibre reinforced composites. Fracture of a mature GRC includes both tensile fractures and pullouts, the highest work of fracture occurs when intermediate/telescopic fractures of the glass strands dominate. Impact resistance, combined with the 'work of fracture' measured as the *area below the stress-strain (or load-deflection) curve , provide a measure of the toughness of the composite.*

Toughness does not normally enter into ordinary structural design, however, with all other parameters being equal (bending and tensile strength etc.) a tougher composite is always preferred. Such a composite exhibits substantial deformations before a final break and a general loss of integrity, a useful warning stage before a failure in practical applications.

5.4. Physical properties [2]

5.4.1. Density

Density (bulk density in kg/m^3) of hardened GRC reflects density of the matrix which depends on content and density of aggregate, and on the degree of compaction of the fresh composite. Free water content and any trapped/entrained air of the hardened GRC also influence the density. Specimens of the composite for assessment of density and for reliable comparisons should be oven-dry, or of identical moisture content. Density of a typical GRC varies from 1800 kg/m3 when dry to 2000 kg/m3 when wet.

5.4.2. Permeability, water absorption and apparent porosity.

A well-compacted GRC has an inherently *low water permeability* and it is considered a waterproof material, suitable for use in water retaining structures. *Permeance by water*, measured on samples of 8mm thick freshly made GRC, is in the range of 0.02 - 0.4 ml/m2/min. *Water absorption* (5-11%) and *apparent porosity* (16-25%) are similar for both the premix and sprayed GRC. Standardised tests, e.g. EN 1170 Pt. 6 or the iGRCA test method [17] are used for the absorption.

5.4.3. Acoustic properties

A typical GRC has high density providing it with an inherently good capacity for *attenuation of noise*. However, the acoustic performance depends not only on the intrinsic capacity of GRC for noise attenuation but also on the geometry and method of fixing of the element. A sheet of GRC 10mm thick, with a density of 2000 kg/m^3 subjected to the sound pressure of 0.2 kPa, will provide a Sound Transmission Class of 5.4.4

5.4.4. Thermal properties

Thermal expansion/contraction is described by a *coefficient indicating 'unit strain' for a one degree of change in temperature,* which has a range of approx. 10-20 x10^{-6} / $^\circ$C for a typical GRC. Minimum values of expansion/contraction occur at high and low Relative Humidities. The maximum value occurs at around 50-80% RH. *Thermal conductivity and thermal resistance* depend on density of the GRC and its moisture content.

► *Thermal conductivity* λ of a typical GRC with density of 1900 - 2100 kg/m^3 is in the range of 0.5 to 1.0 W/m °C. It is a characteristic of the material.

► *Thermal resistance* ***R*** depends on the thickness (t) of the material and its thermal conductivity (λ), $R = t / \lambda$

► *Insulating capability of a building element* ***U*** is an inverse of the resistance R plus the heat losses by convection and radiation. The value of U therefore depends on external factors reflecting the environmental situation of the building element in addition to the thermal resistance of the material.

5.4.5. Hydraulic and abrasion resistance

Smooth mould-side surfaces of GRC make it suitable for elements used in a variety of hydraulic structures. Manning's roughness coefficient, used in the Manning formula for calculation of a hydraulic flow in an open channel is approx. 0.012.

There is no specific guidance for abrasion resistance of GRC. In cases where the abrasion resistance is of significance, an approach similar to that for abrasion resistant concrete may be adopted (abrasion resistant aggregate is used) although there is much less scope to modify the GRC matrix to make it more abrasion resistant (lower aggregate content).

6. Durability

An overall durability of GRC reflects a combination of effects of: (a)alkali-resistance of the fibres, (b) alkalinity of the matrix and (c)resistance of the GRC to degradation caused by exposure to aggressive service environment.

Degradation under (c) includes a range of factors, such as *wetting/drying* and *freeze – thaw cycles exposure to high temperatures (fire).* In addition, durability of a GRC product depends not only on the intrinsic properties of the composite but also on the geometry and durability of fixings adopted for a specific GRC element.

6.1. Frost resistance.

GRC specimens are subjected to cyclic freezing and thawing in a water-saturated state as for ordinary concrete. The effect is similar to long-term ageing: the LOP increases, while the MOR decreases. Due to the very low water/cement ratio of the cementitious matrix, GRC performs as well or better than good frost-resistant concrete, without the need for air-entrainment.

6.2. Fire resistance

Basic constituents of GRC are inorganic and non-combustible. Fire performance classification according to standard test procedure is ***A1*** [18]. GRCs with small amounts of organic admixtures (approx. < 6% b.w.) will not sustain burning but fumes will be emitted. Fire tests classification is ***A2,*** with a rating of ***S1*** for smoke production and ***d0*** for flaming droplet production. The matrix tends to desiccate in fire and undergoes both drying shrinkage and thermal expansion. Thermal movement develops and when it becomes restrained by fixings or the shape of the elements, excessive tensile stresses may be generated, with the subsequent cracking. In practice, performance in fire is very often dominated by the shape & size of the product and when fire resistance is required, appropriate standard *tests for endurance in fire are therefore carried out on prototype full-scale elements.*

6.3. Chemical, biological and other exposure

Carbonation of the cement-rich and relatively dense GRC exposed to air is very slow, reaching only few millimetres after many years in service. With no ferrous reinforcement in GRC and with no effect

IOP Conf. Series: Materials Science and Engineering **246** (2017) 012002 doi:10.1088/1757-899X/246/1/012002

of the glass fibres, any lowering of the alkalinity of the matrix by carbonation actually benefits the GRC by increasing slightly the strength of the matrix.

Effects of acids on the common GRC (OPC based) are similar to those on ordinary concrete, causing an erosion of the cementitious matrix.

There is *no evidence of a biological attack* of significance reported for GRC in practice. Effects of other chemicals have not been studied specifically for GRC, its resistance is generally comparable to that of a very cement-rich ordinary concrete.

No reduction of mechanical properties of GRC has been reported when subjected to *gamma radiation* [19] or another similar exposure.

7. Volume (dimensional) changes

Dimensional changes are of primary significance in design of GRC elements as the systems for attachment/ fixing of GRC elements must be able to accommodate not only the self-weight and imposed loads but also all the movement due to changes in humidity and temperature in-service. Shrinkage strain can generate large movements in large elements, the mix design may be then modified using a 'low-shrinkage' matrix by choosing GRC with the highest practicable content of aggregate (sand) and/or by a modification of the binder.

7.1. Effects of humidity: shrinkage / swelling.

Dimensional changes of GRC are all non-linear and time-dependent as for ordinary concrete. Immersion in water causes an expansion / swelling, drying out causes contraction / shrinkage , which are both reversible, non-linear and their magnitudes are time-dependent. Volume changes due to changes in humidity are assessed by a standard test [20]

For design purposes, the initial, irreversible shrinkage ε_{si} can be estimated at 300 x 10^{-6} with a total long-term shrinkage $\boldsymbol{\varepsilon}_{\mathbf{total}}$ of approx. 1200 x 10^{-6}. Values between 1000-1500 x 10^{-6} are used in practical design of joints and fixings [e.g. 16] , depending on exposure conditions and the GRC mix-design. As the aggregate used is normally not susceptible to drying shrinkage, increasing the content of aggregate lowers the ultimate shrinkage of the composite.

7.2. Creep and fatigue.

Sustained loading of GRC produces an *additional deformation (creep strain)* beyond the initial, instantaneous one. It follows a non-linear relationship, where the increase in creep strain diminishes with age. The instantaneous deformation is largely elastic and reversible, when the stress is less than that of the LOP. Stresses at levels very near to the MOR can cause a delayed, gradual failure of the composite, which is difficult to predict. Cyclic and other types of a sustained load of variable magnitude affect the performance of GRC and generate *fatigue* of the composite. Strength is reduced and the resistance to fatigue is measured by the number of repeats/cycles of the load necessary for a failure to occur.

7.3. Variations in temperature and humidity

Mature GRC reflects changes in ambient temperature in the same way as concrete with a very high cement content. Similarly, the coefficient depend on moisture content of the GRC (matrix), it is lower for very dry and saturated conditions compared to 50% R.H. It is essential to consider dimensional changes due not only to ambient temperature but also due to absorption of sunlight on external GRC surfaces, namely when dark colours were used.

8. e-GRC and environmental performance [2, 21]

The *e*-GRC is a new type of the composite, exploiting *photocatalysis*. The photocatalytic activity of the *e*-GRC is based on a nano-crystalline form of Anatase type of TiO_2 present in the facing layer.

IOP Conf. Series: Materials Science and Engineering **246** (2017) 012002 doi:10.1088/1757-899X/246/1/012002

8.1. Active de-pollution of environment by eGRC

The photocatalytic surface of *e*GRC becomes chemically very active in the presence of daylight (or any other source of UV light) and generates a process in which complex organic molecules of common pollutants (e.g. nitrogen oxides (NO_x), sulphur oxides (SO_x) and numerous volatile organic compounds) are broken down by oxidation and removed from the air. Such pollutants have adverse effects on human health, namely in densely populated urban centres and industrial zones. Trials [22] showed that the photocatalytic de-pollution of air in a city-centre busy street was very significant. It was calculated [27] that applying results from practical trials and covering 15% of visible urban surfaces with products with photocatalytic surfaces in a large city can lead to an overall reduction of the pollution levels by 50%. The European Commission responded with [23] new measures to reduce air pollution.

8.2. Self-cleaning surfaces

Presence of the photocatalytic TiO_2 also affects the hydrophobic properties of the surface. The UV light changes the original hydrophobic surface (repelling water) into a very highly hydrophilic one (contact angle of less than 5 ^{o}), inducing a state of '*super- hydrophylicity*'. A very thin, uniform film of water forms on the exposed photocatalytic surface. The water layer then hinders adhesion of external substances to the surface and helps to maintain the surface clean. A significant level of the 'self-cleaning' capability is established.

8.3. Energy requirements

It is possible to determine the energy requirements for production of individual raw materials making up GRC. A complete assessment of the energy requirement for production of GRC therefore requires the energy input during its manufacture. A modern auto-spray traverse system can produce approx. 50 m2 per hour of GRC about 12mm thick. Energy used is approx. 1.7 kWh per 60 kg batch mix, approx. 29 kWh per tonne of GRC. [24]

8.4. Environmental impact analysis

A typical product, manufactured both in GRC and in traditional reinforced concrete in large numbers and for identical purpose, had been selected to allow a reliable, direct comparison and quantification of the difference in environmental impact to be made [25]. Drainage channels and ducts for communication and other services along railways were examined, as supplied to a typical construction site. Both products were available in traditional reinforced precast concrete and in a premix type of GRC and are already used in very large numbers. On average, the un-weighted impact of traditional concrete unit was 123% higher and the weighted one was 61% higher than that of the GRC. In addition, there are *secondary benefits of GRC products* and a lower environmental impact due to reduced weight of products made from GRC, reduced demands on transport, reduced breakage, reduced amount of site-work and the ease of handling.

9. Summary of benefits

9.1. Economy through a combination of low weight and high strength.

Self-weight of structures decreases when GRC is used and demands on foundations are reduced. GRC cladding is suitable even for very high-rise buildings and offers good performance under seismic loading.

9.2. Freedom of shape.

GRC is easily mouldable into a wide range of shapes, including intricate grilles, panels with double curvature and 3-D objects. The high freedom of shape permits production of structurally very efficient elements. Easily cast, it can produce items with very fine details and reproduce very complex features and elements of both modern and historic buildings.

IOP Conf. Series: Materials Science and Engineering **246** (2017) 012002 doi:10.1088/1757-899X/246/1/012002

9.3. Durability.
Basic reinforcement is non-ferrous and the GRC products are not susceptible to corrosion as in traditional reinforced concrete. Low permeability and a very slow rate of carbonation offer protection against corrosion of steel in adjacent reinforced concrete. GRC has an inherently high resistance to extreme exposure conditions (freeze/thaw, fire etc.)

9.4. Appearance.
An extremely wide range of attractive surface finishes is available, satisfying the highest requirements for an aesthetic appearance of new structures and capable of matching colour and texture of surfaces of existing buildings. Durable and brightly coloured surfaces with enhanced self-cleaning can be achieved in a variety of textures and shapes.

9.5. Environment.
The relatively low weight of GRC products reduces CO2 emissions associated with their transport. There are no VOCs or other pollutants emitted from the material itself, neither in production nor in use.

GRC is fully recyclable into concrete and other applications. In addition, the photocatalytic eGRC reduces directly and significantly the concentration of pollutants in the surrounding air, leading to a better quality of environment, especially in congested urban centres and at a minimal additional cost.

10. Current research

Unlike for other types of fibre reinforced concrete, research into properties and fracture mechanics of GRC in early years had been restricted to in-house research by large companies making the AR glass fibres. Fibres were provided only to licensed GRC producers and not to independent researchers. Combined with inherent difficulties and challenges for research due to the very complex internal structure of GRC, such attitudes almost eliminated independent research for many years. On the other hand, concrete reinforced with other fibres (steel/metal, polymeric fibres, carbon fibres, natural fibres etc.) had been a popular academic research topic for decades. Closer look at the plethora of conferences and multitude of publications on 'fibre-concrete' over the last 40 years reveals that in many of the events, glass fibres were not even mentioned. Situation improved since early 1990s, when AR fibres became freely available, however, current research and development is still confined to very few isolated centres with intermittent projects. Recently, independent research on textile (glass fibre) reinforced concrete (namely in Germany) developed and some of the work has overlapped with GRC.

Alongside the rapid rise in the use of GRC there is now a growing demand for development and for an advance to new versions of GRC of an even higher performance. Manufacturers of AR fibres do maintain some R&D in this area, however, a coordinated, systematic research requires additional support, beyond what the small/medium size producers of GRC can afford. External national/international investment and coordination will be required to set up and operate expensive research infrastructure required to exploit further the potential of GRC.

11. Applications

One of the key features of GRC since its first development has been its versatility in use. However, over the last decade the range of the applications continued to grow and it is now extremely large. The most recent broad review of completed projects [2] included eleven categories, namely *mature structures, civic/public buildings, office and commercial buildings, residential buildings, religious structures, art and recreation, reconstructions and conservation of historic buildings, interior decorations and furniture, architectural building components and civil & environmental engineering.*

It has shown that at one end, GRC is used for mass production of small, simple and unsophisticated items for everyday use (flowerpots, drainage channels, window cills, street furniture etc.). At the other end, it is used by leading international architects for very large scale, high-tech, iconic projects. GRC is able to fulfil demands for highest structural complexity, size of construction elements, freedom of shape to achieve spectacular appearance, durability and the highest quality.

Architects world-wide now enjoy the high degree of design freedom provided by GRC, together with a direct, positive environmental performance it offers. Some of the recently completed projects would have been un-feasible to build just a few years ago. There are many impressive examples, where flowing, curving lines and unusual textures can be identified as common features of the facade structures. Geographically, GRC is no longer limited to the developed world and to a few rich developing countries, it is now found in more remote places such as Bhutan or Georgia.

Two projects have been selected to illustrate the achievements:

Mature structures – projects completed more than 20 years ago: The *office building at 30 Canon Street, London,* UK (Fig 7), remains the best example of a substantial application of the earliest version of GRC. The building is clad with approx. 1900 hollow double-sided GRC panels, produced by an early spray-up process. The facade as shown in Fig 5 has been cleaned and coated.

Public buildings

The *Nanjing Youth Olympic Centre* is a very large development all clad in GRC (Fig 8). A total of 110 000 m^2 of GRC was used to produce 12000 units, which fell into four categories:
40000 m^2 of flat panels from 3m x 2m to 3m x 6m in size, 32000 m^2 of folded panels from basic size of (2+2)m x 2m to (5+2)m x 3m, 30000 m^2 of double-curve panels from 3m x 2m to 6m x 4m, 8000 m^2 of single-curve panels from 3m x 2m to 6m x 4m. The panels were designed for installation in inclined lines both for roofs and walls.

Figure 7. 30 Canon St. , London, UK. Completed 1974. Architect / Structural Engineer: Whinney, Son and Austen Hall , Ove Arup

IOP Conf. Series: Materials Science and Engineering **246** (2017) 012002 doi:10.1088/1757-899X/246/1/012002

Figure 8. Nanjing Youth Olympic Centre. Completed 2014, Architect/Engineer : Zaha Hadid / Nanjing Beilida New materials System Engineering Co.Ltd. GRC was produced by Nanjing Beilida Co. Ltd., China.

12. Conclusions

GRC is already a well established construction material, however, while there has been an enormous amount of research and a great number of publications on concrete reinforced with metallic and other fibres, leading to a disproportionally limited amount of practical applications, it is the other way round in case of GRC.

There is a very large amount of GRC used in practice but very little research and few publications to support it. A strong case exists for widening of the knowledge of what GRC can do and for research required to improve further its properties.

References

[1] Biryukovich K and Biryukovich Yu L 1964 Steklocement Budivelnik (Kiev).
[2] Bartos PJM 2017 Glassfibre Reinforced Concrete: principles, production, properties and applications 176p ISBN 978-184995-326-9 (Whittles Publishing, Dunbeath, UK)
[3] iGRCA 2017 www.grca.org.uk
[4] Brameshuber W et al. Textile Reinforced Concrete 292p ISBN 2-912143-99-3 (Rilem, Cashan, France)
[5] EN 15422:2008 Precast Concrete Products
[6] EN 14649:2005 Test method for strength retention of glass fibres in cement and concrete
[7] Anon. 2014 10 completely 3-D printed houses appear in Shanghai www.3ders.org
[8] Rain C and Kirkegaard PH 2015 Adaptive mould Proc. 16th iGRCA Congr. Dubai 41
[9] Bartos P 1976 Bond Characteristics in fibrous composites with brittle matrices 306p PhD thesis Univ. of Southampton
[10] Bartos P 1985 Effects of changes in fibre strength and bond characteristics due to ageing on fracture mechanism of GRC Proc.Symp.on Durability of Glass Fiber reinforced concrete S. Diamond Ed. (PCI Chicago USA)
[11] Bartos PJM 1987 Brittle matrix composites reinforced with bundles of fibres From Mat. Sci. to Construction Engineering, J Maso Ed. (Chapman and Hall, London)

IOP Conf. Series: Materials Science and Engineering **246** (2017) 012002 doi:10.1088/1757-899X/246/1/012002

[12] Duris M and Bartos PJM 1994 Inclined tensile strength of fibres in a cement based composite Composites V.25 No.10 945
[13] Bartos PJM and Zhu WZ 1997 Cem and Con. Res. V27 No11 1701
[14] Trtik P and Bartos PJM 2000 J.Mat.Sci. V.19 No10. 903
[15] GRCA 2016 Specification for the manufacture, curing and testing of glassfibre reinforced concrete (iGRCA UK)
[16] Jones G 2004 Practical design guide for GRC using limit state theory (iGRCA, Camberley, UK)
[17] GRCA 2016 Methods of testing glassfibre reinforced concrete (iGRCA, UK)
[18] Warrington Fire 2008 Glassfibre reinforced concrete: Classification of reaction to fire (iGRCA, Camberley, UK)
[19] Ridd P 2017 Private communication
[20] EN1170 Pt.7:1998
[21] Bartos PJM 2009 Sklovláknobeton e-GRC: nová cesta ke zlepšení čistoty fasád a kvality ovzduší našich měst. Beton TKS V. 9, No. 2.
[22] Cassar L 2006 Nanotechnology and photocatalysis in cementitious materials Proc. 2nd NICOM deMiguel Y et al. Eds. ISBN 2-912143-87-X
[23] European Commission 2013 Clean air policy package http://ec.europa.eu/environment/air/clean_air_policy html
[24] White I 2016 Private communication
[25] Ferry R and Parrott L 1999 Defining and improving environmental performance in the concrete industry DETR Project 39/3/437 CC 1533 UK

IOP Conf. Series: Materials Science and Engineering **246** (2017) 012003 doi:10.1088/1757-899X/246/1/012003

Replacement of minimum steel bar reinforcement with steel fibres in structural concrete members

F Junker[1], K Holschemacher[1], T Müller[2] and H Kieslich[3]

[1] HTWK Leipzig, University of Applied Sciences, Structural Concrete Institute, Karl-Liebknecht-Strasse 132, 04277 Leipzig, Germany

[2] Baustoffprüflabor Müller & Lobisch GmbH, Am Gewerbepark 8, 04860 Dreiheide, OT Süptitz, Germany

[3] K&K Baugesellschaft mbH, Güterbahnhofstraße 5a, 04860 Torgau, Germany

E-mail: florian.junker@htwk-leipzig.de

Abstract. Using lightweight concrete enables a decrease in dead load and thermal conductivity in the case of the manufacturing of structural concrete members. With the addition of steel fibres in concrete, its properties are altered from brittle to ductile, so that the use of additional minimum reinforcement for securing ductility and crack control can be avoided. This study is aimed at investigating the possibility of replacing conventional minimum steel bar reinforcement with steel fibre reinforcement in lightweight aggregate concrete under flexural loading. Therefore, six full-scaled beams with two different lightweight aggregate concretes (LWAC) (oven-dry densities of <1,200 kg/m^3 and <1,600 kg/m^3) as well as different types of reinforcement were prepared. For each LWAC, a beam with traditional steel bars, a beam with steel fibres and a beam with a combination of steel fibre reinforcement and reduced steel bar reinforcement were produced. The cracking behaviour of the lightweight concrete beams was studied in a four-point bending test. The results of this study show that it is possible to replace a high amount of the conventional mesh or bar reinforcement with steel fibres.

1. Introduction

Lightweight aggregate concrete (LWAC) permits a decrease in dead load. The lighter concrete construction results in a reduced load flow so that successive structural members can have smaller dimensions (especially foundations). Therefore, the costs of the building are reduced. In the precast concrete construction, LWAC offers a cost reduction for transportation [1]. LWAC is suitable for structural members, where the required material strength and required ratio of dead load to live load are reduced [2]. In addition, the lower density improves the thermal insulation.

Despite the low requirements on structural walls, a minimum reinforcement for securing ductility and crack control is demanded by the European Standard (DIN EN 1992-1-1 [3]). Also, further reinforcement is required in areas for openings. With the addition of steel fibres to concrete, its properties are altered from brittle to ductile [4, 5, 6], so the use of additional reinforcement could be avoided. Furthermore, the fibres could minimize the creation of cracks caused by shrinkage of LWAC [7].

This study is aimed at investigating the possibility of replacing conventional minimum steel bar reinforcement with steel fibre reinforcement in lightweight aggregate concrete under flexural loading. Therefore, six full-scaled beams with two different lightweight aggregate concretes

(LWAC) (oven-dry densities of <1,200 kg/m^3 and <1,600 kg/m^3) as well as different types of reinforcement were prepared. For each LWAC, a beam with traditional steel bars, a beam with steel fibres and a beam with a combination of reduced steel bar reinforcement and unmodified steel fibre reinforcement were produced. The cracking behaviour of the lightweight aggregate concrete beams was studied in a four-point bending test.

2. Experimental work

2.1. Materials

A Portland blast-furnace slag cement (conforming to DIN EN 197-1 [8] and ordinary fly ash (per DIN EN 450-1 [9]) were utilized as the cement and mineral addition.

Two grades of expanded glass fine lightweight aggregates (LWF1 and LWF2) and one grade of expanded clay coarse lightweight aggregate (LWC) (conforming to DIN EN 13055-1 [10]) are considered in this investigation. In order to achieve a good gradation, normalweight sand (NWF) was added to all mixtures. More information about the aggregates are provided in table 1.

Hooked steel fibres were utilized. They have a tensile strength of 1550 MPa, a length of 60 mm, a diameter of 0.9 mm (aspect ratio $l/d = 67$), an elastic modulus of 200 GPa and a specific gravity of 7.85 g/cm^3.

Superplasticizer and stabilizer (corresponding to DIN EN 934-2 [11]) were selected to improve the fresh concrete properties. They enhance the workability and inhibit the floating properties of the lightweight aggregates.

Table 1. Aggregate properties

Type	Lightweight			Normal
Designation	LWF1	LWF2	LWC	NWF
Aggregate Size (mm)	0.25–0.50	1–2	2–10	0–2
Particle density (g/cm^3)	0.55	0.35	0.84	2.63
Bulk density (g/cm^3)	0.30	0.22	0.45	1.45
Water absorption (%)[a]	—	—	10	—

[a] Water absorption after 60 minutes

2.2. Methods

Four concrete mixes were considered for this study. The development of the mix designs is shown in [12]. The oven-dry density of hardened concrete was targeted to be 1,200 (D1.2) and 1,600 kg/m^3 (D1.6). Each concrete type was produced without steel fibres and with a steel fibre content of 1.0 vol.-% (F1.0). In this study, a flow table spread of at least 42 cm (16.5") was targeted. Superplasticizer was used at a dosage of 0.5–0.7 % (of the cement content) to achieve this workability.

The coarse lightweight aggregates were added directly (as received) to the mix. The moisture content of the expanded clay aggregates ranged from 4 to 7 % and it was considered in the mix designs.

2.2.1. Production Concretes were mixed in a 1.5 m^3 industrial ready-mixed concrete mixer with the following approach: firstly, dry materials were mixed for 30 seconds; then water was added and mixed for one more minute. Secondly, stabilizer was added and mixed for another minute. Thirdly, superplasticizer was added and mixing was continued for one minute. Finally,

IOP Conf. Series: Materials Science and Engineering **246** (2017) 012003 doi:10.1088/1757-899X/246/1/012003

steel fibres were added and mixed until the total time of 5 and a half minutes had surpassed. The concrete mix compositions are presented in table 2.

Table 2. Concrete mix composition[a]

Components (kg/m^3)	D1.2	D1.2_F1.0	D1.6	D1.6_F1.0
Cement	350	350	390	390
Fly ash	80	80	80	80
Water	185	185	200	200
NWF	283	279	641	631
LWF1	49	48	—	—
LWF2	53	52	—	—
LWC	239	236	307	302
Steel fibres	—	80	—	80
Superplasticizer (%)[b]	0.50	0.50	0.70	0.50
Stabilizer (%)[b]	0.30	0.30	0.30	0.30
Total	1240	1310	1620	1684

[a] Values are rounded

[b] Dosage of additives in percentage of cement content

The concrete specimens were cast in plastic moulds (cubes, cylinders and small beams) and wooden formwork (full-scaled beams) and compacted with an internal vibrator. De-moulding was performed after two days. The cubic and cylindrical specimens were kept in water tanks. The small beam specimens were sealed with foil. Full-scaled beams were left in the outside until they moved to the test site. All tests were carried out after 28 days.

2.2.2. Experimental investigations For all the mixtures, fresh and hardened concrete properties were tested. In the case of the workability of fresh concrete, the flow table test (DIN EN 12350-5 [13]) was conducted. The compressive strength (DIN EN 12390-3 [14]) was tested on 150 mm cubes. The Young's modulus (DIN EN 12390-13 [15]) was conducted on cylinders having a diameter of 150 mm and a length of 300 mm. The flexural strength was carried out on 700 x 150 x 150 mm beams. They were tested using the four-point bending test according to the German Committee for Structural Concrete guideline "Steel Fiber Reinforced Concrete" [16].

The dimensions of the full-scaled beams, the test set-up and the reinforcement drawing are displayed in figure 1. The load was applied on the two cantilevers of the beam, so that a constant flexural moment was generated between the supports. Table 3 lists the designations and the different characteristics of the specimens.

All six test beams had an rectangular cross-section with $w/h = 400/200$ mm. The concrete cover for reinforcements was 20 mm. In the 2500 mm long measuring range were no stirrups placed, so that the crack development was not disturbed by them. Shear reinforcement was only designated in the range of the cantilevers to cover the shear forces. Furthermore, two extra steel bars were added in the upper layer – from the support to the end of the cantilever – to increase the flexural strength in this area.

The quantity and the diameter of the reinforcement bars (upper & lower) in the measuring range of the beam (section A) is given in table 3. For series D1.2–S and D1.6–S, it was calculated for the minimum reinforcement for crack control ($w_{\max} = 0.3$ mm) according to Eurocode 2 [3].

IOP Conf. Series: Materials Science and Engineering **246** (2017) 012003 doi:10.1088/1757-899X/246/1/012003

Table 3. Summary of the experiments

Designation	Concrete type	Reinforcement bars Section A	Steel fibres
D1.2–S	D1.2	4 ø 6 mm	—
D1.2–S+F	D1.2_F1.0	2 ø 6 mm	1.0 vol.-%
D1.2–F	D1.2_F1.0	—	1.0 vol.-%
D1.6–S	D1.6	4 ø 6 mm	—
D1.6–S+F	D1.6_F1.0	2 ø 6 mm	1.0 vol.-%
D1.6–F	D1.6_F1.0	—	1.0 vol.-%

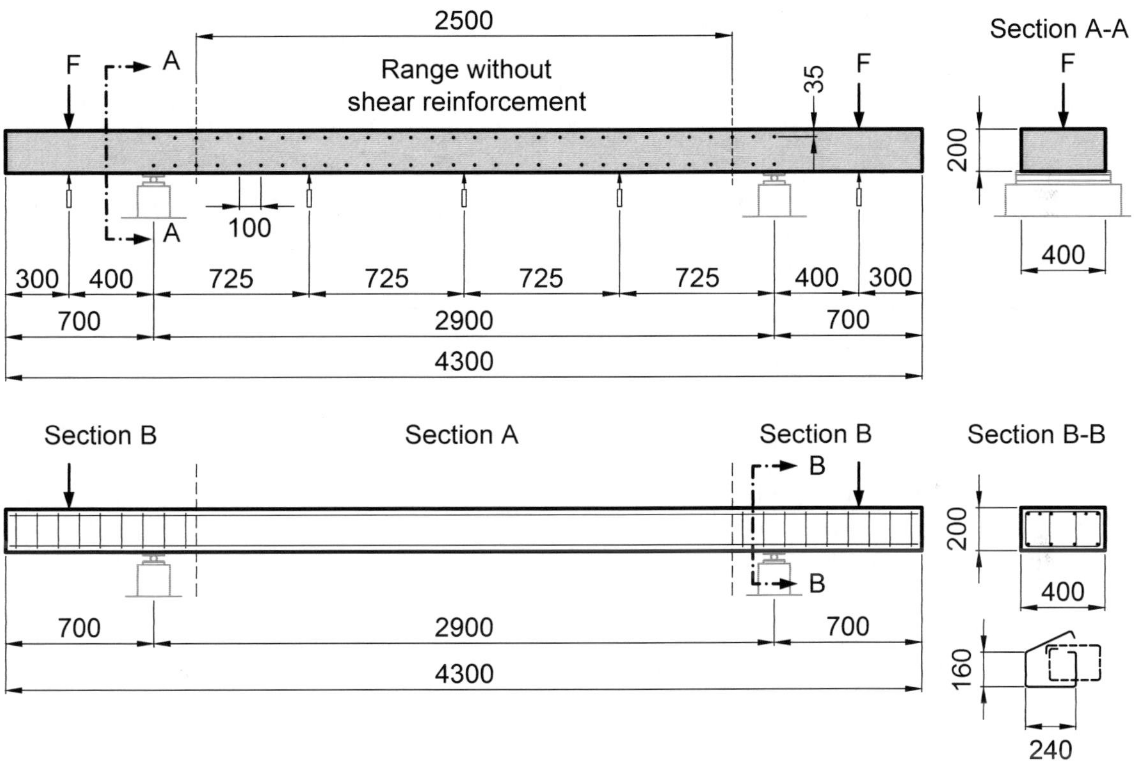

Figure 1. Test set-up and reinforcement drawing of the full-scaled beams

The steel bar reinforcement in the cantilevers (sections B) are for all beams the same:

- upper reinforcement: 6 ø 6 mm
- lower reinforcement: 4 ø 6 mm
- shear reinforcement: 9 (18) ø 6/100 mm.

For the testing of the beams, the load was continuously applied on the cantilevers until the initial crack formation. This also represents the first load step. Then, the load was displacement controlled with fixed load steps on specified mid-span deflections of the beam. During the experiments, the following measurements were conducted (see also figure 2):

- the applied load on the cantilevers with two 500 kN load cells;
- the deflection at five locations with 20 mm LVDT's;
- the crack development on the top side and one side of the beam at each load step;
- the crack widths with a crack width gauge at each load step.

3. Results and discussion

3.1. Fresh and hardened concrete properties

Table 4 presents the results of fresh and hardened concrete properties. Here, the average values from three tests for compressive strength and Young's modulus as well as seven tests for the flexural and residual flexural strength are presented.

The analysis of the residual flexural strength tests is performed with the residual load at deflections of 0.5 mm (serviceability limit state) and 3.5 mm (ultimate limit state). These deflections are related to crack widths and calculated strain conditions, which are expected to occur in the serviceability and ultimate limit states. A deflection value δ of 0.5 mm corresponds to a strain value of $\varepsilon = 3.5$ ‰. The ultimate strain ε of 25 ‰ was specified for the second deflection value $\delta = 3.5$ mm in case of steel fibre reinforced concrete [17].

3.2. Results from four-point bending tests

The moment–midspan deflection relationship obtained from the four-point bending tests is illustrated in figure 2. The diagrams show all three variations of the specimens — steel bar (4 ø 6 mm), steel bar (2 ø 6 mm) with steel fibre as well as steel fibre reinforcement.

For both types of concrete (D1.2 and D1.6), the series with steel bars and steel fibres had an improved residual load-bearing behaviour compared to the other series. The maximum flexural moment was increased by 11 % and 30.5 % for D1.2 and D1.6, respectively.

The beams with steel fibres had an improved performance up to a deflection of circa 5 mm. With further increasing of the displacement, the maximum moment could not be raised. Instead, the predominant crack was quickly opening.

3.3. Replacement of minimum reinforcement

Eurocode 2 [3] requires a minimum reinforcement for ductility and crack control.

Related to the full-scaled beam tests, an increase or stable flexural moment after initial cracking indicates that these requirements are met. This behaviour is independent of the type

Table 4. Fresh and hardened concrete properties

Concrete type	Flow table spread (mm)	Compressive strength (MPa)	Young's modulus (MPa)	Flexural strength (MPa)	Residual flexural strength L1/L2[a] (MPa)
D1.2	530	18.3	11,233	3.2	—
D1.2_F1.0	420	21.0	15,267	3.2	5.0/3.1
D1.6	650	20.8	13,033	4.5	—
D1.6_F1.0	430	30.2	18,267	4.6	6.1/4.5

[a] Performance classes L1 and L2 according to the German guideline for steel fibre reinforced concrete [16]

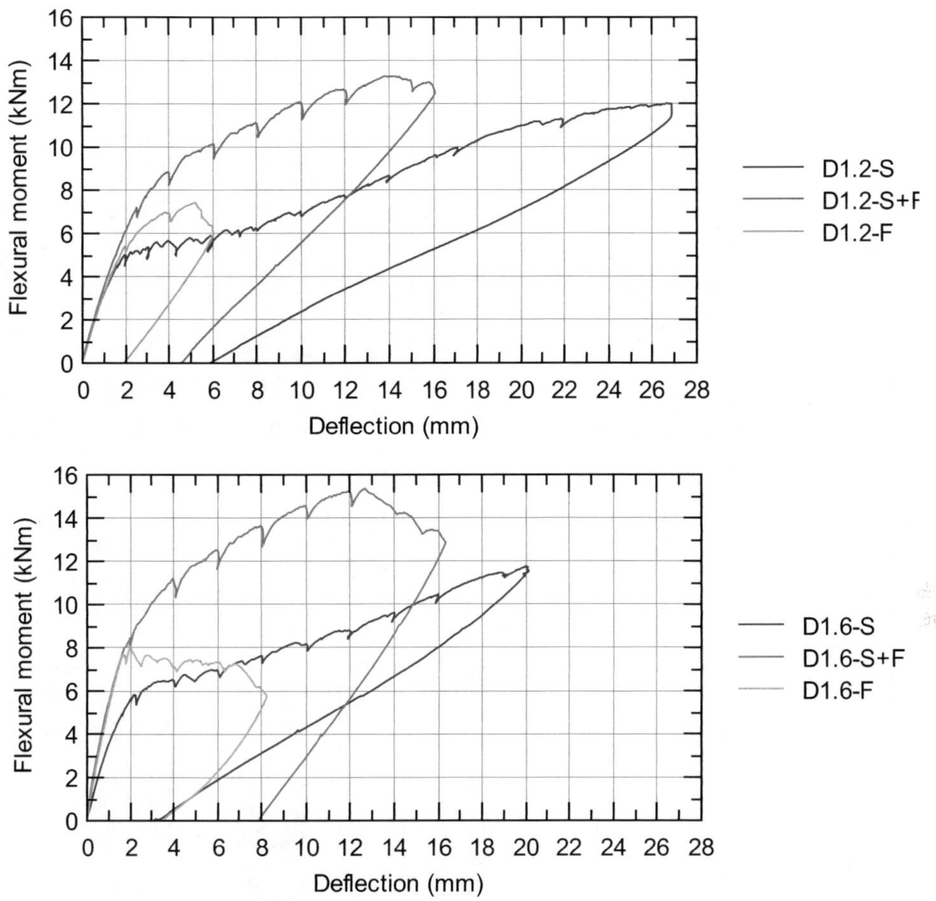

Figure 2. Moment–midspan deflection diagram for concrete series D1.2 (top) and D1.6 (bottom)

of reinforcement (steel bar or steel fibre). For the comparison of the beam tests, the cracking moment was put in relationship with two specified flexural moments. First, the flexural moment at a deflection of $l/500$ (5,8 mm) for securing member ductility and second, the moment at a crack width of 0.3 mm for crack control. If these specified moments are greater or equal to the cracking moment, the requirements for the minimum reinforcement are fulfilled.

The following approach was used for determination of the moment at a crack width of 0.3 mm. First, the measured crack widths were plotted as a function of the deflection for each load step. Second, a regression function of the data points was created for each series. Third, the deflection at a crack width of 0.3 mm of each series was determined using the regression function. And finally, the related moments were obtained using the moment–midspan deflection diagrams. Figure 3 presents the results of this approach. Additionally, the maximum crack width–deflection diagrams shows the faster crack opening with increasing deflection of the series with steel fibres ("S+F" and "F") compared to the series "S".

Figure 4 shows the obtained relationship for both types of concrete (D1.2 and D1.6). If the ratio is bigger than 1.0, the respective reinforcement meets the requirements of the minimum reinforcement. Nearly all series achieve a bigger ratio than 1.0. Only the series D1.6 with steel fibres (D1.6–F) have a slightly lower ratio of 0.93 and 0.96 for the ductility and crack control

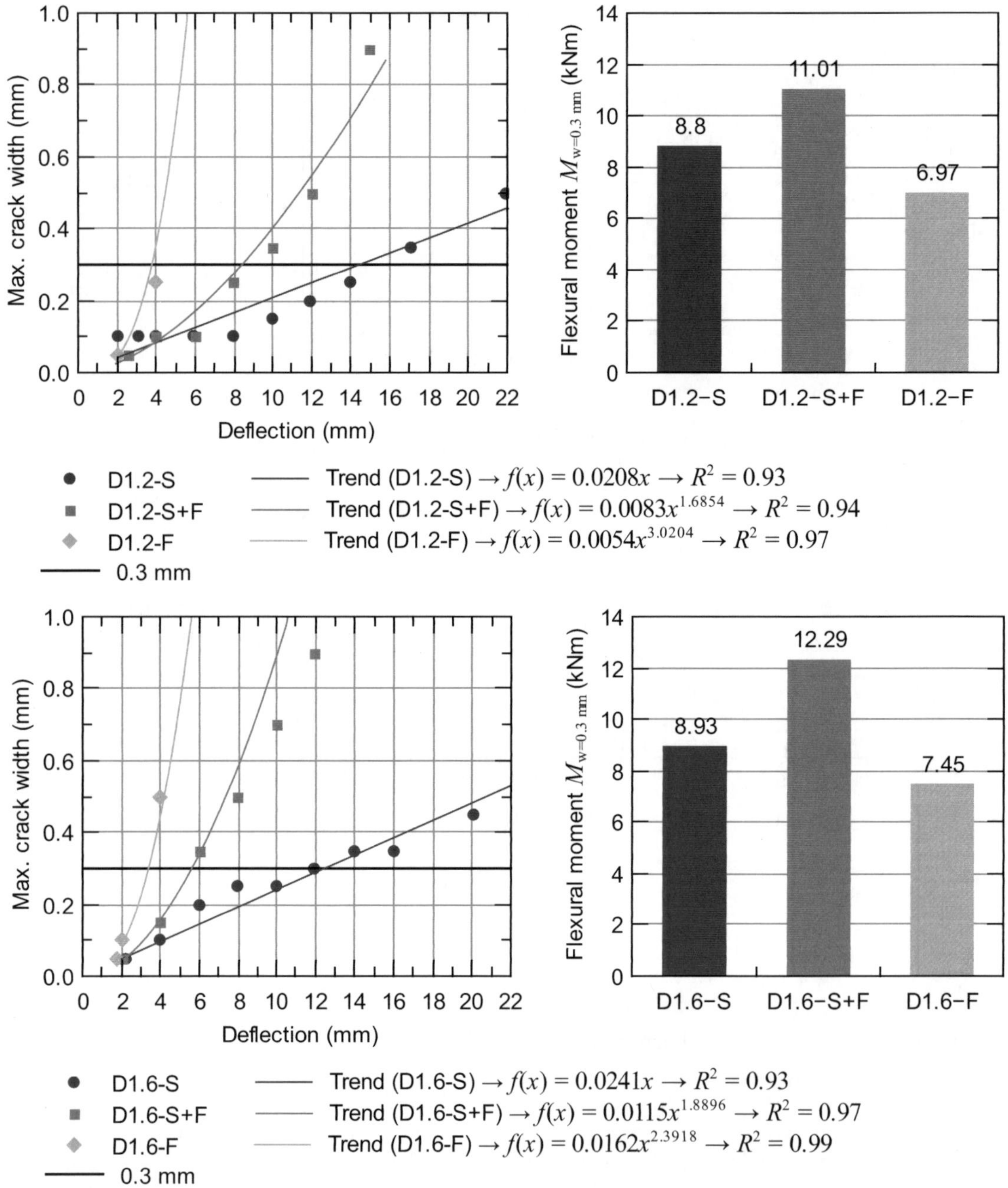

Figure 3. Determination of the flexural moment at a crack width of 0.3 mm — series D1.2 (top) and D1.6 (bottom)

criteria, respectively. The series with combined reinforcement (series "S+F" – 2 ø 6 mm steel bars and fibres) show the highest ratios – except D1.6–S+F for crack control. The series D1.2–F (only steel fibres) has a sufficient load-bearing capacity for securing a ductile member behaviour and crack control and can be utilized for load-bearing walls. The series D1.6–F has a insufficient load-bearing capacity. But it could be possible to improve its performance with another steel

IOP Conf. Series: Materials Science and Engineering **246** (2017) 012003 doi:10.1088/1757-899X/246/1/012003

fibre type (e.g. a higher tensile strength or a higher aspect ratio l/d). Further tests are necessary to proof this hypothesis. For a combined reinforcement with steel bars and fibres, it is possible to reduce the steel fibre content, because of the high capacity ratios.

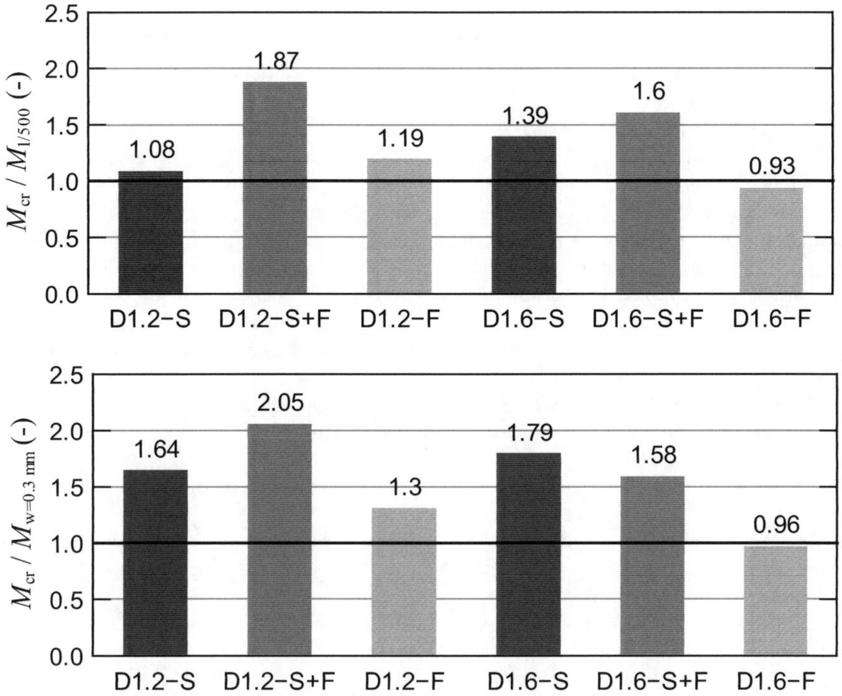

Figure 4. Comparison of the relationship between the cracking moment and the moment at $l/500$ (top) as well as the moment at a crack width of 0.3 mm (bottom)

4. Conclusion

The following conclusions can be drawn from the results of this study:

(1) Addition of steel fibres appreciably increased the ductility of LWAC. A higher load-bearing capacity with combined reinforcement (series "S+F" – steel fibres with reduced steel bars) was conducted.

(2) Specimens with only steel fibres ("F") have an improved performance up to a deflection of approximately 5 mm compared to the series with only steel bars ("S").

(3) Specimens with steel fibres ("S+F" and "F") have a larger crack opening with increasing deflection compared to the series "S".

(4) It is possible to replace the minimum reinforcement for securing ductility and crack control in load-bearing LWAC walls with steel fibres. In this study, a suitable steel fibre reinforced lightweight aggregate concrete (SFRLWAC) with a oven-dry density below 1,200 kg/m^3 was developed.

(5) Another SFRLWAC with a oven-dry density below 1,600 kg/m^3 could be utilized for an application in load-bearing walls with an optimized steel fibre type (e.g. a higher tensile strength or aspect ratio l/d)

Acknowledgments

The authors would like to thank the German Federal Ministry for Economic Affairs and Technology (BMWi) for subsidising the project under the subsidy program "Central Innovation Program SME – Subsidy Module Cooperation Project's".

References

[1] Faust T 2003 *Leichtbeton im konstruktiven Ingenieurbau* Bauingenieur-Praxis (Berlin: Ernst) ISBN 9783433016138 (in German)
[2] Cavaleri L, Miraglia N and Papia M 2003 *Eng. Struct.* **25** 115–125 ISSN 01410296
[3] DIN EN 1992-1-1 Eurocode 2: Design of concrete structures - part 1-1: General rules and rules for buildings; German version EN 1992-1-1:2004 + AC:2010
[4] Campione G 2001 *Mater. Struct.* **34** 201–210 ISSN 1359-5997
[5] König G, Holschemacher K, Weiße D and Schneider H 2004 *Sixth RILEM Symposium on Fibre-Reinforced Concretes (FRC)* RILEM proceedings PRO ed Di Prisco M, Felicetti R and Plizzari G A (Bagneux: RILEM Publications) pp 1321–1330 ISBN 2-912143-51-9
[6] Müller T 2015 *Untersuchungen zum Biegetragverhalten von Stahlfaserbeton und betonstahlbewehrtem Stahlfaserbeton unter Berücksichtigung des Einflusses von Stahlfaserart und Betonzusammensetzung* Ph.D. thesis University of Leipzig (in German)
[7] Hassanpour M, Shafigh P and Mahmud H B 2012 *Constr. Build. Mater.* **37** 452–461 ISSN 09500618
[8] DIN EN 197-1 Cement - part 1: Composition, specifications and conformity criteria for common cements; German version EN 197-1:2011
[9] DIN EN 450-1 Fly ash for concrete - part 1: Definition, specifications and conformity criteria; German version EN 450-1:2012
[10] DIN EN 13055-1 Lightweight aggregates - part 1: Lightweight aggregates for concrete, mortar and grout; German version EN 13055-1:2002
[11] DIN EN 934-2 Admixtures for concrete, mortar and grout - part 2: Concrete admixtures - definitions, requirements, conformity, marking and labelling; German version EN 934-2:2009+a1:2012
[12] Junker F, Holschemacher K, Müller T and Kieslich H 2016 *11th Annual International Concrete Sustainability Conference, At Washington, D.C., USA*
[13] DIN EN 12350-5 Testing fresh concrete - part 5: Flow table test; German version EN 12350-5:2009
[14] DIN EN 12390-3 Testing hardened concrete - part 3: Compressive strength of test specimens; German version EN 12390-3:2009
[15] DIN EN 12390-13 Testing hardened concrete - part 13: Determination of secant modulus of elasticity in compression; German version EN 12390-13:2013
[16] Deutscher Ausschuss für Stahlbeton 2012 *DAfStb-Richtlinie Stahlfaserbeton* (Beuth) (in German)
[17] Deutscher Ausschuss für Stahlbeton 2015 *Commentary on the DAfStb guideline "Steel fibre reinforced concrete"* 1st ed (*Deutscher Ausschuss für Stahlbeton* vol 614(en)) (Berlin and Wien and Zürich: Beuth) ISBN 978-3-410-65279-3

IOP Conf. Series: Materials Science and Engineering **246** (2017) 012004 doi:10.1088/1757-899X/246/1/012004

Probabilistic design of fibre concrete structures

R Pukl[1], D Novák[2], T Sajdlová[1], D Lehký[2], J Červenka[1] and V Červenka[1]

[1] Červenka Consulting, Na Hřebenkách 55, 15000 Prague, Czech Republic
[2] Faculty of Civil Engineering, Brno University of Technology, Veveří 95, 602 00 Brno, Czech Republic

E-mail: radomir.pukl@cervenka.cz

Abstract. Advanced computer simulation is recently well-established methodology for evaluation of resistance of concrete engineering structures. The nonlinear finite element analysis enables to realistically predict structural damage, peak load, failure, post-peak response, development of cracks in concrete, yielding of reinforcement, concrete crushing or shear failure. The nonlinear material models can cover various types of concrete and reinforced concrete: ordinary concrete, plain or reinforced, without or with prestressing, fibre concrete, (ultra) high performance concrete, lightweight concrete, etc. Advanced material models taking into account fibre concrete properties such as shape of tensile softening branch, high toughness and ductility are described in the paper. Since the variability of the fibre concrete material properties is rather high, the probabilistic analysis seems to be the most appropriate format for structural design and evaluation of structural performance, reliability and safety. The presented combination of the nonlinear analysis with advanced probabilistic methods allows evaluation of structural safety characterized by failure probability or by reliability index respectively. Authors offer a methodology and computer tools for realistic safety assessment of concrete structures; the utilized approach is based on randomization of the nonlinear finite element analysis of the structural model. Uncertainty of the material properties or their randomness obtained from material tests are accounted in the random distribution. Furthermore, degradation of the reinforced concrete materials such as carbonation of concrete, corrosion of reinforcement, etc. can be accounted in order to analyze life-cycle structural performance and to enable prediction of the structural reliability and safety in time development. The results can serve as a rational basis for design of fibre concrete engineering structures based on advanced nonlinear computer analysis. The presented methodology is illustrated on results from two probabilistic studies with different types of concrete structures related to practical applications and made from various materials (with the parameters obtained from real material tests).

1. Introduction

The probabilistic approach represents the most accurate method for the safety assessment of civil engineering structures. The accuracy of this approach is much higher if the nonlinear structural analysis is used as a limit state function. The numerical simulation resembles a real testing of structures by considering a representative group of samples, which can be statistically analyzed for the assessment of safety. This concept is supported by the new ***fib*** Model Code 2010 [1] where rational safety assessment approach is presented, which reflects new developments in safety formats based on probabilistic methods. In the Chapter 4 on *Principles of structural design* the probabilistic safety format is introduced as a general and rational basis of safety evaluation. In addition to the partial factor

IOP Conf. Series: Materials Science and Engineering **246** (2017) 012004 doi:10.1088/1757-899X/246/1/012004

format (which remains as the main safety format for most practical cases) a global resistance format is recommended for nonlinear analysis.

In case of fibre concrete structures the design codes are generally not available or not sufficient. Therefore, the computer simulation based on advanced nonlinear finite element analysis has a big potential in design of these structures. The numerical results can be introduced into a suitable engineering safety concept, from which the fully probabilistic analysis is the ultimate tool for design and safety assessment. It is superior to simplified methods because it provides information on the variability of resistance. However, it is computationally demanding and requires good information about random properties of input variables. Therefore, it is applied in special cases, where consequences of failure substantiate the increased effort.

The probabilistic analysis is performed here using software SARA, which integrates ATENA and FReET program tools. The variability of basic properties is described by distribution functions and its parameters (mean, standard deviation, etc.). Probabilistic analysis of the resistance is performed by numerical method such as Latin hypercube sampling (LHS) method. Resulting set of resistance values is approximated by a probability distribution function (PDF) of global resistance, and describes the random properties of the resistance. Finally, for a required reliability index β, or failure probability P_f, a value of the design resistance R_d shall be calculated.

2. Probabilistic nonlinear assessment of engineering structures

Probabilistic analysis is an efficient tool for safety assessment of civil engineering structures, in particular of concrete or fibre concrete structures. In the probabilistic nonlinear approach the structural resistance R_d is calculated by means of the probabilistic nonlinear analysis. The classical statistical and reliability approach is to consider material parameters as random variables with prescribed distribution function. The stochastic response requires repeated analyses of the structure with these random input parameters, which reflects randomness and uncertainties in the input values (see e.g. [2]). In this approach, the resistance function $r(r)$ is represented by nonlinear structural analysis and loading function $s(s)$ is represented by action model. Safety can be evaluated by the reliability index β, or alternatively by failure probability P_f taking into account all uncertainties due to random variation of the input values – material properties, dimensions, loading, and other.

Probabilistic analysis based on the nonlinear numerical simulation includes following steps:

- *Numerical model based on the nonlinear finite element analysis.* This model describes the resistance function $r(r)$ and performs a deterministic analysis of resistance for given set of input variables.
- *Randomization of input variables* (material properties, dimensions, boundary conditions, etc.). This can also include some effects of actions, which are not in the action function $s(s)$ (for example pre-stressing, dead load etc.). Random properties are defined by random distribution type and its parameters (mean standard deviation, etc.). They describe the uncertainties due to statistical variation of resistance properties.
- *Probabilistic analysis of resistance and action.* This can be performed by stratified method of Monte Carlo-type of sampling, such as LHS sampling method. Results of this analysis provide random parameters of resistance and actions, such as mean, standard deviation, etc. and the type of distribution function for resistance.
- *Evaluation of safety* using reliability index β or failure probability P_f.
- Probabilistic analysis can be also used for *determination of design value of resistance function* $r(r)$ expressed as R_d. Such analysis involves repeatedly the first three steps above, and R_d is determined for required reliability β or failure probability P_f.

In order to make the application of the probabilistic nonlinear analysis user-friendly, special software tool has been developed by the authors and their co-workers. The resulting software SARA (Structural Analysis and Reliability Assessment) integrates software tools ATENA and FReET. It is equipped with a user-friendly shell called *SARA Studio*, which leads the user interactively through the modelling and randomization process of the solved problem as described above. All features (or just selected ones) of the involved programs including modeling of deterioration/degradation phenomena can be utilized also in the reliability analysis and performance-based assessment of concrete structures. For this purpose, the interconnectivity between ATENA Engineering as well as ATENA Science (i.e. ATENA input file), with the probabilistic modules was achieved. The program control and data exchange is organized by an efficient and user friendly shell interface (Figure 1).

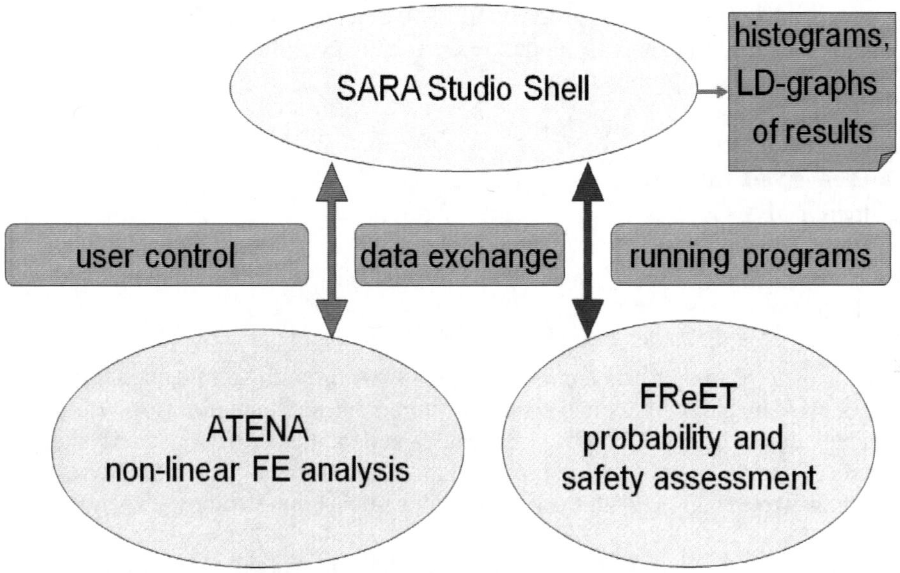

Figure 1. Structure of the SARA system.

3. Material damage modeling

The damage and failure of fiber concrete structures is analyzed in program ATENA [3] for nonlinear analysis of concrete structures. ATENA is capable of a realistic simulation of concrete behavior in the entire loading range with ductile as well as brittle failure modes as shown for instance in [4]. The program was developed for realistic simulation of reinforced concrete structures. It is based on the finite element method with nonlinear material models, and utilized for analysis of beams and girders, plates and shells, bridges, tunnels, dams, composite structures, strengthenings, structural details, fastenings, timber and masonry structures, and also structures made of fibre concrete, e.g. [5], [6].

The ATENA software consists of calculating core ensuring the nonlinear numerical analysis, and a user-friendly graphical interface for an efficient communication between end-user and program core. The numerical core covers the finite element technology, nonlinear material models and nonlinear solution.

The nonlinear material models are based on the orthotropic damage theory and special concrete-related theory of plasticity. As one of the main features the nonlinear fracture mechanics is employed for concrete cracking in tension. Based on the fracture energy approach the tensile cracks are modeled as smeared material damage which enables utilization of the continuum mechanics even for the damaged material. Objectivity of the solution (independency on the finite element mesh) is ensured using crack band method. The material law for concrete exhibits softening after reaching the tensile strength. The behavior of concrete in compression is defined by special theory of plasticity (three-parameter model), with non-associated plastic flow rule and softening. This material model for concrete can successfully reproduce also other important effect, such as volume change under plastic compression or compressive confinement. The constitutive model is described in more detail e.g. in [7]. Nonlinear solution is performed incrementally with equilibrium iterations in each load step.

The native graphical user-interface in ATENA Engineering supports all the specifics of reinforced concrete, e.g. input of discrete reinforcing bars, or evaluation of crack patterns in the damaged structural model. ATENA Engineering is perfectly suitable for static analysis of concrete structures, obtaining their load-displacement response and resistance, crack pattern including crack widths and identification of the failure mode. It can be used for structure optimization, assessment of retrofitting or reinforcement detailing.

In order to extend ATENA potential and features, the recent development combines the calculating core with AtenaStudio runtime and post-processing environment, and a powerful third-party program GiD for pre-processing. The resulting product ATENA Science covers broad range of structural and material behavior in time. It enables to model geometrically complicated shapes and it is suitable for analysis of complex structural problems, such as:

- dynamic implicit analysis
- dynamic eigenvalue analysis
- static stress analysis
- creep analysis
- transport of heat and fluids
- fire analysis

The coupling of the above effects can be often achieved through simultaneous solution of various constitutive models. Thus, dynamic analysis can capture nonlinear material response due to cracking, etc. In the eigenvalue analysis vibration frequencies reflect the stiffness changes due to material damage. In creep analysis, the cracking of concrete and redistribution of stress due to plastic deformations is reflected. In fire analysis material response is strongly dependent on changing temperature fields.

The connectivity to the stochastic analysis and the corresponding SARA Studio is recently available for all levels of the ATENA software – ATENA Engineering (2D and 3D) as well as ATENA Science.

3.1. Modeling of fibre concrete

Typical procedure of modeling a fibre concrete structure by nonlinear finite element analysis consists of several steps. The first one is the estimation of input material parameters based on mixture, contents and type of fibres, etc. The next step is modification of initial input parameters by inverse analysis of material tests, mainly three-point or four-point bending test, to better describe the fibre concrete behavior. The third and last step consist of analysis of the whole structural element using determined material parameters.

The tensile response of fibre concrete material differs from normal concrete not only in the values like tensile strength and especially fracture energy, but also in the shape of tensile softening branch. The original exponential function valid for normal concrete can be used as a first approach, but in most cases a more realistic form of the tensile constitutive law should be employed. Therefore, special material models at macroscopic level are needed for modeling of the fibre concrete material in the numerical simulation of fibre concrete structures as described e.g. in [8].

The most sophisticated and most general model of the fibre concrete material represents an extension to the basic fracture-plastic constitutive law [7] called *3D NLC2 User model*. It describes the tensile behavior according to the material response measured in tests point-wise in terms of the stress-strain relationship. The first part of the diagram is the usual stress-strain constitutive law. After exceeding the localization strain ε_{loc} the material law assumed for the characteristic crack band width L_{ch} is adjusted to the actual crack band width L_t (Figure 2(a)). The characteristic crack band width (characteristic length) is the size (length) for which the defined material law is valid.

The compressive behavior of fibre concrete material model should be also adjusted in order to represent the realistic response. The higher material ductility should be reflected in the compressive material law. Therefore, similar procedure (with eventually different characteristic length) is used for the compression part of the compressive stress-strain law of the mentioned material model shown in Figure (b).

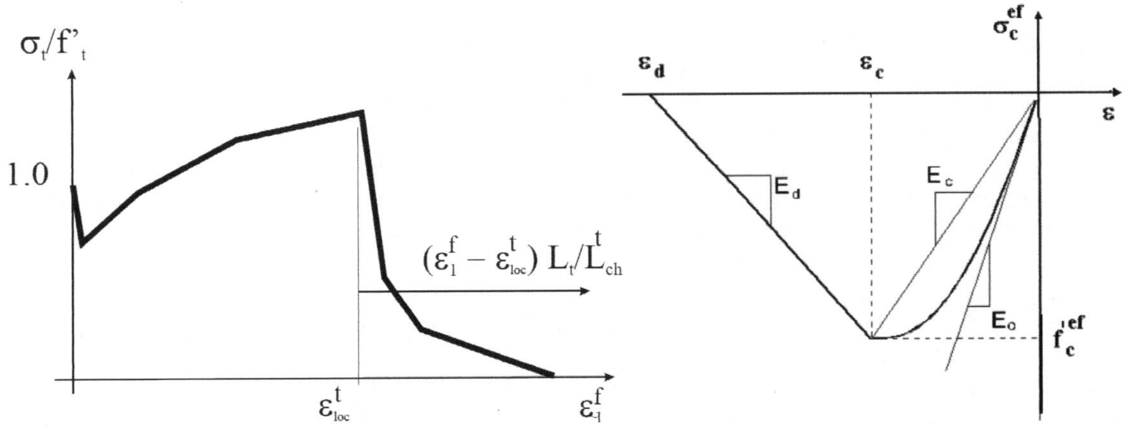

Figure 2. (a) User defined tensile behavior (b) Compressive stress-strain law

The softening law in compression is linearly descending and the end point of the softening curve is defined by plastic displacement w_d (corresponding to ε_d in Figure (b)). By increasing material parameter w_d the contribution of the fibres to the compressive behavior of concrete is considered. Another important compressive parameter for modeling of fibre concrete is reduction of compressive strength due to cracks which says how the strength is reduced while the material is subjected to lateral tension.

It should be noted that the resulting material parameters are appropriate coefficients of the adjusted material model. They don´t necessarily represent the real material properties which would correspond to the material values as described by the property name in the material model.

3.2. Identification of material parameters

The appropriate material parameters such as fracture energy, tensile and compressive strength, and modulus of elasticity of the structural material, are generally not well known, but crucial for a successful computer simulation of the structural response. For identification of these parameters from experimental results an inverse method based on artificial neural networks is applied [9], [10], [11], [12]. Background of the inverse analysis is finite element method model which is used for numerical simulation of fracture test; the model was created in ATENA software.

Subject of identification are the basic three fracture mechanical parameters of the concrete model: modulus of elasticity, tensile strength, fracture energy. These material model parameters are considered as random variables described by probability distribution. The rectangular distribution was chosen as the lower and upper limits represent the bounded range of physical existence. The variables are then simulated randomly based on the small-sample simulation Latin Hypercube Sampling. Multiple calculation of deterministic computational model using random realizations of material model parameters is performed, and statistical set of the virtual structural response is obtained. Random realizations and the corresponding responses from the computational model serve as the basis for the training of the neural network. After the training the neural network is ready to solve the main task: To provide the material parameters for which the numerical simulation will result in the best agreement with the provided experimental data. This task is performed by means of the simulation of the neural network using measured response as an input. It results in a set of identified optimal material input parameters. The last step of the procedure is results verification – calculation of computational model using identified parameters and comparison with the measured data.

Depending on the sample size of the statistical set, the basic statistical characteristics (mean values and standard deviations) of material parameters can be determined using two approaches:

1. "one by one" approach – the parameters of each specimen are identified separately and the final statistics are calculated from the set of all values for each parameter
2. "direct" approach – in the case of a larger statistical set it is more efficient not to identify each specimen one by one but to identify the whole statistical set for all specimens together based on the random response of fracture tests [13]

3.3. Material degradation

Material degradation can be covered by deterministic or stochastic assessment of degradation effects such as concrete carbonation, deterioration, depasivation and reinforcement corrosion. As an extension to the probabilistic software FReET the program FReET-D (FReET Deterioration/Degradation Module, see e.g. [14]) was developed in order to support the life-cycle and performance-based assessment and design of reinforced concrete structures. It represents deterioration module for statistical, sensitivity and reliability assessment of degradation effects in concrete structures in the SARA system. Several recognized models for concrete carbonation and deterioration as well as reinforcement corrosion are included. This deterioration module provides:

- modeling of degradation phenomena in concrete structures
- assessment of service life
- assessment of reliability measure

The results (stochastic or deterministic) can be directly used as inputs in ATENA or SARA programs to account for material deterioration and degradation and to perform life-cycle analysis of concrete engineering structures or structural parts.

4. Stochastic analysis

The stochastic analysis is based on repeated analysis of the prepared model with randomly generated properties. Since each of these samples represent a demanding nonlinear finite element analysis, the number of samples should be kept on moderate level. In the same time the applied methodology should be sufficiently accurate and representative.

The probabilistic software FReET [15] has been developed for stochastic and probabilistic analysis of computationally intensive problems such as nonlinear finite analysis. Stratified simulation technique LHS is used to keep the number of required simulations at an acceptable level. This technique can be used for both random variables and random fields levels. Statistical correlation is imposed by the stochastic optimization technique called simulated annealing. Sensitivity analysis of the input parameters to resulting values is based on nonparametric rank-order correlation coefficients. Procedure can be briefly outlined:

- random input parameters are generated according to their PDF using LHS sampling
- samples are reordered by the Simulated Annealing approach in order to match the required correlation matrix as closely as possible
- generated realizations of random parameters are used as inputs for the analyzed function (computational model)
- solution is performed many times and the results (structural response) are saved.
- simulation process the resulting set of structural responses is statistically evaluated

Main results from the stochastic analysis are:

- estimates of the mean value
- variance
- coefficient of skewness and kurtosis
- empirical cumulative probability density function estimated by an empirical histogram of structural response

This basic statistical assessment is visualized through histograms. This is followed by reliability analysis based on several approximation techniques:

- estimate of reliability by the Cornell safety index β
- curve fitting approach applied to the computed empirical histogram of response variables
- estimate of probability of failure based on the ratio of failed trials to the total number of simulations

State-of-the-art probabilistic algorithms are implemented to compute the probabilistic response and reliability. The main features of the FReET software are:

(a) stochastic model (inputs)
- direct connectivity to the nonlinear analysis input data
- friendly Graphical User Environment (GUE)
- 30 probability distribution functions (PDF), mostly 2-parametric, some 3-parametric, two 4-parametric (Beta PDF and normal PDF with Weibullian left tail)
- unified description of random variables optionally by statistical moments or parameters or a combination of moments and parameters
- PDF calculator
- statistical correlation (also weighting option)
- categories and comparative values for PDFs
- basic random variables visualization, including statistical correlation in both Cartesian and parallel coordinates

(b) probabilistic techniques
- Crude Monte Carlo simulation
- Latin Hypercube Sampling (3 alternatives)
- First Order Reliability Method (FORM)
- Curve fitting
- Simulated Annealing
- Bayesian updating

(c) response/limit state function
- numerical form directly connected to the results of nonlinear FE analysis
- multiple response functions assessed in same simulation run

5. Design and safety of fibre concrete structures

In design of engineering structures sufficient safety must be assured. Comparison of the safety formats by several methods (partial safety factors (PSF), ECOV method, EN 1992-2, see [16], [5]) has been performed for two fibre concrete structures including the probabilistic nonlinear analysis using SARA software system. The results are collected in Table 1.

The first example was a four-point bending beam (Figure 3) made from steel fibre concrete. Model has been created in ATENA 2D (Figure 3(b)). The mean material parameters were tested, calibrated by inverse analysis according to Section 3.2 (Figure 3(c)), the model was loaded until localized failure (Figure 3(d)), and consequently various safety approaches were used to calculate the corresponding design resistances.

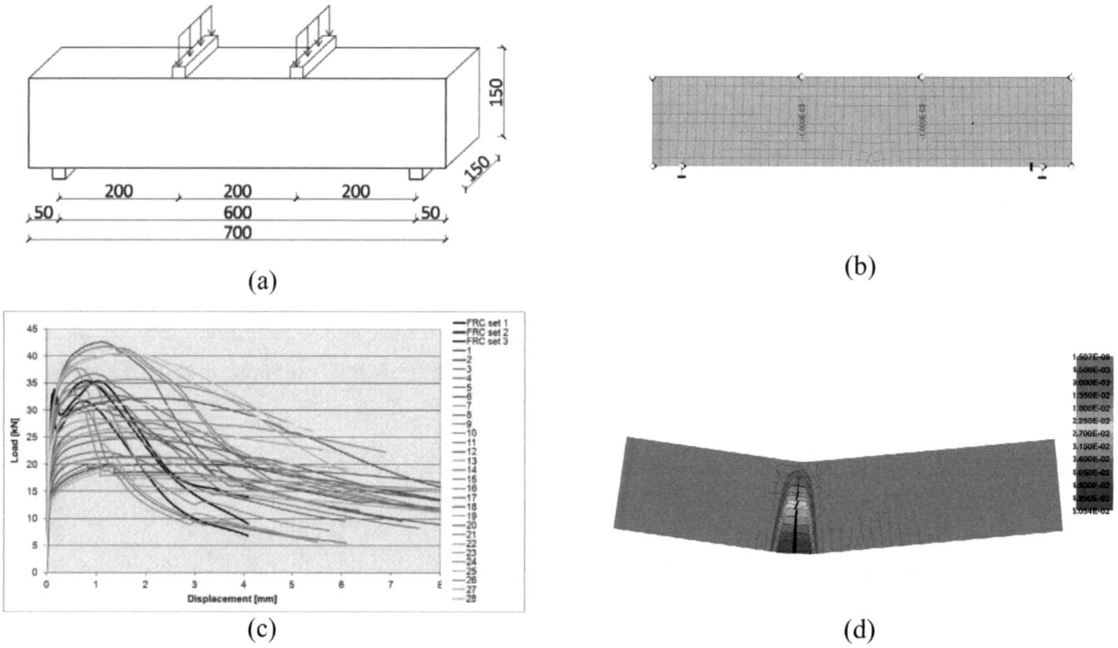

Figure 3. Fibre concrete four-point bending beam.

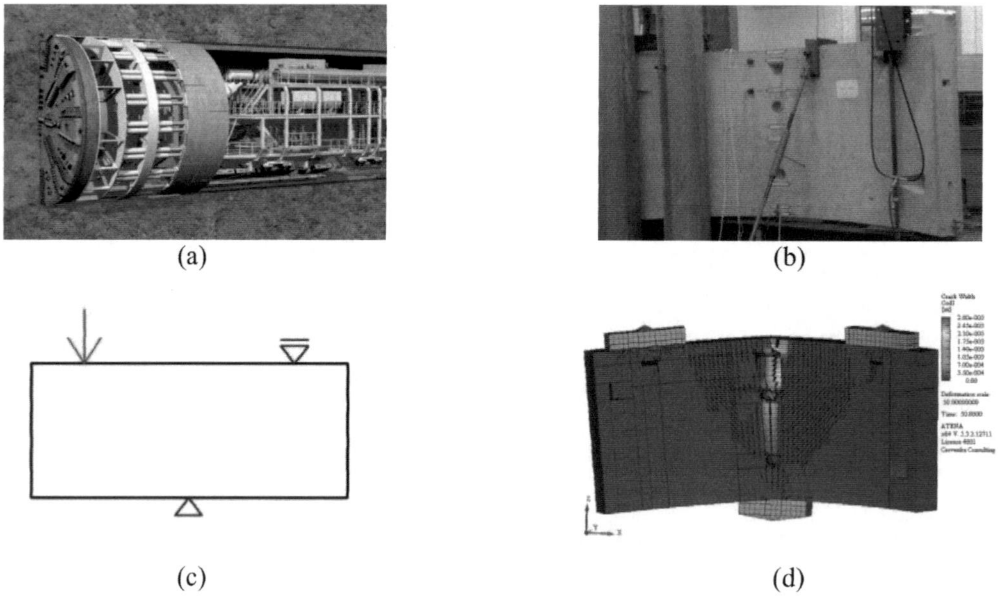

Figure 4. Tunnel tubing made of fibre concrete.

The second example represents a tunnel lining segment (Figure 4) installed by TBM (tunnel boring machine, Figure 4(a)) during tunnel excavation. Laboratory tests of the fibre concrete segments have been performed in Klokner Institute of CTU in order to check their resistance under various loading conditions (Figures 3(b, c)). The numerical model was first validated by experimental data for a loading scenario simulating the action of the TBM machine during the installation and assembly of segments (Figure 4(d)). After that the design resistance was calculated using various safety formats (see Table 1).

Table 1. Safety format results for fibre concrete structures

	R_d/R_d^{PSF}			
	PSF	*Probabilistic*	*ECOV*	*EN 1992-2*
FC bending beam	1	1.16	1.23	0.97
FC tubing	1	1.22	1.27	1.00
Average	1	1.19	1.25	0.99

Note: R_d is normalized with respect to PSF

6. Conclusions

- Authors offer a methodology and user-friendly computer tools for probabilistic design of fibre concrete structures. The utilized approach is based on randomization of the nonlinear finite element analysis of the structural finite element model.
- The input material parameters are represented as random variables or fields. They can be identified from material tests by neural network technology. The input parameters can be reduced in a rational way due to stochastic degradation and deterioration modelling. According to the user request this reduction can be applied in deterministic or stochastic way.
- The results from the nonlinear probabilistic analysis can serve as a rational basis for (performance-based) design of fibre concrete structures.
- The presented methodology is illustrated on results from selected probabilistic studies with different types of fibre concrete structures. The probabilistic analysis resulted here in higher resistance values (by ~20%) with regard to the PSF and to the EN 1992-2 (which is slightly more conservative even regarding to PSF). The ECOV method gives slightly higher resistance values than the probabilistic approach (by ~25% with regard to PSF); however, the presented results do not include model uncertainties which would reduce the calculated structural resistance.
- Due to high variability of fibre concrete material properties it can be recommended to utilize preferably the probabilistic methods, where the actual material variability can be accounted for the evaluation of structural performance, safety and reliability. The probabilistic nonlinear safety analysis is a demanding, but the most suitable method for design and performance assessment of fibre concrete structures.

Acknowledgement
Presented results are partially based on research performed within projects LO1408 AdMaS UP – Advanced Materials, Structures and Technologies (National Sustainability Programme I, Ministry of Education, Youth and Sports of the Czech Republic) and 16-04132S Epistemic uncertainty of crack models in reinforced concrete structures (GAČR).

References

[1] ***fib*** 2013 Model Code for Concrete Structures 2010 (Berlin: Wilhelm Ernst & Sohn)
[2] Bergmeister K, Novák D, Pukl R and Červenka V 2009 Structural assessment and reliability analysis for existing engineering
[3] Červenka V, Červenka J and Pukl R 2002 ATENA – A tool for engineering analysis of fracture in concrete, *Sadhana* **27** (4) pp 485-492
[4] Červenka V 2002 Simulating a response, *Concrete Engineering International* **4** (4) pp 45-49
[5] Červenka V 2011 Design of fibre reinforced concrete structures based on nonlinear analysis *Proc. Fibre Concrete 2011 - Technology, Design, Application* ed. A. Kohoutková (Prague: CTU) pp 11-12
[6] Marek M, Gramblička M, Baláž D, Pukl R, Staš M and Schneider J 2016 Steel Fiber Reinforced Concrete Precast Tunnel Segments: Design Using Nonlinear Analysis. *Proc. Underground Construction Prague 2016 / EETC 2016* (Prague) paper 5/15
[7] Červenka J and Papanikolaou V K 2008 Three Dimensional Combined Fracture-Plastic Material Model for Concrete *Int. Journal of Plasticity* **24** (12) pp 2192-2220
[8] Pukl R, Červenka J, Červenka V, Novák D, Vořechovský M and Lehký D 2005 Deterministic and statistical models for nonlinear FE-analysis of FRC-based structures *Proc. 1st Central European Congress on Concrete Engineering – Fiber Reinforced Concrete in Practice* (Graz) pp 130-133
[9] Lehký D, Keršner Z and Novák D 2014 FraMePID-3PB software for material parameter identification using fracture tests and inverse analysis *Advances in Engineering Software* **72** pp 147–154
[10] Novák D and Lehký D 2006 ANN Inverse Analysis Based on Stochastic Small-Sample Training Set Simulation *Engineering Application of Artificial Intelligence* **19** pp 731-740
[11] Novák D, Pukl R 2007 Modeling fibre concrete: from experiment to reliability assessment *Proc. Fibre Concrete 2007 - Technology, Design, Application* ed. A. Kohoutková (Prague: CTU) pp 137-142
[12] Pukl R and Sajdlová T 2013 Multi-level Optimization of Input Parameters for Modeling of Fibre Reinforced Concrete *Proc.* ***fib*** *Symposium* (Tel Aviv)
[13] Lehký D, Novák D 2011 Statistical material parameters identification of fibre-reinforced concrete. *Proc. 9th International Symposium on High Performance Concrete – Design, Verification & Utilization (HPC 2011)* (Rotorua, New Zealand) pp 38 + CD (8p).
[14] Novák D, Vořechovský M and Teplý B 2014 FReeT: Software for the statistical and reliability analysis of engineering problems and FReET-D: Degradation module *Advances in Engineering Software* **72** pp 179-192
[15] Novák D, Vořechovský M, Rusina R 2003 Small-sample probabilistic assessment – software FREET *Proc 9th Int. Conf. on Applications of Statistics and Probability in Civil Engineering – ICASP 9* (Rotterdam: Millpress) pp 91-96
[16] Červenka V 2013 Reliability-based non-linear analysis according to fib Model Code 2010 *Structural Concrete* **14** pp 19-28

IOP Conf. Series: Materials Science and Engineering **246** (2017) 012005 doi:10.1088/1757-899X/246/1/012005

Comparison of experimental and theoretical data for steel fiber reinforced concrete slabs on ground under a point load according to design codes of the Republic of Belarus and the Russian Federation

A Valodzin[1]

[1] MEng, PhD student, Civil Engineering Department, Belarusian National Technical University, Minsk, Republic of Belarus

E-mail: alexander.volodin@bk.ru

Abstract. Fiber reinforced concrete is a perspective construction material. But there is still a lack of calculation methods for that material. Some calculation methods for steel fiber reinforced concrete structures are offered in the Republic of Belarus and the Russian Federation design codes. The comparison has been made for experimental and theoretical data concerning steel fiber reinforced concrete slabs on ground under a point load. Mean value correction factors and coefficients of variation of the error terms have been determined. Determination of those values is a first step to further reliability index determination.

1. Introduction

Fiber reinforced concrete is one of the perspective construction materials. Such a concrete belongs to a vast class of composite materials which are widely used in different industries. In fiber reinforced concrete fibers are uniformly distributed inside the concrete matrix. Fibers themselves can be either metallic or nonmetallic, either mineral or organic. The name of the material depends on the fibers origin — steel fiber reinforced concrete, glass fiber reinforced concrete and so on.

Steel fibers inside the concrete matrix are randomly distributed. In plain concrete microcracks unite into macrocracks which leads to failure or loss of important qualities, such as water permeability, freeze-thaw resistance, resistance to aggressive substances and so on. In case of steel fiber reinforced concrete the opening of microcracks is restricted with steel fibers. Random fiber distribution makes it possible to control cracks in various directions.

2. Research objectives

The main research objective is to estimate the reliability of current calculation methods for steel fiber reinforced concrete (SFRC) slabs on ground (industrial floors).

In order to achieve the objective it is necessary to fulfill the following tasks:

1. To collect information concerning current calculation methods for SFRC slabs on ground and analyze them.
2. To collect information concerning experimental data for SFRC slabs on ground.
3. To perform analysis of bearing capacity of SFRC slabs on ground according to the chosen calculation methods.

4. To make a comparison between theoretical and experimental bearing capacity values.

3. Calculation methods

In this paper SFRC slabs on ground are analyzed according to methods offered in the following codes:

— TCP 45-5.03-300-2015 "Steel fiber concrete products and structures. Design rules" (Republic of Belarus) [1];

— SP 29.13330.2011 "Floors" (Russian Federation) [2].

3.1. TCP 45-5.03-300-2015 "Steel fiber concrete products and structures. Design rules" calculation method

The code is used for SFRC cross section design. The sequence of calculations needed to determine the ultimate bending moment is described hereafter.

In order to determine the SFRC tensile strength f_{fctd} it is essential to define the failure variant.

Fiber anchorage length:

$$l_{f,an} = \frac{\eta d_f f_{fk}}{f_{ck}}, \tag{1}$$

where d_f – fiber diameter, f_{fk} – characteristic fiber tensile strength, f_{ck} – characteristic concrete compressive strength.

The η coefficient has a tabulated value.

If $l_{f,an} < \frac{l_f}{2}$ then failure occurs due to fibers being torn apart (failure variant 1). In that case the SFRC tensile strength is calculated according to 7.2.6 [1]:

$$f_{fctd} = 1.1 f_{cd}\left[K_T k_{or}^2 \rho_{fv} \cdot \frac{f_{fd}}{f_{cd}} \cdot \left(1 - \frac{l_{f,an}}{l_f}\right) + 0.08 - 0.1 \cdot \sqrt{2\rho_{fv} - 0.005}\right], \tag{2}$$

where f_{cd} – concrete compressive strength, K_T – tabulated coefficient, k_{or} – fiber orientation coefficient for cross sections in tension, ρ_{fv} – fiber volume percentage, f_{fd} – fiber tensile strength, l_f – fiber length.

If $l_{f,an} \geq \frac{l_f}{2}$ then failure occurs due to fibers being pulled out (failure variant 2). In that case the SFRC tensile strength is calculated according to 7.2.7 [1]:

$$f_{fctd} = 1.1 f_{cd} \cdot \left(K_T \cdot \frac{k_{or}^2 \rho_{fv} l_f}{8\eta d_f} + 0.08 - 0.5\rho_{fv}\right). \tag{3}$$

The SFRC compressive strength is defined as follows:

$$f_{fcd} = f_{cd} + k_n^2 \varphi_f \rho_{fv} f_{fd}, \tag{4}$$

where k_n – fiber orientation coefficient for cross sections in compression, φ_f – fiber confinement reinforcement efficiency coefficient.

Coefficients k_{or} and k_n have tabulated values.

The fiber confinement reinforcement efficiency coefficient:

$$\varphi_f = \frac{5 + L}{1 + 4.5L}, \tag{5}$$

where

$$L = \frac{k_n^2 \rho_{fv} f_{fd}}{f_{cd}}. \tag{6}$$

According to 7.2.12 [1] the SFRC compressed zone characterictics:

$$\omega = 0.8 - 0.008 f_{fcd}. \tag{7}$$

The relative compressed zone height limit:

$$\xi_{\lim} = \frac{\omega}{1 + \frac{\sigma_{sR}}{\sigma_{sc,u}}\left(1 - \frac{\omega}{1.1}\right)}. \quad (8)$$

Compressed zone height:

$$x = \frac{f_{fctd} h}{f_{fcd} + f_{fctd}}, \quad (9)$$

$$\xi = \frac{x}{h} \leq \xi_{\lim}, \quad (10)$$

where h – cross section height.
The ultimate bending moment:

$$M_{Rd} = 0.5 f_{fcd} bhx. \quad (11)$$

3.2. SP 29.13330.2011 "Floors" calculation method

The code offers the theory to determine forces acting in the floor slab under loads of different configurations and also the bearing capacity of the slab.

Hereafter is the calculation method for a floor slab under a point load.

The ultimate bending moment:

$$M_{ult} = f_{fctd} \frac{h^2}{3.5}, \quad (12)$$

where f_{fctd} is SFRC tensile strength.
The ultimate point load:

$$P_{ult} = \frac{M_{ult}}{K_1}. \quad (13)$$

The K_1 coefficient has a tabulated value and depends on the load area dimensions and the radius of relative stiffness.

4. Experimental data

The database contains 11 specimens. The data used is presented in table 1.

Table 1. Experimental data.

No.	Reference	Authors	Slab dimensions (mm)	Fiber dosage (kg/m^3)	Modulus of subgrade reaction (N/mm^3)	Failure load (kN)
1	[3]	Meda, Plizzari	3000x3000x150	30	0.0785	265
2				30	0.0785	238.6
3				30	0.0785	252.3
4				30	0.0785	246.2
5				45	0.0785	231.9
6	[4]	Beckett	3000x3000x150	20	0.035	350
7				20	0.035	390
8				30	0.035	340
9				30	0.035	345
10	[5]	Faulkner	3000x3000x150	20	0.025	240
11				30	0.050	380

5. Calculation results and comparison of experimental and theoretical data

Eleven specimens were analyzed according to the following schedule:

1. Determination of ultimate bending moments according to [1] and [2].
2. Determination of ultimate point load according to [2].

Thus, for each specimen two ultimate point load values were obtained.

The results are presented in table 2.

Table 2. Calculation results (theoretical data).

No.	Ultimate point load according to [2] (kN) Ultimate bending moment calculated according to [1]	Ultimate bending moment calculated according to [2]
1	267.51	165.09
2	269.30	166.28
3	273.37	168.74
4	233.94	144.27
5	264.73	163.13
6	290.18	180.53
7	293.49	182.78
8	291.58	180.91
9	297.13	184.65
10	262.42	163.30
11	272.82	169.42

Visual representation of theoretical values compared with experimental values can be seen on figure 1.

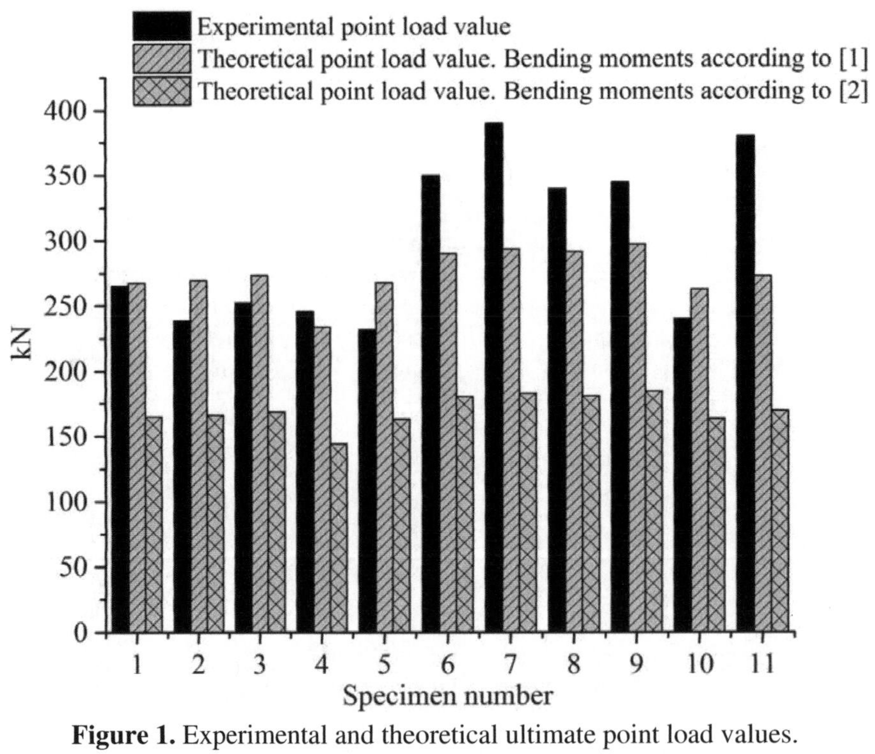

Figure 1. Experimental and theoretical ultimate point load values.

IOP Conf. Series: Materials Science and Engineering **246** (2017) 012005 doi:10.1088/1757-899X/246/1/012005

Theoretical ultimate point load values calculated with the use of ultimate bending moments according to [2] are much lower than experimental values. The average difference between these values equals to 41%. At the same time the use of ultimate bending moments calculated according to [1] leads to more accurate results. The average difference between experimental and theoretical values in that case equals to 5%.

In order to provide more accurate statistical description of obtained results the mean value correction factor b and the coefficient of variation of the error terms V_δ can be calculated according to [6].

Table 3. Mean value correction factors and coefficients of variation of the error terms.

	Ultimate point load according to [2]	
	Ultimate bending moment calculated according to [1]	Ultimate bending moment calculated according to [2]
b	1.093	1.764
V_δ	0.166	0.164

6. Conclusions

Design codes [1] and [2] provide methods for calculating ultimate point loads for SFRC slabs on ground. The point load itself is calculated according to [2] using ultimate bending moments for SFRC cross sections determined according to [1] or [2]. The best convergence between experimental and theoretical data is reached when ultimate bending moments are calculated according to [1].

This work is the first step to determine the reliability index of described methods of calculation of SFRC slabs on ground under a point load.

References

[1] TCP 45-5.03-300-2015 "Steel fiber concrete products and structures. Design rules". Republic of Belarus Architecture and Construction Ministry, 2015, 48 pp.

[2] SP 29.13330.2011 "Floors". Russian Federation Regional Development Ministry, 2011, 68 pp. Sorelli L G, Meda A, Plizzari G A 2006 Steel Fiber Concrete Slabs on Ground: A StructuralMatter *ACI Structural Journal* 103-S58 551-558.

[3] Beckett D 1990 Comparative Study on Plain, Fabric Reinforced and Steel Fiber Reinforced Concrete Ground Slabs *Concrete* **3** 43-45.

[4] Faulkner H, Huang Z, Teutsch M 1995 Comparative Study of Plain and Steel Fiber Reinforced Concrete Ground Slabs *Concrete International* **1** 45-51.

[5] EN 1990:2002. Eurocode – basis of structural design, 119 pp.

IOP Conf. Series: Materials Science and Engineering **246** (2017) 012006 doi:10.1088/1757-899X/246/1/012006

Dramix® Steel fibres in residential foundation slabs in Czech Republic: design approach

S Pouillon[1]

[1] N.V. Bekaert S.A., Bekaertstraat 2, B-8550 Zwevegem, Belgium

E-mail: Steven.pouillon@bekaert.com

Abstract. Steel fibres are used more and more in structural concrete elements, like residential foundation slabs. The development of high performant steel fibres (e.g. Dramix® 4D and 5D), the evolution in standardization and the demand coming from construction companies and investors are pushing this trend. Engineers are using the yield line method to design steel fibre concrete structures in an easy and economical way, especially if the structure has a regular load layout mainly containing wall loads and point loads. The load configuration, soil characteristics and material characteristics determine the final solution.

1. Introduction

Dramix® steel fibres become more and more the preferred concrete reinforcement material in a wide range of residential and industrial structural applications in Czech Republic. Bekaert expanded his steel fibre product range with Dramix® 4D and 5D fibres, especially developed to reinforce structural concrete elements. Already for more than 20 years, steel fibres are used to reinforce most of the Czech concrete industrial floors on grade. But very recently, engineers, contractors and investors are often opting for this material to also reinforce their concrete structural elements: residential and industrial foundation slabs, industrial slabs on piles and crack width controlled slabs.

2. From well-known steel fibre application to structural steel fibre applications

Various developments have boosted the use of steel fibre concrete in structural applications. As a result of 3 years fundamental research, Bekaert launched a range of high performing steel fibres (4D and 5D). These steel fibres make it possible to design structural applications with steel fibre reinforced concrete (SFRC), economically and technically safe. Secondly, various standardization committees have created a technical and juridical framework for engineers to design structural steel fibre concrete structures in a safe and legal way. And last but not least, the practical and economic advantages of steel fibre reinforcement – as they are experienced in floors and other standard applications – remain valid and very appreciated by all stakeholders in the new applications.

2.1. Outperforming new Dramix® steel fibres

Steel fibres increase the ductility of the concrete by bridging cracks and transferring tensile stresses over these cracked zones. Many different parameters influence the performance of the steel fibre concrete: the *tensile strength* of the steel wire, type of *anchorage*, *ductility* of the wire, the *dosage* of steel fibres and of course the *concrete* class and quality. Bekaert has grouped its Dramix® products into 3 families: 3D, 4D and 5D (More information about the different characteristics can be found in

IOP Conf. Series: Materials Science and Engineering **246** (2017) 012006 doi:10.1088/1757-899X/246/1/012006

[1] and [2]). The steel fibres belonging to the 3D family are the standard ones (e.g. 3D 80/60BG, 3D 65/60BG and 3D 45/50BL) and are mainly used to reinforce industrial floors on grade (jointed floors, jointless floors), tunnels and precast elements. The newly developed Dramix® steel fibres, the families 4D and 5D, are especially designed to reinforce structural concrete elements. The main difference between the 4D/5D family and the 3D family is the significantly increased performance in serviceability limit state (4D) and ultimate limit state (5D). The combination of type of anchorage (more end hooks), stronger wires (up to 2200 MPa) and more ductile wires (up to 7% of elongation) are key in this increased performance.

Figure 1. The logos and shape of the 3 Dramix® fibre families

2.2. Evolution in standardization [3]

Steel fibre reinforced concrete has been the subject of much research over the last 40 years. In line with the knowledge built up, guidelines and standards have been developed gradually.

In the early 1980's, the SFRC test standards – JSCE-SF4 [4] and ASTM C1018 [5] – were in place and widely used. These standards – still topical – describe how SFRC can be tested in order to determine the performance of this material under bending conditions. During such a test, a SFRC prism is tested under bending, until a crack is formed with a specific crack opening. The performance is expressed as "flexural toughness" and represents the energy dissipated during the test. An equivalent post-crack flexural stress is derived from the flexural toughness. Many European countries developed their own national SFRC test standards, based upon the two pioneer standards. Actually, the EN14651 [6] test standard is mostly used in Europe, but also widely accepted outside Europe. In some countries, like Czech Republic, local standards (ČSN P 73 2452 [7]) are as well valid. The performance of the SFRC in EN14651 is expressed directly in flexural post crack stresses at specific crack mouth openings. The term crack mouth opening – and not crack opening - is used because a notch is made at the center bottom of the beam. This, in order to force the beam to crack in the middle. In 1995, the first design guideline was published: "Dramix Design Guideline". This document contained the first basic design rules in order to design concrete structures with steel fibres. This document was a result of a cooperation between N.V. Bekaert S.A., the Belgian universities KULeuven and UGent and the Belgian building research institute WTCB. This design guideline served as the basis for some national design recommendations. A deeper investigation was done by a group of international universities brought together by RILEM (RILEM TC162-TDF [8]). This cooperation resulted in a test and a design method. A structured approach – material characterization, performance indicators, constitutive law – was worked out and fully backed up by theory and experiments. Also overseas, in 2008, ACI 318 [9] incorporated steel fibres as a structural element to take up shear forces in concrete elements.

Different countries (e.g. Germany, Sweden and Italy) also published in the meantime national design guidelines and/or codes, often as an add-on of EN1992-1-1 [10]. A new milestone was reached in 2014 when the final version of the Model Code 2010 [11] was published. Steel fibres are fully incorporated in this state-of-the-art concrete design guideline. This document was also the trigger for the launch of TC 250/SC2/WG1/TG2(CEN); this task group works on the integration of steel fibres into the future EN1992-1-1. The ModelCode2010 and the DAfSTb-guideline [12] serve as the source code for this work.

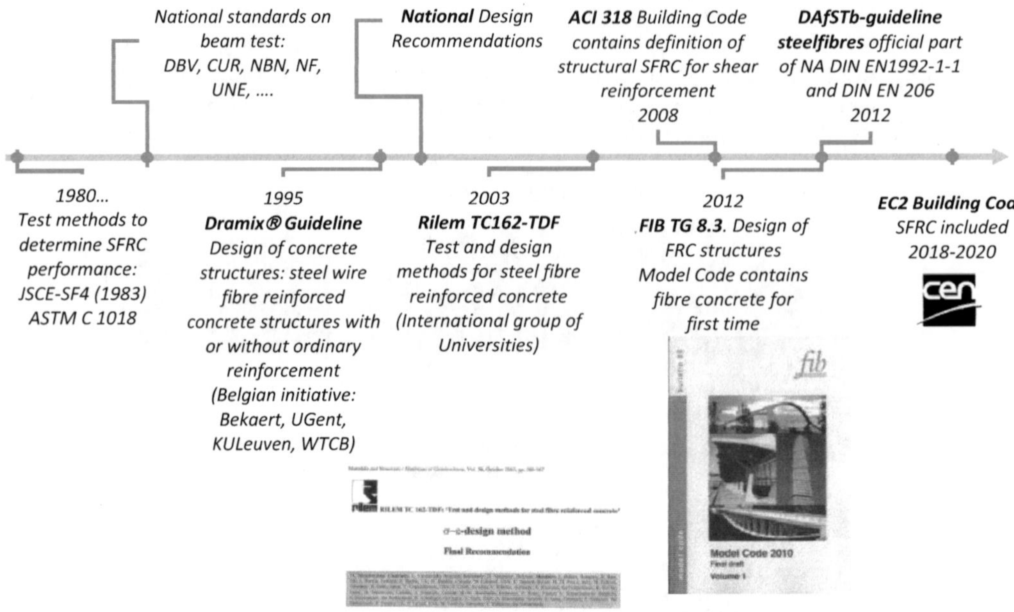

Figure 2. Timeline: Evolution of design standards and guidelines for SFRC

2.3. *Practical and economic advantages of steel fibre concrete in structural applications*

The practical and economic advantages are mainly a consequence of the low reinforcement complexity that can be obtained when using steel fibres and the fact that the 'reinforcement' is pre-mixed in the concrete.

The first is very clear when a fibre only solution is proposed. In that case, there is no need for complex drawings and for long site preparations, which is reducing the cost in each stage of the project (design and execution). Furthermore, errors in the drawings and errors during the reinforcement placing are almost eliminated. When a fibre only solution is technically not possible, engineers will always try to combine standard meshes – and avoid (thick) diameter bars – combined with steel fibres. The latter optimizes the crack control (less total reinforcement steel needed compared with a traditional solution) and increases the bending moment capacity significantly. Additionally, due to the strong shear and punching resistance of steel fibre concrete, shear and punching reinforcement can often be omitted.

The second is clearly facilitating the logistics. Instead of transporting to and storing the reinforcement on site, the steel fibres are delivered together with the truck mixer. Direct pouring from the truck mixer or using a pump is still possible with steel fibres.

3. Residential foundation slabs

Different type of residential foundation slabs exist: small (e.g. 50m²) to big (e.g. 1000m² or more), regularly versus irregularly shaped, lightly loaded versus highly loaded, on pile foundations or at grade… In particular – but not exclusively – the lightly to medium loaded residential foundation slabs on grade are very suitable to reinforce with (only) steel fibres. The advantages are evident:

- Fast execution (no mesh/rebar installation, little detailing work)

IOP Conf. Series: Materials Science and Engineering **246** (2017) 012006 doi:10.1088/1757-899X/246/1/012006

- Easy logistics (no storage of meshes on small construction sites)
- Easy design
- Economic

In general 2 load types are acting on a regular residential foundation slab: wall loads ("line load") and column loads ("point loads"). These loads can be positioned in the middle of the slab or at the edge/corner of the slab. Other important input parameters for a design are the soil characteristics (bedding modulus and bearing capacity - which can be assumed homogeneous over the small surface) and the characteristics (long-term E-modulus, thickness and maximum long-term compressible strength) of the insulation under the complete slab or only at the perimeter (if applicable).

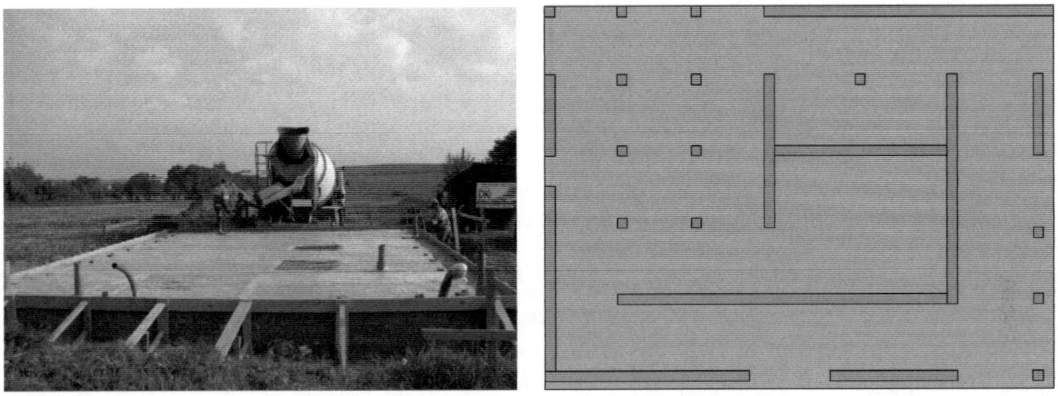

Figure 3. Example of the preparation and construction of a residential foundation slab (left) and a typical load configuration on a such slab (right)

4. Design of a foundation slab on grade

A typical SFRC design follows below mentioned flow:

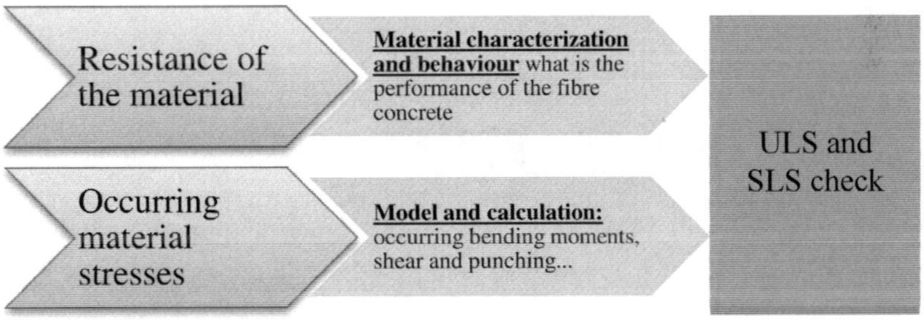

Figure 4. Design flow of SFRC structures

4.1. Material characterization and behavior

SFRC can be characterized using different methods or following different standards. Common tests standards in Europe are the EN14651 [6] (3-point bending test with notch) and the DAfStb [12] (4-point bending test). Also the Czech standard ČSN P 73 2452 [7] is suitable to follow when determining the performance of SFRC. It is although important that the material characterization and the model used to describe the behavior (constitutive law) are in correspondence with each other. The constitutive law and rules to derive the stress-strain curve from residual flexural strength values are inseparably linked with the type of test used to develop them. So e.g. is a 4-point bending test the basis for a design according the DAfStb Steel Fibre Concrete Guideline and the 3-point bending test EN 14651 the basis for a design according the ModelCode2010 [11] or Rilem TC162-TDF [8].

Important considerations in the material characterization of SFRC are the number of samples and the way how the characteristic residual strength values are determined. 6 beams is generally regarded as the absolute minimum, taking the natural spread in these type of beam tests into account; it is common to obtain a coefficient of variance of 25% (Reference can be made to SS 812310 [13]). In order to calculate the characteristic value, reference can be made to the international standard ISO 12491 [14]. The estimation of the lower mean value (when the standard deviation of the population is unknown) is given by:

$$\bar{x} - t_p \cdot s/\sqrt{n} \qquad (1)$$

where

s is the sample standard deviation;
t_p is the fractile of the t-distribution for $v = (n-1)$ degrees of freedom;
p is the probability (close to 1) given in table 3 or ISO 12491.

Other methods are described in DAfSTb Steel Fibre Concrete Guideline, Annex O.6 for example. Here a log-normal distribution is assumed.

The constitutive law expresses the stress-strain relationship of the SFRC in the tension zone. The characteristic residual flexural stress vs. crack width results of the beam tests are converted in stress – strain values. Together with a material safety factor (1,5 is common for SFRC in structural applications, like concrete in compression), the derived material behaviour can directly be used in the design of the bending moment capacity of a section, shear/punching capacity and in the calculation of crack openings.

The main parameters in the determination of the stress-strain curve are the conversion factor (bending vs. axial tension), the size factor, the orientation factor and the material safety factor.

4.1.1. The conversion factor. The characteristic residual flexural strength values need to be converted in uniaxial tensile strength values. Typical values are presented in the table pod.

Table 1. Main conversion factors (or range) for 3 SFRC design guidelines/standards

Strain	Rilem TC162-TDF	DAfStb SFRC Guideline	ModelCode 2010
At small strain	$\sigma_2 = 0{,}45\, f_{R1,k}$	$\sigma_2 = 0{,}40\, f_{cfl,L1}$	$\sigma_2 = 0{,}45\, f_{R1k}$
At ultimate strain	$\sigma_3 = 0{,}37\, f_{R3,k}$	$\sigma_3 = 0.25 - 0.44\, f_{cfl,L2}$	$\sigma_3 = \sigma_2 - \frac{w_u}{CMOD_3}(\sigma_2 - 0{,}5 f_{R3,k} + 0{,}2 f_{R1,k})$

σ_1 is the characteristic tensile strength of unreinforced concrete, σ_2 is the characteristic tensile strength of SFRC at small strain (typically between 0 ‰ and 3,5 ‰), σ_3 is the characteristic tensile strength of SFRC (typically around 20‰ - 25‰. For the definition of the other parameters, reference is made to the selected standards/guidelines.

4.1.2. The size factor. In many applications, a much bigger cracked section is present than tested in the standard beams. To account for the much smaller standard deviation expected in the actual structure, a size factor is introduced. For bending situations in slabs, following formula can be used in accordance with the DAfStb:

$$\kappa_G^F = 1{,}0 + A_{ct}^f \cdot 0{,}5 \leq 1{,}7 \qquad (2)$$

with A_{ct}^f area of tension zones in the concrete of the cracked cross-sections or plastic hinges in m².

IOP Conf. Series: Materials Science and Engineering **246** (2017) 012006 doi:10.1088/1757-899X/246/1/012006

4.1.3. The orientation factor. In many applications, a random fibre orientation can be expected (like in general slabs). For some applications (due to their shape, way of casting, concrete consistency…), a significant fibre alignment can be expected (advantageously or disadvantageously). To account for this an orientation factor should be used.

4.1.4. The material safety factor. A typical material safety factor for SFRC in tension is 1,5 (DAfStb: 1,25/0,85, ModelCode2010: 1,5).

4.2. Design of the structure (in U.L.S.)

The design of a residential foundation slab can be economically performed by using the yield-line theory. Yield-line theory requires to determine the relevant yield-line mechanism together with the corresponding load case.

For residential foundation slabs two major mechanisms are governing:

- folded plate pattern
- fan pattern

While the folded plate is relevant for both line loads (walls) and point loads positioned in a line (row of columns), the fan pattern only applies to point loads.

Shear and punching need to be verified in addition whereby both design situation are usually not governing for steel fibres concrete without additional reinforcement.

This rather simple approach is valid under specific conditions:

- no ground water pressure
- a continuous support with suitable ground
- no shear walls, frames or cores present

4.2.1. Line load. A line load (or a line of point loads) at the inner side of the slab will typically create 2 negative yield lines (m'_{Ed}) and 1 positive yield line (m_{Ed}), as shown in the picture pod:

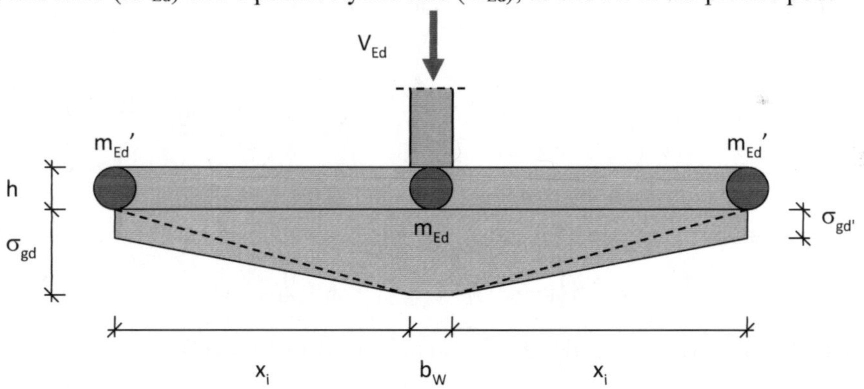

Figure 5. The yield line system for a line load (inner side of the slab)

The situation in case of a line load at the edge is slightly different and shown in Figure 6.

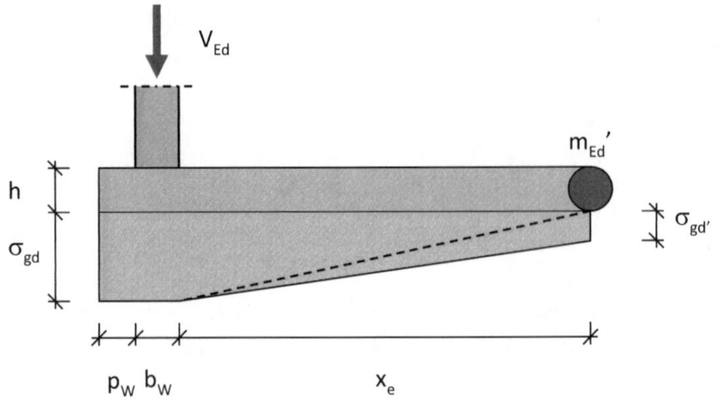

Figure 6. The yield line system for a line load (edge side of the slab)

The principle is that the load V_{Ed} must be in balance with the soil pressure distribution (described by σ_{gd} and $\sigma_{gd'}$) and (excluding the slab death weight), not exceeding the soil bearing capacity $\sigma_{gd,max}$ (equilibrium of vertical forces). The position of the negative yield lines depends on the elastic length[1] for the minimum distance (between $0{,}5l_{el}$ and l_{el}) and the distance to the edge and/or other loads for the maximum distance (to avoid intersection with other yield lines). The respective occurring plastic bending moment ($m'_{Ed} + m_{Ed}$) in the cracked sections (yield lines) can easily be found by the principle of moment equilibrium:
For an inner line load:

$$m_{Ed} + m'_{Ed} = \sigma_{gd'} \cdot \left(\frac{x_i}{2} + \frac{b_w}{2}\right) \cdot x_i + \left(\sigma_{gd} - \sigma_{gd'}\right) \cdot \left(\frac{x_i}{3} + \frac{b_w}{2}\right) \cdot \frac{x_i}{2} + \sigma_{gd} \cdot \frac{{b_w}^2}{8} - V_{Ed} \cdot \frac{b_w}{8} \quad (3)$$

For an edge line load:

$$m'_{Ed} = V_{Ed} \cdot \left(x_e + \frac{b_w}{2}\right) + \sigma_{gd}\left(\frac{{p_w}^2}{2} + p_w \cdot b_w + p_w \cdot x_e + \frac{{b_w}^2}{2} + b_w \cdot x_e + \frac{{x_e}^2}{3}\right)$$

$$-\sigma_{gd'} \cdot \frac{{x_e}^2}{6} \quad (4)$$

The shear forces can be calculated as follows:

$$V_{Ed} = \left[\left(\frac{\sigma_{gd} - \sigma_{gd'}}{2}\right) \cdot \frac{(x_i - d)}{x_i} + \sigma_{gd'}\right] \cdot (x_i - d) \quad (5)$$

with d = the slab thickness for SFRC, d = d_{eff} for combined reinforcement)

4.2.2. Point load. The governing yield line system for point loads is the fan pattern. Reference is made to Losberg [15]. The circumferential crack will occur at the top (negative yield line) and the diagonal cracks will occur at the bottom (positive yield lines).

---

[1] Elastic length: $l_{el} = \sqrt[4]{\frac{E_c \cdot h^3}{12 \cdot (1 - v^2) \cdot k}}$, with E_c the Youngs Modulus of concrete, h the slab thickness, v the Poisson ratio, k the bedding modulus.

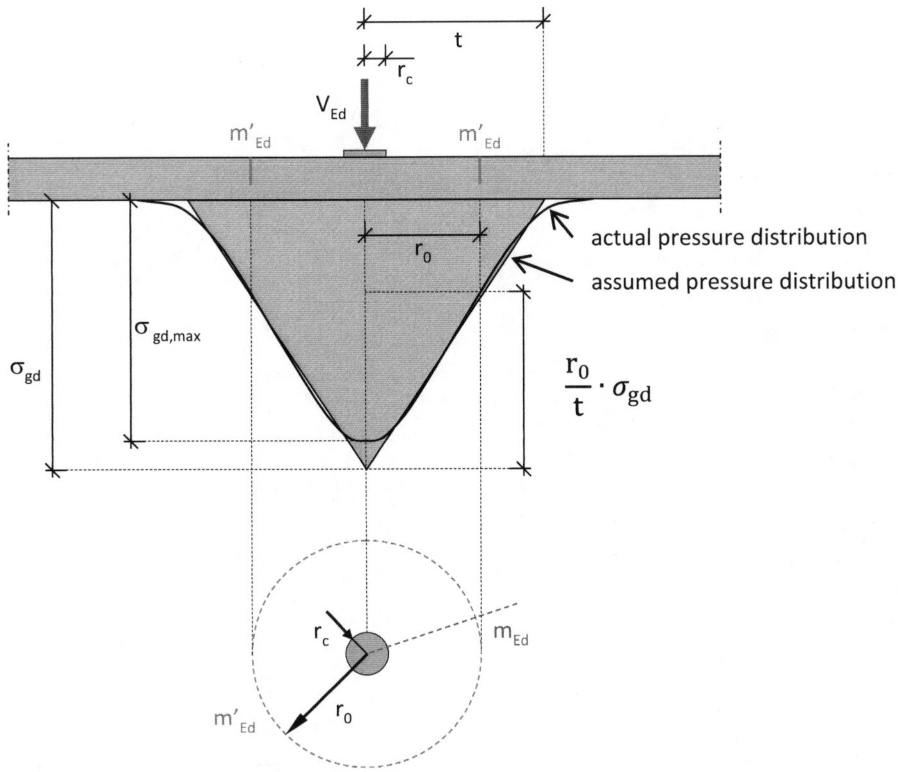

Figure 7. The yield line system for a point load (inner side of the slab). Reference is made to Losberg for all equations.

4.3. Design of the structure (in S.L.S.)

Due to the nature of concrete, crack free structures cannot be guaranteed and are normally not required either. Serviceability of a foundation slab can thus be achieved in different ways, depending on the actual requirements, the actual way of constructing and the intended use:

- avoiding of cracks (e.g. applying construction methods to avoid cracks)
- dealing with cracks (e.g. procedure defined prior to construction)
- limiting of cracks (e.g. controlling cracks using combined reinforcement)

5. Example

This paragraph presents an example of a residential foundation slab in Czech Republic. The load analysis (reaction forces from the wall loads and column loads) is shown in Figure 8.

IOP Conf. Series: Materials Science and Engineering **246** (2017) 012006 doi:10.1088/1757-899X/246/1/012006

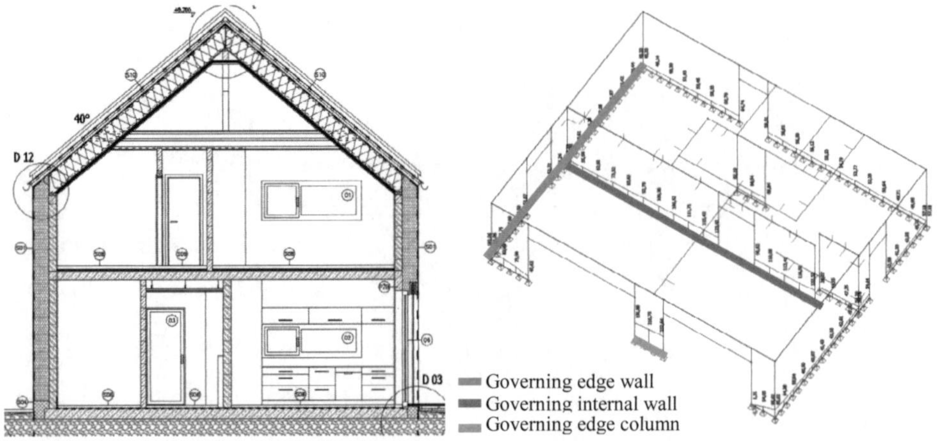

Figure 8. A cross section of the house (left) and the reaction forces from the wall loads and column loads (in U.L.S.) (right)

5.1. Input parameters

Table 2. Input parameters

Soil	Bedding modulus	k	0.03 N/mm³
	Maximum bearing capacity	σ_{gd}	200 kN/m²
Inner wall	Line load	V_{Ed}	132.4 kN/m
	Wall width	b_w	175 mm
	Closest other load	s	1925 mm
Edge wall	Line load	V_{Ed}	101.1 kN/m
	Wall width	b_w	175 mm
	Closest other load	s	3875 mm
Edge column	Point load	V_{Ed}	208,4 kN
	Column dimension	$a \cdot b$	175 mm · 1000mm
	Closest other load	s	3000 mm

5.2. U.L.S. design: bending moments

As solution, following slab characteristics were determined: 240mm, C30/37 + 30 kg/m³ 5D 65/60BG.

Table 3. Results

Bending moment capacity	According to DAfStb	m_{Rd} m'_{Rd} $m_{Rd} + m'_{Rd}$	28.4 kNm/m 28.4 kNm/m 56.8 kNm/m
Occurring bending moments	Inner wall	$m_{Ed}+m'_{Ed}$	15.1 kNm/m
	Edge wall	m'_{Ed}	10.5 kN/m
	Edge column	$m_{Ed}+m'_{Ed}$	55.4 kNm/m

IOP Conf. Series: Materials Science and Engineering **246** (2017) 012006 doi:10.1088/1757-899X/246/1/012006

6. Conclusion

Dramix® steel fibres become more and more the preferred concrete reinforcement material in a wide range of structural applications in Czech Republic. Well established standards are guiding engineers in their designs of structural concrete elements with steel fibers. Contractors and investors are experiencing each day the practical and economic advantages of the newly developed Dramix® steel fiber families by Bekaert. Residential foundation slabs, taken as a design example in this article, are highly suitable to be reinforced with the new Dramix® fibre families 4D and 5D; from technical and economic point of view.

References

[1] New applications for steel fibre reinforced concrete and combined reinforcement, Guirguis Philipp, *Procedia Egineering (2013), Concrete and Concrete Structures 2013*
[2] Nové typy ocelovych vlakien pre aplikacie do nosnych prkov, Dojcak Juraj, *SAVT Patria Beton 2013 Konferencia*
[3] Fibre reinforced concrete in structural applications, Pouillon Steven, *Fibre Concrete 2015, Prague, Czech Republic*
[4] JSCE-SF4, Method of tests for flexural strength and flexural toughness of steel fiber reinforced concrete
[5] ASTM-C1018, Standard Test Method for Flexural Toughness and First-Crack Strength of Fiber-Reinforced Concrete (Using Beam With Third-Point Loading)
[6] EN14651, Test method for metallic fibered concrete – Measuring the flexural tensile strength (limit of proportionality (LOP), residual)
[7] ČSN P 73 2452, Fibre-reinforced concrete – Testing of hardened fibre-reinforced concrete
[8] RILEM TC 162 TDF, Design of Steel fibre reinforced concrete – *Method, recommendations, Material and Structures, March 2001*
[9] ACI 318, 318-14: Building Code Requirements for Structural Concrete and Commentary
[10] EN1992-1-1, Eurocode 2: Design of concrete structures - Part 1-1: General rules and rules for buildings
[11] ModelCode2010, The fib Model Code for Concrete Structures 2010
[12] DAfStb-Richtlinie, Stahlfaserbeton, Ausgabe November 2012
[13] SS 812310:2014, Fiberbeton – Dimensionering av fiberbetongkonstruktioner
[14] ISO 12491, Statistical methods for quality control of building materials and components
[15] Anders Losberg, Design Method for structurally reinforced concrete pavements, 1961

IOP Conf. Series: Materials Science and Engineering **246** (2017) 012007 doi:10.1088/1757-899X/246/1/012007

On the residual properties of damaged FRC

R Zerbino[1], M C Torrijos[1] and G Giaccio[1]

[1] Department of Civil Engineering, La Plata National University. CONICET. LEMIT-CIC, 52 entre 121 y 122, 1900 La Plata, Argentina

E-mail: zerbino@ing.unlp.edu.ar

Abstract. A discussion on the residual behaviour of Fibre Reinforced Concrete (FRC) is performed based on two selected cases of concrete degradation: the exposure at High Temperatures and the development of Alkali Silica Reactions. In addition, and taking in mind that the failure mechanism in FRC is strongly related with the fibre pull-out strength, the bond strength in damaged matrices was shown concluding that the residual bond strength is less affected than the matrix strength. As the damage increases, the compressive strength and the modulus of elasticity decrease, being the modulus of elasticity the most affected. There were no significant changes produced by the incorporation of fibres on the residual behaviour when compared with previous experience on plain damage concrete. Regarding the tensile behaviour although the first peak decreases as the damage increases, even for a severely damage FRC the residual stresses remain almost unaffected.

1. Introduction

The construction of concrete structures of extended service life represents a key point to improve the sustainability of a country's infrastructure. In addition, the repair and strengthening of existing structures has become an important and urgent issue around the world.

Fibre Reinforced Concrete (FRC) is a high performance concrete particularly efficient for the construction of structures exposed to severe conditions as well as for reinforcement and retrofitting. Fibres incorporation controls cracking processes in concrete, resulting in great improvements of the material toughness and the structures durability. Steel and Polymer FRC have been used for decades in slabs on grade, pavements and bridge decks, tunnels, precast elements among others.

Concrete is a composite material where inclusions of different size and shape, the aggregates, are surrounded by a more or less continuous matrix that acts like a binder. The failure mechanism of concrete is closely related to the initiation and propagation of cracks, coarse aggregates generate weaker zones (interfaces) in which the development of the cracks begins. Furthermore, macrodefects (big pores, air bags) generated during compaction or bleeding may concentrate around the bigger aggregates and also cracks will appear due to differences in stiffness, shrinkage or thermal expansion coefficients between the cement paste and the aggregates. Significant cracking at micro and macro levels can also appears when concrete is exposed to more severe conditions, affecting the functionality and reliability of the structures.

As well as plain or conventional reinforced concrete, FRC can be used in structures in contact with aggressive environments or exposed to severe external conditions that produce damage in the internal structure of the material.

Frequently, for the assessment of the capacity in service or for taking decisions about replacement, reinforcement or repairing damaged structures, the evaluation of the properties of the existing concrete is required. Usually, the evaluation of the structures is made by means of the compressive strength measured on drilled cores; however other properties as stiffness, tensile strength or energy of fracture can change in a very different magnitude. The same may occur with the transport properties.

This paper presents a discussion based on previous works on the residual properties of FRC after severe degradation processes. The exposure at High Temperatures (HT) and the development of Alkali Silica Reactions (ASR) were selected as cases of study; the effect degradation processes on the pull out strength of individual fibres is also considered.

It is known that another very important case of degradation is the bars corrosion in reinforced concrete structures. The incorporation of fibres also can lead to benefits through their effect on the control of crack widths. However, the analysis of this pathology exceeds the scope of this work.

2. Residual properties of damaged plain concrete

There is extensive literature on the behaviour of damaged plain concrete. A summary of the main knowledge regarding the residual properties for the two studied cases is presented.

2.1. High temperatures (HT)

An important case of damage of the internal structure of concrete appears when it is exposed to high temperatures (HT) [1-7]. Initially, the increase in temperature results in the elimination of the pore's water and the consequent contraction of the paste with crack formation. Temperature rise over 500 ºC produces non-reversible effects, as the loss of chemical bounded water takes place. Microcracking at the interfaces appears due to differences between thermal dilatation coefficients of paste and aggregates, which increases the size and closeness of the internal cracks.

The alteration of concrete structure affects the failure mechanism, and this is reflected in the shape of the stress-strain curves. In damaged concrete the growth and propagation of matrix cracks starts earlier and there is an important decrease in the period of stable matrix crack growth as the degradation is more severe. In compression, there were reported decreases in ultimate stress, increases in final deformations and rapid and important reduction in Poisson's ratio. The tensile strength shows a great variability, and it is more sensitive than the compressive strength to the changes produced by HT. While damage produces a drop in tensile strength the fracture energy is less affected due to the greater branching in crack propagation in the post-peak regime [3, 5].

Large reductions in ultrasonic pulse velocity and in dynamic modulus of elasticity as temperature increases were observed in HT damaged concrete. The modulus of elasticity was more affected than the compressive strength, thus the residual values of velocity can diminish much more than those corresponding to strength and are very effective to detect damaged zones [4].

The temperature level, the exposure time and the cooling conditions are the main factors affecting the mechanical properties [1, 2]. The type of aggregate represents an important factor meanwhile the type of binder seems to be not so significant. Considerable changes can be produced by moisture content of concrete, in some cases the vapour pressure in the pores increase during heating and spalling takes place. Many times visual inspection showed that specimens exposed to temperatures higher than 500 °C had significant surface cracking, which was more important when they were cooled quickly.

Finally, temperatures lower than 200 °C, do not have a significant effect on the compressive strength, but can affect the transport properties mainly the external layer of concrete, promoting significant changes in concrete durability. In HT damaged concrete the residual velocity and capacity of capillary absorption grew with the crack density until a maximum is reached and then decreased, although the density and width increased. On the other hand the water permeability showed a direct relationship with the level of damage, increasing as the crack density and width increase. It can be mentioned that the transport properties can also be affected by less severe conditions as the presence of microcracking due to drying shrinkage [8].

IOP Conf. Series: Materials Science and Engineering **246** (2017) 012007 doi:10.1088/1757-899X/246/1/012007

2.2. *Alkali-silica reaction (ASR)*

Alkali-silica reaction (ASR) can induce loss in serviceability of concrete structures. ASR occurs in concretes with reactive aggregates, when sufficient alkalis (K_2O, Na_2O), and moisture are available. Different levels of damage and cracking appear in concrete microstructure according to the kinetic of ASR. Rapid reaction rates induce internal stresses at the interfaces and cement pastes, producing micro and macrocracks. On the other hand in concretes with strained quartz, the reactions are localized inside the aggregates in reactive zones (intercrystals), where pore solution can reach. This process takes place very slowly and the attack is not generalized all around the aggregate surface. [9–12].

Usually linear expansions are used to evaluate the material degradation; nevertheless the residual mechanical properties could be different for a same expansion. ASR strongly affects concrete tensile strength and modulus of elasticity. The reductions in compressive strength are always lower than those observed in stiffness, being in some cases compressive strength even not affected.

The failure mechanism of concrete in compression is also clearly affected by ASR. The shape of the stress–strain curves reflects the presence of internal fissures. The growth and propagation of matrix cracks tends to start earlier due to defects produced by ASR. The period of stable crack propagation is less affected than the period of unstable crack growth which is widely extended, showing that the capability of controlling crack propagation decreases leading to premature failure. In tension, the differences in the crack pattern of sound and damaged concretes are reflected in the shape of the load deflection curves, both in the pre- and post-peak regime. Damaged concretes show an increased non-linearity before the peak and a more gradual softening.

Also in this case, the changes on transport properties strongly depend on the characteristics of the crack pattern [8].

3. Residual bond strength of individual fibres in damaged matrix

It is well known that the post peak capacity of FRC depends on the type and content of fibers. The fibers mainly act after a crack appears in the matrix being the pull out the most important mechanism involved. Although fiber – matrix bond increases with the compressive strength, the increments are more significant in the later. This behavior was verified for synthetic macrofibres (M) and hooked-end steel fibres (S) [13].

To study the effect of ASR on bond strength pull out tests were performed. Four mortars with w/c 0.50 were prepared with reactive sand, two incorporated a high content of alkalis (D) and the others were used as reference (C). Reactive and reference mortars incorporating synthetic microfibers were also done with the aim of evaluating if there is any effect in controlling matrix microcracking and consequently in bond strength; they are identified as Cm and Dm. Specimens of 40 x 20 mm section and 70 mm length with a single fibre embedded were used; the bond length was half of the fibre length. In addition, prisms of 25 x 25 x 300 mm were cast to measure the expansions evolution.

Figure 1.a shows the evolution of mortar expansions. Mortars D and Dm that incorporate high content of alkalis show important expansions over 0.6 %. As expected, mortars C and Cm do not expand. Figure 1.b plots the results of bond strength vs. mortar compressive strength obtained with hooked end steel fibres (S) and different types of macrosynthetic fibres (M). It can be seen that mortars D and Dm show a clear decrease in compressive strength, which is slightly lower in the later. However, the variations in pull out strength are significantly lower. In this case, when the matrix is damaged by ASR the mortar compressive strength decreased near 40 %, while the bond strength is only marginally affected specially fibre S.

The same tendency was verified when the residual properties were evaluated in matrices damaged by HT. The variations in compressive strength are usually higher than those observed in bond. In a recent study [14] pull out tests were performed on concrete specimens incorporating steel fibres exposed at 20, 300, 375 and 475 °C; no change in pull-out response up to 375°C was observed, and only a reduction in load capacity of near 20 % appeared for 475°C. The corresponding concrete compressive strengths were 63.7, 55.7, 54.4 and 44.2 MPa respectively.

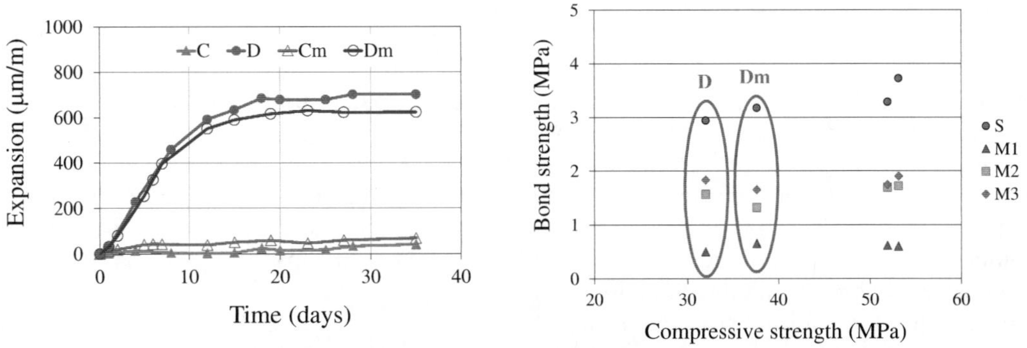

Figure 1. a (left) Evolution of the mortar expansions; b (right) bond strength vs. mortar strength.

4. Residual properties of FRC affected by HT

The residual mechanical behaviour of thermally damaged high strength FRC incorporating different types and contents of hooked end steel fibres were studied [15]. Two exposure conditions were used, 1 hour at 500 °C and 24 hours at 150 °C; reference specimens at 20°C were also tested.

The residual stress strain curves in compression of plain (P) and FRC incorporating 40 kg/m^3 (40H) or 80 kg/m^3 (80H) of High carbon steel fibres and 40 kg/m^3 (40L) of Low carbon steel fibres are given in Figure 2. The axial, lateral and volumetric strains are plotted. It was found that as cracking increases the lateral/axial strains ratio increase at lower stresses and more rapidly.

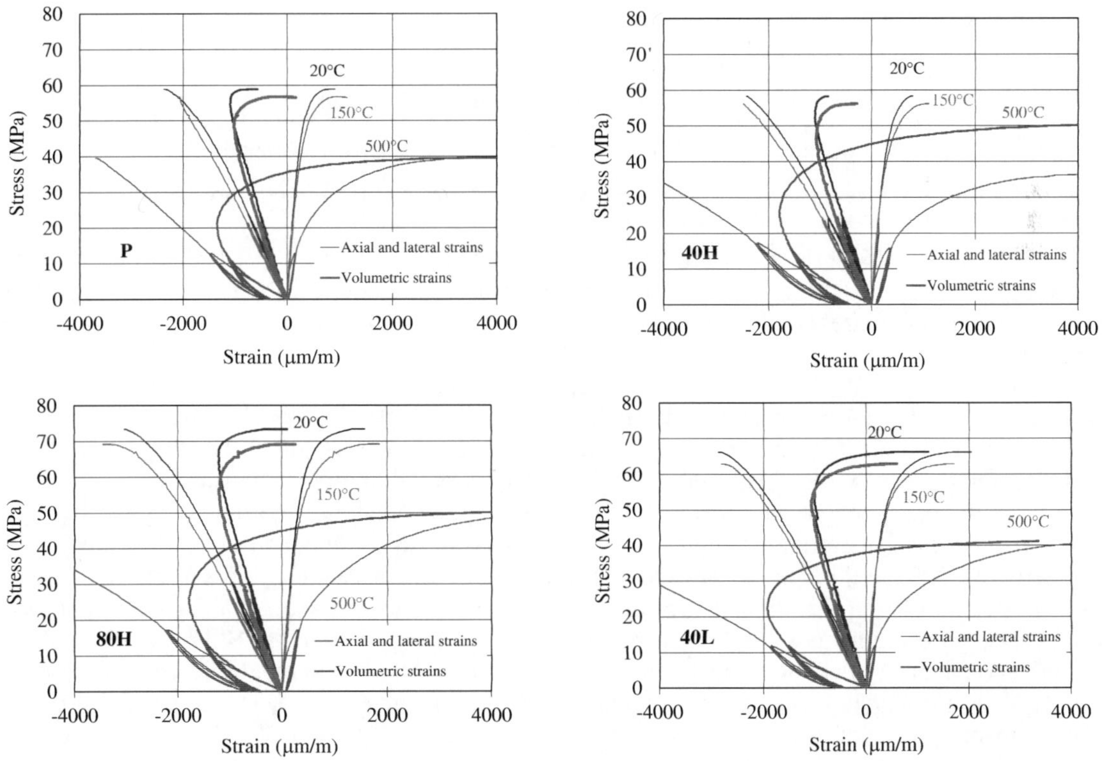

Figure 2. Stress vs. axial, lateral and volumetric strain curves in compression.

Regarding the failure mechanism in compression, Figure 3 presents the initiation stress (f_{init}, associated with the start of matrix cracking), the critical stress (f_{crit}, associated with the onset of unstable crack propagation) which corresponds to the peak volumetric strain [16], and the compressive strength (f´c). It can be seen that these parameters tend to increase as fibre content increases and when High–carbon steel fibres were used. When concretes are exposed to 500 °C, f_{init} and f_{crit} decrease, being the decrease in f_{crit} more significant. This means that, due to the cracking generated by high temperature, the growth and propagation of matrix cracks started earlier. The period of stable crack propagation is reduced and, although the period of unstable crack growth is extended the capability of controlling crack propagation decreases. The fibres increase the load-carrying capacity in post-cracking regime when compared with plain concrete (see P vs. 80H).

Summarizing, FRC follow similar residual compressive behaviour as the plain concrete, but the presence of fibres lead to slight increase in strength and in the stress at which cracks initiate.

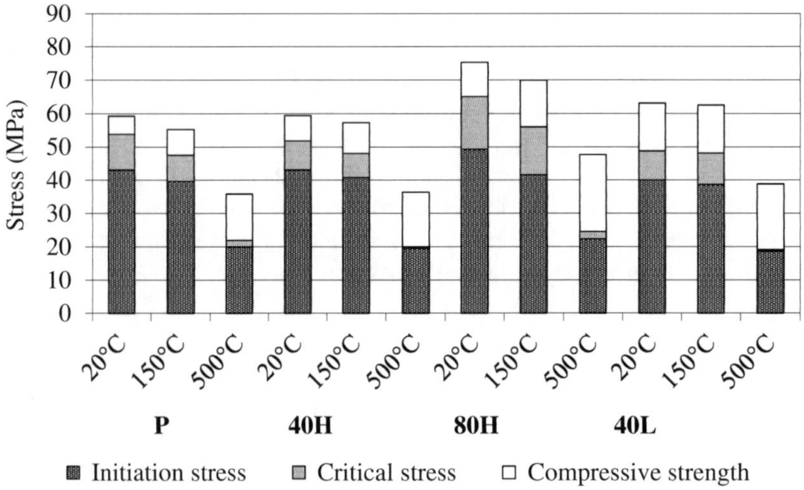

Figure 3. Stress vs. axial, lateral and volumetric strain curves in compression.

The main benefits of fibre incorporation in concrete are related to the tensile behaviour and the corresponding increments in toughness; in this way the characterization of FRC is based on their post peak capacity (residual stress). Thus the study of the post peak loading capacity in damage concrete becomes relevant.

To evaluate the tensile behaviour flexural tests on notched beams were performed. Figure 4 presents typical load – deflection curves in bending. As it was expected, the matrix degradation leads to a reduction of the residual mechanical properties of concrete. It was found that the shape of the load-deflection curves in FRC exposed to 150 °C was similar to the undamaged concrete. The reductions in flexural strength were lower in FRC than in plain concrete, and the post-peak strength was less affected than first-crack strength. For the most severe exposure condition the degradation of the material is reflected by an increased non-linearity of the ascending branch of the curves, nevertheless some FRC still exhibited strengthening type behaviour and kept an almost constant load bearing capacity during the post-peak. Concrete prepared with high-carbon steel fibres tends to show rather higher residual values. Some differences can be attributed to the number of fibres as the low-carbon type has the same aspect ratio but greater length. It is interesting to note that in FRC 40L the post-peak loading capacity was very similar between 20 °C, 150 °C and 500 °C, the degradation specially affects the first crack strength.

Many times Non-destructive tests (NDT) are used for the evaluation of concrete structures. Nevertheless in damaged concretes some particular aspects must be taken into account. In thermally damaged plain concrete, ultrasonic pulse velocity (UPV) strongly differs from the usual values measured in sound concrete; the UPV method cannot be used to estimate strength but represents a useful tool to identify damaged zones [4]. Figure 5 plots the variation of the static elastic modulus with the compressive strength and the relationship of the static modulus o elasticity with the UPV; the results of concretes P, 40H, 80H and 40L previously presented, exposed to 20, 150 and 500 ªC are included. In addition, the residual properties of other fibre concrete exposed to the same conditions are shown, all fibre concretes are identified as FRC. As it occurs in plain concrete, there is not a good correlation between the residual modulus of elasticity and the residual compressive strength in FRC, but concretes prepared with different materials, types and contents of fibres, and exposed to different grades of alteration show a very good relationship between the modulus of elasticity and the UPV.

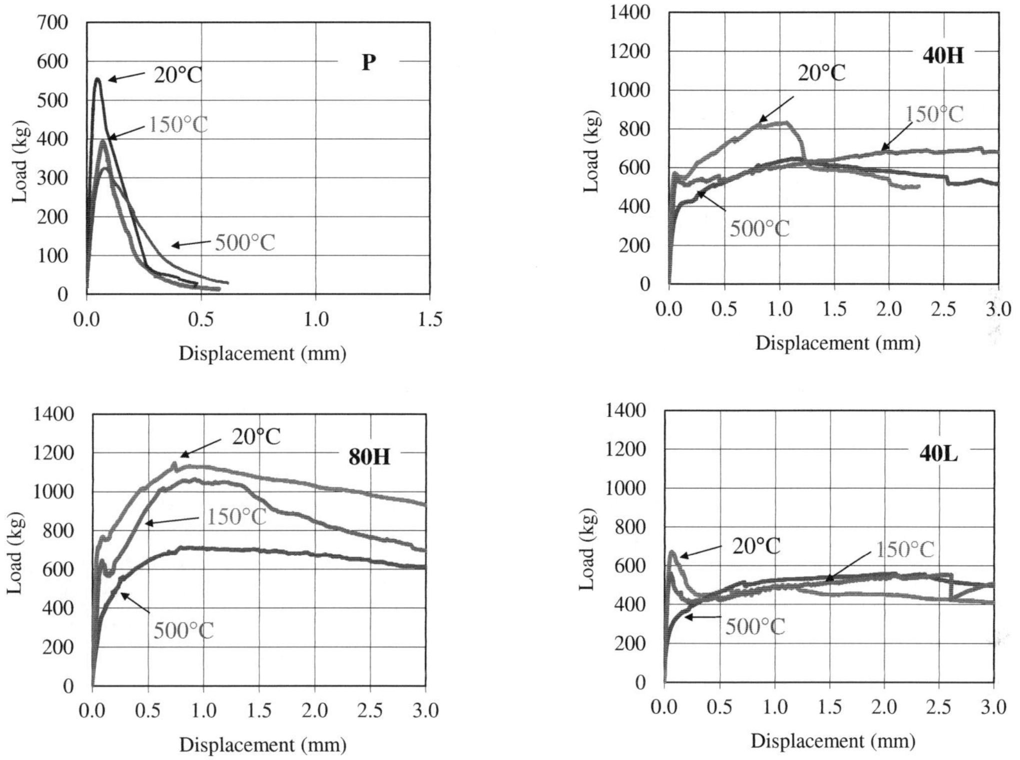

Figure 4. Load - deflection curves on notched prisms.

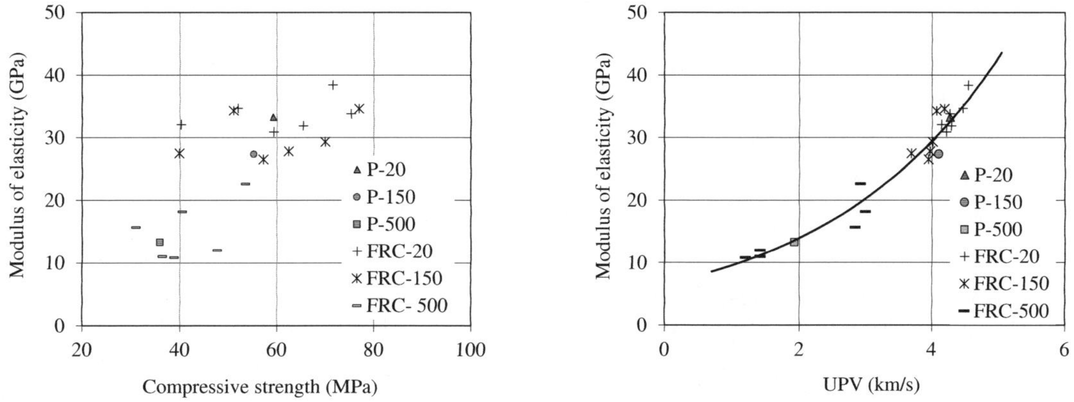

Figure 5. Left: Compressive strength vs. elastic modulus. Right: UPV vs. elastic modulus.

5. Residual properties of FRC affected by ASR

A reference concrete (P) and two FRC incorporating 40 kg/m^3 of hooked-end steel fibres (S) or 3 kg/m^3 of synthetic macrofibres (M) were done. All of them were prepared with similar mixture proportions and incorporate a 19 mm MSA highly reactive crushed quartzitic sandstone as part of the coarse aggregate to promote the development of ASR. The total alkali content in concrete was equal to 4 kg/m^3 (NaOH was added in the mixing water). The expansion along time, as representative of the reaction level, and the compressive strength, the modulus of elasticity and the flexural behaviour, as characteristic residual mechanical properties, were evaluated [17].

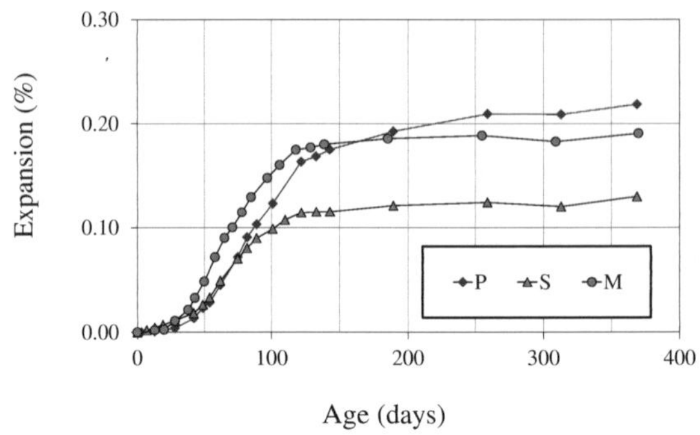

Figure 6. Linear expansions.

The linear expansions of the different concretes are given in Figure 6. A high rate of expansion can be seen until near 150 days and then it markedly decreases, achieving values near 0.2 %. Concrete S shows lower expansions; at one year the expansion is near 0.13 %.

To evaluate the residual mechanical properties compression tests were performed in standard cylinders at different ages from 28 days to 1 year. Test at 28 days were used as a reference value as at that age the ASR was not significant (see Figure 6). Similarly, bending tests on notched prisms were performed at different ages up to 1 year according to the guidelines of the EN 14651 standard [18]; in this case the Limit of Proportionality (f_L, that correspond to the first peak stress, representative of the matrix strength) and the residual stress f_{R3} (that corresponds to the nominal stress for a Crack Mouth Opening Displacement, CMOD, of 2.5 mm) were selected for the analysis.

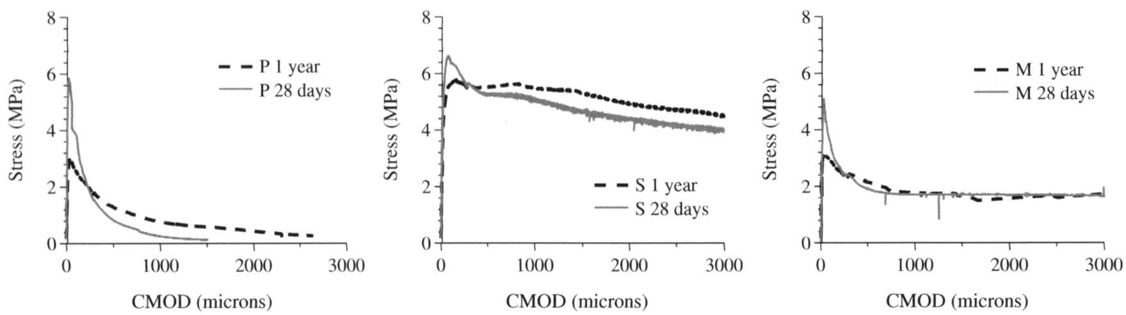

Figure 7. Stress - CMOD curves in bending.

Figure 7 compares typical Stress – CMOD curves of the three concretes at 28 days and 1 year. Each FRC shows a different post peak response, in accordance with the type of fibre incorporated. It can be seen than in plain concrete as the damage increases the softening branch of the curves become more extended, and they even show some residual load capacity. This behaviour is attributed to the possibility of greater branching and meandering of cracks [10]. In concrete M, with synthetic macrofibres, after achieving the maximum value the loads significantly decrease and they remain almost constant until the end of the test. The steel FRC presents a very gradual descending branch, with high stress values. In both FRC, although the first peak load clearly decreases at 1 year (severely damaged), the post peak loading capacity remains almost constant. It can be mentioned that, an extensive crack pattern was seen in 150 mm side prisms of these concretes [17].

Figure 8 summarizes the effect produced by the cracking developed in each concrete on their mechanical response; the results of compressive strength, modulus of elasticity, first-peak strength (f_L), and the residual strength (f_{R3}) versus measured expansions are plotted. As in damaged plain concrete, in FRC the compressive strength and the modulus of elasticity decrease as the deleterious processes advances; the presence of internal cracking leads to greater reductions in stiffness than in strength [8, 10]. In tension, the strength decreases as the expansion increases but the presence of fibres, particularly the steel ones, improves the concrete behaviour. When the post peak strength capacity is considered a quite different response was found; the residual stress remains almost constant, even for very high expansions. Consequently, FRC incorporating steel of synthetic macrofibres conserve their residual loading capacity when severe ASR damaged has taken place. This is in accordance with previous studies that showed minor changes in fibre-matrix bond strength in mortars affected by ASR [13].

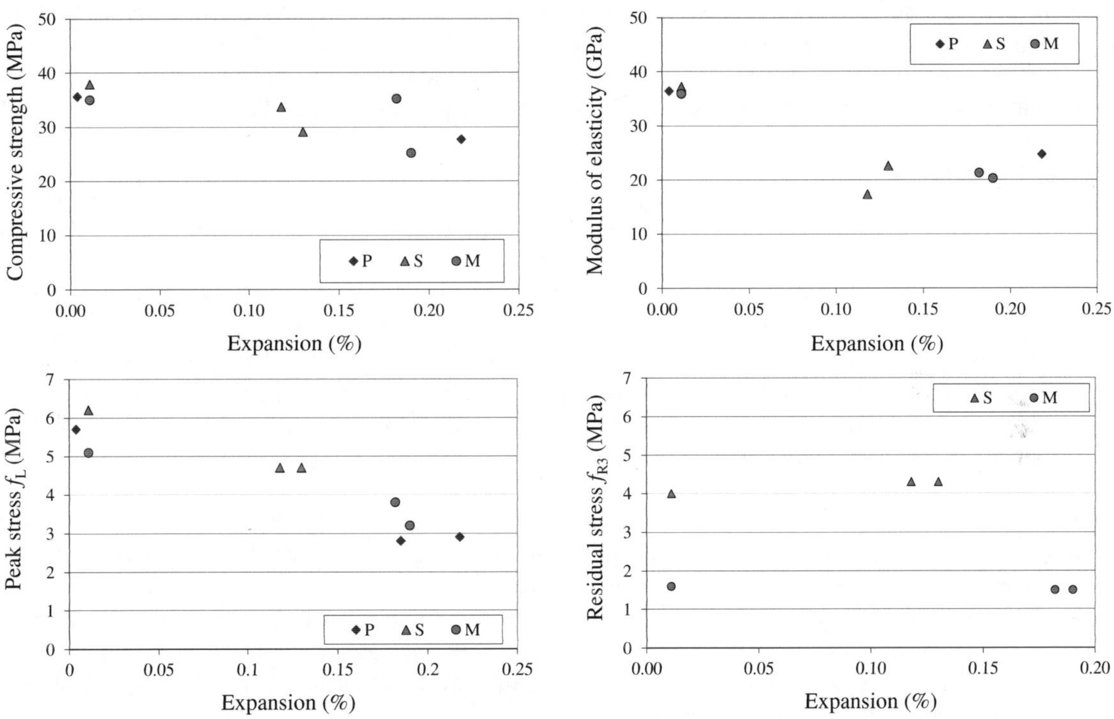

Figure 8. Relationship between mechanical properties and linear expansions: a) compressive strength, b) modulus of elasticity, c) first-peak strength, d) residual stress for a CMOD of 2.5 mm (f_{R3}).

Several codes estimate concrete stiffness from the results of compressive strength. The modulus of elasticity vs. compressive strength relationship proposed by the *fib* Model Code 2010 [19] is represented in Figure 9. The values obtained on concretes P, S and M at 28 days and at 6 and 12 months are plotted. It can be seen that when there is very few cracking, tests performed at 28 days, the measured modulus of elasticity is close to the prediction curve. When concrete is damaged the conventional expressions that estimate the modulus of elasticity from the compressive strength are no longer valid, the experimental values of stiffness are significantly overestimated.

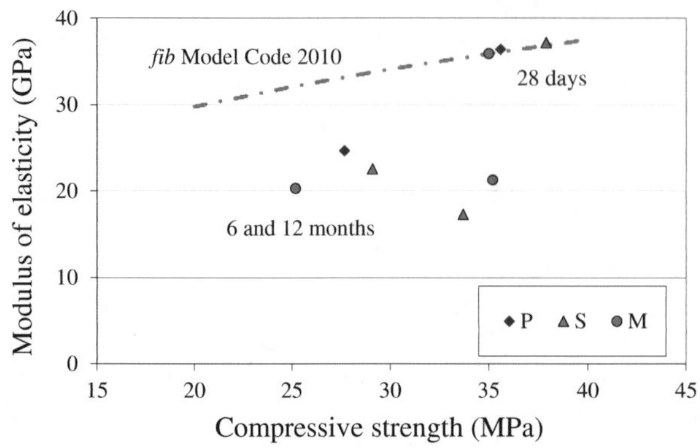

Figure 9. Relationship between compressive strength and modulus of elasticity.

Although the incorporation of fibres does not avoid ASR, some fibres may be useful to reduce in some extent the expansion rate and level, as well as the induced crack sizes [20-21]. Regarding the mechanical performance, it should be emphasized that, FRC maintain their residual properties without significant modifications even when severe matrix damage occurred. Then, the use of FRC is a useful tool to extend the structures service life as they conserve their residual mechanical properties and crack control capacity even if unexpected significant damage processes take place.

6. Concluding remarks

The residual properties of FRC after exposed to different severe conditions have been analysed in this paper, the cases of exposure at High Temperatures and the development of Alkali Silica Reactions were considered. The main findings are pointed as follows:

- No significant differences in the residual behaviour in compression have been found between plain and FRC. The compressive strength is less affected than the elastic modulus and the failure mechanism is clearly affected by the internal damage. The shape of the stress–strain curves reflects the presence of internal fissures, the growth and propagation of matrix cracks tends to start earlier, the period of stable crack propagation is less affected than the period of unstable crack growth which is widely extended, showing a decrease in the capability of controlling crack propagation.
- As well as in plain concrete, the tensile strength decreases as the damage increases but residual post-peak strength capacity due to the presence of steel or macrosynthetic fibres is mainly no modified, even for very severe damage.
- The capacity of FRC to conserve a significant part of its post-peak strength capacity is in accordance to pull out tests of single fibres, which indicate that in damaged concrete the reductions in fibre-matrix bond strength are significantly lower than the decreases measured on the matrix compressive strength.

References

[1] Bazant Z Kaplan M, 1996 *Concrete at High Temperatures* 1st Ed (Longman, Essex, UK).
[2] Mohamedbhai GTG 1986 Effect of exposure time and rates of heating and cooling on residual strength of heated concrete *Mag. of Concrete Research* 38 (136) 151-158
[3] Barragán B, Giaccio G, Zerbino R 2001 Fracture and failure of thermally damaged concrete under tensile loading *Materials & Structures* 34 (239) 312-319
[4] Di Maio A, Giaccio G, Zerbino R 2002 Non-destructive tests in the evaluation of concrete exposed at high temperatures *ASTM J Cement, Concrete and Aggregates* 24 (2) 58-67
[5] Baker G. 1996 The effect of exposure to elevated temperatures on the fracture energy of plain concrete *Materials & Structures* 29 (190) 383-388.

[6] Luo X, Sun W, Chan SYN 2000 Residual compressive strength and microstructure of high performance concrete after exposure to high temperature *Materials & Structures* 33 (229) 294 -298
[7] Luo X, Sun W, Chan SYN, 2000 Effect of heating and cooling regimes on residual strength and microstructure of normal strength and high-performance concrete, *Cement and Concrete Research* 30 (3) 379-383
[8] Torrijos MC, Giaccio G. Zerbino R. 2010, Internal cracking and transport properties in damaged concretes, *Materials & Structures* 43 (1) 109-131. DOI 10.1617/s11527-010-9602-z
[9] Fournier B, Berubé M, 2000 Alkali-aggregate reaction in concrete: a review of basic concepts and engineering implications, *Can. J. Civil Eng*. 27 (2) 167–191
[10] Giaccio G, Zerbino R, Ponce JM, Batic OR 2008 Mechanical behavior of concretes damaged by alkali silica reaction, *Cement and Concrete Research* 38, 993-1004.
[11] Jones AE, Clark LA, 1996 A Review of the Institution of Structural Engineers Report: structural effects of Alkali-silica reaction (1992), Proc. *10th ICAAR*, Melbourne Australia, 394–401
[12] Takemura K. Ichitsubo M, Tazawa E, Yonekura A, 1996 Mechanical performance of ASR affected nearly full-scale reinforced concrete columns, Proc. *10th ICAAR*, Melbourne Australia, 410–417.
[13] Bossio ME, Torrijos MC, Zerbino R, Giaccio G, 2012, Pull out behaviour of macro synthetic fibres: effects of fibre type, matrix strength and microcracking, in: Proc. of *Bond in concrete 2012: Bond, anchorage, Detailing, Fourth Int Symp*, Brescia, Italy, Vol 2, ISBN 978-88-907078-3-4, Ed. JW Cairns et al. 901-906
[14] Ruano G, Isla F, Luccioni B, Zerbino R. Giaccio G. 2016 Influencia de altas temperaturas en el comportamiento mecánico del hormigón reforzado con fibras *Mecánica Computacional XXXIV*, Eds. S. Giusti et al., Córdoba, Argentina, , 2463-2481
[15] Giaccio G, Zerbino R, 2005 Mechanical behaviour of thermally damaged high-strength steel fibre reinforced concrete, *Materials & Structures* 38, 335-342
[16] Shah SP. Chandra S. 1968 Critical stress, volume change, and microcracking of concrete, *ACI Materials Journal*, 65 (9) 770-781
[17] Giaccio G, Bossio ME, Torrijos MC, .Zerbino R. 2015 Contribution of fiber reinforcement in concrete affected by alkali–silica reaction". *Cement and Concrete Research* 67 310-317
[18] EN 14651 2005 Test method for metallic fibre concrete European Standard
[19] Fédération Internationale du Béton 2012 *fib* Model Code for Concrete Structures 2010 Vol 1 350 pages ISBN 978-2-88394-105-2; Vol 2 370 pages ISBN 978-2-88394-106-9
[20] Pires de Carvalho MR., de Moraes Rego Fairbairn E, Dias Toledo Filho R, Chagas Cordeiro G, Hasparyk NP, 2010 Influence of steel fibers on the development of alkali-aggregate reaction, *Cement and Concrete Research* 40 598-604
[21] Haddad RH, Smadi MM, 2004 Role of fibers in controlling unrestrained expansion and arresting cracking in Portland cement concrete undergoing alkali–silica reaction, *Cement and Concrete Research* 34, 103–108

IOP Conf. Series: Materials Science and Engineering **246** (2017) 012008 doi:10.1088/1757-899X/246/1/012008

Crack control in concrete members reinforced by conventional rebars and steel fibers

G Tiberti[1], I Trabucchi[2], M AlHamaydeh[3], F Minelli[4] and G Plizzari[5]

[1] Assistant Professor, [2] Ph.D. Student, [4] Associate Professor, [5] Professor, Department of Civil Engineering, Architecture, Land, Environment and of Mathematics, University of Brescia, Italy

[3] Corresponding Author, Associate Professor, Department of Civil Engineering, American University of Sharjah, Sharjah, United Arab Emirates

E-mail: malhamaydeh@aus.edu

Abstract. Tension stiffening is still a matter of discussion within the scientific community. The ability of resisting tensile stresses by un-damaged concrete portions that spans in between cracks is significant and could be improved with adoption of tougher material as Fiber Reinforced Concrete (FRC). In addition, FRC may provide noticeable residual stresses at a crack location, linking the two adjacent faces due to the bridging effect provided by the fibers. Within this framework, this paper aims at further investigating the ability of Steel Fiber Reinforced Concrete (SFRC) in reducing the crack spacing and width by utilizing SFRCs with high post-cracking residual strengths.

1. Introduction

The employment of Fiber Reinforced Concrete (FRC) has gained considerable attention in recent years, as numerous researches have demonstrated its effectiveness in many structural applications. Moreover, a number of physical, semi-empirical and empirical models have been recently developed toward the formulation of appropriate design procedures that are useful for practitioners; especially for strain-softening materials. The recent inclusion of FRC in the fib Model Code 2010 [1], referred to as MC2010 herein, in national codes as well as the organization of FRC-devoted conferences [2, 3, 4] confirms this positive development.

FRC has been particularly used in structural elements when crack propagation control is of primary importance, e.g. in precast tunnel segments [5] or in beams where little or no shear reinforcement is provided [6]. In several of these applications, the reinforcement generally consists of a combination of conventional rebars and fibers [7].

FRC significantly improves the behavior at the Serviceability Limit State (SLS), with respect to crack and deflection control. It is well known that, after cracking, tensile stresses are induced in the concrete between cracks and hence they stiffen the response of a reinforced concrete member subjected to tensile stresses. This stiffening effect is referred to as "tension stiffening". Several authors already studied this mechanism in traditional Reinforced Concrete (RC) elements [8, 9, 10]. In fibrous RC elements, the transfer of non-negligible residual stresses at crack locations provides an additional significant mechanism that influences the member response. The combination of these two mechanisms (tension stiffening and the post-cracking residual stresses provided by fibers at any crack, referred to as "residual strength" or "tension softening" herein) results in a different crack pattern, characterized by a reduced crack spacing and crack width.

A number of research studies have been carried out so far on the tensile behavior of FRC members since late 90's. Mitchell [11] presented one of the first studies; more recently, Bischoff [12, 13] performed monotonic and cyclic tests and included shrinkage effects in the analysis.

Within this framework, the present paper describes the main results obtained from 17 tests on tension-ties to the aim of evaluating crack formation and development in structures reinforced by SFRC having high post-cracking residual strength. Besides a description of the tests, particular attention was devoted to the influence of fibers in terms of crack formation and development: the crack initiation, the crack formation stage, the crack spacing and its progression will be evaluated. The results will be also compared against the formulation proposed by fib MC 2010 [1].

2. Experimental program

In previous research works carried out at the University of Brescia [14, 15] a broad experimental program was developed and about one-hundred uniaxial tension Reinforced Concrete (RC) and Steel Fiber Reinforced Concrete (SFRC) members containing one central steel rebar were tested. Several key-parameters were investigated such as element size (b), rebar diameter (ϕ), ϕ/ρ_{eff} ratio and SFRC post-cracking properties. These fibrous and non-fibrous members will be identified as RC and SFRC tensile ties, respectively. Note that the effective reinforcement ratio (ρ_{eff}) is the conventional reinforcing rebar area divided by the area of concrete in tension surrounding the reinforcement: in the presented samples, $\rho=\rho_{eff}$.

In order to further improve the knowledge on the cracking behavior of SFRC members, 17 SFRC tensile ties were cast within a collaborative program with the American University of Sharjah. The principal aim was to investigate the effects of high-performance SFRCs, presenting noticeable high post-cracking nominal residual strengths f_{Ri}, according to EN 14651 [16].

2.1. Uniaxial tension SFRC test specimen configurations

The geometry of tested specimen is shown in Figure 1. Five square cross sections were selected: 80, 120, 180 and 200 mm in size (b, Figure 1). Samples having 80 mm size are 950 mm long and have been tested by means of a hydraulic universal servo-controlled machine [15]. All the other specimens are 1500 mm long; holed steel plates were welded at both of the rebar ends. The latter were connected by pins in vertical position to the strong floor and to a steel reaction frame (for more details, see [15]).

Reinforcing bars having a diameter of 10, 20 and 30 mm (B450C steel, according to European standard EN 10080 [17]), corresponding to a reinforcement ratio (ρ) varying from 0.98% to 2.23%, were employed (Figure 1). The properties of the reinforcing bars used in the tie elements are reported in Table 1. The concrete cover was, in all cases, at least 2-3 times the bar diameter to prevent splitting phenomena during the tests [18, 19].

Specimens were reinforced by hooked-end-steel fibers having a length of 30 mm, a diameter of 0.50 mm (aspect ratio of 60) and tensile strength of 3100 MPa. The fiber content was 40 kg/m^3 (fiber volume fraction, V_f is about 0.5%).

All samples were cast with the same concrete matrix designed to obtain a C35/45 concrete, according to Eurocode 2 [20]. Specifically, a cement content of 390 kg/m^3; water-to-cement ratio of 0.46; sand content (0-4 mm) of 648 kg/m^3; coarse aggregate content (4-20 mm) of 1202 kg/m^3; and superplasticizer content of 3.3 l/m^3.

Table 2 reports the mechanical properties of used concrete and the volume fraction of steel fibers. Each combination of fiber reinforcement, member dimension and steel reinforcement ratio defines a specific set of tests, whose repetitions and notations are listed in Table 3, in which the entire experimental program is reported in detail. Three specimens were in general tested for each material combination. In Table 3 the mean experimental crack spacing s_{rm} is reported as well. The mean crack spacing of a single specimen was evaluated by measuring the distance between visible cracks on the surface. Furthermore, the mean crack spacing of each set of samples (s_{rm}) was calculated as the mean value of the measured mean values of each single specimen.

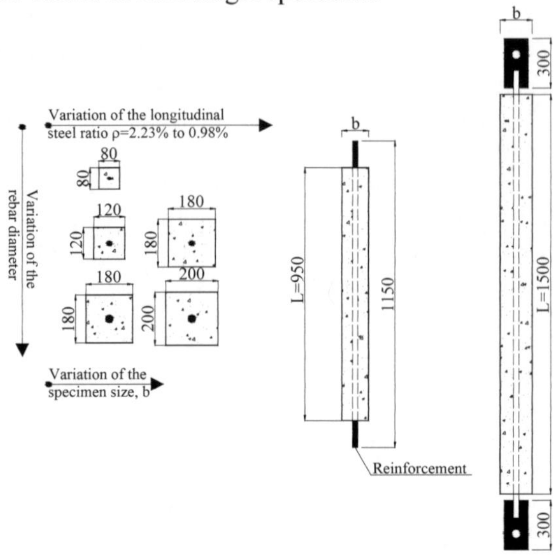

Figure 1. Geometry and reinforcement details of specimens.

Table 1. Properties of steel reinforcing bars.

Rebar	A_s [mm^2]	ϕ [mm]	E_s [GPa]	f_y [MPa]	ε_{sh} [x10^{-3}]	f_{ult} [MPa]
ϕ10	77	10	193	511	-	720
ϕ20	305	20	192	548	14.72	633
ϕ30	711	30	207	510	11.52	621

Table 2. Mechanical properties of concrete and fiber content.

Batch ID	Days after casting	$f_{cm,cube}$ [MPa]	f_{cm} [MPa]	f_{ctm} [MPa]	E_c [GPa]	Fibers , V_f [% - kg/m^3]
SFRC 0.5MH	149	53.1 (0.06)	44.1 (*)	3.27 (†)	35.2 (†)	0.50 - 40
Coefficient of variation (CV) was reported in round brackets.						
* Calculated by assuming f_{cm}=0.83·$f_{cm,cube}$.						
† Calculated as suggested by fib Model Code 2010 [1].						

Table 3. Experimental program, specimen notation and mean crack spacing.

Rebar	Batch ID	b [mm]	Length, L [mm]	Reinf. Ratio (%)	Clear cover [mm]	Specimen ID	No. of Spec's	s_{rm} [mm]
ϕ10	0 Plain*	80	1000	1.25	35	N 80/10 – 0	3	144
	0.5MH					N 80/10 – 0.5MH	4	111
ϕ20	0 Plain*	120	1500	2.23	50	N 120/20 – 0	3	170
	0.5MH					N 120/20 – 0.5MH	3	127
ϕ20	0 Plain*	180	1500	0.98	80	N 180/20 – 0	3	358
	0.5MH					N 180/20 – 0.5MH	4	215
ϕ30	0 Plain*	180	1500	2.23	75	N 180/30 – 0	3	232
	0.5MH					N 180/30 – 0.5MH	3	172
ϕ30	0 Plain*	200	1500	1.80	85	N 200/30 – 0	3	310
	0.5MH					N 200/30 – 0.5MH	3	180
* Reference plain concrete specimens which were tested in a previous experimental campaign (2nd phase, [15]).								

IOP Conf. Series: Materials Science and Engineering **246** (2017) 012008 doi:10.1088/1757-899X/246/1/012008

The main data of reference RC tensile ties were also reported in Table 3. In fact, RC samples tested in a previous broad experimental campaign [15] were assumed as reference for analyzing results obtained from sets of the tensile ties that are labeled as SFRC 0.5MH. It is worthwhile underlining that the mechanical properties of steel reinforcing conventional bars (ϕ20 and ϕ30) reported in Table 1 are similar to those exhibited by rebars adopted in the previously mentioned research work [15], making the comparison of results valid and consistent.

2.2. Material properties

A number of tests were conducted in order to determine the material properties. Standard tests on 150 mm cubes were carried out for the determination of the concrete compressive strength ($f_{cm,cube}$). The corresponding values with their coefficient of variations are reported in Table 2.

The cylindrical tensile strengths and Young modulus were determine according to MC2010 [1], as reported in Table 2. Similarly, the cylindrical compressive strengths are reported as well. The latter were determined from the experimental cubic strengths by adopting the relationship $f_{cm}=0.83 \cdot f_{cm,cube}$.

In addition, among the many standards that are available for the material characterization, all SFRC were characterized according to the European Standard EN 14651 [16] and MC2010 [1], which require that three point bending tests (3PBT) to be performed on small notched beams (150·150·550 mm). Experimental curves for the entire series SFRC 0.5MH, concerning the nominal stress vs. Crack Mouth Opening Displacement (CMOD) are depicted in Figure 2. The mean of the individual specimens' curves is also depicted. It can be noticed that the post-cracking flexural behavior of series SFRC 0.5MH is generally exhibiting the desirable high residual strength and toughness performance, as well be highlighted below.

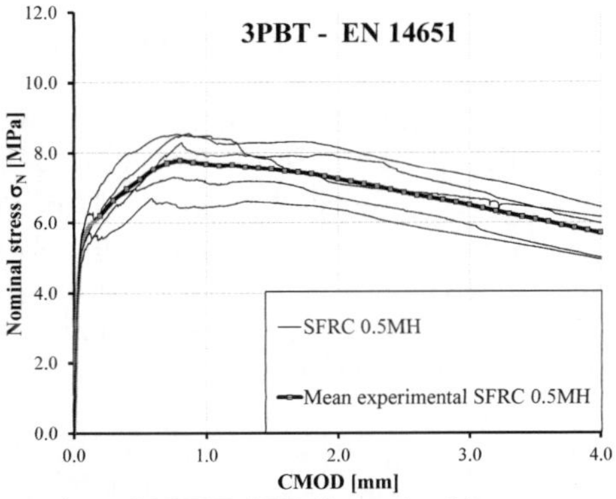

Figure 2. Experimental results of 3PBT SFRC notched beams according to EN 14651 [16].

Based on these curves, the residual strengths $f_{R,j}$ (evaluated at 4 different CMOD values, i.e. 0.5, 1.5, 2.5 and 3.5 mm [16]), and the flexural tensile strength (limit of proportionality) f_L were calculated, as listed in Table 4 (mean values). Also, the corresponding coefficient of variations are reported in Table 4. It is worthwhile underlining that both series present noticeable high residual strengths (i.e. toughness). In a previous experimental campaign [15], it was proven that the higher the f_{R1m} (corresponding to CMOD equal to 0.5 mm) is, the smaller the mean crack spacing (s_{rm}) exhibited by tensile ties. In this regard, it can be observed that the series SFRC 0.5MH present a f_{R1m} of 7.24 MPa, which is much higher than the f_{R1m} shown by SFRCs tested in the aforementioned study [15].

Table 4. Fracture parameters of the SFRCs according to EN 14651 [16].

Batch ID	f_{Lm} [MPa]	f_{R1m} [MPa]	f_{R2m} [MPa]	f_{R3m} [MPa]	f_{R4m} [MPa]
SFRC 0.5MH	5.08 (0.04)	7.24 (0.09)	7.54 (0.09)	6.86 (0.11)	6.09 (0.12)
Coefficient of variation (CV) was reported in round brackets.					

2.3. Set-up and instrumentation

Samples having a length of 950 mm were tested by means of a hydraulic universal servo-controlled (closed-loop) universal testing machine. The test was carried out with stroke-control, while monitoring the specimen behavior up to the onset of the rebar strain-hardening [15]. The deformation rate varied from 0.1 to 0.2 mm/min up to the rebar yield; then the rate was progressively increased up to 1 mm/min, the latter is beyond an average member strain of approximately 2%.

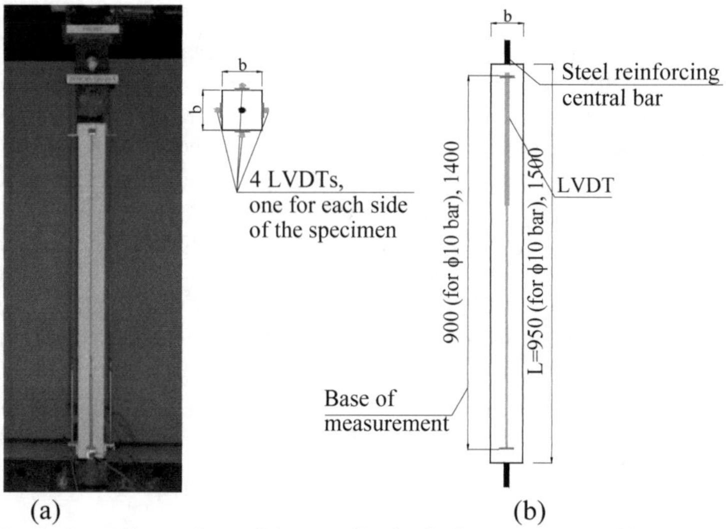

Figure 3. Typical configuration of the tensile tie during test (a) and instrumentation (b).

For the longer specimens (b=120, 180, 200 mm, Figure 1), tests were carried out by means of a steel reaction frame conveniently modified for the scope (for more details see [15]). Two previously machined steel plates, with bolt-holes, were welded at the ends of the specimen (Figure 1), which was connected by pins in vertical position to the strong floor and to the steel frame. The upper end of the specimen was connected to an electromechanical screw jack, with a maximum capacity of 1500kN. Note that the plates and the corresponding welding did not change the behavior of the bare bar as preliminary verified through uniaxial tests on the welded joints. Tests were carried out under stroke control and by assuming the same load-procedure of the first phase.

All specimens were stored in a fog room (R.H. > 95%; T=20 ± 2°C) until 2 or 3 days before testing; then they were air dried in the laboratory. Shrinkage strains were measured by means of free shrinkage prisms: since the measured strains were negligible, no-shrinkage offset strains were herein applied.

A typical instrumented specimen is shown in Figure 3a: four Linear Variable Differential Transformers (LVDTs, one for each side), were employed to measure the mean deformation of the specimen over a length ranging from 900 mm to 1400 mm, the former in the 950 mm long specimens, the latter for the remaining experiments (Figure 3b).

3. Results and discussion

In this Section, the behavior of SFRC tensile ties will be deeply investigated in terms of crack development by analyzing the evolution of mean crack spacing and average member strain with the progressive increment of the applied tensile forces.

3.1. Typical tensile tie behaviour: benefits of fiber reinforcement in combination with rebars

The diagrams reported in Figure 4a and in Figure 4b provide typical axial load vs. average tensile member strain (ε_{sm}) plots of SFRC and RC specimens, for rebars having a diameter of 20 mm and 30 mm, respectively. The average member strain ε_{sm} was calculated as the mean elongation of the 4

LVDTs, divided by the length of the base measurement. In both the diagrams, a comparison between the corresponding typical RC (plain, 2nd phase, [15]) and SFRC members are provided. In addition, the response of the bare bar is reported.

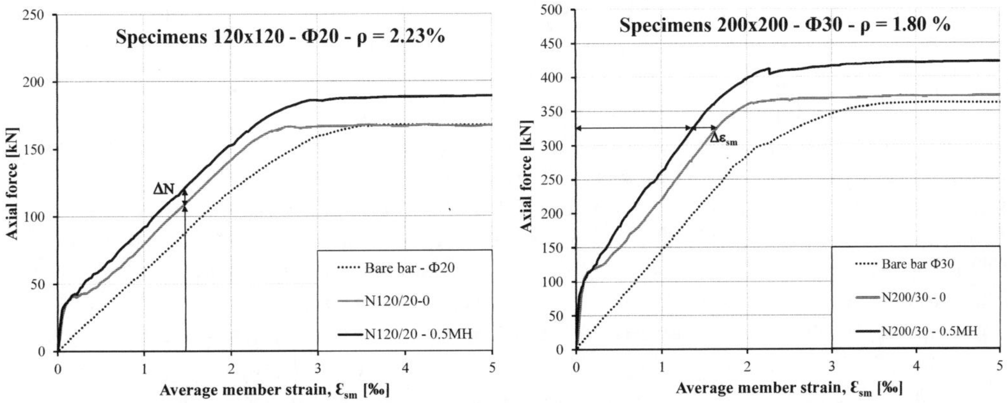

(a) (b)

Figure 4. Typical response of tie elements: (a) specimen N120/20 (b) specimen N200/30.

Even though tests were conducted well beyond that value, the results are plotted up to a maximum average strain of $5\cdot10^{-3}$mm/mm in order to appropriately assess the tensile behavior at SLS and capture the behavior at first yielding. By inspection of the diagrams, it is evident that fibrous and non-fibrous samples experience first cracking at approximately the same load level. This is attributed to the fact that fibers generally do not affect the concrete tensile strength, at least for volume fractions lower than 1%, for strain-softening materials. By further increasing the applied load, the phase where no new cracks occur, and those existing widen, is denoted as the "stabilized crack stage": it takes place for ε_{sm}=$0.87\cdot10^{-3}$mm/mm and $0.47\cdot10^{-3}$mm/mm in groups N 120/20 and N 200/30, respectively (Figure 4). One should notice that there is no clear evidence to support a possible relationship between strains at the crack stabilized stage and fiber toughness. Counterintuitively, the higher number of cracks reported in SFRC elements, in fact, form in the same range of average strains as in their RC counterpart elements.

The typical responses of FRC tensile ties reported in Figure 4 enable to emphasize one of the two main advantages related to the combination of rebars and fibers; that is the global stiffness increase caused by the transmission of noticeable residual stress across cracks. This tendency can be recognized also during the crack formation stage but it is especially clear during the stabilized crack stage. In fact, as schematically depicted in Figure 4a, by referring to a certain value of average strain (ε_{sm}), the improved toughness due to fibers determines an increment of the average tensile stress of the undamaged concrete portions within cracks, resulting in an increment of the applied global force ΔN. Hence, the tension stiffening increases with respect to that of the RC counterpart specimens. This is consistent with reported improvements of incorporating of fiber into different types RC materials and structural elements [21-24]. In the same way, by referring to a certain axial load level, a considerable reduction to the average strain occurs (denoted as $\Delta\varepsilon_{sm}$ in Figure 4b). The effect is referred to in literature as "tension stiffening strain".

3.2. Evolution of the mean crack spacing and of the tension stiffening strain

The expected mean crack width of a SFRC member diminishes because of the post-cracking toughness of SFRC, involving two important contributes: a reduction of both the mean crack spacing (s_{rm}) and the average member strain (ε_{sm}). The former aspect has been investigated in the diagrams reported in Figure 5a, while the latter in Figure 5b, for series N 120/20.

The mean crack spacing of a single specimen was evaluated by measuring the distance between visible cracks on the surface. Furthermore, for each set of samples, s_{rm} was calculated as the mean value of the measured mean values of each single specimen. In Figure 5a, the evolution of the mean crack spacing s_{rm} is plotted as a function of the average strain up to the end of the crack formation stage for specimens N 120/20. These diagrams report the average curves of the 3 specimens belonging to series SFRC 0.5MH. The s_{rm} vs. ε_{sm} response of the corresponding control RC counterpart specimens are also reported for reference. The latter diagrams were obtained in a previous experimental study [15].

From this plot, the reduction of the mean crack spacing can be readily noticed. The residual post-cracking strength provided by steel fibers (at any crack) contributes to the reduction of the transmission length (l_t) necessary to transfer tensile stresses in concrete through bond. This effect can be considerable even if it is assumed, as a first approximation, that the bond stresses between rebar and surrounding concrete are not affected by fibers. This assumption is currently also supported by MC 2010 [1]. Considering the entire sets of tests, it was estimated an average percentage reduction of mean crack spacing for series SFRC 0.5MH with respect to the plain RC counterparts of about 31.

The average member strain of fibrous and non-fibrous tensile ties (basically the average strain of the rebar embedded in a prismatic tie, ε_{sm}) is provided as a function of the bare bar strain ε_s. Consider first a range of strains up to $3x10^{-3}$mm/mm, corresponding to the stabilized crack stage up to the onset of yielding. By referring to a given applied force, represented by a certain bare bar strain (ε_s), in plain RC elements, the average member strain ε_{sm} is lower (tension-stiffening strain). Moreover, in the presence of fibers, a further increase of this phenomenon can be observed as in Figure 5b. From this plot, a reduction of about 28% is noted between SFRCs and the control RC series at a bare bar strain ε_s of $1\cdot10^{-3}$mm/mm. This reduction generally depends on the longitudinal steel ratio (ρ_{eff}): the lower the ratio, the higher the percentage reduction is. Moreover, it strictly depends on the concrete toughness, without any influence of the tensile strength. By considering all the samples, series SFRC 0.5MH presents a reduction of about 42% with respect to the control samples.

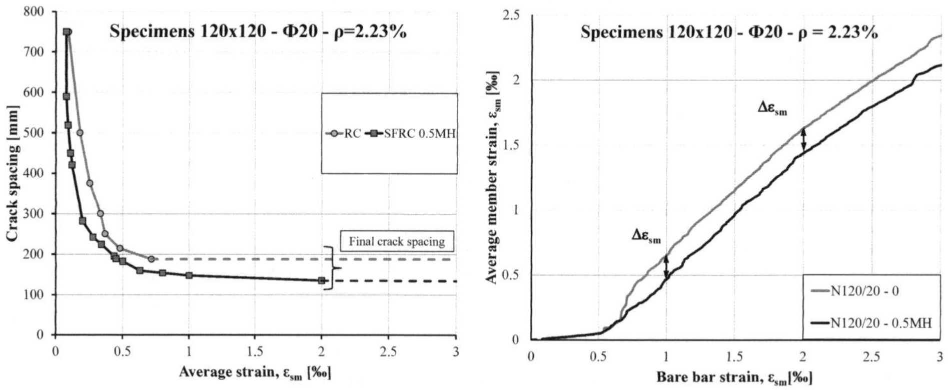

(a) (b)

Figure 5. SFRC and RC series: evolution of the crack spacing (a) and of the tension stiffening strain (b).

In Figure 6a, the mean crack spacing (s_{rm}) is plotted versus the key parameter ϕ/ρ_{eff}, which is generally included in many building codes for the prediction of s_{rm}. In particular, the experimental results are plotted for the plain RC specimens as well as the SFRC 0.5MH. The diagrams reported in Figure 6a confirm the tendency previously observed in Figure 5a: for the same value of ϕ/ρ_{eff}, the use of fibers results in a considerable reduction of the mean crack spacing due to the enhanced material toughness. Since several formulations proposed in the literature and in design codes define the crack spacing to be linearly proportional to the parameter ϕ/ρ_{eff}, a linear regression was utilized in order to evaluate the dispersion of the experimental results. As depicted in Figure 6a, the coefficient of correlation R^2 is equal to 0.92 for RC (plain control) samples whereas it is 0.97 for SFRC 0.5MH members. Basically, a possible linear relationship between s_{rm} and ϕ/ρ_{eff} can be found appropriate. Even though, for SFRC, that should be considered with the addition of a further built-in parameter taking into account the post-cracking residual strength provided by fibers.

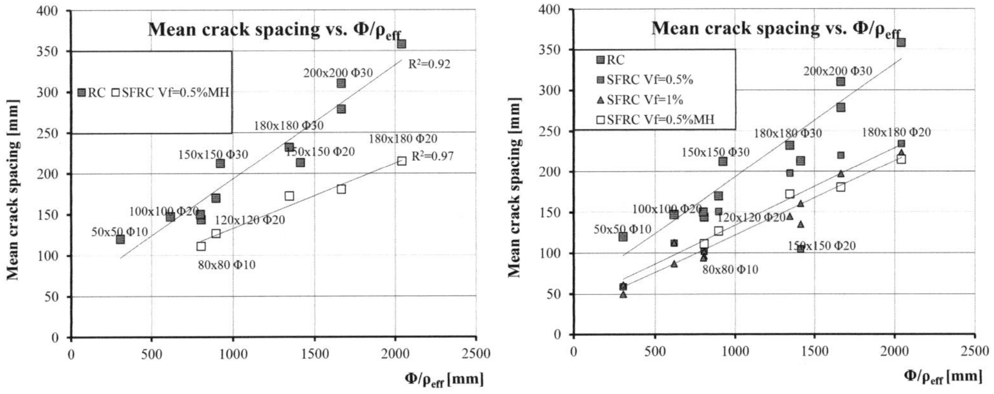

(a) (b)

Figure 6. Mean crack spacing vs. ϕ/ρ_{eff} non-fibrous series/fibrous series: (a) SFRC 0.5MH; (b) comparison with samples of previous experimental campaigns [15].

In Figure 6b, the comparison between s_{rm} and ϕ/ρ_{eff} was plotted by considering results from the SFRC 0.5MH samples investigated in the research work presented herein and those of a previous study [15]. It can be observed a rather similar behavior of previously tested SFRCs series, whose global mean reductions of s_{rm} was equal to 30% for SFRC with V_f=0.5% and 37% for SFRC with V_f=1%. The latter are rather similar to the average reductions exhibited by series SFRC 0.5MH (31%), even though the SFRC 0.5MH series demonstrate much higher post-cracking residual strengths (f_{R1m}). Based on these results, it seems that the increment of f_{R1m} beyond typical ranges of 4-5 MPa does not provide a corresponding reduction of mean crack spacing. This is evidenced by the fact that the SFRC 0.5MH group had an f_{R1m} in excess of 7 MPa.

Recently, the MC2010 [1] proposed a relationship for estimating the mean crack spacing, in which the fiber resistant contribution is taken into account by means of a reduction (f_{Ftsm}) applied to the introduction length ($l_{s,max}$, generally assumed for RC elements), which includes the FRC toughness at SLS (f_{Ftsm}=0.45 f_{R1m}). Accordingly, the following expression can be derived:

$$s_{rm} = 1.17 \cdot l_{s,\max} = 1.17 \cdot \left[k \cdot c + \frac{1}{4} \cdot \frac{\phi}{\rho_{eff}} \cdot \frac{(f_{ctm} - f_{Ftsm})}{\tau_{bm}} \right] = 1.17 \cdot \left[k \cdot c + \frac{1}{4} \cdot \frac{\phi}{\rho_{eff}} \cdot \frac{(f_{ctm} - 0.45 \cdot f_{R,1m})}{\tau_{bm}} \right] \quad (1)$$

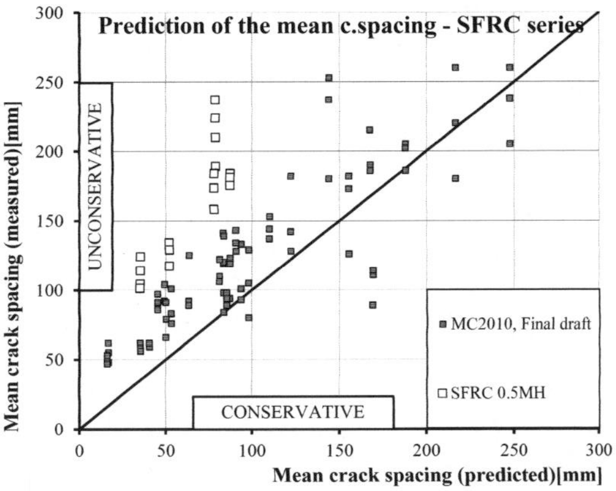

Figure 7. Crack spacing prediction for fiber series according to MC2010 [1].

Eq.1 has been applied based on the fracture parameters (mean values) reported in Table 4. The factor k was assumed equal to 1, while the bond stress (τ_{bm}) over the concrete tensile strength (f_{ctm}) ratio was assumed equal to 1.8 even for SFRC specimens. The predictions of Eq.1 are reported in Figure 7 while indicating the regions where it is conservative or otherwise. It is important to underline that the Mean Absolute Percentage Error (MAPE) for the SFRC 0.5MH series, is about 54%, as a result of un-conservative estimation for most of the samples. Hence, it can be seen that the current MC2010 [1] formulation does not consider that beyond 4-5 MPa values for residual strengths (f_{R1m}), the crack spacing may not be further reduced.

4. Concluding remarks

In the present paper, an experimental study was presented aiming at evaluating the cracking behavior of RC and SFRC ties, with special focus on the enhancement of crack control due to the addition of fibers. A series of 17 tension-tie specimens' tests have been carried out on SFRC samples exhibiting high post-cracking residual strengths. Based on the presented results and findings, the following main conclusions emerge:

- SFRC positively influences the behavior of tension-ties at SLS due to two main aspects: in SFRC elements, the tension-stiffening significantly increases and the mean crack spacing (s_{rm}) reduces in comparison to plain RC members. A crack spacing reduction of around 31% was seen in SFRC elements with V_f=0.5%.
- An increase in reinforcement ratio ρ decreases the mean crack spacing of both SFRC and plain RC elements.
- For a given reinforcement ratio, an increase in the conventional reinforcing bar diameter increases the mean crack spacing of both SFRC and non-fibrous concrete (with similar trend).
- SFRCs with rather high post-cracking strengths (f_{R1m} higher than 7 MPa) do not present a noticeably lower mean crack spacing with in comparison to SFRCs with f_{R1m} around 4-5 MPa. This phenomenon is not currently accounted for by the crack spacing formulation of MC2010 [1].

Acknowledgments

A special acknowledgment goes to M.Sc. Eng. Alberto Bossini, and to the technicians Mr. Andrea Delbarba and Mr. Luca Martinelli for their valuable support in performing the tests and in the data processing.

References

[1] Model Code 2010 2012 *Final Complete Draft* (fib bulletins 65 and 66) March 2012-ISBN 978-2-88394-105-2 and April 2012-ISBN 978-2-88394-106-9

[2] Di Prisco M., Felicetti R. and Plizzari G.A. 2004 *Fibre Reinforced Concrete* (BEFIB 2004, Bagneaux, France) RILEM Publications S.A.R.L., PRO39

[3] Gettu R. 2008 *Fibre Reinforced Concrete: Design and Applications* (BEFIB 2008, Bagneaux, France) RILEM Publications S.A.R.L., PRO60

[4] Barros J.A.O et al. 2012 *Fibre Reinforced Concrete: challenges and opportunities* (CD & Proceeding book of abstract of the eight RILEM international Symposium BEFIB 2012, Guimarães, Portugal) Publisher: RILEM Publications S.A.R.L., ISBN: 978-2-35158-132-2; e-ISBN: 978-2-35158-133-9

[5] Plizzari G.A. and Tiberti G. 2007 *Structural Behaviour of SFRC tunnel segments* (Proceedings of the 6th International Conference on Fracture Mechanics of Concrete Structures, FraMCos 2007 Vol 3) Editors Carpinteri A., Gambarova P., Ferro G., Plizzari G.A. (Catania, Italy) pp. 1577-1584 ISBN 978-0-415-44066-0

[6] Minelli F. and Plizzari G.A. 2013 *On the effectiveness of steel fibers as shear reinforcement* (ACI Structural Journal Vol 110 No 3) pp. 379-390 ISSN 0889-3241

[7] Tiberti, G., Minelli, F., Plizzari, G. (2014). *Reinforcement optimization of fiber reinforced concrete linings for conventional tunnels*, Composites Part B: Engineering, Vol. 58, March 2014, ISSN: 1359-8368, pp. 199-207, doi: http://dx.doi.org/10.1016/j.compositesb.2013.10.012
[8] Beeby A. W. 1971 *The prediction of Cracking in Reinforced Members* (PhD Thesis University of London)
[9] Beeby A. W. and Scott R. H. 2005 *Cracking and deformation of axially reinforced members subjected to pure tension* (Magazine of Concrete Research Vol 57 No 10) pp 611-621
[10] Borosnyói A. and Balázs G. L. 2005 *Models for flexural cracking in concrete: the state of the art* (Structural Concrete Vol 6 No 2) pp. 53-62 ISSN: 1464-4177 E-ISSN: 1751-7648
[11] Mitchell D. and Abrishami H. H. 1997 *Influence of steel fibers on tension stiffening* (ACI Structural Journal Vol 94 No 6) pp 769-773
[12] Fields K. and Bishoff P. H. 2004 *Tension stiffening and cracking of high strength reinforced concrete tension members* (ACI Structural Journal Vol 101 No 4) pp. 447-456
[13] Bischoff P.H. 2003 *Tension stiffening and cracking of steel fibre reinforced concrete* (Journal of Material in Civil Engineering, ASCE, Vol 15 No 2) pp 174-182
[14] Tiberti G., Minelli F., Plizzari G.A., Vecchio F.J. 2014 *Influence of concrete strength on crack development in SFRC members,* Cement and Concrete Composites, Vol. 45, January 2014, ISSN: 0958-9465, pp. 176-185, doi: http://dx.doi.org/10.1016/j.cemconcomp.2013.10.004
[15] Tiberti G., Minelli F., Plizzari G.A., 2015 *Cracking behavior in reinforced concrete members with steel fibers: A comprehensive experimental study,* Cement and Concrete Research, Vol. 68, February 2015, ISSN: 0008-8846, pp. 24-34, doi: http://dx.doi.org/10.1016/j.cemconres.2014.10.011
[16] EN 14651 2005 *Test method for metallic fibre concrete - Measuring the flexural tensile strength* (limit of proportionally (LOP), residual) European Committee for Standardization 18 pp
[17] UNI EN10080 2005 *Steel For The Reinforcement Of Concrete – Weldable Reinforcing Steel* 25 pp
[18] Mitchell D. and Abrishami H. H. 1997 *Influence of steel fibers on tension stiffening* (ACI Structural Journal Vol 94 No 6 pp 769-773
[19] Bigaj van Vliet A. 2001 *Bond of Deformed Reinforcing Steel Bars Embedded in Steel Fiber Reinforced Concrete – State of the Art* (Technical report TU-Delft) pp 65
[20] Eurocode 2 EN 1992-1-1 2005 *Design of Concrete Structures, general rules and rules for bilding*
[21] Yehia S., AlHamaydeh M., Alhajri R., Abdelsalam A., Farid A. 2011 *Steel Fiber SCC High Strength Lightweight Concrete with Local Available Materials, Proc. Central European Congress on Concrete Engineering (CCC 2011)*, Balatonfüred, Hungary, September 22-23.
[22] Yehia S., AlHamaydeh M., Abdelsalam A. 2013 *Development and Evaluation of Self-Consolidated High Strength Lightweight Steel Fiber Concrete in UAE, Proc. Central European Congress on Concrete Engineering (CCC 2011), IABSE Symposium Report*, 99 (16), 1068-1074, Rotterdam, Netherlands, May 6-8.
[23] AlHamaydeh M., Jarallah H., Ahmed M. 2017 *Punching Shear Capacity of Two-Way Slabs Made with Macro Synthetic Fiber-Reinforced Concrete, Proc. 11th Int. Conf. on Composite Science and Technology (ICCST-11)*, Sharjah, UAE, April 4-6.
[24] AlHamaydeh M., Orabi M., Ahmed M., Mohamed S., Jabr A., Al Hariri M. 2017 *Punching Shear Capacity of Macro Synthetic Fiber-Reinforced Concrete Two-Way Slabs with GFRP Rebars, Proc. 11th Int. Conf. on Composite Science and Technology (ICCST-11)*, Sharjah, UAE, April 4-6.

IOP Conf. Series: Materials Science and Engineering **246** (2017) 012009 doi:10.1088/1757-899X/246/1/012009

Production of footbridge with double curvature made of UHPC

J Kolísko[1], D Čítek[1], P Tej[1] and M Rydval[1]

[1] Klokner Institute CTU in Prague, Czech Republic

E-mail: jiri.kolisko@cvut.cz

Abstract. This article present a mix design, preparation and production of thin-walled footbridge made from UHPFRC. In this case an experimental pedestrian bridge was design and prepared. Bridge with span of 10 m and the clear width of 1.50 m designed as single-span bridge. Optimization of UHPFRC matrix and parameters of this material leads to the design of very thin structures. Total thickness of shell structure 30 - 45 mm. Bridge was cast as a prefabricated element in one piece. Self-compacting character of UHPFRC with high flowability allows the production of the final structure. Extensive research was done before production of footbridge. Experimental reached data were compared with extensive numerical analysis and the final design of structure and UHPFRC matrix were optimized in many details. Two versions of large scale mock-ups were casted and tested. According to the complexity of whole experiment a casting technology and production of formwork were tested and optimized many times.

1. Introduction

The UHPC is very promising material and its very favourable material properties allow to design very thin structures. Experiences with the real UHPFRC applications and results of many experimental research programs in Czech Republic lead to design modern footbridge using new principles and knowledge [2-5]. This article is focused on experimental research in field of material optimization, testing of material properties and preparation of the preliminary mock-up of the part of the construction, testing of load bearing capacity of thin walls and on final mock-up 1:1 of the whole footbridge. Each part of this extensive experimental research is presented in more detail in this article. Design and numerical analysis of this footbridge is part of another article presented at the conference [1].

The span of the footbridge is 10 m. The bridge is designed with double curvature – horizontal and vertical. In the vertical and transverse direction it is a circular arc with a camber of 0.4 m. The cross section of the bridge has a width of 1.5 m. The bridge load bearing structure consists of a bottom deck with only 45 mm thickness and 30 mm thick side walls. The side walls serve as handrails. Handrails height is 1.1 m in the middle of the bridge span and 1.5 m at the support area. In place of the support area a bridge deck is a little bit thicker because of anchors zones. These zones are reinforced with ordinary reinforcement for reliable transfer of the shear and anchor forces from the supports. The rest of the bridge has no conventional reinforcement and it is reinforced only by the steel fibre reinforcement.

Figure 1. Visualization of the bridge

The main shape of the footbridge ("U" shape) was taken partly from steel construction and partly from shell construction. Optimized UHPFRC mixture should be able to be cast in such a thin and difficult shape. The main outstanding features of this material are the high compressive strength, high tensile (direct and flexural) strength and very high durability also [1-4]. Favourable durability is caused by a very high density of the cement matrix and by very low level of porosity with unconnected pores. This fact is provided by very fine particles (slag, silica fume) and very low water-cement coefficient. High levels especially of flexural strength are provided by using of the steel fibres. These perfect material properties should be supplemented by high workability[6-9].

Production of "U" shaped UHPFRC footbridge is very complex problem. Whole construction need to be cast in one part, one time and design of the structure need be optimized with respect to all factors during casting, demoulding, transporting and using.

2. "U" profile mock-up

Because of very complex design of the footbridge a preliminary mock-up need to be prepared. The footbridge was casted in the upside-down position because of double curvature and more several reasons that will be describe in more details in this article. The support part of the construction was chosen as a highest part with high of 1.5 m. The wooden formwork of preliminary mock-up was prepared and everything was managed at the same way as for final footbridge was designed. The main reasons of preparing of the mock-up were: verification of the casting process, surface conditions of the element, verification of right consistency of the fresh UHPFRC mixture and ability of the mould to be completely filled. The crucial factor for the element was the time of demoulding, respectively the time of releasing of the formwork. UHPFRC has very high shrinkage that was developed at a very early age of concrete. Developing of the shrinkage cracks is than caused by fixed formwork which is relatively stiff and does not allowed any movement of the casted element. All this factors need to be considered. Casting of the mock-up was in September 2016.

2.1. Preparation of the mock-up

The wooden formwork was designed and prepared to withstand pressure of the fresh concrete which is the main load during casting (Figure 2). The technology of the casting was the main issue during whole process of design. Because of double curvature of the bridge and very thin deck and side walls the bridge had to be casted in upside-down position. There was also problem with the surface of the construction which is not in the mould during casting. The formwork was closed from all sides and also the bridge deck (in this position the top deck) was closed. Only three openings were made on the top desk used for pouring of the concrete. Smooth surface of the concrete was achieved in all visible surfaces of the bridge. Only upper side of the casted element in closed formwork has always many pores because of air coming from the UHPC matrix upward from the walls and deck. In this case the position is advantageous because the surface with pores is on the nonvisible bottom part of the bridge in final position. Total volume of used UHPFRC for this element was 0.4 m^3.

Figure 2. Formwork of the preliminary mock-up

Figure 3. Preliminary mock-up after the demolding

2.2. Optimization of UHPFRC mixture

UHPFRC mixture, used in this construction, need to fulfil very high requirements. Extensive research in the field of material optimization was carried out in Klokner Institute. The key parameter for UHPFRC mixture was high flowability. Because of the upside – down pouring of the footbridge, concrete need to flow inside by the 30 mm thin area and fill both thin walls and upper deck. Usually consistency of the mixture can be adjusted by volume of added water and superplasticizer. In this case (especially for UHPFRC) added water means lower mechanical properties and higher shrinkage. Because of these problems a new UHPFRC mixture were designed and tested. Black pigment was added to the mixture according to requirement of the architect. Many mixtures were tested with a different volume of pigment in dependence to flowability and mechanical properties. The final colour shade was discussed regarding to the material parameters. Final mixture consist of the cement CEM II 52,5 N, fine aggregate with maximum size of 2 mm, slag, silica fume and steel fibres (0.2 mm thick and 13 mm long). Volume fraction of the fibres was 1.5%. The volume of water and superplasticizer depended on the temperature conditions with regard to the workability.

2.3. Material properties of UHPFRC

During casting of the preliminary mock-up many specimens for testing of the material properties were prepared. Because of early demoulding age, many specimens were tested for compressive and flexural strength. First tests were carried out 5 hours after casting. Next set of specimen was tested after 2 hours and the tests continued until the demoulding of the element. The development of material properties at early age is shown in Figure 4. The crucial value of the compressive strength was set to 30 MPa. At this point the formwork should be released and removed. Preliminary mock-up was demoulded after 11-12 hours after casting. Formwork was completely removed and surfaces of the element were cured by special fluid wax. The surface quality was checked. Element is shown in Figure 4 after demoulding. No shrinkage cracks and no empty areas without UHPFRC were observed. Only the top surface under top wooden desk was covered by the layer with many pores.

Development of material properties at first 10 days after casting shows in Figure 5.

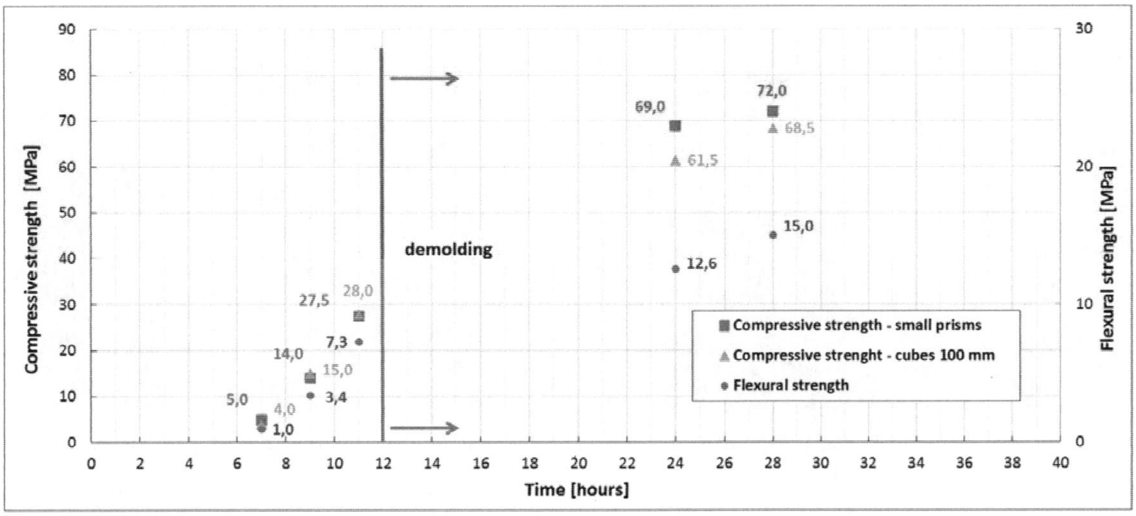

Figure 4. Development of compressive and flexural strength at first 40 hours after casting

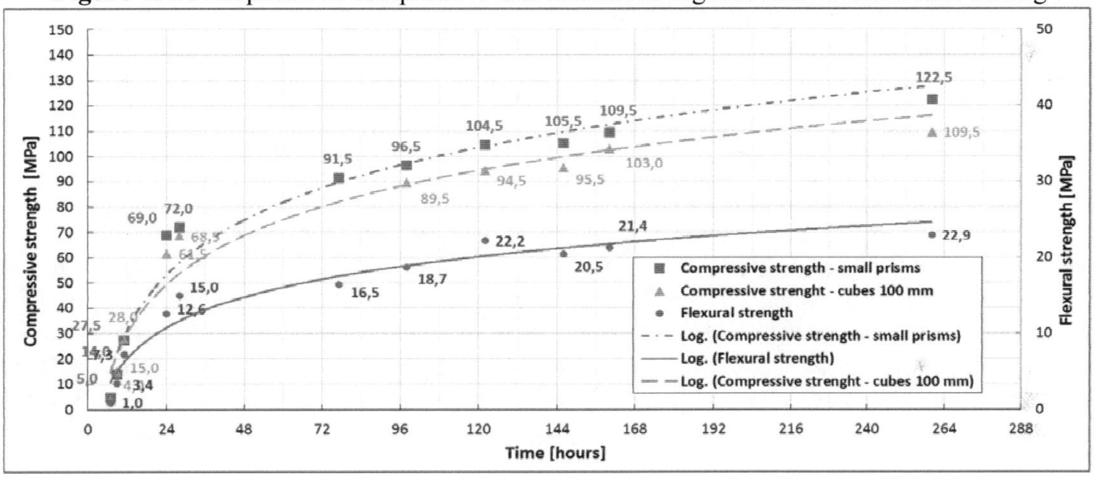

Figure 5. Development of compressive and flexural strength at first 10 days after casting

2.4. Loading tests of the thin walls

After 7 days after demoulding the element were turned to the final position and transported on the truck to Klokner Institute. Load bearing capacity of the thin wall was tested by horizontal force at a height 1.3 m. The applied axial force was distributed by steel profile to entire width of the wall. Tests were performed after 35 days after casting. Element was fixed by concrete block to prevent horizontal deflection of whole element. Horizontal deflections were measured by three potentiometric sensors. The first sensor was placed at the height of the applied force, second at the half of the height and third at the bottom deck. This measurement served as control that there was no horizontal deflection of whole element.

At first the load bearing capacity of the right thin wall in inward direction were evaluated (Figure 6). Maximum horizontal force 3.1kN was reached. It means that force divided by width of the tested wall (1.5 m) was 2.1kN/m′. The second test was performed in opposite direction (a way out direction – Figure 7) and the maximum force was 3.5 kN (2.3 kN/m′). It was concluded that load bearing capacity of the walls (handrails) is sufficient.

Figure 6. Test of load bearing capacity of the right thin wall – an inward direction

Figure 7. Test of load bearing capacity of the left thin wall – a way out direction

3. "U" profile full size footbridge mock-up

The final formwork was prepared after optimization of many details from preliminary mock-up. Double curvature of the bridge was applied on the top deck and side walls. Two openings for pouring UHPFRC were placed on the edges of the top deck on the ends of the bridge. Development of the material properties and knowledge from casting of the preliminary tests were used for

3.1. Preparation of the final formwork

The formwork was prepared in Klokner Institute. Because of length of the bridge it had to be divided into the four parts. Very complex design of the formwork needed to be optimized in case of fast and easy releasing and demoulding. UHPFRC has very high early shrinkage so the fast releasing of the screws and desks is needed to enable some movement of construction and prevent cracks development. Whole final construction of the formwork was prepared at KI premises and transported after control check by designer. Each part of the formwork was transported individually to precast plant Steti, where the bridge was casted (Figure 8). All parts of the formworks were joined together, the release agent was applied to the all surfaces and mould was closed. All formwork beams were reinforced by additional wooden beams (Figure 9) for safety reasons and to withstand of pressure of the fresh concrete.

Preparation of the formwork and casting of the footbridge was in very cold temperatures during winter time. Temperature on the precast plant was between 1-8°C, so the additional heating was necessary after casting of the UHPFRC. The crucial effect was to accelerate hydration of cement and increase early age strength of the UHPFRC. Two heaters were placed inside the "tunnel" from both sides and two and two heaters were placed outside the side walls. Temperature was controlled by extensive measurement. Two temperature sensors were placed into the concrete on the edges (pouring openings) and several sensors were placed on the formwork to control a temperature on the mould surfaces. Sensors placed into a concrete showed development of temperature of the mixture depending on heating and increasing temperature during hydration.

3.2. Casting, demoulding and transport of the footbridge

The prepared mould was poured in January 2017. The final volume of the UHPFRC was 1.4 m^3. Three batches were mixed and transported by crane. The mould was poured from the top of two edges of the bridge, where the final bearing will be installed. No additional vibrations were applied because of the self-compacting character of the mixture.

IOP Conf. Series: Materials Science and Engineering **246** (2017) 012009 doi:10.1088/1757-899X/246/1/012009

Figure 8. Transport of the ¼ of the formwork

Figure 9. Formwork prepared for the casting

Because of very low temperatures in the hall an additional heating of the formwork was provided. Maximum temperature outside of the formwork did not reach 40°C but it was still sufficient for increasing early strength of UHPFRC. Casting of the whole element took approximately one hour. Consistency of the fresh concrete was measured in several times after mixing. Cold temperature had a positive effect on consistency so there was not a problem with cold joint between different batches. After the end of casting, whole mould were covered by polystyrene plates and PVC foil to keep the same temperature conditions.

Development of the compressive and flexural strength was tested at the same way as during preliminary mock-up. The time of releasing of the formwork was set to 11-12 hours after casting. Additional heating was provided for the next 5 days. After 4 days a temperature inside concrete was settled to the same temperature as on wooden desk surfaces, so the formwork should be removed with no additional temperature shock. Final footbridge in the upside-down position is shown in Figure 10 and 11. All sides of the thin walls and especially top deck had very smooth surface with good quality.

Construction was removed from hall to the storage place outside the hall, after 7 days from casting. It had to be stiffened by four inside wooden frames (Figure 10) to prevent damage of the thin shell in torsion during transport. Element was transported by two cranes (Figure 12) and placed on polystyrene desks to allow volumetric changes caused by shrinkage, creep and changes caused by temperature (Figure 13). Temperature measurement was kept during storage.

The next step of the program is to turn the footbridge to the final position. This action is planned to end of February 2017. Turning will be provided by two cranes. Footbridge in final position will be placed on prepared concrete blocks and anchored on the edges. There are two prepared anchor elements at the deck. Because of horizontal curvature in opposite sides of load bearings a tensile stress can be observed and must be reliable transferred.

Final step will be a loading test of the final footbridge and verification of the numerical analysis. Test will be performed in the precast plant and the results of the test will be presented at the conference and other publications.

IOP Conf. Series: Materials Science and Engineering **246** (2017) 012009 doi:10.1088/1757-899X/246/1/012009

Figure 10. Demoulding of footbridge

Figure 11. Final footbridge after demoulding

Figure 12. Transport of final 1:1 mock-up of the footbridge in upside-down position

Figure 13. Storing of final 1:1 mock-up of the footbridge in upside-down position

4. Conclusion

Actual paper presents very complex experimental program of the design, casting and testing of the "U" shaped thin walled UHPFRC footbridge. Single-span bridge with span of 10 m and the clear width 1.50 m with the total thickness of shell structure 30 - 45 mm was cast as a prefabricated element in one piece. Self-compacting character of UHPFRC with high flowability allows the production of the final structure. An extensive research carried out in Klokner institute was done before final construction casting. Based on the requirements of architect and designer a new UHPFRC mixture was developed. Material properties were tested especially at early age of concrete because of releasing of the formwork and demoulding of the mould after 11-12 hours after casting. An extensive numerical analysis was done. Results of numerical modelling are presented at second paper focused on design of the footbridge. Preliminary mock-up was produced and technology of casting and many details of formwork and mould were optimized based on experiences gained. Casting of the full scale model was performed during winter time so an additional heating of the form was necessary. After demoulding the whole footbridge was transported to storage place. Turning to the final position and testing of the load bearing capacity on full scale footbridge is scheduled for spring 2017 and will be presented on the conference.

Acknowledgements

This contribution is the result of the research supported by the research project GAČR 17-22796S. The tests were carried out in laboratory of Klokner Institut, CTU in Prague.

References

[1] Tej P., Čítek D. and Kolísko J.: – Design of thin-walled U-profile footbridge made of UHPC. UHPFRC 2017, Montpellier, France, Oct.2-4,2017

[2] JSCE-USC; "Recommendations for Design and Construction of Ultra-High Strength Fiber-Reinforced Concrete Structures – Draft."

[3] AFGC/SETRA: Ultra High Performance Fibre-Reinforced Comcretes, Recommandations (06/2013). Paris CEDEX.

[4] Kolísko J., Rydval M., and Huňka P.: UHPC – Assessing the Distribution of the Steel Fibre and Homogeneity of the Matrix. Tel Aviv, Israel: fib Symposium Tel Aviv, 2013.

[5] Vítek, J. L., Coufal, R., Čítek, D.: UHPC - Development and Testing on Structural Elements. Concrete and Concrete Structures 2013, Žilina, 2013, University of Žilina, pp. 218-223

[6] Abbas, S., Nehdi, M.L., Saleem, M.A.: Ultra – High Performance Concrete: Mechanical Performance, Durability, Sustainability and Implementation Challenges, International Journal of Concrete Structures and Materials, Vol.10, No.3, pp. 271-295, September 2016

[7] C. Shi, Z. Wu, J. Xiao, D. Wang, Z. Huang, Z. Fang; Review on ultra high performance concrete: Part I. Raw materials and mixture design, Concstruction and Building Materials 101, p. 741-751, 2015.

[8] D. Wang, C. Shi, Z. Wu, J. Xiao, Z. Huang, Z. Fang; Review on ultra high performance concrete: Part II. Hydratation, microstructure and properties, Concstruction and Building Mat. 96, p. 368-377, 2015

[9] Duque, L.F.M, Varga,I., Graybeal, B.A.: Fiber Reinforcement Influence on the Tensile Response of UHPFRC, First International Interactive Symposium on UHPC -2016, Des Moines, IOWA, Jul.18 – 20, 2016

IOP Conf. Series: Materials Science and Engineering **246** (2017) 012010 doi:10.1088/1757-899X/246/1/012010

Application of BCN test for controlling fiber reinforced shotcrete in tunnelling works in Chile

S Carmona[1] and C Molins[2]

[1] Universidad Técnica Federico Santa María, Valparaíso Chile
[2] Universitat Politècnica de Catalunya, Barcelona, Spain.

E-mail: sergio.carmona@usm.cl

Abstract. In many tunnelling projects currently under construction in Chile, the quality of fiber reinforced shotcretes (FRS) is controlled by means of its energy absorption capacity determined by testing squared panel following the EFNARC recommendation. Nevertheless, this test requires large and heavy specimens, which have to be filled when concrete is sprayed into the tunnel and does not allow testing the concrete actually placed onto the support. Due to these difficulties, the quality of fiber reinforced shotcretes used in some projects has been controlled by means of the Barcelona (BCN) test, which is an indirect tension test, using cores drilled from the hardened tunnel support. To apply the BCN test, the dissipated energy measured by means of this test has been correlated with the energy absorption capacity of the fiber reinforced shotcretes using experimental data obtained from works. The aim of this paper is presenting this correlation and it application.

1. Introduction

In different tunnel construction projects in Chile, fiber reinforced shotcretes (FRS) used in the supports are controlled by their energy absorption capacity determined by means of the square panel tests according to EFNARC of 1996 [1]. However, this test requires large size and weight test specimens, which must be filled during the projection of the concrete into the tunnel. This often causes the panel to present defects or damages that disturb the results, increasing their scatter, with coefficients of variation (CoV) between specimens of the same sample greater than 20%.

In view of these difficulties, the use of the Barcelona (BCN) test [2],[3], has been proposed as an alternative for controlling fiber reinforced shotcrete, because this test is characterized for requiring a small specimen, with a large specific surface of cracking, which reduces the scatter of the results, and the specimens can be obtained by drilling cores from the hardened tunnel supports.

In order to use the Barcelona test in the control of the fiber reinforced shotcretes in the tunnel construction works of different projects, a correlation between the energy absorption capacity determined by EFNARC panels and energy dissipated in the BCN tests was developed. For that, at points where 14 samples of panels were filled point where panels were filled, were drilled 28 cores which were tested by BCN tests.

The aim of this paper is to present the results of this correlation and also its application to the quality control of the fiber reinforced shotcrete in the construction of tunnels in Chile.

IOP Conf. Series: Materials Science and Engineering **246** (2017) 012010 doi:10.1088/1757-899X/246/1/012010

2. Absorption energy capacity of fiber reinforced shotcrete

According to the project specifications, the energy absorption capacity of fiber reinforced shotcrete was determined by the square panel test, according to the EFNARC recommendation, establishing a minimum value of 1000 J.

In this test, a square panel of 600 × 600 mm and 100 mm thick, supported on its four edges, is loaded at the center by a contact surface of 100 × 100 mm, as can be seen in Figure 1. Following the EFNARC recommendations, the rough face of the specimen shall be on the bottom during the test and test is conducted under deformation control at a rate of midspan deflection of 1.5 mm/min.

The load – deformation curve shall be continuously recorded until a deflection of 25 mm is achieved at the center point of the panel. Using this response, the energy absorption capacity, E_{25}, can be calculated as:

$$E_{25}(\delta = 25) = \int_{0}^{25} P(\delta)\ d(\delta) \tag{1}$$

where$P(\delta)$ is the load at deflection $\delta = 25$ mm.

Figure 1. View of EFNARC panel test.

3. Energy dissipated in BCN test

The BCN test is an indirect tension test in which a cylindrical specimen of FRC is subjected to a double punching compression load by means of two cylindrical steel punches placed at the center of the upper and the lower faces, respectively, as shown in Figure 2. The cylinder has a diameter (*d*) equal to its height (*h*), and punches have a diameter (2*a*) equal to 0.25 of the specimen diameter. During the test, the applied load, *P*, and the specimen Total Crack Opening Displacement (*TCOD*), measured at half of it height; have to be recorded continuously, in order to obtain the *P* - *TCOD* curve.

During the test, the applied load produces a conical volume under triaxial compression stress beneath the punches, increasing the cylinder diameter and producing tensile stresses perpendicular to the radial lines of specimen. Due to this tensile stress with cylindrical symmetry, when the stress exceeds the tensile strength of concrete, cracks perpendicular to the field propagate through the specimen. This allows that the compression cone penetrates into the cylinder, as can be seen in Figure 3, increasing the specimen radius and producing two or more cracks. Then, the final state of the specimen present two aligned cracks or three cracks arranged approximately at 120° or, sometimes, four perpendicular cracks [4].

IOP Conf. Series: Materials Science and Engineering **246** (2017) 012010 doi:10.1088/1757-899X/246/1/012010

Figure 2. BCN test setup.

When the specimen cracks, the circumferential dilatation corresponds to the total cracks opening displacement (TCOD) and the energy dissipated can be calculated as:

$$E_{BCN,x} = \int_0^{Rx} P(Rx)\ d(Rx) \tag{2}$$

where $E_{BCN,x}$ is the energy dissipated at a certain total circumferential deformation R_x value. According to the standard UNE 83515 [2], the energy has to be determined at 2.0 mm, 2.5 mm, 4.0 mm and 6.0 mm.

4. Equivalence between $E_{BCN,6}$ and E_{25}

From an experimental research, in which three types of fiber reinforced concretes were tested, Carmona and Molins [5] proposed a correlation between dissipated energy at a circumferential deformation of 6.0 mm, $E_{BCN,6}$, and the energy absorbed by the panel at a center point deflection of 25 mm, E_{25}. These values were used taking into account that the panel test measures the energy absorption capacity at deflection of 25 mm, when cracks are widely open and the specimens exhibit a high degree of damage, allowing fibers to develop their full reinforcement capacity, these absorption energy will be compared with the energy dissipated in the BCN until a *TCOD* of 6 mm, which is the maximum deformation established in the standard.

Using a nonlinear regression analysis, theses author obtained and equation of the following form:

$$E_{25} = a \times \left(E_{BCN,6}\right)^b \tag{3}$$

where a, and b, are experimental parameters which depend on the type and content of fibers in the concrete. In the research conducted by Carmona and Molins [5], detailed in Table 1, the value of the experimental parameters obtained were $a = 37.02$ and $b = 0.68$, with a coefficient of determination $r^2 = 0.998$, and this fit is plotted in Figure 3.

IOP Conf. Series: Materials Science and Engineering **246** (2017) 012010 doi:10.1088/1757-899X/246/1/012010

Table 1. Summary of Carmona and Molins [5] research.

Material (kg/ m³)	Concrete		
	FRS – 4	FRS – 8	FRS – 12
Cement type IP	420		
Sand 0/10	1655		
Super plasticizer admixture	2.10		
Active admixtures	2.94		
Water	215		
Synthetic fiber content	4	8	12
Concrete Properties			
Compressive strength (MPa)	45	47	36
Volumetric substitution (%)	0.44	0.88	1.32
$E_{BCN,6}$ (J)	95.4 (9.4)	132.8 (11.8)	179.6 (12.1)
E_{25} (J)	803.6 (11.7)	1064.9 (9.9)	1252.4 (10.7)

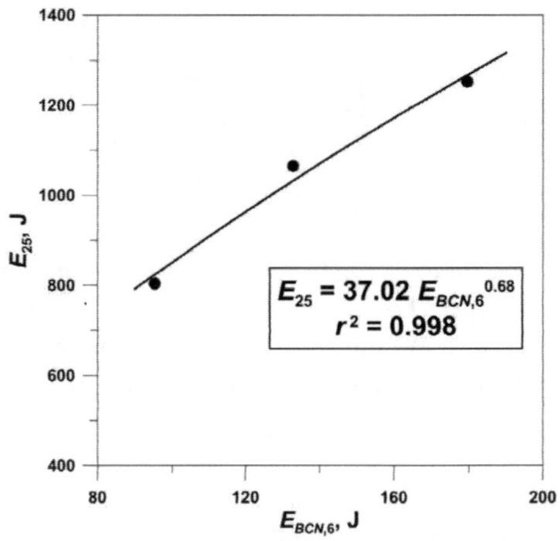

Figure 3. Correlation between energy dissipated and energy absorption capacity obtained by Carmona and Molins [5].

5. Development of an equivalence between both tests for FRS control at works

To replace the EFNARC panels by the BCN test in FRS control in the construction of mining tunnels in Chile, the following procedure was proposed to establish equivalence with the results of tests performed during the control at works:

- Identify 15 points where panels were sampled and whose intra-sample absorbed energy values have a CoV lower than 10%.
- Drill at least 2 cores of 150 mm diameter from the shotcrete of the tunnel support at each of the selected points.
- Obtain the dissipated energy at $TCOD = 6.0$ mm of each core by mean of BCN test.
- Perform the statistical analysis and establish the correlation between the energy absorption capacity of the EFNARC square panel and the energy dissipated by the FRS in the BCN tests.

In this project, the concrete was reinforced with 6 kg/m^3 of synthetic fibers. The fibers incorporated in the concrete were Barchip 54, with a length of 54 mm, a tensile strength of 640 MPa, a specific weight of 9.1 kN/m^3 and a Young Modulus of 10 GPa.

Following the proposed procedure, the 15 EFNARC panel sampling points given in Table 2 were selected, along with the values of the energy absorption capacity obtained during the control.

Table 2. Values of absorbed energy used in this research.

Sample	Panel 1	Panel 2	E (J)	CoV (%)
1	1005	967	986	2,7
2	1036	1025	1031	0,8
3	986	979	983	0,5
4	1009	998	1004	0,8
5	996	977	987	1,4
6	990	986	988	0,3
7	1023	910	967	8,3
8	1000	1001	1001	0,1
9	940	1004	972	4,7
10	1020	1004	1012	1,1
11	1025	1014	1020	0,8
12	948	931	940	1,3
13	978	1028	1003	3,5
14	970	953	962	1,3
15	986	1022	1004	2,5

At the same sampling points showed in Table 2, two cores of diameter $d = 150$ mm and length $h = 150$ mm were drilled for the BCN test. The tests were performed in a hydraulic system of 300 kN of capacity under displacement control, following the specifications and configuration given in standard UNE 83 515 (Figure 2). The circumferential deformation was measured at half of the height of the specimen with a circumferential extensometer with a total range of 12 mm. Data were recorded by Hewlett Packard model 7500 - XVI system.

The final state of the specimen can be seen in Figure 4, along with the typical experimental $P - TCOD$ and $TCOD - E_{BCN,6}$ curves. The results for each sample are given in Table 3.

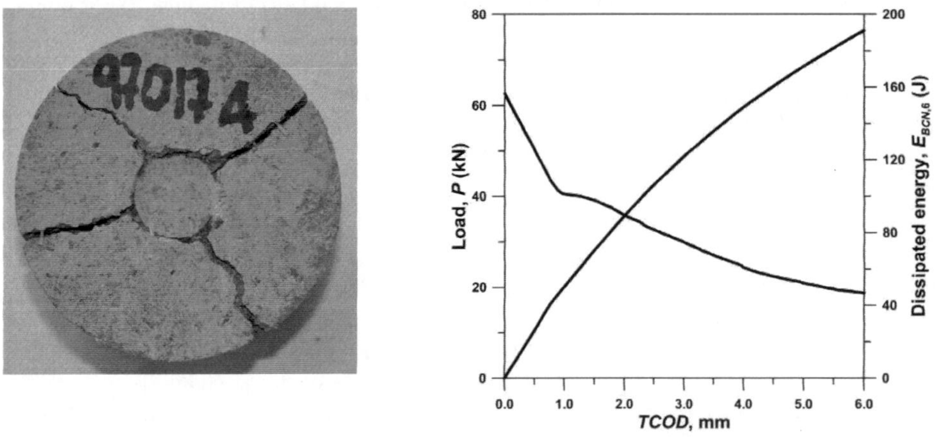

Figure 4. State of specimens after BCN tests; (b) Typical curves obtained with BCN tests.

Table 3. Dissipated energy, $E_{BCN,6}$, obtained with BCN tests.

Sample	BCN-1 (J)	BCN-2 (J)	$E_{BCN,6}$ (J)	CoV (%)
1	218	237	228	5,9
2	**231**	**349**	**290**	**28,8**
3	292	258	275	8,7
4	305	305	305	0,0
5	**244**	**208**	**226**	**11,3**
6	**172**	**300**	**236**	**38,4**
7	**236**	**200**	**218**	**11,7**
8	230	220	225	3,1
9	191	219	205	9,7
10	287	295	291	1,9
11	**254**	**177**	**216**	**25,3**
12	322	311	317	2,5
13	276	292	284	4,0

As can be seen in Table 3, two samples (number X and Y) were discarded because they failed suddenly when cracking load was reach, due to low fiber content. On the other hand, the CoV of four samples, displayed in bold, are higher than 10% and, then, these results were also discarded.

With the aim of removing outliers from the data, both $E_{BCN,6}$, and E_{25} values were analysed using Grubbs and Dixon criteria using a statistical software. The results are shown in Table 4., It can be seen that sample from point 12, displayed in bold, ha to be discarded because the value of E_{25} was an outlier. Finally, only 8 values were used to obtain the correlation between $E_{BCN,6}$, and E_{25}.

Table 4. Results of outliers analysis.

Sample Point N°	E_{25} (J)	Criteria		$E_{BCN,6}$ (J)	Criteria	
		Grubbs	Dixon		Grubbs	Dixon
1	986	0,039	0,03889216	228	-0,7921524	-0,7921524
3	983	-0,114	-0,11424572	275	0,34170413	0,34170413
4	1004	0,805	0,80458155	305	1,05782404	1,05782404
8	1001	0,673	0,67332051	225	-0,85182906	-0,85182906
9	972	-0,574	-0,57365935	205	-1,32924233	-1,32924233
10	1012	1,176	1,17648783	291	0,72363475	0,72363475
12	**940**	**-1,974**	**-1,9737771**	317	1,34427201	1,34427201
13	1003	0,783	0,78270471	284	0,52471255	0,52471255

By means of statistical software, a nonlinear regression analysis was developed obtaining the following equation:

$$E_{25} = 639.4 \times \left(E_{BCN,6}\right)^{0.08} \quad (4)$$

This correlation has a coefficient of determination $r^2 = 0.5378$. The experimental data and the equation (4) are plotted in Figure 5a, and the percent difference between experimental data and the estimated values are shown in Figure 5b, in which can be seen that the equation (4) has a good fit.

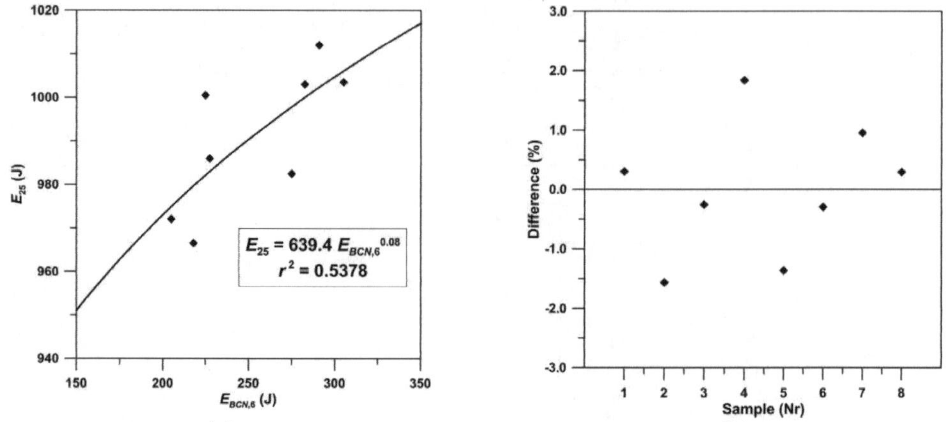

(a) (b)

Figure 5. (a) Experimental values and equation (4); (b) Percent difference between experimental values of E_{25} and equation (4).

In order to prove the goodness of equation (4), the values of E_{25} were estimated using all the experimental data available in Table 3, achieving the result showed in Figure 6, where differences are lower than 5% for the most of the experimental data.

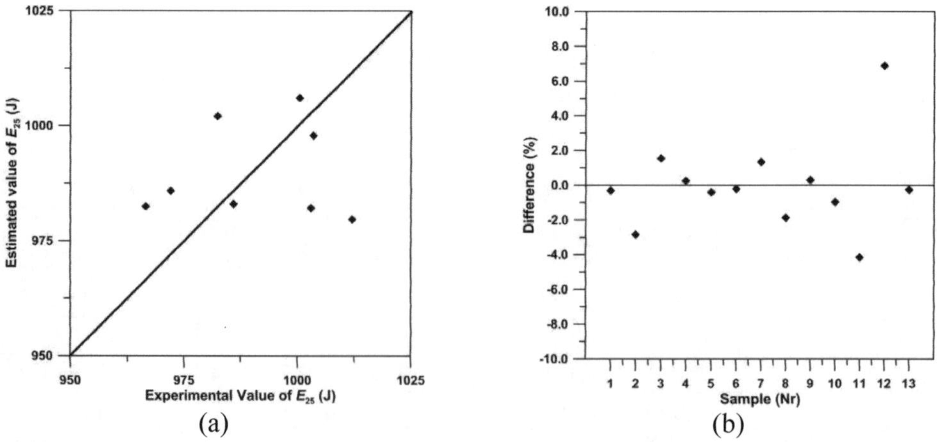

(a) (b)

Figure 6. (a) Comparison between experimental values and equation (4) using all then available data; (b) Percent difference between experimental values of E_{25} and equation (4).

IOP Conf. Series: Materials Science and Engineering **246** (2017) 012010 doi:10.1088/1757-899X/246/1/012010

6. Conclusions

Taking in account that at laboratory level the BCN test has proved to be an adequate experimental procedure for systematic quality control of FRS at works due to its simplicity and low scatter, the results of this research allowed to to replace the EFNARC panel test in a mining tunnel work.

To apply the BCN test instead of EFNARC panel, a correlation based on the dissipated energy has been establish by using the values of energy absorption capacity determined during the ordinary control and the results of BCN tests conducted on cores drilled from the existing support in the tunnel.

The correlation proposed has differences lower than 5% for the most of experimental data available for this research and is now in use at work.

Additional refinements of this procedure will take into account the actual strength of concrete at the age of testing. Panel tests were developed at 28 days while the BCN test on the drilled cores were developed at 1000 days in average.

Acknowledgements

This research was supported by the Fondecyt Project "Use Of The Generalized Barcelona Test For Characterization And Quality Control Of Fiber Reinforced Shotcretes In Underground Mining Works", N°1150881.

References

[1] EFNARC. European specification for sprayed concrete, 1996.
This reference has two entries but the second one is not numbered (it uses the 'Reference (no number)' style.

[2] AENOR. UNE 83515 – Hormigones con fibras. Determinación de la ersistenciaa fisuración, tenacidad y resistencia residual a tracción. Método Barcelona. AEN/CTN 83 – Hormigón, Madrid (in Spanish), 2010.

[3] Molins, C., Aguado, A., Marí, A. Quality control test for SFRC to be used in precast segments. Tunn Undergr Space Technol 2006, 21 (3), 423 – 424.

[4] Carmona, S., Aguado, A., Molins, C. Generalization of then Barcelona test for the toughness control of FRC. Mater Struct 2012, 45 (7), 1053 – 1069.

[5] Carmona, S., Molins, C. Use of Barcelona test for controlling fiber reinforced shotcrete in tunnelling. Proceedings of 13th International Conference Underground Construction Prague 2016, 7 pp.

IOP Conf. Series: Materials Science and Engineering **246** (2017) 012011 doi:10.1088/1757-899X/246/1/012011

Influence of steel fibres on the blast response of normal-strength and high-strength reinforced concrete columns

A Hammoud[1] and H Aoude[1]

[1]Department of Civil Engineering, University of Ottawa, Ottawa, ON, K1N5S5, Canada

E-mail: haoude@uottawa.ca

Abstract. This paper examines the influence of steel fibres on the blast performance of normal-strength concrete and high-strength concrete columns. As part of the study, four normal-strength and high-strength concrete columns built with and without steel fibres are tested under simulated blast loads using the shock-tube facility at the University of Ottawa. The specimens include two columns built with plain concrete and two columns built with steel fibre-reinforced concrete. The results show that the addition of steel fibres in reinforced concrete columns leads to important enhancements in blast performance, with improved control of mid-span displacements at equivalent blasts and increased damage tolerance.

1. Introduction

Availability of constituent ingredients, ease of construction, and design flexibility are among the many qualities that make concrete the material of choice for the design and construction of modern infrastructure. Despite its many advantages, concrete's deficiency is in its poor tensile resistance. The addition of steel fibres is a practical solution to this problem, and allows concrete to have increased tensile resistance, post-cracking capacity and toughness [1]. In reinforced concrete structures such as beams, the provision of fibres leads to enhanced shear resistance and flexural ductility [2-5]. In structures subjected to seismic load-reversals, the use of fibres can be used to relax transverse reinforcement requirements and increase energy-dissipation capacity [6]. The toughness and fragmentation resistance of steel fibre-reinforced concrete (SFRC) also make it well-suited for use in impact and blast applications [7-10].

This paper presents the results from an ongoing experimental program examining the effect of steel fibres on the blast behavior of reinforced concrete columns. The paper presents the results from four normal-strength and high-strength concrete columns, built with and without steel fibres, which are tested under simulated blast loads using a high capacity shock-tube at the University of Ottawa. The effect of steel fibres on blast behaviour is examined by comparing mid-span displacements and damage tolerance of companion columns at equivalent blasts.

2. Experimental program

2.1. Specimen designs

A total of four reinforced concrete columns are included in this study. Table 1 summarizes the design details of the specimens and includes information on concrete type, fibre properties and reinforcement

IOP Conf. Series: Materials Science and Engineering **246** (2017) 012011 doi:10.1088/1757-899X/246/1/012011

details. As shown Figure 1, the columns had cross-sectional dimensions of 152 mm x 152 mm, with a total length of 2440 mm. Longitudinal reinforcement in the columns consisted of 4 - 10M Canadian size bars (d_b = 11.3 mm, A_s=100 mm^2), while transverse reinforcement was provided using square ties made from 6.3 mm steel wire (A_s=31 mm^2), spaced at s = 75 mm over the column height. In all cases the clear concrete cover was kept constant at 5 mm. Two concrete types were considered in the test program. The first mix which is described as "NSC" in the specimen nomenclature, had a target strength of 50 MPa, while the second consisted of a high-strength concrete (HSC) mix with a target strength of 100 MPa. For each concrete type, one column was built with plain concrete and one column was cast with fibre-reinforced concrete having a volumetric ratio of 1% of steel fibres (80 kg/m^3) (the use of "F0" and "F1" in the specimen names refers to the use of 0% and 1% fibres, respectively). The hooked-end fibres had a length (L_f) of 30 mm, a diameter (D_f) of 0.55 mm, with tensile strength of 1350 MPa. The concrete strengths for each column were obtained by testing 100 mm x 200 mm cylinders in compression and are summarized in Table 1. Sample stress-strain curves for the plain and fibre-reinforced HSC are shown in Figure 2 as examples. The 10M bars used in the NSC columns had an average yield strength of 483 MPa, while those in the HSC columns had yield strength of 474 MPa. The steel wire used for the transverse reinforcement had average yield strength of 604 MPa and 542 MPa in the NSC and HSC columns, respectively. Sample stress-strain curves for the steel reinforcement are shown in Figure 2.

Table 1. Column test matrix.

Beam I.D.	Concrete Mix type	Concrete Strength f'_c (MPa)	Steel fibre properties L_f / D_f (mm/mm)	Steel fibre properties V_f (%)	Steel reinf. properties Transverse Steel	Steel reinf. properties Long. Steel
NSC-F0-10M-75	NSC	51.6	-	-	6.3 mm square ties @ s = 75 mm	4 - 10M
NSC-F1-10M-75	NSC	46.1	30/0.55	1.0		
HSC-F0-10M-75	HSC	93.9	-	-		
HSC-F1-10M-75	HSC	93.3	30/0.55	1.0		

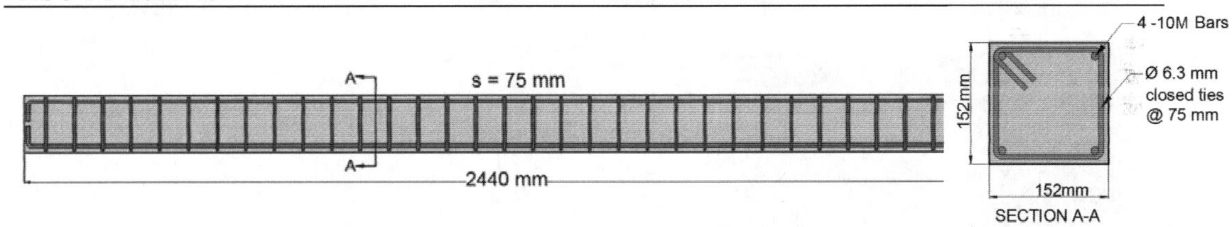

Figure 1. Dimensions and reinforcement properties for reinforced concrete columns.

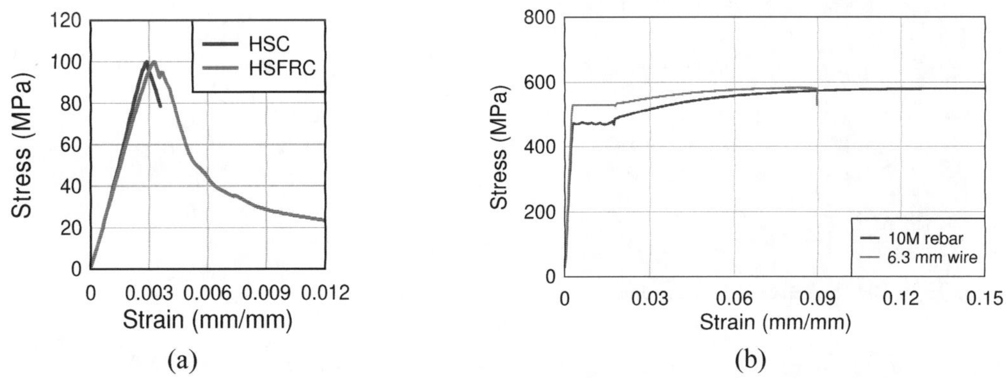

Figure 2. Sample stress-strain curves (HSC series): (a) concrete in compression and (b) steel reinforcement in tension.

2.2. Test setup

The columns in this study were tested under simulated blast loads using the University of Ottawa shock-tube. As shown in Figure 3a, the shock-tube consists of four main parts: (1) a variable length driver section, (2) a spool section, (3) an expansion section and (4) a rigid end test frame with a 2 m x 2m square opening. The shockwave is produced by rapidly releasing compressed air from the driver section into the expansion section using a double-diaphragm firing system in the spool section. For non-planar elements, such as beams and columns, a load transfer device (LTD) is used to collect the shockwave pressure at the shock-tube opening and redirect it onto the test specimens. In this study, the LTD consists of a light gauge steel metal sheet connected to a series of steel beams which resulted in the application of blast loading as an approximately uniformly distributed load. Figure 3b shows a typical column prior to testing. The columns, which had a clear span of 1980 mm, were fixed to the shock-tube frame using partially-fixed supports. All columns were tested under combined axial loading and transverse blast loads, with axial load (P = 300 kN) applied using a hydraulic jack placed at the bottom of the specimens.

The columns were subjected to different simulated blast pressure–impulse combinations using the shock-tube which tested them under elastic (Blast 1), yield (Blast 2), and ultimate conditions (Blast 3). Sample shockwaves for Blasts 1 to 3 are shown in Figure 3c. The average reflected impulses under Blasts 1-2-3 were approximately 175, 420 and 765 kPa*msec, respectively.

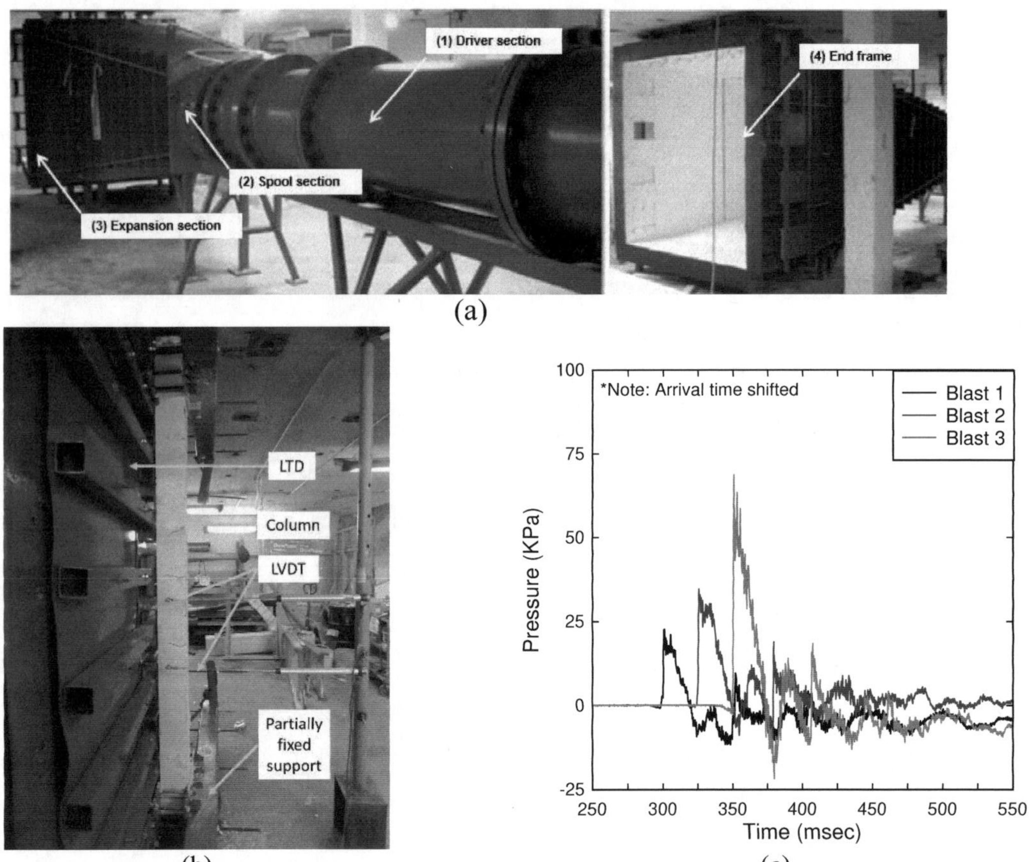

Figure 3. (a) Shock-tube, (b) column setup and (c) sample shockwaves for Blasts 1-3.

3. Results

Table 2 summarizes the experimental results for the four column specimens tested in this study, including the positive phase shockwave impulse (I_r) imparted on each column at each blast, and column response in terms of maximum & residual mid-span displacements (d_{max} & d_{res}), and maximum support rotations. Table 3 shows the damage in the columns at the end of testing, while Figure 4 compares the displacement response of the columns at selected blasts.

Table 2. Blast test results.

Column I.D.	Blast #	Shockwave Impulse	Displacement		Support rotation
		I_r (kPa*ms)	d_{max} (mm)	d_{res} (mm)	θ (°)
NSC-F0-10M-75	1	130	6.4	1.8	0.37
	2	381	29.9	9.7	1.73
	3	781	126.2	108.6	7.26
NSC-F1-10M-75	1	127	5.1	0.6	0.30
	2	411	23.3	5.4	1.35
	3	767	87.7	47.6	5.06
HSC-F0-10M-75	1	233	8.4	1.3	0.49
	2	429	33.6	14.7	1.94
	3	779	132.6	118.7	7.63
HSC-F1-10M-75	1	216	7.1	0.5	0.41
	2	453	19.4	6.7	1.12
	3	734	83.8	52.9	4.84

3.1. Column NSC-F0-10M-75

Column NSC-F0-10M-75 was the first specimen tested in the research program. It was constructed with plain normal-strength concrete, 4-10M bars and 6.3 mm ties spaced at 75 mm. The column was undamaged after Blast 1, while Blast 2 resulted in some transverse cracking at the levels of the ties along the column height. Failure of this column occurred under Blast 3 loads, with the column recording maximum & residual displacements of 126 mm and 109 mm, with a maximum support rotation of 7.3°. As shown in Table 3 failure of the column was associated with important damage, with crushing of concrete, spalling of the cover and buckling of the compression steel reinforcement at midspan.

3.2. Column NSC-F1-10M-75

The design of column NSC-F1-10M-75 is identical to that of the previous specimen, with the exception being the addition of steel fibres at a volumetric ratio 1.0% in NSC-F1-10M-75. As with its companion Blast 1 and Blast 2 resulted in limited damage and deformations. Failure of the column would occur under Blast 3 loading. Some improvements are noted both in terms of control of displacements and control of damage when compared to the previous specimen. Maximum and residual displacements after Blast 3 were 88 mm and 48 mm, with a maximum support rotation of 5.1°. As shown in Table 3 the addition of fibres led to minimal spalling and crushing, with prevention of bar buckling.

3.3. Column HSC-F0-10M-75

Column HSC-F0-10M-75 was the first of two specimens built with high-strength concrete. This control column was built without fibres. As in the previous series the relatively moderate blasts at the first two shots resulted in only minor cracking of concrete. Under Blast 3, the column suffered failure

IOP Conf. Series: Materials Science and Engineering **246** (2017) 012011 doi:10.1088/1757-899X/246/1/012011

with concrete crushing and spalling at midspan as shown in Table 3. The maximum & residual displacements were recorded to be 133 mm and 119 mm at this blast, with a support rotation of 7.6°.

3.4. Column HSC-F1-10M-75

Column HSC-F1-10M-75 was the last column tested in this research study. It is a companion to specimen HSC-F0-10M-75, and was built with high-strength concrete and 1% steel fibres. Blasts 1 and 2 resulted in small deformations and limited damage in this specimen. Failure of the column would once again occur under Blast 3 loading, however the use of fibres results in improved deformation control, with maximum and residual displacements of 84 mm and 53 mm, and maximum support rotation of 4.8°. As shown in Table 3, damage in the column was more controlled when compared to its plain HSC companion, with limited crushing and spalling. Failure in this column occurred due to pullout of the fibres at a critical crack at midspan.

Table 3. Comparison of damage in columns after Blast 3 loading

Column ID			
NSC-F0-10M-75	NSC-F1-10M-75	HSC-F0-10M-75	HSC-F1-10M-75
		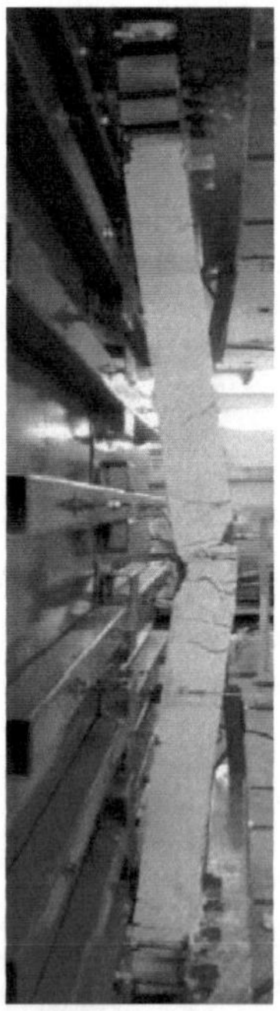	
		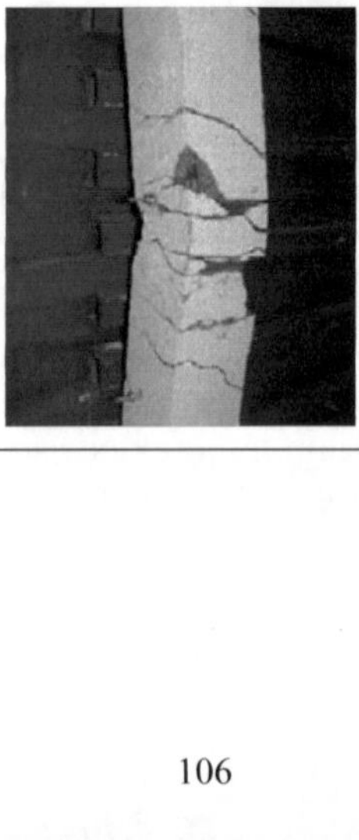	

4. Discussion of results

Comparison between the behaviour of companion columns in the NSC and HSC series allows for a study of the effect of fibres on the blast performance of normal-strength and high-strength concrete columns. All four columns in the study failed at Blast 3, however some enhancements in performance are observed in terms of the ability of fibres to improve displacement control under equivalent loading.

The specimens in the NSC series showed similar deformations after Blast 1. However, as shown in Figure 4, the maximum and residual displacements were reduced by 20% and 44% for column NSC-F1-10M-75 (with fibres) when comparing to NSC-F0-10M-75 (plain concrete), after Blast 2. The effect of fibres on controlling displacements is even more clear at Blast 3, with the provision of fibres allowing for reductions of 31% and 56% in maximum and residual displacements.

Similar trends are observed in the high-strength concrete series. As with the previous set, both HSC columns showed similar response under Blast 1. However, as shown in Figure 5 control of maximum deformations was improved for the HSFRC column (HSC-F1-10M-75) at Blast 2 and 3, with reductions of 42% and 37% when compared to the plain HSC column (HSC-F0-10M-75). Residual displacements were also better controlled for column HSC-F1-10M-75 at these blasts, with reductions of 54% and 55% when compared to the plain HSC companion.

The steel fibres also had a significant influence on the damage in the columns. While all four specimens failed at Blast 3, damage was clearly better controlled when fibres were provided (see the close-up photos of damage in Table 3). While the plain NSC and HSC columns show prominent damage with significant crushing and spalling of concrete at mid-span, crushing is more limited and spalling is well controlled in the in the fibre-reinforced companions. High-speed video shows this translated into a significant reduction in the amount of secondary blast fragments, with the high fragmentation in the plain concrete columns minimised when fibres were provided (see example in Figure 6).

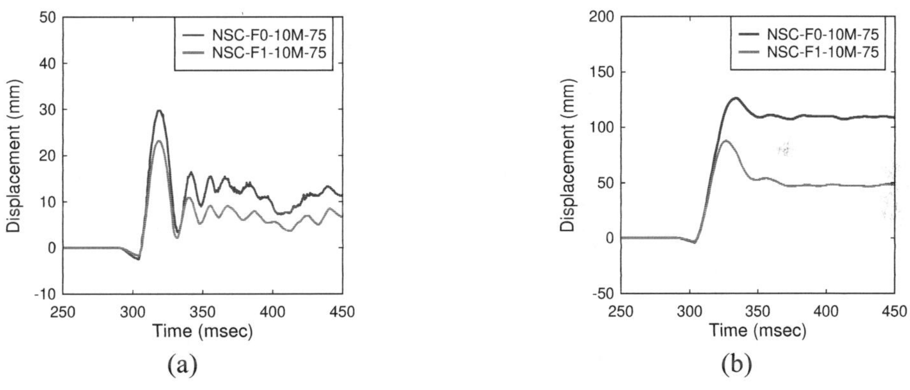

Figure 4. Mid-span displacement time-histories for NSC columns: (a) Blast 2 and (b) Blast 3.

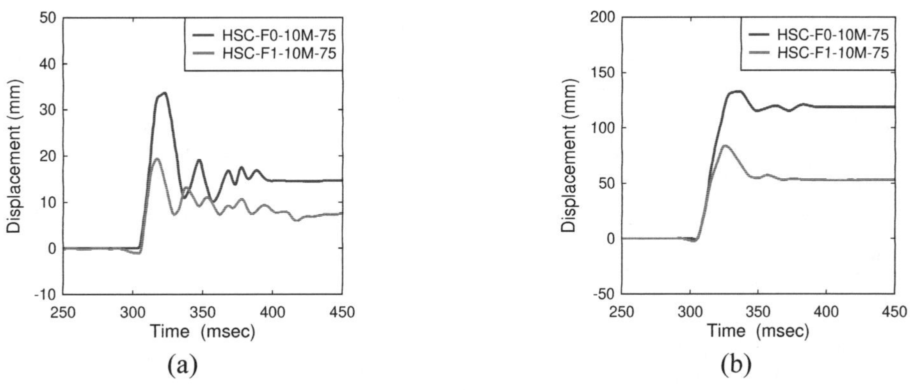

Figure 5. Mid-span displacement time-histories for HSC columns: (a) Blast 2 and (b) Blast 3.

(a)

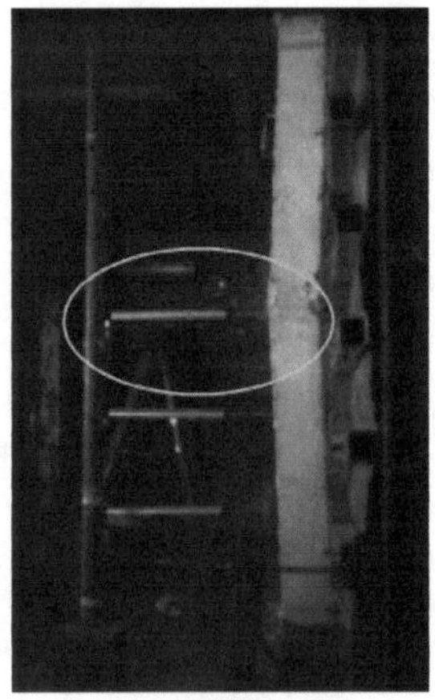

(b)

Figure 6. Example of secondary fragments at Blast 3: (a) NSC-F0-10M-75, (b) NSC-F1-10M-75.

5. Conclusions

This paper presented the results from four reinforced concrete columns, built with and without steel fibres, and tested under blast loading. The following conclusions can be summarized from this study:

- The use of steel fibres resulted in enhanced control of maximum and residual displacements in the NSC and HSC columns, although blast capacity was not increased;
- The use of steel fibres improved damage tolerance in the columns, and resulted in reduced concrete crushing and better control of cover spalling;
- The addition of steel fibres significantly reduced the amount of secondary blast fragments at column failure.

References

[1] Li V S 2002 Large volume, high-performance applications of fibers in civil engineering, *J. of Applied Polymer Sc.* **83(3)** 660-86

[2] Barros J A and Figueiras J A 1999. Flexural behavior of SFRC: testing and modeling. *J. of Mat. in Civil Eng.* **11(4)**, 331-39.

[3] Meda A, Minelli F, Plizzari G A and Riva P 2005 Shear behaviour of steel fibre reinforced concrete beams. *Mat. and Struct.* **38(3)** 343-51.

[4] Parra-Montesinos G J 2006 Shear strength of beams with deformed steel fibers. *Concr. Int.*, 28(11), 57

[5] Di Prisco M, Plizzari G and Vandewalle L 2009 Fibre reinforced concrete: new design perspectives. *Mat. and Struct.* **42(9)** 1261-81.

[6] Parra-Montesinos G J and Chompreda P 2007 Deformation capacity and shear strength of fiber-reinforced cement composite flexural members subjected to displacement reversals. *J. of Struct. Eng.* **133(3)** 421-31.

[7] Banthia N, Mindess S and Trorrier J E 1996 Impact resistance of steel fiber reinforced concrete. *ACI Mat. J.* **93** 472-79.

[8] Gopalaratnam V S and Shah S P 1986 Properties of steel fiber reinforced concrete subjected to impact loading *ACI J.* **83(1)** 117-26

[9] Lok, T S and Xiao J R 1999 Steel-fibre-reinforced concrete panels exposed to air blast loading *Struct. and Build.* **134(4)** 319-31

[10] Magnusson J, Hallgren M and Ansell A 2010 Air-blast-loaded, high-strength concrete beams - Part I: experimental investigation. *Mag. Concrete Res.* **62(2)** 127-36

IOP Conf. Series: Materials Science and Engineering **246** (2017) 012012 doi:10.1088/1757-899X/246/1/012012

Influence of mixing procedure on mechanical properties of high-performance concrete

J Fládr[1] and P Bílý[1]

[1]Faculty of Civil Engineering, Czech Technical University in Prague, Thákurova 7, 166 29 Praha 6, Czech Republic

E-mail: josef.fladr@fsv.cvut.cz

Abstract. Mechanical parameters of cementitious composites are primarily determined by the composition of cement matrix. In case of high performance concrete (HPC), high-quality compounds and increased amounts of cement are exploited to obtain the best possible properties of the material. As these components are expensive, the aim is to use them with maximum efficiency. With optimized procedure of homogenization of compounds, parameters of concrete can be improved without changing the composition of the mix. Several mixing procedures were applied to a selected HPC mix. Compressive and flexural strength of particular batches were compared. The results showed that the compressive strength of HPC made of the same compounds can vary by almost 30 % just due to amendments of the mixing procedure, while the flexural tensile strength was almost indifferent to the homogenization technique used.

1. Introduction

The exploitation of high-performance concrete (HPC) in real applications becomes more common in the recent years thanks to their extraordinary mechanical properties and durability. The limiting factor is still the unit price of the material which is in average four times higher than in case of normal strength concrete, mainly due to increased amount of cement and the use of high-quality compounds. Mechanical parameters of concrete can be influenced not only by the composition of the mix, but also by the way the compounds are mixed. The parameters of homogenization procedure, such as the type of the mixer, mixing speed, air pressure in the mixer, sequence of addition of the compounds, mixing times or temperature, can also affect the properties of both fresh and hardened concrete significantly, as shown in the detailed review carried out by Dils et al [1].

The main objective of the research presented in this paper was to compare the effect of selected parameters of the mixing procedure, namely the order of addition of the components and the mixing times, on compressive and flexural tensile strength of a selected HPC mix.

2. Experimental program

The experiments were carried out on 40x40x160 mm prismatic specimens that were subjected to four-point un-notched bending test according to ČSN EN 12390-5 standard [2]. The remnants of the specimens were used for compressive strength test according to ČSN EN 12390-3 [3]. The tests were performed in the age of 28 days. All the samples were made of the same mix (see table 1) and

IOP Conf. Series: Materials Science and Engineering **246** (2017) 012012 doi:10.1088/1757-899X/246/1/012012

prepared in the same mixer (standard laboratory concrete mixer, nominal volume of 80 liters, mix prepared at atmospheric pressure and room temperature).

Table 1. Composition of concrete mix.

Component	Amount (kg.m^{-3})
Cement CEM I 42.5	650
Water	152
Aggregate 0-4 mm	1710
Aggregate type	basalt
Superplasticizer (polycarboxylate)	27
Silica fume	80

Mixing procedures for mixes M1 to M5 differed according to tables 2 to 6. For mix M1, standard mixing procedure for HSC was used. At first, aggregate was homogenized, than cement was added, followed by silica fume and water with superplasticizer. In mix M2, silica fume was added before cement and mixing time of silica fume was extended from 180 s to 300 s. In case of mix M3, silica fume was added as the last component (after the water with superplasticizer). The mixing procedure of M4 mix was the same as standard (M1), but the mixing time of silica fume was increased from 180 s to 600 s. In the last mixture (M5), the effect of aggregate pre-wash (removal of dust by washing the aggregate by water) was tested. The mixing procedure was the same as standard with extended mixing time of silica fume (300 s).

Table 2. Mixing procedure – mix M1.

Step	Mixing time (s)
Aggregate	10
Cement	60
Silica fume	180
Water + superplasticizer	30

Table 3. Mixing procedure – mix M2.

Step	Mixing time (s)
Aggregate	10
Silica fume	300
Cement	60
Water + superplasticizer	30

Table 4. Mixing procedure – mix M3.

Step	Mixing time (s)
Aggregate	10
Cement	60
Water + superplasticizer	30
Silica fume	300

Table 5. Mixing procedure – mix M4.

Step	Mixing time (s)
Aggregate	10
Cement	60
Silica fume	600
Water + superplasticizer	30

Table 6. Mixing procedure – mix M5.

Step	Mixing time (s)
Aggregate (pre-washed)	10
Cement	60
Silica fume	300
Water + superplasticizer	30

3. Results and discussion

Average flexural tensile strength and compressive strength of particular mixes is given in table 7. Plot in figure 1 presents the mean load-deflection curves from the bending test.

Table 7. Experimental results.

Mix	Flexural tensile strength (MPa)	Flexural tensile strength relative to M1 (%)	Compressive strength (MPa)	Compressive strength relative to M1 (%)
M1	14.5	100.0	156.9	100.0
M2	14.6	100.7	164.3	104.7
M3	15.4	106.2	142.0	90.5
M4	15.5	106.9	119.0	75.8
M5	14.4	99.3	121.1	77.2

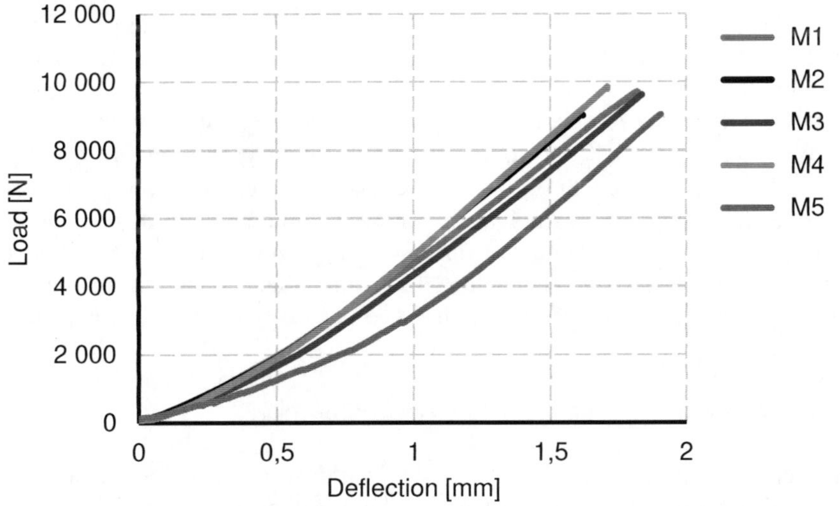

Figure 1. Load-deflection curves from the bending test.

The flexural behavior of materials was almost unaffected by the mixing procedure. Mix M5 had slightly lower elastic modulus than the remaining ones, but the variance of flexural strength of all the mixes was insignificant (less than 7 %).

IOP Conf. Series: Materials Science and Engineering **246** (2017) 012012 doi:10.1088/1757-899X/246/1/012012

In case of compressive strength, the process of homogenization was found to have significant effect on the results. Mix M2 reached the best results. Comparison with mix M1 shows that it is more suitable to add the silica fume before cement. A possible explanation is that the silica particles covered the surface of aggregate grains and functioned as microfiller, leading to higher density and strength of the interfacial transition zone (ITZ) between the aggregates and cement paste. ITZ is the weakest part of the composite, therefore the improvement of ITZ may significantly enhance the overall characteristics of the composite. Another contributing factor might have been a phenomenon described by Mazanec et al. [4]. The gaps between the coarser cement grains and aggregate particles are normally filled by water. If small silica fume particles are present, they fill the gaps and the water which would otherwise have filled the gaps is used to reduce the friction between the particles. This leads to better compaction of the mix, higher bulk density and higher compressive strength.

When silica fume was added to the wet mix M3, compressive strength was lower by 10 % compared to M1, probably due to the fact that the silica fume could not be properly dispersed in the wet mix and clusters of silica fume grains were created. This result was anticipated by the authors, but the decrease in compressive strength was lower than expected.

Extending the mixing time of silica fume from 180 s to 600 s (mix M5) led to deterioration of compressive strength of hardened concrete, the decrease was almost 25 % in comparison with the basic mixing procedure (M1). This results was surprising, it was expected that the strength will be higher due to more thorough dispersion of the components in the mix, or the same at worst. The authors assume that longer mixing allowed silica fume particles to adsorb more water; this resulted in lack of water for the hydration of cement. Another possible explanation can be deduced from the finding of Schitzel et al. [5] who observed that the number of fine particles in the mix is raised by increased abrasion of the coarse aggregate when mixing time is extended. As a consequence, specific surface of aggregate is increased, leading to higher adsorption of water to the surface of aggregates, lower amount of water available for hydration of cement and reduced compressive strength. In any case, it can be clearly stated that extending the mixing time of silica fume above 300 s has no point as it only not only decreases the mechanical properties, but increases the consumption of energy, wear of the mixer and time required for preparation of fresh concrete at the same time.

In case that aggregate pre-wash was applied to remove the fine particles (mix M6), compressive strength was decreased. The authors expected an increase in compressive strength, but the effect of fines content in aggregate on concrete properties is a subject of constant debates. The fines content would increase the surface area and thus the water demand, but also increase the packing density. Hence, the fines content has both undesirable and desirable effects [6]. The mineralogy of fines is also important – if they are of a clay nature, their influence is negative, but silt type fines can function as mineral filler with positive effect [7]. Another important factor is the amount of fines; Topçu and Uğurlu [7] have shown that the content of 7 – 10 % of mineral filler (less than 0.15 mm) that is not of a clay type increases the compressive strength of concrete, while at 15 % dosage the effect becomes negative. The mineralogy and exact content of fines in the aggregate used in the experimental program presented in this paper were determined neither before nor after the aggregate pre-wash. This step should be undertaken in the future to clarify the observed negative effect of aggregate pre-wash.

4. Conclusions

The main conclusions of the work can be summarized as follows:

- Flexural tensile strength of HPC of the given composition seems to be almost indifferent to the mixing procedure applied, the variance of the results was 7 %.
- Compressive strength was significantly affected, the difference between the best and the worst results was almost 30 %.
- It was found that addition of silica fume before cement can increase compressive strength compared to the standard procedure when silica fume is added after cement.

- Extension of mixing time of silica from 300 s to 600 s led to significant reduction of compressive strength.
- Aggregate pre-wash did not lead to the expected improvement of material parameters. This result will be further analyzed.

Acknowledgements
This paper was prepared thanks to the support of the Science Foundation of the Czech Republic (GAČR), project "Analysis of the relations between the microstructure and macroscopic properties of ultra-high performance concretes" (no. 17-19463S).

References
[1] Dils J, De Schutter G, Boel V 2012 Influence of mixing procedure and mixer type on freshand hardened properties of concrete: a review *Materials and Structures* **45** pp 1673-1683
[2] Český normalizační institut 2001 Testing hardened concrete – Part 5: Flexural strength of test specimens *(ČSN EN 12390-5)*
[3] ÚNMZ 2009 ČSN EN 12390-3 Testing hardened concrete – Part 3: Compressive strength of test specimens *(ČSN EN 12390-3)*
[4] Mazanec O, Lowke D, Schiessl P 2009 Mixing of high performance concrete: effect of concrete composition and mixing intensity on mixing time. *Materials and Structures* **43** pp 357-365
[5] Schiessl P, Mazanec O, Lowke K 2007 SCC and UHPC effect of mixing technology on fresh concrete properties. *Advances in construction materials part VI* pp 513–522
[6] Kwan A K H, Ng P L, Huen K Y 2014 Effects of fines content on packing density of fine aggregate in concrete *Construction and Building Materials* **61** pp 270-277
[7] Topçu I B, Uğurlu A 2003 Effect of the use of mineral filler on the properties of concrete *Cement and Concrete Research* **33** pp 1071-1075

IOP Conf. Series: Materials Science and Engineering **246** (2017) 012013 doi:10.1088/1757-899X/246/1/012013

Utilisation of Fibre Reinforced Sprayed Concrete for Primary Lining of the Považský Chlmec Tunnel

M Baláž[1], M Uhrin[2], P Růžička[3] and M Hilar[4]

[1]FCE CTU in Prague and HOCHTIEF CZ a. s.
[2]SUDOP PRAHA a.s.
[3]HOCHTIEF CZ a. s.
[4]FCE CTU in Prague and 3G Consulting Engineers s.r.o.

E-mail: martin.balaz@hochtief.cz

Abstract. The aim of this paper is evaluation of the possibility to utilise fibre reinforced sprayed concrete for the primary lining of the Považský Chlmec Tunnel and comparison of such alternative design with the realised standard solution. The paper also includes a short description of the encountered geotechnical conditions and performed numerical modelling, which was necessary for determination of internal forces in the designed underground structure. The evaluation of fibre reinforced tunnel lining capacity is based on international standards. The paper also contains additional information about tunnel linings from fibre reinforced sprayed concrete such as standards and guidelines in the Czech Republic and abroad, project experience from the Czech Republic and abroad, details about design and construction or relevant laboratory testing.

1. Introduction

Internationally, it has become common practice to utilise fibre reinforced sprayed concrete for the lining of tunnels if the geotechnical and other project conditions allow so. The structural fibre reinforcement is ideally meant to replace a conventional reinforcement comprising lattice girders, mesh or re-bar, and thus achieve simpler, faster and overall more efficient construction. [9]
However, in the Czech Republic and Slovakia there have so far been very few attempts in this area. An indirect consequence of such development is that local legislation, technical standards, standards of large client organizations and common practice of design, construction and supervision are „unprepared" for this construction system.

Hochtief CZ considered use of fibre reinforced sprayed concrete for the Považský Chlmec Tunnel project during the construction preparation stage based on two key factors: First, very favourable ground conditions. And second being the fact that the project delivery was contracted under "yellow fidic" contract conditions that allowed design and specification changes by the contractor to some extent. Overall, the motivation was construction, cost and programme efficiency.

In the end the tunnel was constructed using the conventional reinforcement system mostly due to lack of time to fully prepare the alternative fibre reinforcement system including all required formal approvals. However, considerable effort was spent on preparation of the competitive alternative system and the solution was developed to a high level of detail. This paper details the lessons learned during this exercise and presents key issues encountered on the way.

IOP Conf. Series: Materials Science and Engineering **246** (2017) 012013 doi:10.1088/1757-899X/246/1/012013

2. Basic information about the Považský Chlmec Tunnel

The Považský Chlmec Tunnel is situated on the D3 motorway section on the northern side of town Žilina in Slovakia. The alignment with the aggregated length of 4250 m first overcomes the Hričov dam reservoir on a dual carriageway viaduct. The route then smoothly passes to the western portal of the Považský Chlmec Tunnel. It emerges from it after another 2 kilometres and passes to the dual carriageway bridge over the Kysuca River.

The excavation of the twin-tube tunnel started in January 2015 and was finished in August 2016. The tunnel was excavated according to the NATM principles with horizontal excavation and lining sequence (top heading, bench and invert). Cut-and-cover sections were situated not only in the areas of the eastern and western portal but also at the mid-point construction pit. Special feature of the Považský Chlmec Tunnel is the fact that the tight construction schedule required work at up to six headings. Taking into consideration the tunnel length, this fact is undoubtedly a rarity. HOCHTIEF CZ was the contractor for the excavation of the tunnel from the mid-point construction pit. TuCon was its sub-contractor for driving the tunnel from the western and later eastern portal (Figure 1).

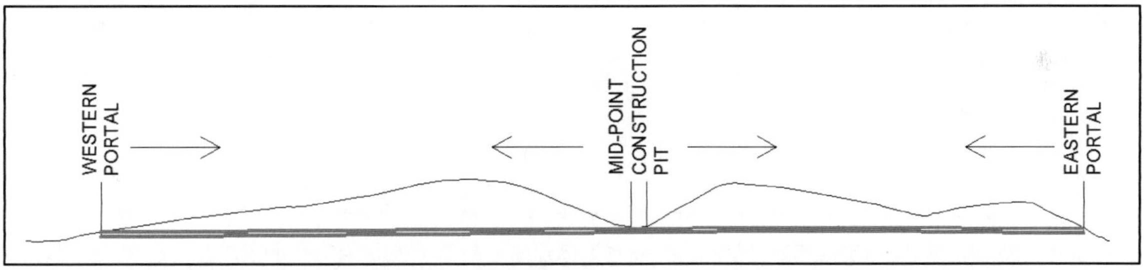

Figure 1. Construction sequence schematic shown in longitudinal section

3. Encountered geological conditions

The tunnel was mostly excavated in relatively strong ground with high overburden (in some parts almost 120 m). A thin sprayed concrete layer of primary lining in these geotechnical conditions performs especially a safety function.

The area along the tunnel route is formed by flysch series of claystones, sandstones and layers of conglomerate. The rock mass was affected by folding and tectonics. Several typical excavation conditions were defined along the tunnel route for design purposes (strength of rock mass, properties of discontinuities, permeability etc.). For further analysis of mined sections the rock mass was divided within the framework of the engineering geological survey into four quasi-homogeneous blocks (M1, M2, M3 and the worst M4).

4. Calculation of internal forces

Numerical analyses were carried out to assess the feasibility of the considered structural systems for the primary lining. Internal forces in the primary lining were calculated by FEM software Plaxis 2D. The predicted internal forces depended mostly on the geotechnical conditions (height of overburden, properties of rock mass) and stiffness of the primary lining (thickness, modulus of elasticity and rockbolts).

4.1. Examined cross sections

For excavation there were designed 8 support classes (ESC 4.1, 4.2, 5.1, 5.2, 6.1, 6.2, 6.3 and MP1). The differences between the classes are thickness of the primary lining, number of wire mesh layers, number and length of rockbolts, length of advance, maximum distance between workplaces (top heading, bench and invert) and reinforcement type of invert.

The assessment of fibre reinforced concrete lining was examined for three excavation support classes in three cross sections of the tunnel. Maximum values of internal forces were calculated for support classes 4.2, 5.1 and 6.2 with different thickness of primary lining (150 mm, 200 mm and 250 mm). For each examined cross section there was created one calculation model with different input parameters (Figure 2).

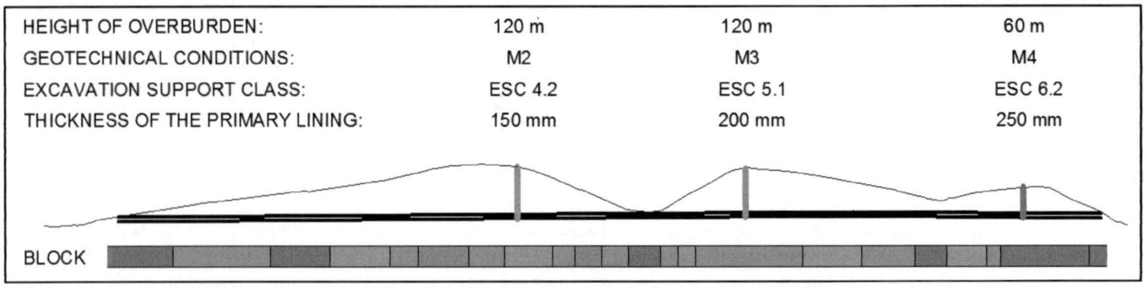

Figure 2. Locations and basic parameters of analysed FRSC sections

4.2. Load cases

Calculation load cases represent the main steps of the construction process (Figure 3). In the performed calculations there was taken into account also the relaxation effect of the ground. The relaxation of ground represents the three-dimensional effect of rock mass behaviour (the value of relaxation depends on length of advance, size of excavation profile, geotechnical conditions and other factors).

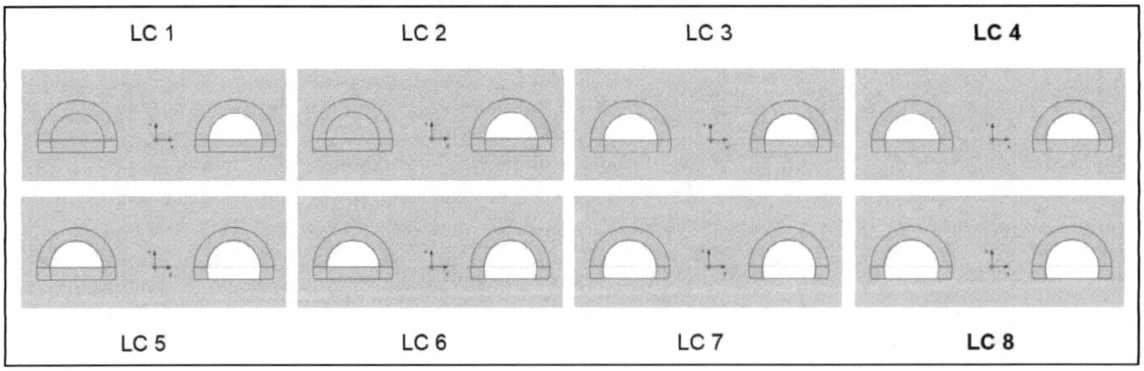

Figure 3. Load case schematics for the excavation support class 5.1

IOP Conf. Series: Materials Science and Engineering **246** (2017) 012013 doi:10.1088/1757-899X/246/1/012013

4.3. Calculated internal forces in the primary lining

Values of normal forces and bending moments in the primary lining were used as basic parameters for further assessment of the lining performance. The results with maximum lining stress values from the three FEM models for examined load cases are shown in Figure 4.

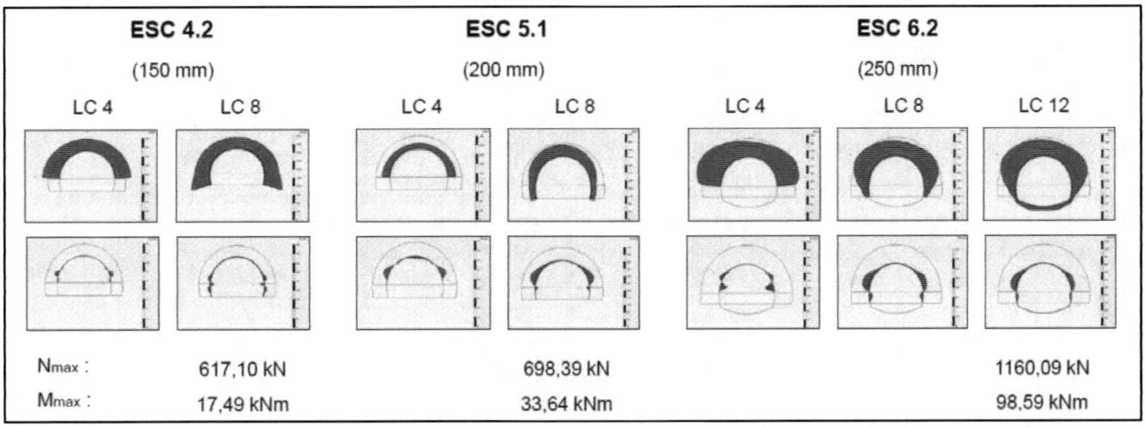

Figure 4. Calculated internal forces in the primary lining for the particular load cases (LC)

The whole primary lining in the examined excavation support classes is loaded in compression. Most parts of the lining are also loaded in bending but with relatively low values (especially in classes 4.2 and 5.1). These results are not surprising for tunnel construction in similar geological conditions (hard rock with high overburden). The calculated combinations of internal forces are appropriate for the design of fibre reinforced sprayed concrete lining which has a limited load bearing capacity compared to standard solution with wire mesh.

5. Assessment of load bearing capacity of fibre reinforced primary lining

Recent civil contracts typically require design according to the currently valid standards. The basic guidelines for the design of structures in the Czech Republic and Slovakia are Eurocodes. Unfortunately, Eurocode 2 (the standard for design of concrete structures) gives no guidance for dimensioning of fibre reinforced concrete.

For a comparison the following steps were carried out: First, calculation of load bearing capacity of standard reinforced primary lining with wire mesh (Figure 5a). Secondly, load bearing capacity of plain concrete without positive effect of fibres was determined (Figure 5b) and in the last step was considered effect of fibres according to foreign guidelines (Figure 5c). In the exercise the load bearing capacity of FRSC primary lining was assessed first according to the international standard Model Code 2010 (results being presented in this paper) and, as additional independent check, according to Austrian guidelines (Richtlinie Spritzbeton and Richtlinie Faserbeton). The main difference between these two standards is in the approach to determine the residual tensile strength of the fibre reinforced concrete.

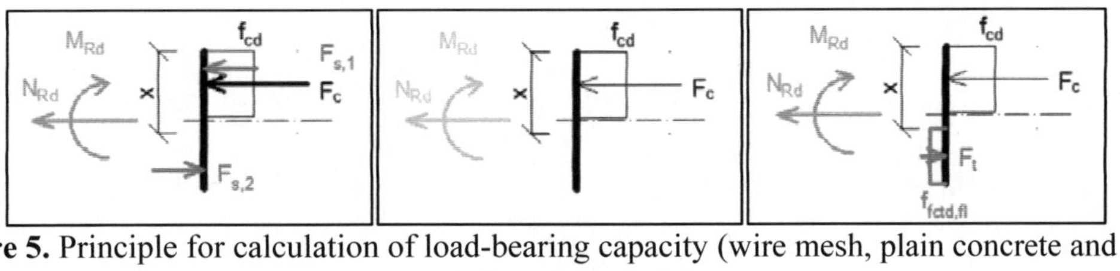

Figure 5. Principle for calculation of load-bearing capacity (wire mesh, plain concrete and fibre reinforced concrete)

5.1. Load bearing capacity without positive effect of fibres

The objective of the second calculation was determination of baseline for assessment of the impact of fibres on the overall flexure strength. The calculation of the load bearing capacity without positive effect of fibres is similar to standard case of reinforced concrete, but only without compressive and tensile forces in reinforcement (Figure 5a and 5b). The compressive strength was determined according to the strength class of the specified concrete (fibres have no significant impact on compressive strength).

The proposed FRSC lining satisfied the condition of load bearing capacity without the effect of fibres for the excavation support classes 4.2 and 5.1. In the case of support class 6.2 the load exceeded the capacity of unreinforced case interaction chart (Figure 6).

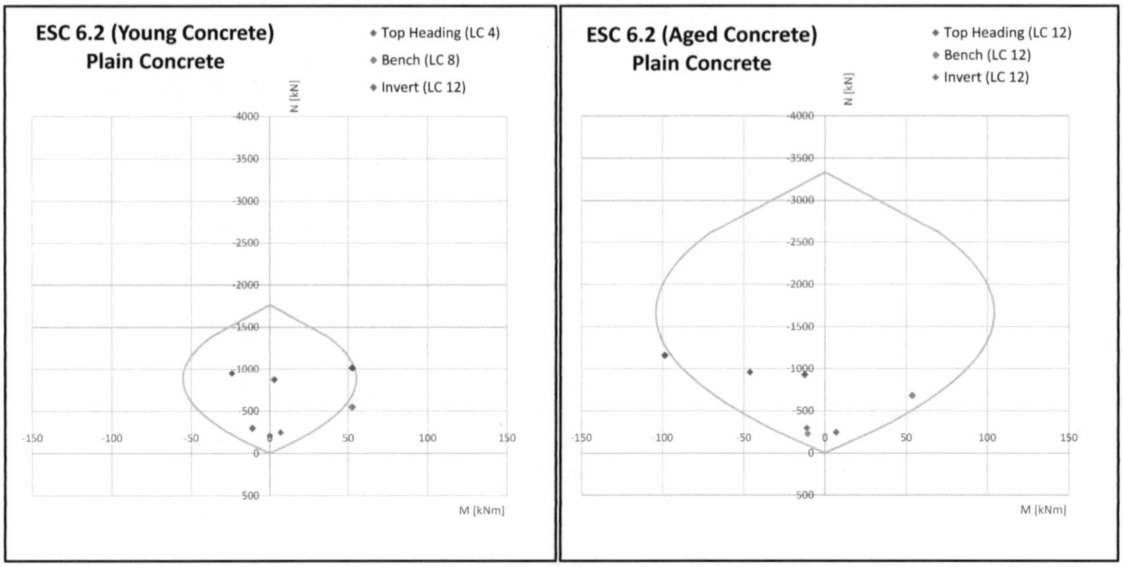

Figure 6. Assessment of load bearing capacity **without** positive effect of fibres (ESC 6.2)

5.2. Load bearing capacity with positive effect of fibres

Assessment of load bearing capacity with positive effect of fibres was performed according to the principles of Eurocodes with additional application rules adopted from international guidelines. Load bearing capacity was calculated according to Model Code 2010 and Austrian guidelines. In both cases the performed calculations led to the similar results.

Addition of fibres in the concrete mix influences mechanical properties of the final product. FRSC is a composite material with residual tensile strength (the value of it depends on the type of fibres, dosage, cohesion of a cement matrix and discrete fibres etc.). The specified fibre reinforced primary lining with residual flexural tensile strength 1.5 MPa satisfied the condition of load bearing capacity (Figure 7).

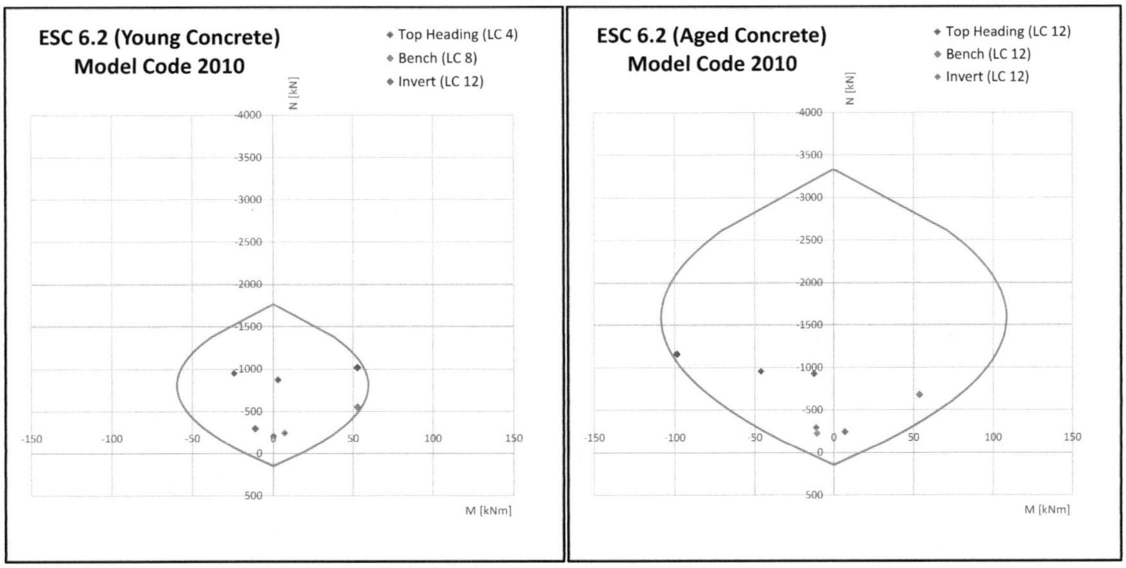

Figure 7. Assessment of load bearing capacity **with** positive effect of fibres (ESC 6.2)

6. Flexural tensile strength and toughness of fibre reinforced shotcrete samples

Mechanical properties of fibre reinforced sprayed concrete can be determined by loading tests. Besides common concrete tests the fibre reinforced sprayed concrete samples can be also tested for determination of energy absorption (on panels) and flexural tensile strength (on beams).

The first group of tests on panels practically simulate the area of primary lining between rockbolts which is loaded by unstable rock block. The second group of tests on beams is typical for European guidelines like RILEM or Model Code 2010 (three-point load test) and Austrian Richtlinie (four-point load test).

Figure 8 shows the results of three point load tests according to EN 14651 from a foreign tunnel project. In underground constructions the common dosage is 25 – 45 kg/m^3 of steel fibres or 4 – 6 kg/m^3 of structural plastic macro fibres.

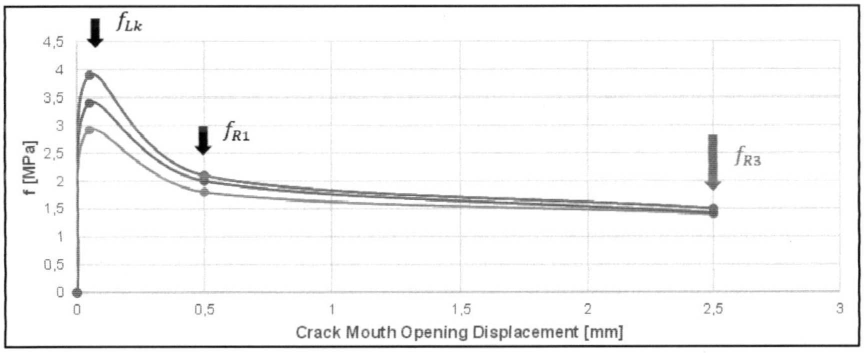

Figure 8. Results of three point load test (flexural tensile strength vs crack mouth opening)

On the basis of known results from foreign tunnel projects for the purpose of initial evaluation on the Považský Chlmec Tunnel there was specified a sprayed fibre reinforced concrete with residual flexural tensile strength of 1.5 MPa (sprayed concrete class **C20/25 D1S2.0 D3S1.5** according to EN 14487-1). This value was used in chapter 5.2 for determination of load bearing capacity of the fibre reinforced primary lining.

Besides assessment of the load-bearing capacity, it was also necessary to assess the toughness of the fibre reinforced structure (Table 1). The toughness check was assumed to represent equivalent of code minimum reinforcement check for RC structures. The assessment of toughness was carried out also according to Model Code 2010 (equations (1) and (2)).

$$\frac{f_{R1}}{f_{Lk}} > 0{,}4 \tag{1}$$

$$\frac{f_{R3}}{f_{R1}} > 0{,}5 \tag{2}$$

Table 1. Assessment of toughness of FRSC samples

Test no.	Measured Flexural Tensile Strength [MPa]			Assessment of Toughness according to Model Code 2010			
	f_{LK}	f_{R1}	f_{R3}	f_{R1}/f_{LK}	f_{R3}/f_{R1}	$f_{R1}/f_{LK} > 0{,}4$	$f_{R3}/f_{R1} > 0{,}5$
1	2.9	1.8	1.4	0.6	0.8	OK	OK
2	3.4	2.0	1.4	0.6	0.7		
3	3.9	2.1	**1.5**	0.5	0.7		

7. Time savings

The performed calculations proved that it was possible to design the primary lining of FRSC with the same thickness as the standard solution in all excavation support classes (150 mm, 200 mm and 250 mm). On the basis of these results we can compare both solutions in terms of time savings of NATM operations. On the Figures 9 and 10 are shown average times of NATM operations. In the calculations was considered the same time for spraying conventionally reinforced and fibre reinforced concrete.

ESC 5.1 - Top heading (average advance speed 3,54 m/day)		
Average time of NATM operations	[min]	[%]
Drilling	97	14,0
Charging and blasting	77	11,1
Ventilation	28	4,0
Loading and hauling, scaling and geological mapping	156	22,5
Construcion of wire mesh and lattice girders	78	11,3
Shotcrete	125	18,1
Rock bolts	45	6,5
Machinery failures	20	2,9
Downtime	37	5,3
Other	29	4,2
Σ	**692**	100

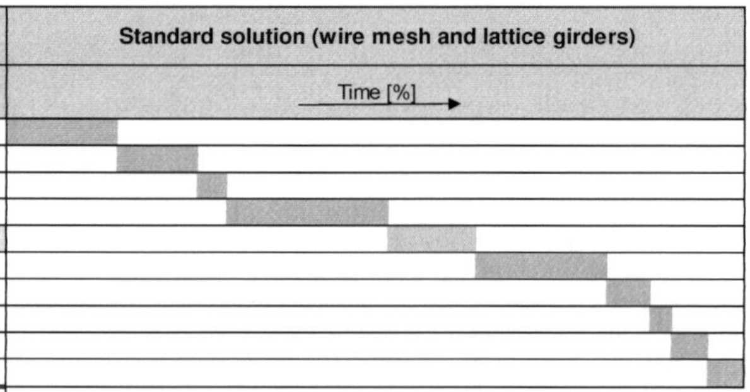

Figure 9. Average time of NATM operations and advance speed of tunnelling in the case of **standard solution** (wire mesh and lattice girders)

IOP Conf. Series: Materials Science and Engineering **246** (2017) 012013 doi:10.1088/1757-899X/246/1/012013

ESC 5.1 - Top heading (average advance speed 3,99 m/day)		
Average time of NATM operations	[min]	[%]
Drilling	97	15,8
Charging and blasting	77	12,5
Ventilation	28	4,6
Loading and hauling, scaling and geological mapping	156	25,4
Construcion of wire mesh and lattice girders	0	0,0
Fibre reinforced shotcrete	125	20,4
Rock bolts	45	7,3
Machinery failures	20	3,3
Downtime	37	6,0
Other	29	4,7
Σ	**614**	100

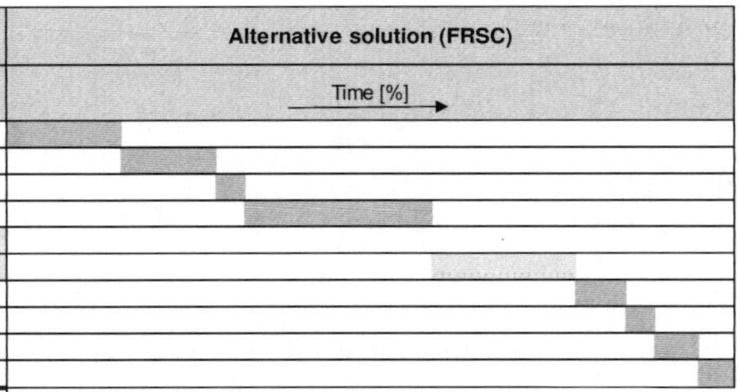

Figure 10. Average time of NATM operations and advance speed of tunnelling in the case of **alternative solution** (fibre reinforced sprayed concrete)

8. Conclusion and comparison of alternative design with realised standard solution

Time and economic advantages of the utilisation of FRSC in underground construction were studied. According to performed calculations it was found that the alternative was more efficient in this specific case. The main advantages of FRSC were considerable time savings (approx. 16 %) and lower production cost (over 6 %).

On the other hand, in some cases the primary lining without wire mesh and lattice girders may not be appropriate. Especially in adverse geological conditions the addition of fibres in the concrete cannot fully substitute the effect of wire mesh (the influence of added fibres in concrete on load bearing capacity of primary lining was calculated approx. 5 % in this case). It is always necessary to make a study for a specific project. Equally important is the availability of material, experience of technicians and workers with new technologies, attitude of project owner towards innovative solutions etc. The main advantages and disadvantages are summarized in Table 2.

Table 2. Summary of advantages and disadvantages of the alternative solution

Fibre Reinforced Sprayed Concrete for Primary Lining of the Považský Chlmec Tunnel	
Advantages	Disadvantages
Time savings (milestones, competitive offer, possible production cost savings)	Higher material costs
Construction safety (workers do not have to work under unsupported ground to install wire mesh and lattice girders)	In adverse geological conditions cannot fully substitute standard primary lining with steel bar reinforcement
Advantageous for complicated shapes (cross passages and tunnel crossings)	In case that construction of secondary lining is planned it necessitates a higher profiling accuracy
Higher quality of sprayed concrete layer (shadowing behind bar reinforcement)	In case of steel fibres it is necessary to provide a thin cover layer under waterproofing
Construction material savings (environmental considerations)	Additional requirements of loading tests (panels or beams)

IOP Conf. Series: Materials Science and Engineering **246** (2017) 012013 doi:10.1088/1757-899X/246/1/012013

9. Acknowledgements
Financial support by the research grants SGS17/057/OHK1/1T/11 and TACR TE01020168 is gratefully acknowledged.

References
[1] Baláž, M.: *Utilisation of SFRC for Primary Lining of the Považský Chlmec Tunnel*, Thesis, FCE CTU, 2017
[2] *EN 14487-1: Sprayed concrete, Part 1: Definitions, specifications and conformity*, CEN, 2005
[3] *EN 14651: Test method for metallic fibered concrete – Measuring the flexural tensile strength (limit of proportionality (LOP), residual)*, CEN, 2005
[4] Mařík, L.: *Považský Chlmec Tunnel on Žilina (Strážov) – Žilina (Brodno) Section of the D3 Motorway – from Design to Realisation*, Tunel 3/2015
[5] Matúška, T.: *Tunnel Excavation and Steel Fibre Reinforced Shotcrete*, Thesis, TUKE FBERG, 2016
[6] *Model Code 2010 – Volume 1*, CEB FIP, 2010
[7] *Model Code 2010 – Volume 2*, CEB FIP, 2010
[8] RILEM TC: *Test and design methods for steel fibre reinforced concrete*, Materials and Structures, Vol. 36, 2003
[9] Růžička, P.; Uhrin, M.; Píšová, B.: *Design and Application of Fibre Reinforced Sprayed Shotcrete in Convetional Tunnelling, 43th Conference Foundations*, Brno 2015

IOP Conf. Series: Materials Science and Engineering **246** (2017) 012014 doi:10.1088/1757-899X/246/1/012014

Recent development in blast performance of fiber-reinforced concrete

R Hajek[1], M Foglar[1] and A Kohoutkova[1]

[1] Department of Concrete and Masonry Structures, Czech Technical University in Prague, Faculty of Civil Engineering, Thákurova 7/2077, Prague, Czech Republic

E-mail: marek.foglar@fsv.cvut.cz

Abstract. The paper presents an overview of the recent development in blast performance of fiber reinforced concrete. The paper builds on more than ten years' history of the research in this field by the team of the Department of Concrete and Masonry Structures of the Faculty of Civil Engineering of the Czech Technical University in Prague.

1. Introduction

1.1. Motivation

Research on the resistance of structures to extreme loading has gained considerable attention in recent years, reflecting the interest in improving the resistance of the civil and transport infrastructure to blasts or impacts. Due to the current geopolitical situation, there has been a considerable increase in the number of terrorist acts. Due to the raised probability of exposure to blast loading, it has become necessary to design government buildings, buildings of significant symbolic value, transport infrastructure, etc., with enhanced safety for users.

This paper presents the findings gathered by the research team from Department of Concrete and Masonry structures, Faculty of Civil Engineering, Czech Technical University in Prague, Czech Republic since the beginning of the research in the field of blast loading on concrete structures in 2006.

The research is based on long-term experimental program of testing the resistance of concrete slab to blast in full-scale. Full-scale experimental program of this magnitude is unparalleled.

Numerical modeling was also performed, but mainly as a tool for evaluating the experimental data and for identifying unmeasurable processes occurring inside the specimen during the experiment. The author's approach and results in field of computer modeling of blast effects on cementitious composites are also included in the paper.

1.2. The route towards blast-resistant concrete

Concrete is one of the most commonly used building materials due to its high durability, resistance to aggressive environments and relatively low financial cost. It usually has sufficient compressive strength to replace costlier construction steel but its tensile strength is low (only 10% of compressive strength) compared to steel with essentially the same compressive and tensile strength. In past decades many attempts were made to improve the characteristics of concrete so it could be used to replace steel in many other applications. The improvement is usually done by introducing another material into the concrete matrix normally consisting of aggregate and Portland cement chemically reacted with water creating a hard matrix binding the other materials together. The first and still the most commonly used approach is to use steel reinforcing bars to provide tensile strength to the structure. However, this approach does not improve the tensile strength of the material. Tensile stresses are transferred to the reinforcement and the concrete surrounding the material usually fails in tension and cracks. The reinforcement holds cracked concrete parts together.

Dispersed reinforcement in form of fibers of various materials was introduced to improve the tensile properties of concrete matrix and it is now being commonly used in structures to reduce risk of cracking of concrete; mainly to ensure required durability of the structure. Concrete commonly used in structural engineering has limited performance when subjected to extreme loads with higher strain rates. This is mainly because concrete is a brittle material with tensile strength much lower than its compressive strength. Fiber reinforced concrete has mechanical properties that enhance its suitability for use in structures subject to high strain rates, e.g. blast or impact loading [1, 2, 3]. The fracture energy value is the decisive material characteristic for an assessment of damage to concrete structures caused by high strain rate loading.

Fracture energy is defined as the amount of energy required to open a unit crack surface area. Fracture energy is a material property, and it does not depend on the size of tested specimen nor the layout of the testing method. Fracture energy of concrete is usually derived from bending tests to total collapse of the specimen. The fracture energy value is equal to the area under the force-deflection diagram (F-δ) recorded during the bending test.

Since 2010, blast experiments have been conducted with basically the same arrangement, using various types and compositions of concrete, ranging from normal strength concrete without distributed reinforcement (NSC) to steel and polypropylene fiber-reinforced concrete (SFRC and PFRC). The results of earlier experiments are presented in [1] and [3].

The focus of the authors then shifted to the use of concrete with even greater strength. The experiments presented in [4] tested two pairs of specimens made of high strength concrete (HSC); both sets were tested with a varying content of high-strength steel (HSS) fibers. Two additional specimens were made of ultrahigh-performance fiber-reinforced concrete (UHPFRC). One of UHPFRC specimens featured a basalt fiber mesh placed in a concrete cover between the soffit of the specimen and the bottom reinforcement. This arrangement, utilizing a mesh commonly used in retrofitting existing structures was used as a basis for the authors' planned future research in blast retrofitting of concrete structures.

Last sets of experiments conducted in 2015 and 2016 introduced new materials into the concrete matrix of full scale specimens creating extremely heterogeneous structure of the specimen. To the date of issue of this paper, the results from those latest experiments were not yet published elsewhere but this paper.

1.3. Theoretical background to blast loading

To study the behavior of a material subject to a blast loading, it is essential to understand the basic principles of blast mechanics. Blast is a pressure disturbance caused by the sudden release of energy. The energy release may have a number of sources. The most known and usually the most harmful is the detonation of an explosive charge.

The blast loads are created by the rapid expansion of an energetic material, creating a blast wave propagating away from the explosion source. For a distant blast, the flow of air particles caused by expansion of the blast wave creates a pressure load analogous to wind or acoustic load. For analogy of pressure and acoustic load in bridge design refer to [5]. Structures subjected to close-in and contact blast are primarily damaged by the blast wave passing through the structure and reflecting from any material interface.

All the types of blast loading were studied by the authors' research team. Special attention was paid to the propagation of shock waves through air and interaction between passing shock wave and concrete blast barriers, see [6, 7, 8].

Further and more detailed description of mechanics of blast loads as well as the physics description of blast phenomena can be found in [9] and [10].

2. Experimental program

Full-scale testing is the most accurate method for determining the blast resistance of a complex heterogeneous material such as concrete or other cementitious composites. Although the current state-of-the-art numerical modeling can provide reliable data on blast loading, it is difficult to calibrate a material model without a prior experiment and subsequent optimization. The authors used their experience in field testing, and decided to conduct full-scale experimental program focused on the blast resistance of concrete bridge decks.

The main full-scale experimental program gradually evolved in correspondence with the shift of author's focus from the study of standard RC structures to the study of behavior of complex composite materials. Experiments conducted in 2015 and 2016 tested highly heterogeneous concrete-based materials with promising results.

A reduced-scale experimental program was designed to supplement the costly full-scale experimental program and to characterize the effect of size of aggregate relative to size of a structure and the effect of reinforcement ratio of the structure. The reduced-scale experimental program was extensive. Multiple types of concrete were tested.

2.1. Full scale experimental program

The experiments were performed in the Boletice military training area, in cooperation with the Armed Forces of the Czech Republic. The location and the layout of the experiment remained practically same during the years, though some improvements were gradually made to the layout.

The specimens were concrete slabs 6 m in length, 1.5 m in width and 0.3 m in thickness. The steel reinforcement remained the same in all the experiments. It consisted of 11 pcs 16 mm dia. reinforcing bars every 140 mm on both surfaces, 10 mm dia. bars every 150 mm as an outer transverse reinforcement. Shear reinforcement was provided by 8 mm dia. links (9 pcs/m2). Concrete cover of 50 mm was taken to the outer surface of the transverse reinforcement. All full-scale specimens were reinforced with standard steel bars with characteristic yield stress of 500 MPa.

2.1.1. Layout of the experiment. The scheme and the layout of the experiment are shown in figure 1 and figure 2. The ground beneath the slabs was excavated to a depth of 2 m to avoid the results being influenced by the shock wave bouncing off the ground. 25 kg TNT charges were placed on steel chairs in the middle of each slab. The chairs provided 450 mm standoff from the slab.

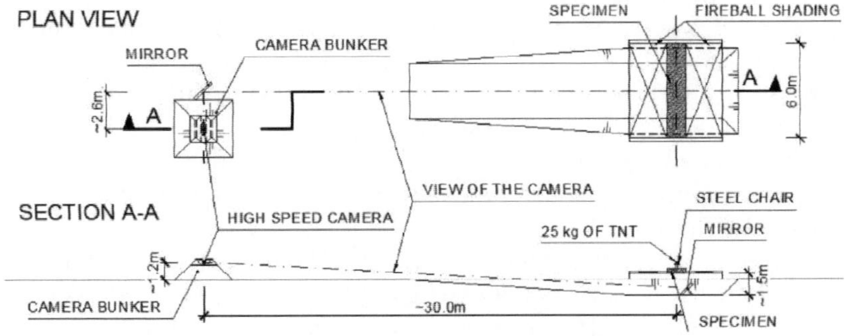

Figure 1. Layout of the experiment

2.1.2. Instrumentation of the experiment. The response of the specimen was recorded with a high-speed camera with a frame rate of 15,000 frames per second (fps). The additional lighting of the scene required for high-speed photography was provided by aluminum powder ignited by a small charge of plastic explosive. The camera observed the spall formation on the soffit of the specimen through an angled mirror (see figure 3). This thesis is not focused on instrumentation of the experiment. For detailed information on the instrumentation used during the experiment refer other author's papers [11, 12, 13].

Figure 2. Concrete specimen ready for the experiment

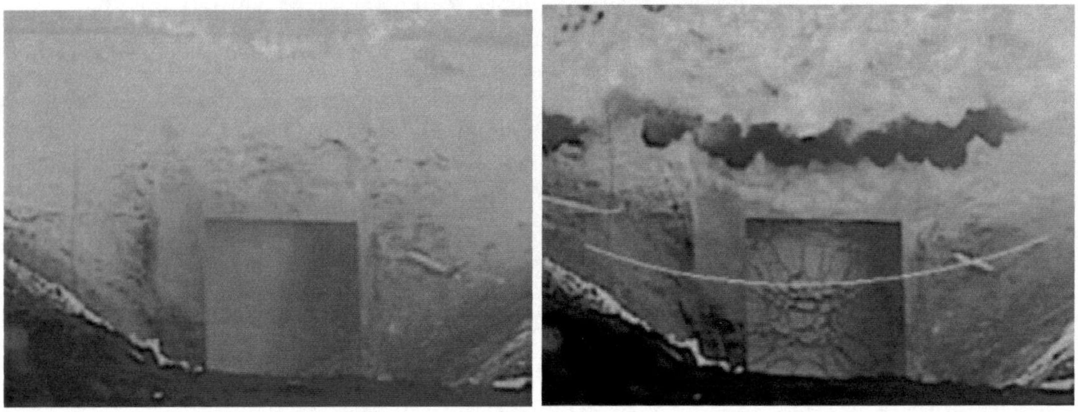

Figure 3. High speed camera image - time 0.7 ms (left), time 2.8 ms (right)

2.1.3. Materials of the specimens. Many variations of composition of concrete were tested during the years. The experiments started in 2010 with specimens made of standard concrete class C30/37 according to Eurocode. In 2011, dispersed reinforcement in form of polypropylene (PP) fibers was used. Concrete strength reached up to 80 MPa.

Low strength steel fibers (yield strength 350 MPa) were added to the concrete matrix of tested specimens in 2013 to study the influence of increased tensile strength and fracture energy of concrete on its resistance to blast effects.

To further improve the behavior of concrete, high strength steel fibers (yield strength 2200 MPa) were used in 2014. Compressive strength of concrete varied from 60 MPa to 130 MPa depending on the amount of fibers, amount of cement and overall composition of the matrix.

To follow the route of gradual improvement of performance of concrete, recent experiments introduced additional materials to the concrete matrix to dramatically increase the heterogeneity of the material. In addition to the dispersed steel fiber reinforcement, a further reinforcement in the form of a basalt mesh (30 x 30 mm) was placed in the concrete cover at 25 mm distance from the soffit of some specimens. Other specimens included multiple layers of the basalt mesh evenly spaced along the depth of the specimen.

The melting point of the basalt fibers is 1350 °C. The tensile strength of the basalt fabric was about 4200 MPa with a tensile modulus of elasticity of about 85 GPa. The unit weight of the mesh is 250 g/m2 with weight density of 2.67 g/cm3. The minimal tensile strength of the basalt fiber is 700 mN/tex, the minimal tensile strength of the unit width of the mesh is 50 kN/m. Each string of the mesh consists of multiple fibers so the diameter of each string cannot be accurately defined, but required for FEM modelling. From parameters presented above, an effective area of each string equal to 1.4 mm2 was calculated instead.

The most recent specimens tested in 2015 and 2016 also included a different approach to increasing the heterogeneity of concrete. Recycled textile sheets were introduced to create disturbance to the propagation of shock waves through the specimen. The goal was to try to increase the ability of concrete to dissipate blast energy. A typical concrete element in form of a hollow-core prestressed panel was also tested with the same boundary conditions to provide the direct comparison of behavior of specimens specially designed to withstand effects of a blast and the behavior of a typical concrete structure. Cross sections of the specimens with drastically increased heterogeneity are shown in figure 4 to figure 6.

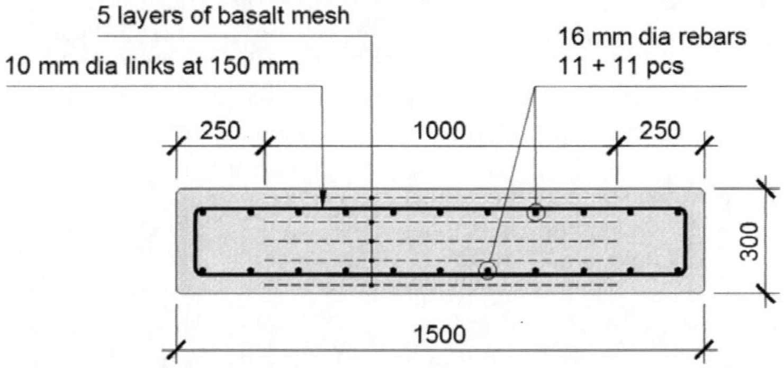

Figure 4. Cross section of Specimen No. 18

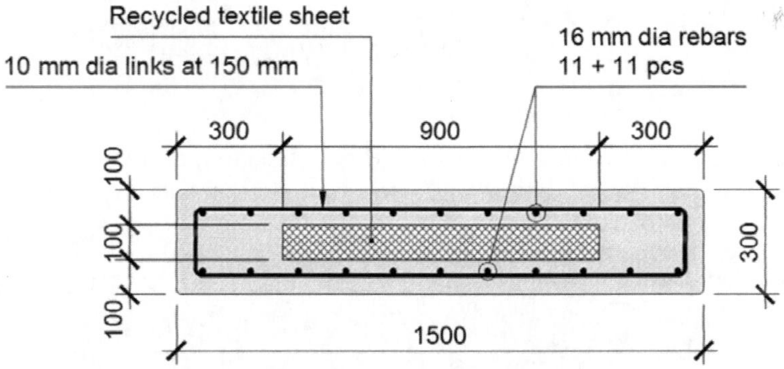

Figure 5. Cross section of Specimen No. 19

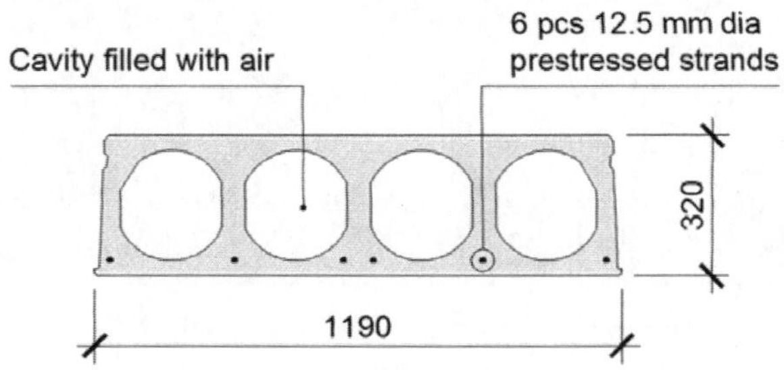

Figure 6. Cross section of Specimen No. 20

IOP Conf. Series: Materials Science and Engineering **246** (2017) 012014 doi:10.1088/1757-899X/246/1/012014

2.1.4. Results of the experiment. The effects of blast loading on the top and bottom surfaces of the specimens after each test were documented. Residual deflection of each specimen was also measured. Damage of selected specimens is shown in following figures. For detailed description and discussion of results refer to a corresponding authors' published paper.

Figure 7. Damaged specimen No. 1 after the blast (60.1 MPa compressive strength, no additional reinforcement) – top view (left) + bottom view (right)

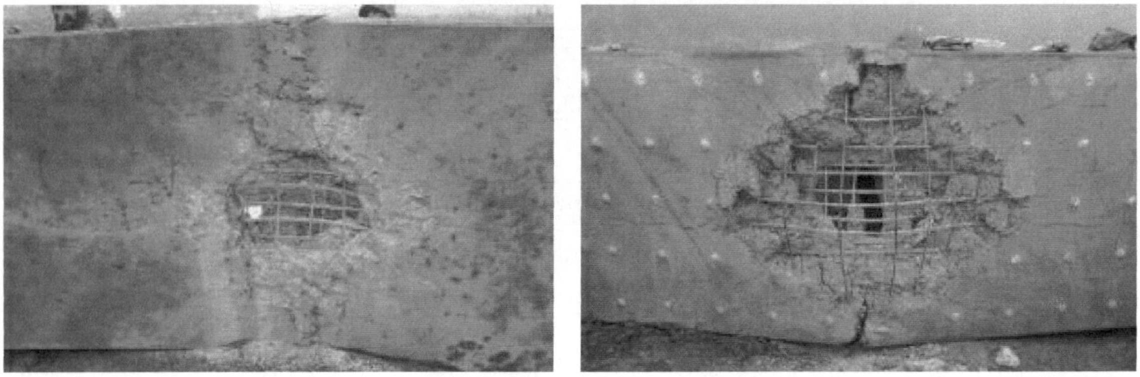

Figure 8. Damaged specimen No. 15 after the blast (76.1 MPa compressive strength, 80 kg/m^3 of high strength steel fibers 30 mm long) – top view (left) + bottom view (right)

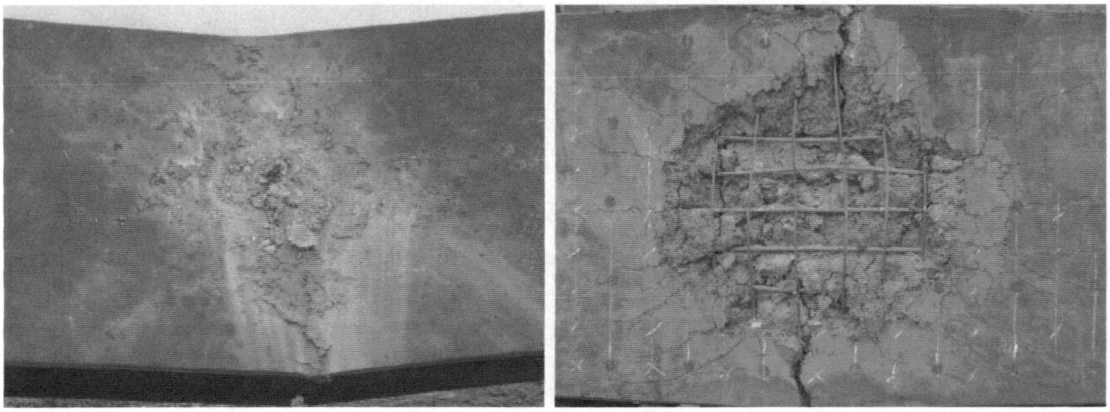

Figure 9. Damaged specimen No. 16 after the blast (129.5 MPa compressive strength, 160 kg/m^3 of high strength steel fibers 13 mm and 30 mm long) – top view (left) + bottom view (right)

2.2. *Reduced scale experimental program*

Full scale experiments are very expensive and time & space demanding. It was decided to verify, whether the material properties influencing the blast resistance of a specimen can be studied on reduced scale experiments. Some of the results presented in this section were already published in [14].

2.2.1. Materials of the specimens. Two sets of reduced scale specimens were created. First set used standard concrete mixture, the second set used micro-concrete. The same amount of steel reinforcement was used for both materials. The amount of steel bar reinforcement of reduced scale specimens was derived by reducing the cross sectional area of the longitudinal reinforcement roughly by the scaling factor.

The use of micro-concrete model for simulation of behavior of concrete structures during the earthquake simulation test seems to be common in earthquake experiments. Micro-concrete seems to have the ability to simulate standard concrete quite well in elastic stage, but because of the difference in material properties between standard concrete and micro-concrete, especially the nonlinear stress-strain relationship, the prediction of nonlinear behavior of the structure remains questionable.

Micro-concrete, to the authors' best knowledge, has never been used in reduced scale experiments to simulate the behavior of full scale concrete structure exposed to blast loading. The use of micro-concrete, if proven applicable, could allow testing the blast resistance of concrete structure in reduced scale and thereby greatly reduce cost of the experiment. Compared to an earthquake, the effects of blast on structure are much more complex. The strain rate dependent non-linear behavior is influenced by propagation of the shock wave through the structure and its reflections from interfaces between the heterogeneous materials within the concrete matrix, i.e. cement mortar, aggregate and reinforcement. Because of this phenomenon the comparison of performance of micro-concrete and full scale concrete subjected to blast loads has to be studied in detail.

2.2.2. Results of the experiment. At first glance, the results of full scale experiment (for example Fig. 4.11) show certain similarities to the reduced scale experiment, but there are also multiple key differences in the specimen damage. A further study of the influence of the scale reduction and composition of concrete on the results is definitely needed.

Typical damage of a standard concrete specimen is shown in as well as typical damage of a micro-concrete specimen is shown in figure 10. The results obtained from micro-concrete specimens show very similar damage mode as previous results. There are differences in spalled areas of each specimen. The size of spalled area varies depending on the amount of steel reinforcement. The amount of reinforcement in the soffit of the specimen seems to have the greatest impact on specimen damage. The amount of top reinforcement seems to have no significant effect.

Figure 10. Blast damage to standard concrete specimen (left) and micro-concrete specimen (right) after the reduced scale experiment.

The comparison of results of full scale and reduced scale experiments with standard concrete specimens shows that the basic mode of damage is present in both cases. But closer look suggests that the behavior of concrete between reinforcing bars is different. The reason of the difference is the use of the non-scaled concrete matrix. It is necessary to maintain the same ratio between the reinforcement and aggregate to obtain results that correspond.

3. Numerical simulations

Computer modelling was widely used during authors' research throughout all the topics that the authors have studied. It is documented in authors' published papers.

3.1. Purpose of the computer modeling

Computer modeling of shock wave propagation in air served as a tool to design effective blast barriers for use in public buildings, see [6], [7] and [8]. Computer modeling of shock wave propagation in cementitious composites was conducted mainly to supplement experimental results and to identify and quantify various physical phenomena that cannot be measured directly during the experiment. For further information, refer [1], [3] and [4]. FEM model was mainly used to identify and visualize the processes occurring in the specimen directly after the detonation. Currently, it is impossible to directly capture the behavior of specimen in the short timeframe of a blast load. Even high-speed camera recording is problematic due to numerous difficulties caused by the rapid release of blast energy.

Numerical models of the full-scale experiments, mainly focused on the effect of material heterogeneity of the structure, were developed for evaluating the experimental results. The model set-up process is described in the following text. The numerical model and the material models of the concrete were calibrated according to the outcomes of the experiments described in the previous section.

3.2. Approach to the computer modeling

The LS-DYNA solver was used for non-linear analysis of the experiment. The software is based on the finite element method (FEM) either in implicit form or in explicit form. It also allows the creation of finite element meshes of elements defined by various mathematical formulations and their combinations to create mathematical models for various types of materials and loadings. The layout of some typical computational models and results is shown in the following figures.

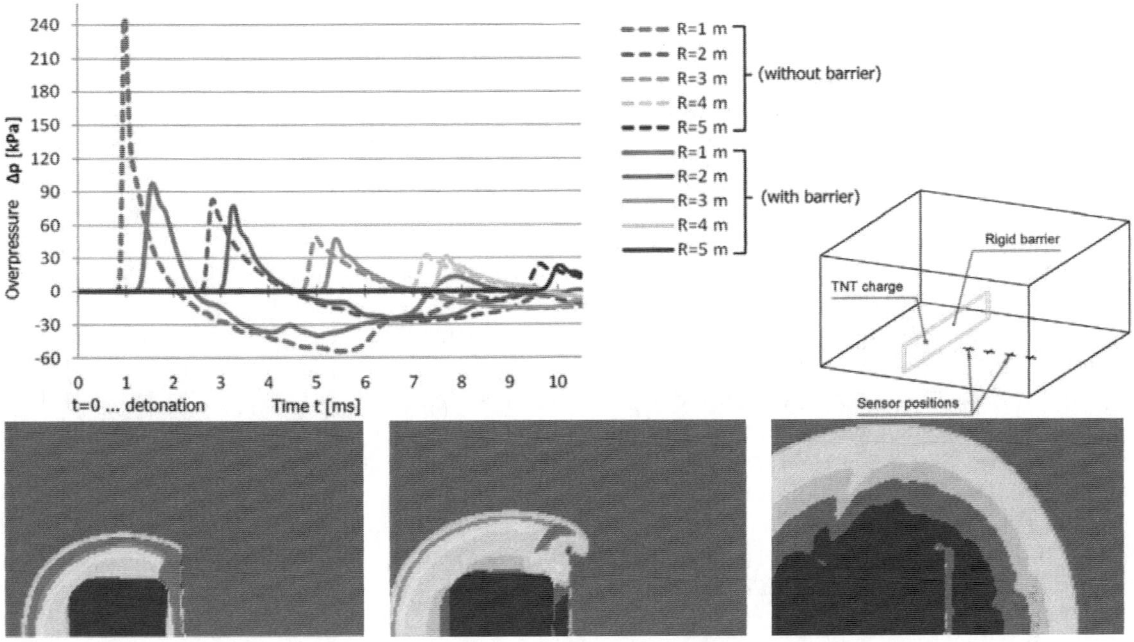

Figure 11. Overpressure at the front of passing shock wave at multiple standoff distances with and without a barrier

Figure 12. Pressure contours at the cross section of the full scale specimen 0.25 ms after the detonation. See the apparent pressure concentration near the stirrup caused by the reflection of shock wave of the interface between two materials with significantly different densities.

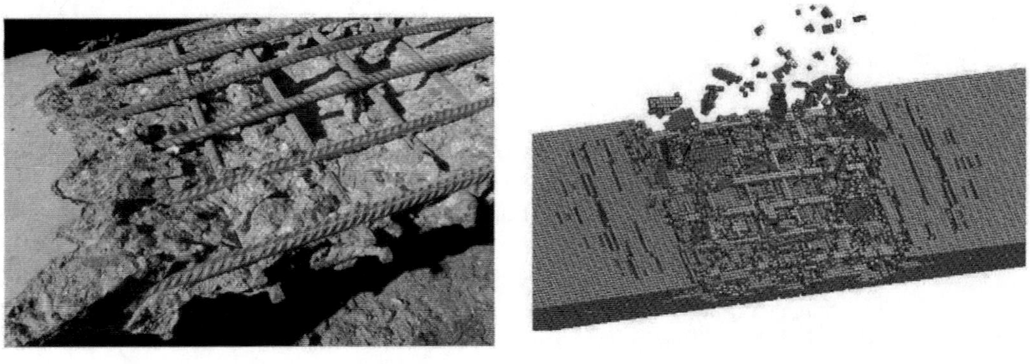

Figure 13. Reduced scale FEM model damage compared to the experiment.

4. Conclusions

Conclusions from each of the sub-topics studied by the authors' research team can be found in corresponding published papers. This section includes only the most significant findings and remarks for future research in the field of blast resistance of concrete structures.

The blast performance of reinforced concrete specimens increases with the addition of high-performance steel fibers and with an increased volume of high-performance steel fibers. By contrast, the addition of low-performance steel fibers has no significant effect. Increasing the fiber content and the compressive strength of fiber concrete up to ultrahigh-performance fiber concrete further enhances its blast performance. None of the UHPFRC specimens was noticeably breached during blast loading, whereas all the SFRC, PFRC and RC specimens were breached.

A basalt mesh inserted into the concrete cover at the soffit of the UHPFRC specimen improves its blast performance, as expressed by the area of spalling and by the volume of debris. The UHPFRC specimen with a basalt mesh experienced a larger extent of internal damage than the regular UHPFRC specimen. This phenomenon was studied numerically, and it was proved that the cause is the internal rebound of the blast overpressure, which interacts with further propagating blast overpressure waves.

In addition, numerical modelling identified the importance of another effect: the stress concentration on the interfaces between materials of different densities. This is caused by the rebound of the shock wave. In the case of a standard bar reinforcement, the stress concentration may weaken the bond between the concrete and the reinforcement, and may seriously affect the overall behavior of the structure.

The effect of stress concentration may also contribute to the delamination of a composite material. This delamination is not necessarily an issue; it can provide benefits to the overall blast performance of the specimen. For example, when the entire cover (50 mm in thickness) becomes disconnected from the specimen after the explosion, it is ejected to the surroundings and can cause damage to people or equipment. When a basalt mesh is used within the cover layer, it can reduce the volume of debris ejected from the specimen to zero, or it can at least decrease its initial velocity and thus reduce its range and kinetic energy. In fact, the increased delamination and increased deformation of the concrete cover can be very effective in dissipating the energy of the blast wave. The formation of a massive horizontal crack across a wide area requires a lot of energy. In combination with the subsequent plastic deformation of the cover layer restrained by the basalt mesh, a large amount of blast energy is mitigated within the structure of the panel.

Based on the results obtained so far, the research team is currently focusing on similar topics based on the observations made during the research presented in this paper, including:

- Behavior of highly heterogeneous concrete composites with controlled heterogeneity subject to close-in explosion
- Behavior of structures made of concrete and concrete-based composites to extreme loading caused by contact explosion
- The study of combined effect of multiple extreme loads acting together, e.g. explosion with subsequent fire, fire with subsequent explosion, high speed impact followed by explosion and fire etc.

References

[1] Foglar M and Kovar M 2013 Conclusions from experimental testing of blast resistance of FRC and RC bridge decks *Int. J. Imp. Eng.* **59** 18

[2] Kovar M and Foglar M 2015 An analytical description of the force-deflection diagram of FRC *Composites: Part B* **69** 550

[3] Foglar M, Hajek R, Kovar M and Stoller J 2015 Blast performance of RC panels with waste steel fibers *Const. Build. Mat.* **94** 536

[4] Foglar M, Hajek R, Fladr J, Pachman J and Stoller J 2017 Full-scale experimental testing of the blast resistance of HPFRC and UHPFRC bridge decks *Const. Build. Mat.* **145** 588

[5] Foglar M, Göringer J 2013 Influence of the structural arrangement of bridges on the noise induced by traffic *Eng. Struct.* **56** 642
[6] Hajek R and Foglar M 2014 The reduction of peak overpressure using concrete blast barriers *WIT Trans. Built Env.* **141** 265
[7] Hajek R and Foglar M 2015 Numerical and Experimental Analysis of the Effect of Rigid Barriers on Blast Wave Propagation *ASCE J. Struct. Eng.*
[8] Hajek R, Foglar M and Fladr J 2016 Influence of Barrier Material and Barrier Shape on Blast Wave Mitigation *Const. Build. Mat.* **120** 54
[9] Dusenberry D O 2010 *Handbook for Blast-Resistant Design of Buildings* (Hoboken, New York: Wiley & Sons)
[10] Henrych J 1979 *The dynamics of explosion and its use* (New York: Elsevier)
[11] Foglar M, Pachman J, Pelikan V, Kunzel M and Kovar M 2014 The structural response of a reinforced concrete specimen subjected to adjacent blast loading, *WIT Trans. Built Env.* **141** 171
[12] Hajek R, Kovar M, Foglar M, Pachman J and Stoller J 2015 Field testing of concrete members subjected to contact and adjacent blast *Adv. Mat. Res.* **1106** 164
[13] Kovar M, Foglar M, Hajek R. 2014 Blast performance of real-scale reinforced concrete specimens with varying fiber type and content *WIT Trans. Built Env.* **141** 159
[14] Stohr J, Hajek R, Foglar M 2015 Blast performance of reduced scale reinforced concrete specimens *Civil-Comp Proc.* **108**

Effects of aggregate grading on the properties of steel fibre-reinforced concrete

M Acikgens Ulas[1], K E Alyamac[2] and Z C Ulucan[3]

[1] Fırat University, Faculty of Architecture, Department of Architecture, 23119, Elazig, TURKEY
[2] Purdue University, Faculty of Engineering, School of Civil Engineering, West Lafayette, USA
[3] Fırat University, Faculty of Engineering, Department of Civil Engineering, 23119, Elazig, TURKEY

E-mail: merveacikgenc@firat.edu.tr

Abstract. This study investigates the effects of changing the aggregate grading and maximum aggregate size (D_{max}) on the workability and mechanical properties of steel fibre-reinforced concrete (SFRC). Four different gradations and two different D_{max} were used to produce SFRC mixtures with constant cement dosages and water/cement ratios. Twelve different concrete series were tested. To observe the properties of fresh concrete, slump and Ve-Be tests were performed immediately after the mixing process to investigate the effects of time on workability. The hardened properties, such as the compressive, splitting tensile and flexural strengths, were also evaluated. In addition, the toughness of the SFRC was calculated. Based on our test results, we can conclude that the grading of the aggregate and the D_{max} have remarkable effects on the properties of fresh and hardened SFRC. In addition, the toughness of the SFRC was influenced by changing the grading of the aggregate and the D_{max}.

1. Introduction

The term steel fibre-reinforced concrete (SFRC) is used to describe concrete containing hydraulic cement, water, fine or fine and coarse aggregates with discontinuous discrete steel fibres. SFRC is designed to enhance the ductility of the concrete by delaying the onset of tension cracks or blocking the propagation of cracks [1-3]. Furthermore, the addition of steel fibres in plain concrete can increase the tensile strength and toughness [4,5]. As the steel fibres affect the mechanical properties of the concrete positively, they also affect workability negatively. Since the main components of the SFRC are similar to conventional concrete, the fresh and hardened properties are affected by the cement content, the water/cement ratio and the aggregate properties other than steel fibres [1,6]. The water/cement ratio is one of the most widely researched parameter because of its major influence on the workability and concrete strength. The influences of aggregate parameters on the properties of the SFRC, however, are generally not as well appreciated [1,3,7]. Since approximately 75% of the concrete volume is occupied by aggregates, it is known that the aggregate properties greatly affect the performance of the concrete. Numerous parameters related to the aggregates, such as chemical and mineral compositions, shape, roughness, degree of weathering, specific gravity, hardness, strength, physical and chemical stability and pore structure, affect the properties of the concrete. Moreover, particle size distribution known as grading or gradation and the maximum aggregate size (D_{max}) have significant effects on the fresh and hardened properties of the concrete [2,4,5,7,8].

The grading describes the distribution of aggregate particles of various sizes. The D_{max} can vary, but in any aggregate composition, particles of different dimensions will be present. The grading of a given sample can be used to tell if an aggregate composition is too coarse, too fine or lacing in a particular size. Aggregate gradations also reflect the amount of voids that must be filled by cement paste and the surface area of aggregates that needs to be coated with cement paste. Moreover the surface area of the aggregates determines the amount of water necessary to wet all of the solids. Thus, grading is important for determining the workability and tendency for segregation. Grading can also influence the economy of the mixture by determining the amount of cement paste. In addition, grading become important because the hardened properties of concrete cannot be fully realized if the concrete is unworkable and difficult to compact [3,9]. Thus, the grading of the aggregates affects both the fresh and hardened properties of the concrete.

Well-graded aggregates can give workable mixes that can be properly transported, placed and compacted [9]. The American Concrete Institute (ACI) committee 544 recommends combined aggregate gradations for SFRC. Compared SFRC to conventional concrete, some SFRC mixtures are characterized as having higher cement and finer aggregate for workability decreasing with fibre content. Therefore, increasing the fibre volume fraction (V_f) and D_{max} can also increase the tendency for clustering of the fibres and improving bond strength between steel fibres and concrete [2,10]. To prevent the clustering of steel fibres, the coarse aggregate content should be less than 55% of the total combined aggregate content by absolute volume. To provide better workability, SFRC requires increasing the paste content, limiting the D_{max} between 10 mm and 38 mm, and optimization of the grading of the aggregate [2,6]. According to Yang et al. [11], considering that the sand particles were more uniform and spherical in shape, the concrete likely had a lower void content, which improved its workability and uniformity of strength [11]. Thus, the composition of the aggregates has effects on both fresh and hardened SFRC. Chenkui and Guofan [12] investigated the hardened properties of SFRC produced with two different D_{max} values, four different fine to coarse aggregate ratios and different types of steel fibres with different V_f. SFRC containing larger amounts of coarse aggregate had good performance, and all strength values showed improvements with the addition of steel fibres.

Most standards related to the grading of concrete aggregates recommend gradations expressed as percent by mass passing a specified set of test sieves. Moreover, standards give charts or logarithmic grading curves which provide recommended limits for gradations [2,13-15]. In this way, it is possible to easily determine the different sizes of aggregates needed to obtain the best concrete mixture. In this paper, rather than investigating only aggregate ratios in a narrow range, different grading curves with different D_{max} values are considered. We experimentally investigate the properties of both fresh and hardened SFRC. Dvorkin et al. recommended that the SFRC composition can only be optimized by determining the coarse and fine aggregate fractions with the best fresh properties [16]. With this purpose, this paper deals with the effects of 4 different aggregate gradations with 2 different D_{max}.

2. Experimental Program

2.1. Material Properties

CEM I 42.5 N type ordinary Portland cement (PC) was used in all mixtures. Specific gravity and specific surface of the cement were respectively 3.1g/cm^3 and 3749 cm^3/g. This cement type complies with the requirements of the TS EN 197-1 [17] standard.

The workability of a cement matrix generally decreases with the addition of steel fibres. With the use of chemical admixtures, however, it is possible to maintain the workability without adding extra water [2,18]. For all mixtures, a polycarboxylate-based super plasticizer (SP) admixture was used. The specific gravity of the admixture was 1.1 g/cm^3.

The steel fibres added to the SFRC mixtures were hooked end-type. The tensile strength of the steel fibres was 1250 MPa and the modulus of elasticity was 200 GPa. The fibre length (l) was 60 mm and the fibre diameter (d) was 0.75 mm; therefore, the aspect ratio (l/d) was 80.

Two groups of crushed calcareous aggregates were used, the first group has D_{max}=16 mm, and the second group has D_{max}=31.5(32) mm. These two groups of aggregates were sieved and divided according to the basic sieve series which is given by the TS 706 EN 12620 [19] standard. Aggregates, divided into the sieve sizes, were added to concrete mixtures with the percentages of designed gradations. 8 types of gradations were calculated on the basis TS 802 [15] standard. The specific gravity of coarse (31.5~8mm), medium (8~2mm) and fine (1mm~filler) aggregates were respectively 2.68, 2.60 and 2.58 g/cm^3. The water absorption rate of coarse (31.5~8mm), medium (8~2mm) and fine (1mm~filler) aggregates were 0.3, 2.2 and 3%, respectively.

2.2. *Mixture Proportions*

All of the SFRC mixtures were prepared with constant cement dosage to provide normal strength values. The *W/C* ratio and the amount of SP were also kept constant. Moreover, the volume fractions (V_f) of the steel fibre were 0% (for reference or plain concrete) and 1%. The proportions used in the SFRC mixtures are listed in table 1, with the aggregate contents given according to sieve sizes. For example, $\frac{16}{8}$ means that the aggregates passed through a 16 mm square mesh sieve and were retained on an 8 mm square mesh sieve. Furthermore, the numbers 16 and 31.5 (32) in the mixture codes refer to the D_{max} values of the mixtures. The letters before the numbers define the grading curves of the mixtures. The mixture code with R means that the mixture was produced without steel fibres (reference series) (table 1).

The volumes of the aggregates were the same for all of the mixtures. Thus, the volume of cement paste was also constant for all of the mixtures. The aggregate compositions were prepared for the grading curves, which are shown in figure 1.

Table 1. SFRC mixture proportions.

Mixture Proportions (kg/m^3)													
					Aggregates divided by sieve sizes (mm)								
Mix Code	PC	W/C	Steel Fibre	SP	$\frac{32}{16}$	$\frac{16}{8}$	$\frac{8}{4}$	$\frac{4}{2}$	$\frac{2}{1}$	$\frac{1}{0.5}$	$\frac{0.5}{0.25}$	$\frac{0.25}{0.125}$	<0.125
R-A16	350	0.5	0	7.0	0	282	420	402	310	181	109	91	36
A16	350	0.5	78.5	7.0	0	282	420	402	310	181	109	91	36
A32	350	0.5	78.5	7.0	282	696	274	201	128	91	72	54	54
R-AB16	350	0.5	0	7.0	0	216	365	374	310	208	154	118	82
AB16	350	0.5	78.5	7.0	0	216	365	374	310	208	154	118	82
AB32	350	0.5	78.5	7.0	216	621	265	210	146	118	100	100	72
R-B16	350	0.5	0	7.0	0	151	310	347	310	235	199	145	127
B16	350	0.5	78.5	7.0	0	151	310	347	310	235	199	145	127
B32	350	0.5	78.5	7.0	151	546	256	219	164	145	127	145	91
R-C16	350	0.5	0	7.0	0	19	201	292	310	308	290	217	181
C16	350	0.5	78.5	7.0	0	19	201	292	310	308	290	217	181
C32	350	0.5	78.5	7.0	19	414	237	219	201	199	181	217	145

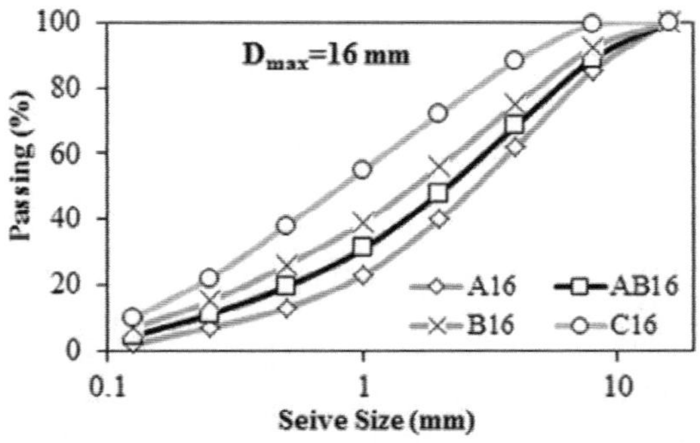

(a)

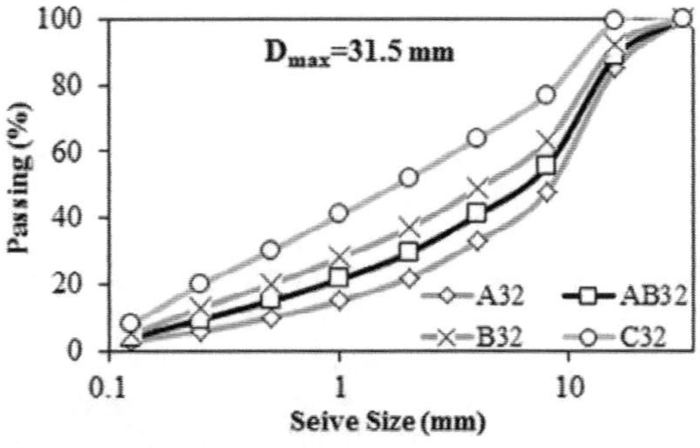

(b)

Figure 1. Gradation curves of aggregates for (a) D_{max}=16 mm and (b) D_{max}=31.5 mm.

The grading curves were designed based on the grading recommendations of TS 802 [15]. In figure 1, grading curves A, B and C, which are given by TS 802 [15], are the limits for the gradations of aggregates that produce economically feasible and workable concrete. According to the standard, the section between A and B represents the most workable and economical concrete mixture (i.e., favorable section). The section between B and C (serviceable section) can also be used. As the curve approaches the 100% line, the aggregate composition becomes finer, whereas as the graphic approaches the 0% line, the aggregate composition becomes coarser [3]. Thus, TS 802 [15] recommends that the limits A and C should not be exceeded.

In this study, two different D_{max} values and four different grading curves were used as A, B, C and AB (figure 1). For each D_{max}, the AB curve was the average of the A and B curves. The *W/C* ratio and SP content were determined to provide a 10~15 cm slump value for the R-AB16 mixture because adding steel fibres reduced the workability of the fresh concrete.

It was considered that the fineness modulus (*k*) and the fine to coarse aggregate ratios were varied for the 8 different aggregates. Before the testing procedures, it was helpful to calculate the parameters of the aggregates (table 2).

Table 2. Calculated parameters of aggregate gradings.

D_{max}=16 mm					D_{max}=31.5 mm				
Code	Fineness Modulus (*k*)	Fine Aggregate (%)	V_{void} (%)	Fine/Coarse (%)	Code	Fineness Modulus (*k*)	Fine Aggregate (%)	V_{void} (%)	Fine/Coarse (%)
A	4.7	62	38	1.61	A	5.8	32	37	0.48
AB	4.3	68	36	2.14	AB	5.4	40	35	0.68
B	3.9	75	35	2.96	B	4.9	48	33	0.94
C	3.2	88	34	7.28	C	4.1	63	31	1.73

The fineness modulus is used to characterize and evaluate a grading curve. This parameter is tied to the particle size distribution and can be obtained by adding the cumulative percentages of materials retained on each of the standard sieves and dividing the sum by 100. The larger fineness modulus is, the coarser the aggregate is, whereas the lower the fineness modulus is, the finer the aggregate is [9, 19]. In addition, considering coarse aggregates as particles larger than 4 mm and fine aggregates as particles smaller than 4 mm, according to TS 706 EN 12620 [19] standard, the fine/coarse aggregate ratios can be calculated by the mass of each aggregate group.

The void volume or content (V_{void}) of the aggregate particles are presented in table 2. The space between aggregates is another important parameter. To measure the V_{void} the aggregates are placed in a container of known volume. The weight is then measured and the volume is calculated by dividing the aggregate weight by the specific gravity. The V_{void} of the aggregate particles is obtained using,

$$V_{void} = \left[1 - \frac{V_a}{V_c}\right] \times 100 \tag{1}$$

where V_{void} is the void volume or content of the aggregate particles (%), V_a is the volume of the aggregates (dm^3) and V_c is the volume of the container (dm^3).
As shown in table 2, larger D_{max} values make the gradations coarser. In addition, the V_{void} decreases. As mentioned before, the volume of cement paste needed to fill between the aggregate particles is constant for all mixtures. Thus, variable V_{void} for different gradations might result in incomplete filling. It is known that unfilled V_{void} in concrete can adversely affect the properties of the concrete.

2.3. Test Procedure

Firstly, to determine the fresh properties of concrete mixtures, Slump and Ve-Be tests were performed after immediately casting according to the TS EN 12350-2 [20] and the TS EN 12350-3 [21], respectively.

Hardened concrete tests were performed on concrete specimens cured in water for 28 days. To determine mechanical properties, 150×150×150 mm cube specimens were used for compressive and splitting tensile strength as recommended by respectively the TS EN 12390-3 [22] and the TS EN 12390-6 [23]. Moreover, to determining flexural strength and toughness, 4-point flexural strength test was performed on 150×150×500 mm prismatic beam specimens, according to the TS 10515 [24] standard. During the flexural strength test, changing of load and deflection of beam specimens were recorded, and the load-deflection curves of specimens were obtained.

3. Results and Discussions

3.1. Fresh Concrete Test Results

Conventional slump measurements cannot determine the workability of SFRC mixtures adequately because compaction by mechanical vibration is recommended in most SFRC applications. The Ve-Be test is more suitable for assessing the workability of SFRC mixtures [1-3]. Thus, in this study, Ve-Be tests were performed in addition to slump tests.

Figure 2 provides the respective slump values. As expected, the addition of steel fibres significantly reduced the workability of the fresh mixtures. Most of SFRC slump values were zero, especially for SFRC mixtures with D_{max}=31.5 mm (figure 2).

The Ve-Be times are shown in figure 3. The Ve-Be times increased with the addition of steel fibres, with increasing D_{max} values compared to the reference mixtures. Because increasing the Ve-Be time means that the workability decreases, it can be said that steel fibre addition reduces the workability. Thus, changing the grading of the aggregate has a noticeable effect on the workability.

In figure 3, the workability of the reference mixtures are not very different from each other. The mixtures graded B were the most workable mixtures compared to the other gradations (figure 3).

Mixtures graded C performed poorly in terms of workability. The C grading was the finest gradation used in this study (the lowest fineness modulus in table 2). As the grading becomes finer, the aggregate surface are becomes higher leading to decreased workability [1,3,9]. The A grading was the coarsest gradation used in this study (the fewest fine aggregates and the highest fineness modulus in table 2). The workability of A mixture was very low, especially for SFRC. It is known that very coarse aggregates produce harsh and unworkable concrete mixtures [3]. Metha and Monteiro [3] content that aggregate grading and D_{max} can be specified due to the reasons such as mixing, placing, etc. Moreover, for mixtures with D_{max}=16 mm and D_{max}=31.5 mm (figure 3), the workability of the mixtures tends to decrease with higher D_{max}. This proves that coarser aggregates reduce the workability. Furthermore, especially with the addition of steel fibres, the SFRC mixtures tend to clustering due to the D_{max} and grading of aggregates [1,2]. This is why the ACI 544-1R [2] report specifies the D_{max} and grading of aggregates for SFRC mixtures and recommends a lower volume fraction of steel fibres to be used with larger D_{max} values. Here, we emphasize that from A to C, finer gradations of aggregates positively affect the workability to a certain degree.

The Ve-Be data were more variable than the slump data. Thus, Ve-Be tests are obviously more suitable for measuring the workability of SFRC mixtures in this study. Most of the slump values for the SFRC mixtures were 0, whereas the Ve-Be times varied (figures 2 and 3).

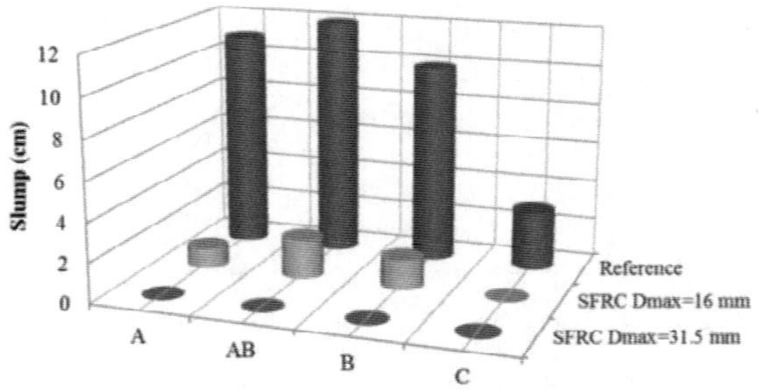

Figure 2. Slump values of the mixtures.

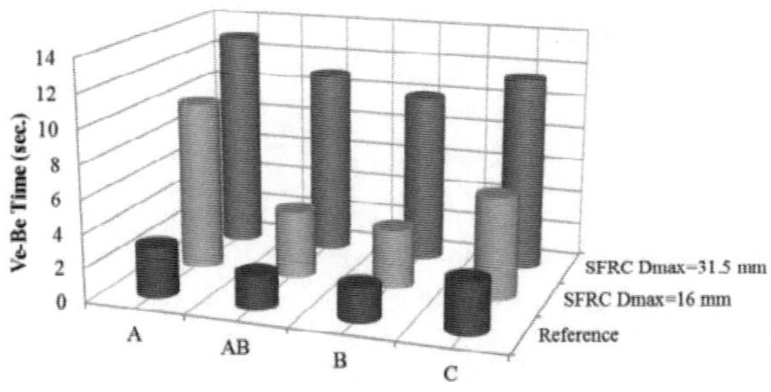

Figure 3. Ve-Be times of the mixtures.

3.2. Hardened Concrete Test Results

Figure 4 shows the changes in compressive strength (f_c) of the specimens. The aggregate grading influenced the compressive strength of both SFRC and reference concretes. The effect of steel fibres on compressive strength is negligible and the compressive strength of concrete increases with increasing D_{max} [3,7,25,26]. Although the differences between compressive strength of SFRC and reference were not very large, changing the gradation, and especially the D_{max}, affected the compressive strength with a constant cement dosage and W/C ratio. Increasing the D_{max} increased the compressive strength (figure 4). It can be seen in figure 4 that the grading is more influential than the presence of steel fibres on the compressive strength, especially when increasing the coarse aggregate content with D_{max}.

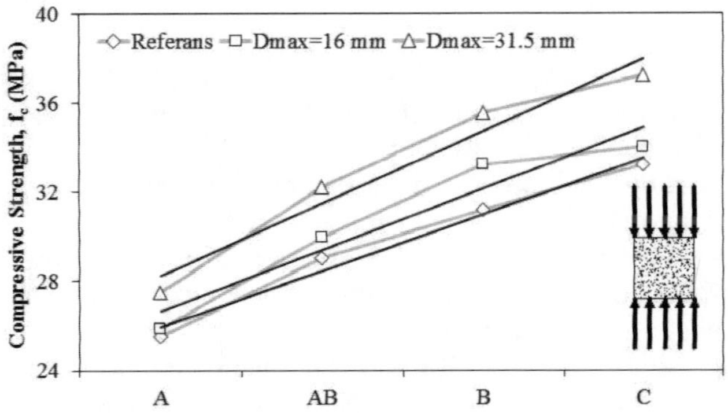

Figure 4. Compressive strengths of specimens.

It is also seen in figure 4 that the compressive strength of both SFRC and the reference concrete increases as the fineness modulus and V_{void} of aggregate particles decrease from A to C (table 2). As explained before, all of the mixtures had the same volume of cement paste. It is well known that unfilled voids by the cement paste decrease the mechanical strength of concrete. In addition, these voids can also increase by the lower workability [1,3].

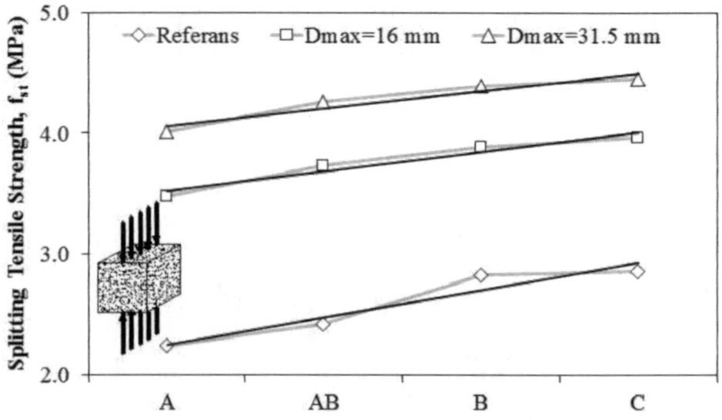

Figure 5. Splitting tensile strengths of specimens.

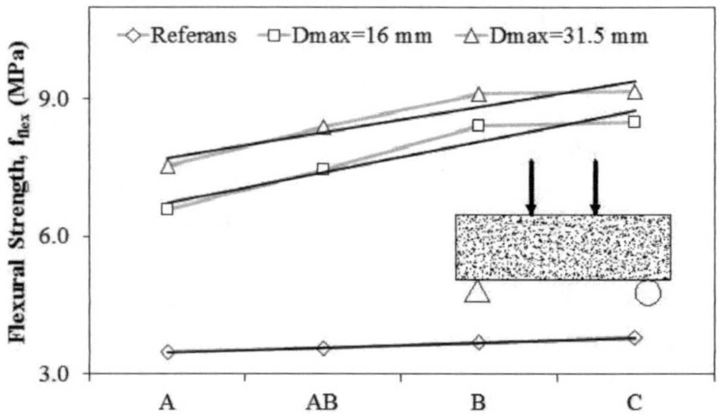

Figure 6. Flexural strengths of specimens.

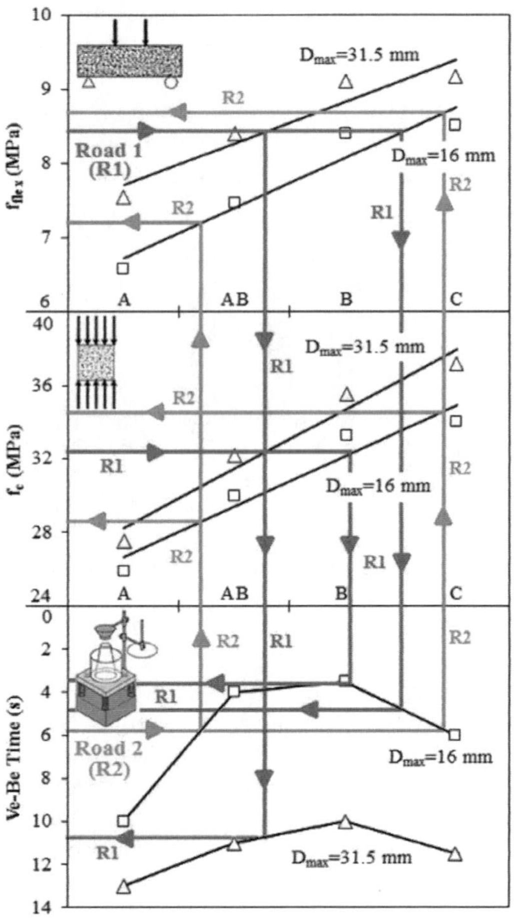

Figure 7. Relationship between Ve-Be times, compressive and flexural strengths.

The compressive strengths of specimens with the finest grading C were higher than specimens with the coarsest grading A (figure 4). Considering a constant cement dosage and *W/C* ratio, it is possible to increase the compressive strength by almost 30% for D_{max}=16 mm and 35% for D_{max}=31.5 mm by changing the aggregate grading. The compressive strength of concrete has the greatest influence on the cost of structures. The economic value tends to increase with increasing strength. Cement with the most expensive production unit cost (almost 10 to 15 times as much as the price of aggregates) has the largest influence on both the economic value and the compressive strength of the concrete [3,27,28]. Here, on account of maintaining a constant cement dosage, *W/C* ratio, amount of SP and V_f. Thus, the costs of the mixtures were almost the same. If the cost of one SFRC mixture is considered as constant, the cost per unit compressive strength (1 MPa) of the SFRC mixtures varies. Changing the gradation from A to C reduces the cost per unit compressive strength (1 MPa) while maintaining the mixture proportions. In this way, more economical SFRC mixtures can be provided by changing the aggregate gradation (figure 4).

Figure 5 and 6 gives the splitting tensile (f_{st}) and flexural strengths (f_{flex}) under ultimate loads. Aggregate grading affected the splitting tensile and flexural strength for both SFRC and plain concrete. Changing the gradation and the D_{max} affected the splitting tensile strength values for constant cement dosages and *W/C* ratios (figure 5). Steel fibres have a remarkable effect on the splitting tensile and flexural strength other than the compressive strength. Because steel fibres make concrete less brittle and more ductile, adding steel fibres in plain concrete increases the tensile strength of the

concrete [4,5,8,12]. As expected, the steel fibres enhanced the splitting tensile and flexural strengths of the concrete specimens.
For both plain concrete and SFRC, the tensile and flexural strength increased with decreasing V_{void} for both D_{max} values of the A to C gradations (table 2 and figures 5 and 6). It was explained previously that voids decrease the mechanical strength. As with the compressive strength, the influence of larger D_{max} on the tensile and flexural strength of SFRC was greater (figure 6). It is also known that coarse aggregates with higher D_{max} values enhance the ultimate flexural strength of SFRC [12].
Figure 7 provides a general comparison among gradations for Ve-Be times and f_c and f_{flex} of SFRC. As is seen in figure 7, there is a relationship between strength values and Ve-Be time. For the SFRC mixtures, changing the grading had a variable effect on the workability. This variability can be compared with the compressive and flexural strengths of the SFRC specimens. If Road 1 is followed, it is seen that a larger D_{max} produces concrete with lower workability for the same f_c and f_{flex}. If Road 2 is followed, it is seen that the finer grading produces concrete with higher fc and f_{flex} for the same workability and D_{max}. Thus, for a constant cement dosage and W/C ratio, aggregate grading alone can be determinative factor for both fresh and hardened SFRC.

3.2.1. Toughness of SFRC. Plain concrete was exceeded at once the deflection that corresponds to the ultimate flexural strength; on the other hand, the SFRC continues to carry on notable loads even at the deflections considerably in excess of the fracture deflection of the plain concrete. Because the failure occurs due to fibre debonding which is depend on bond stresses occurred between steel fibre and concrete matrix, the SFRC does not break suddenly after the first crack unlike plain concrete. Bond stress depends on concrete quality as much as it depends on fibre characteristics. The toughness (T) value, defined as the area under the load-deflection curve, reflects the energy absorption of deflected specimen to the specified deflection value [1,29,30]. It is known that D_{max} has a slight effect on the toughness [12] although studies in the literature are insufficient which investigated the relationship between the grading and the toughness.

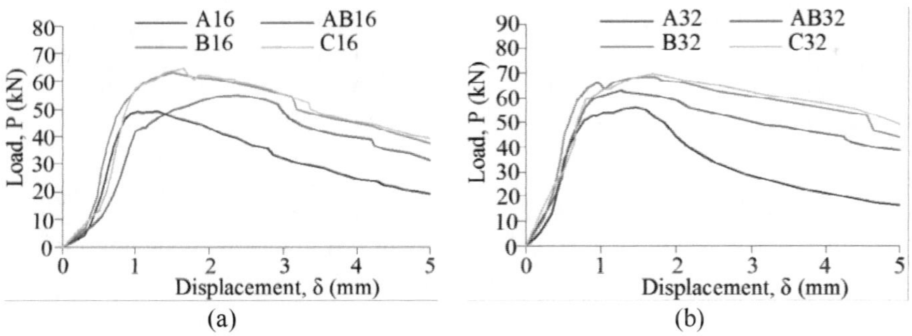

Figure 8. Load – displacement curves of SFRC specimens for (a) D_{max}=16 mm and (b) D_{max}=31.5 mm.

Figure 8 gives the load (P)-deflection (δ) curves of SFRC specimens. Figure 9 provides the toughness of the SFRC specimens. From A to C gradation, finer grading affected toughness of SFRC positively for both D_{max} values. It can be said that larger D_{max} increased 10% toughness of SFRC (figure 9).

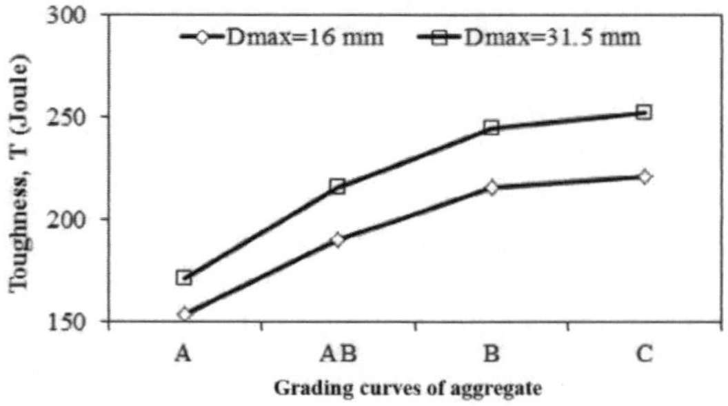

Figure 9. Toughness of SFRC specimens.

4. Conclusions

Based on our experimental results on the effects of gradation and D_{max} on SFRC properties with constant cement dosages and *W/C* ratios, the following conclusions can be drawn.

In addition to reduce of the workability by using 1% steel fibres, most of the slump values of the SFRC mixture were 0. The results of the Ve-Be tests were more variable than the slump tests for SFRC. Changing the aggregate grading had a noticeable effect on the workability of both SFRC and reference concrete. The finest and the coarsest grading showed low workability for both D_{max} values. SFRC mixtures with smaller D_{max} were more workable.

The compressive, splitting tensile and flexural strengths of the hardened specimens were investigated experimentally. Improved mechanical test results were observed with increasing D_{max}. All of the mechanical properties were increased with finer gradation for both D_{max} values. Unfilled voids negatively affected the mechanical properties. In terms of the toughness of the SFRC specimens, increasing the D_{max} slightly increased the toughness of SFRC. The toughness increased with finer gradations.

For the future studies, gradation curves may be defined with more parameters instead of with just one fineness modulus parameter since the volume of voids between aggregate particles was affected the properties of SFRC this much. Finally, it is possible to improve both the workability and mechanical properties of SFRC by changing the aggregate grading and D_{max} while maintaining a constant cement dosage and *W/C* ratio. In addition, by choosing the right grading and D_{max}, SFRC mixtures can be produced economically in terms of the cost per unit strength.

Acknowledgements

This research was supported by Fırat University Scientific Research Projects Unit (Grant No. MF.13.02). Authors acknowledge to Kemerli Firm (Turkey) for their material supports.

References

[1] Açıkgenç M 2015 A Graphic Based Approach for the Mix Design of Steel Fiber Reinforced Concrete. In: *Civil Engineering Department,* (Elazig, Turkey: Fırat University) p 186
[2] ACI 544.1R-96 2002 *State-of-the-art report on fiber reinforced concrete (Reapproved 2005),* (Michigan: Manual of Concrete Practice, American Concrete Institute)
[3] Mehta P K and Monteiro P J M 2006 *Concrete: Microstructure, Properties and Materials* (New York: McGraw-Hill)
[4] Banthia N and Sappakittipakorn M 2007 Toughness enhancement in steel fiber reinforced concrete through fiber hybridization *Cem.Concr. Res.* **37** 1366-72
[5] Gopalaratnam V S and Gettu R 1995 On the characterization of flexural toughness in fiber reinforced concretes *Cem. Concr. Compos.* **17** 239-54
[6] ACI 544.3R-93 2005 *Guide for specifying, proportioning, mixing, placing, and finishing steel fiber reinforced concrete (Reapproved 1998),* (Michigan: Manual of Concrete Practice, American Concrete Institute)

[7] Meddah M S, Zitouni S and Belâabes S 2010 Effect of content and particle size distribution of coarse aggregate on the compressive strength of concrete *Constr. Build. Mater.* **24** 505-12
[8] Nataraja M C, Dhang N and Gupta A P 1999 Stress-strain curves for steel-fiber reinforced concrete under compression *Cem. Concr. Compos.* **21** 383-90
[9] Alexander M and Mindess S 2005 *Aggregates in concrete* (London: Taylor & Francis Group)
[10] Kim J J, Kim D J, Kang S T and Lee J H 2012 Influence of sand to coarse aggregate ratio on the interfacial bond strength of steel fibers in concrete for nuclear power plant *Nuclear Eng. Des.* **252** 1-10
[11] Yang S L, Millard S G, Soutsos M N, Barnett S J and Le T T 2009 Influence of aggregate and curing regime on the mechanical properties of ultra-high performance fibre reinforced concrete (UHPFRC) *Constr. Build. Mater.* **23** 2291-8
[12] Chenkui H and Guofan Z 1995 Properties of steel fibre reinforced concrete containing larger coarse aggregate *Cem. Concr. Compos.* **17** 199-206
[13] ASTM C 33 2003 *Standard specification for concrete aggregates,* (USA: Philadelphia: Annual Book of American Society of Civil Engineers (ASTM) Standards)
[14] BS EN 12620 2009 *Aggregates for concrete,* (London: British Standards Institution)
[15] TS 802 2009 *Design concrete mixes* (Ankara: Turkish Standards Institutions)
[16] Dvorkin L, Dvorkin O, Zhitkovsky V and Ribakov Y 2011 A method for optimal design of steel fiber reinforced concrete composition *Mater. Des.* **32** 3254-62
[17] TS EN 197-1 2005 *Cement. Part 1: compositions and conformity criteria for common cements,* (Ankara: Turkish Standards Institutions)
[18] Uygunoğlu T 2011 Effect of fiber type and content on bleeding of steel fiber reinforced concrete *Constr. Build. Mater.* **25** 766-72
[19] TS 706 EN 12620 2009 *Aggregates for concrete,* (Ankara: Turkish Standards Institutions)
[20] TS EN 12350-2 2002 *Testing fresh concrete – part 2: slump test,* (Ankara: Turkish Standards Institutions)
[21] TS EN 12350-3 2002 *Testing fresh concrete – part 3: vebe test,* (Ankara: Turkish Standards Institutions)
[22] TS EN 12390-3 2010 *Testing hardened concrete - Part 3 : Compressive strength of test specimens* (Ankara: Turkish Standards Institutions)
[23] TS EN 12390-6 2010 *Testing hardened concrete - Part 6: Tensile splitting strength of test specimens,* (Ankara: Turkish Standards Institutions)
[24] TS 10515 1992 *Concrete–steel fibre reinforced-test method for flexural toughness,* (Ankara: Turkish Standards Institutions)
[25] Altun F, Haktanir T and Ari K 2007 Effects of steel fiber addition on mechanical properties of concrete and RC beams *Constr. Build. Mater.* **21** 654-61
[26] Olivito R S and Zuccarello F A 2010 An experimental study on the tensile strength of steel fiber reinforced concrete *Compos. Part B: Eng.* **41** 246-55
[27] Kim T, Tae S and Roh S 2013 Assessment of the CO2 emission and cost reduction performance of a low-carbon-emission concrete mix design using an optimal mix design system *Renewable and Sustainable Energy Reviews* **25** 729-41
[28] Narasimhan H and Chew M Y L 2009 Integration of durability with structural design: An optimal life cycle cost based design procedure for reinforced concrete structures *Constr. Build. Mater.* **23** 918-29
[29] Bentur A and Mindess S 2005 *Fiber Reinforced Cementitious Composites* (New York: Elsevier Science Publishing)
[30] Ferrara L, Park Y D and Shah S P 2008 Correlation among fresh state behavior, fiber dispersion, and toughness properties of SFRCs *J. Mater. Civ. Eng. (ASCE)* **20** 493-501

IOP Conf. Series: Materials Science and Engineering **246** (2017) 012016 doi:10.1088/1757-899X/246/1/012016

Non-Destructive Testing of Steel Fibre Reinforced Concrete

D Corbett[1] and G Tronca[1]

[1] Proceq SA, Ringstrasse 2, CH-8603 Schwerzenbach, Switzerland

E-mail: david.corbett@proceq.com

Abstract. For standard reinforced concrete, there are several non-destructive test (NDT) methods available for measuring the concrete cover and for locating subsurface objects and defects. Whether or not these methods may also be applied to fibre reinforced concrete has been the subject of a recent study. The results and a recommendation for the most suitable technology for use with fibre reinforced concrete is the topic of this paper.

1. Introduction

One of the major design parameters for ensuring a minimum service life of a structure is the concrete cover. Together with the concrete quality this determines how long it will take for aggressive agents such as chlorides and carbonation layers to reach the steel reinforcement. All major standards bodies have recommendations for the concrete cover based on the environmental exposure class to which the structure is subjected. The concrete quality is typically specified prescriptively in terms of compressive strength class, minimum cement content and water / cement ratio. The concrete cover can be checked at the time of construction by non-destructive methods.

The two most common technologies used for measuring concrete cover are magnetic eddy current and ground penetrating radar. A study [1] has recently been carried out at HSR "Hochschule für Technik" Rapperswil in Switzerland to investigate whether or not these technologies along with several others could be applied to fibre reinforced concrete.

As developers and manufacturers of NDT equipment, Proceq were invited to carry out their own tests on the specially dedicated test blocks to reach their own conclusions independently. This paper presents the results of that testing.

2. The Test Objects

Three test objects were made available by HSR. Each test object was a 150x50x20cm concrete prism with eight different rebars and a plastic tube embedded at different depths (Figures 1 and 2).

Figure 1. Test blocks

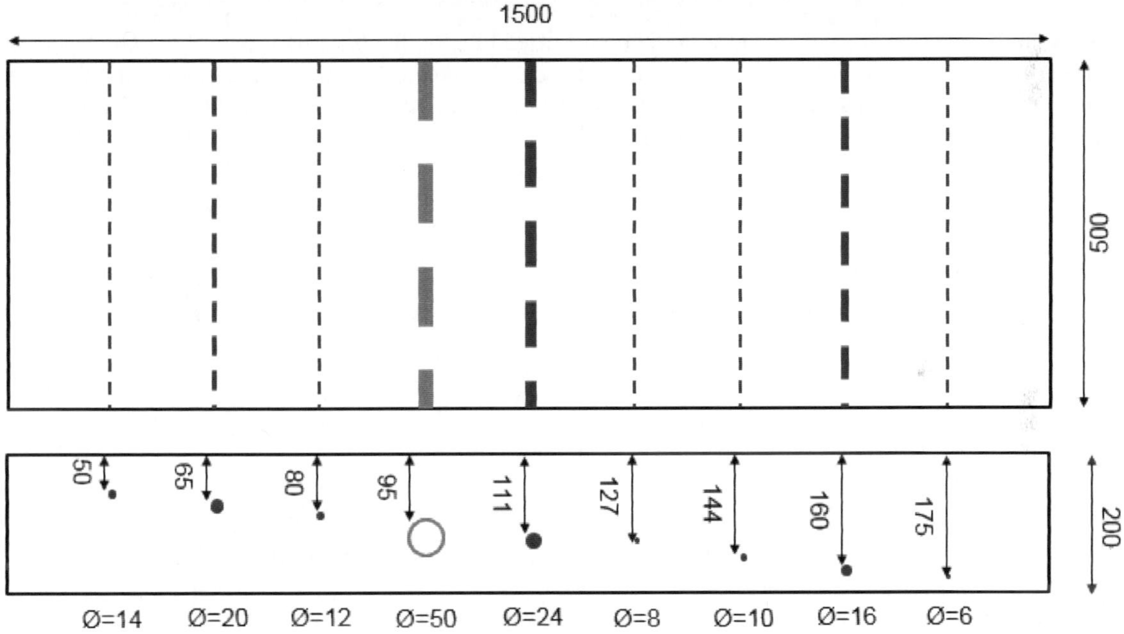

Figure 2. Test Block Dimensions and Layout

Although identical in size and layout the difference between the test objects was in the fibre content of the concrete used. Test block 1 was constructed with standard concrete with no fibre content. Test block 2 had a fibre content of 30 kg/m^3 and test block 3 had a content of 60 kg/m^3. Dramix steel fibres 1 x 50mm were used. The concrete used had a concrete compressive strength class of C25/30 with a maximum aggregate size of 32mm.

IOP Conf. Series: Materials Science and Engineering **246** (2017) 012016 doi:10.1088/1757-899X/246/1/012016

3. Pulse Eddy Current Testing

The instrument used for this test was the Proceq Profometer 650 AI (pulsed eddy current), Figure 3. This is a typical cover meter working on magnetic eddy current principles. This technology is generally recognized as providing the most accurate assessment of cover on site.

Figure 3. Profometer 650 AI Cover Meter

Longitudinal single line scans were made on all three blocks. Such scans record the signal strength as the scan cart is moved along the block and also detects the location of rebars automatically. The signal strength is converted into concrete cover based on the rebar size. Figure 4 shows the result from test block one which acts as the reference block.

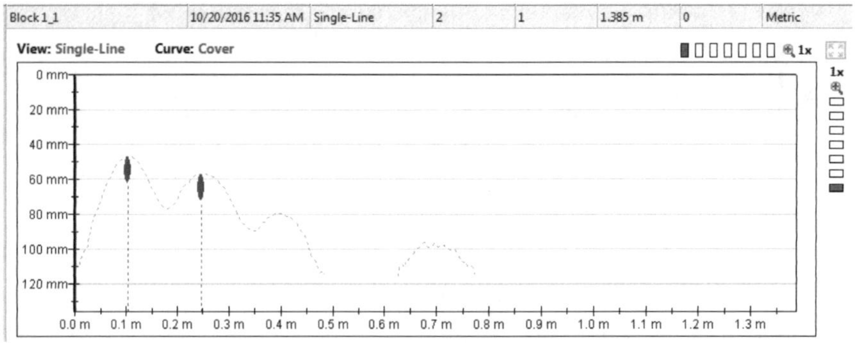

Figure 4. Block 1 Eddy Current Single-Line Scan

The first two rebars can be easily detected as they lie within the standard measuring range. After that the signal strength becomes too small for automatic detection, although from the cover single line it is easy to see the location of the 3rd rebar and the 5th rebar. The 4th object is a hollow plastic pipe which cannot be detected by magnetic eddy current instruments.

Now compare this with the scan of block 2 containing 30 kg/m^3 of steel fibres.

IOP Conf. Series: Materials Science and Engineering **246** (2017) 012016 doi:10.1088/1757-899X/246/1/012016

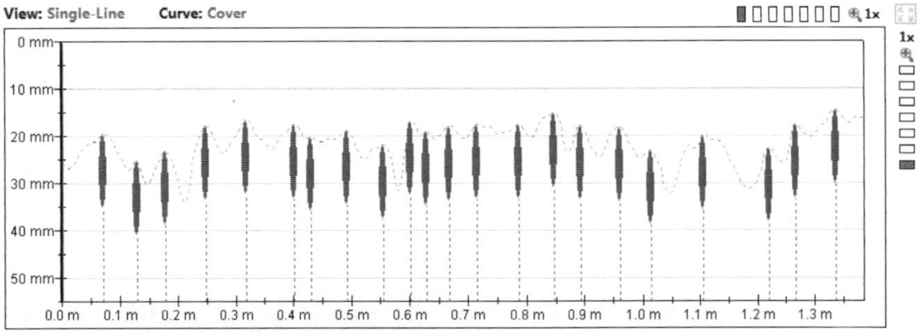

Figure 5. Block 2 Eddy Current Single Line Scan

No rebar can be clearly detected. The signal curve was always affected by large numbers of local peaks, completely masking the actual position of the rebars. The same occurred when testing on block 3 with the higher steel fibre content.

4. Ground Penetrating Radar (GPR) Testing

Two instruments were used for this test, the Hilti PS-100 and the GSSI Mini Structure Scan XT. Once again scans were carried out on the reference block and then on the blocks containing fibres. The results from each instrument can be seen in Figures 6 and 7.

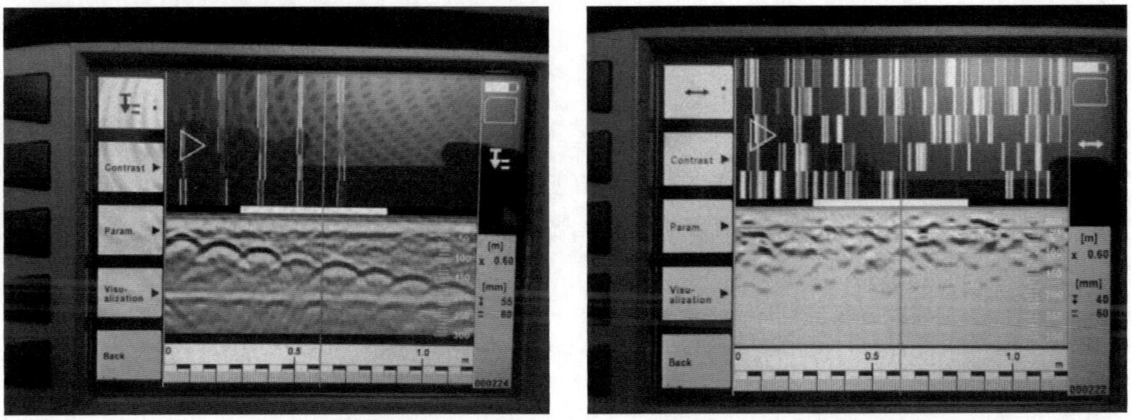

Figure 6. Hilti GPR scan of block 1 (left) and block 2 (right)

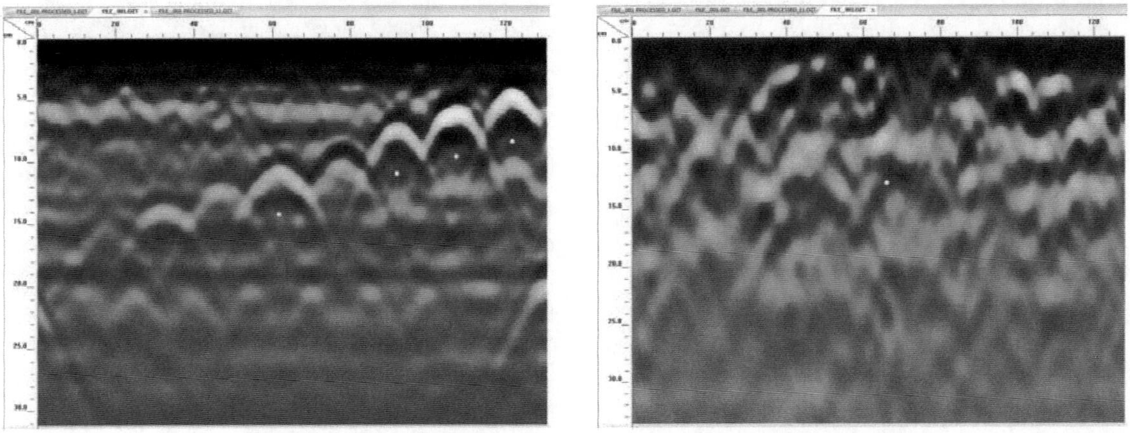

Figure 7. GSSI GPR scan of block 1 (left) and block 2 (right)

In both examples, it can be seen that the rebars and also the hollow pipe can be clearly detected on block 1. The back wall is also clearly identifiable. However, on block 2 the results are meaningless. It is impossible to detect the objects.

5. Ultrasonic Pulse Echo Testing

The instrument used for this test was the Proceq Pundit 250 Array.

Figure 8. Pundit 250 Array Ultrasonic Pulse Echo Instrument

Panorama B-scans were carried out on all three blocks. The results are shown in Figure 9.

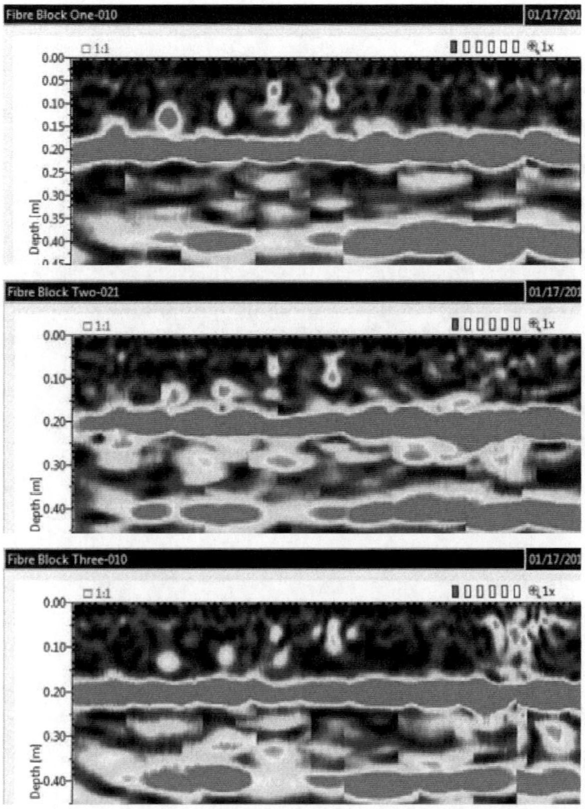

Figure 9. Ultrasonic Pulse Echo Scans – Block 1 (top), Block 2 (middle), Block 3 (bottom)

It can be clearly seen here that there is no real difference between the scan on the control block and the scans on blocks 2 and 3 with increasing steel fibre content. This means that it is quite feasible to use this technology for assessing cover in fibre reinforced concrete elements.

5.1. Determination of the Cover

When the ultrasonic pulse echo testing was carried out the test blocks had been flipped over when compared with figure 2. The actual orientation can be seen in Figure 10

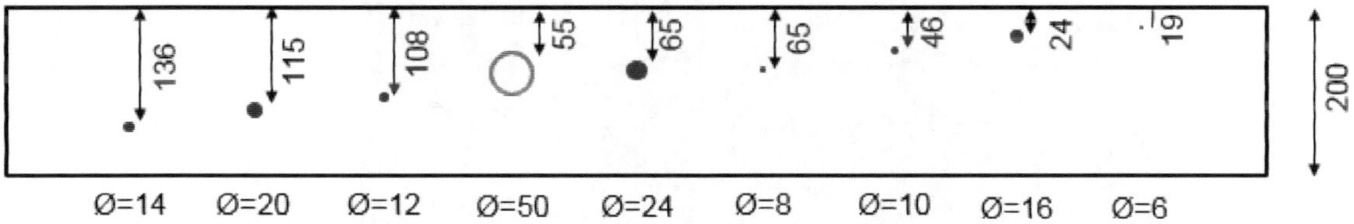

Figure 10. Test Block Orientation for Pulse Echo Tests

In order to obtain accurate depth information with pulse echo testing, it is necessary to calibrate the pulse velocity. In the Pundit 250 Array instrument this is done by detecting the multiple back wall echoes and calibrating this against the slab thickness information. This can be seen in Figure 11.

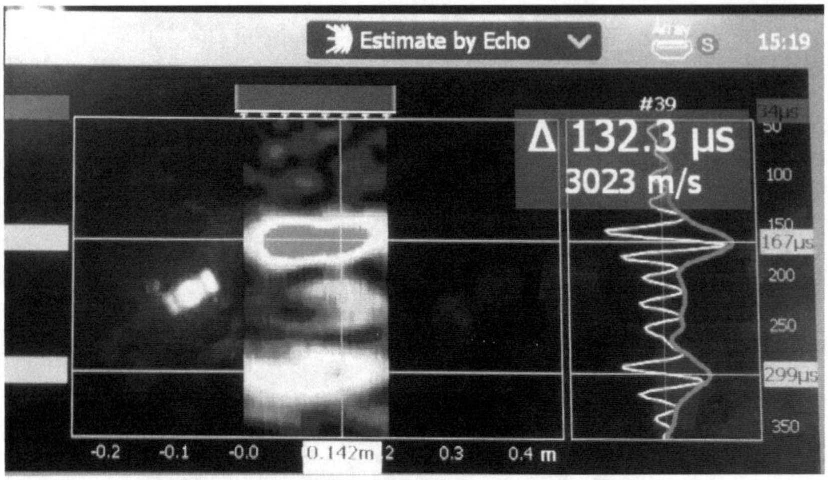

Figure 11. Calibration of the Pulse Velocity

The A-scan peaks are used to align the cursor to the first and second back wall echoes. The known thickness of the slab is then used to calculate the pulse velocity. The pulse velocity determined on each of the three slabs was as follows:

- Block one – no fibre content – 2939 m/s
- Block two – 30 kg/m^3 fibre content – 3117 m/s
- Block three – 60 kg/m^3 fibre content – 3023 m/s

It may be expected that with increasing fibre content, the pulse velocity would likewise increase due to the presence of more steel which obviously has a higher pulse velocity than concrete. It was thought that perhaps the pulse velocity determination could also be correlated to steel fibre content. However, as can be seen from the results this is not the case. There is an explanation for this in a study by Gebretsadik published in 2013 [2], who concluded that increased amount of fibre content led to the formation of additional voids in the concrete matrix which has the opposite effect of reducing the pulse velocity.

Once the pulse velocity has been determined the depth of the objects within the structure can be measured. An example is shown in figure 12.

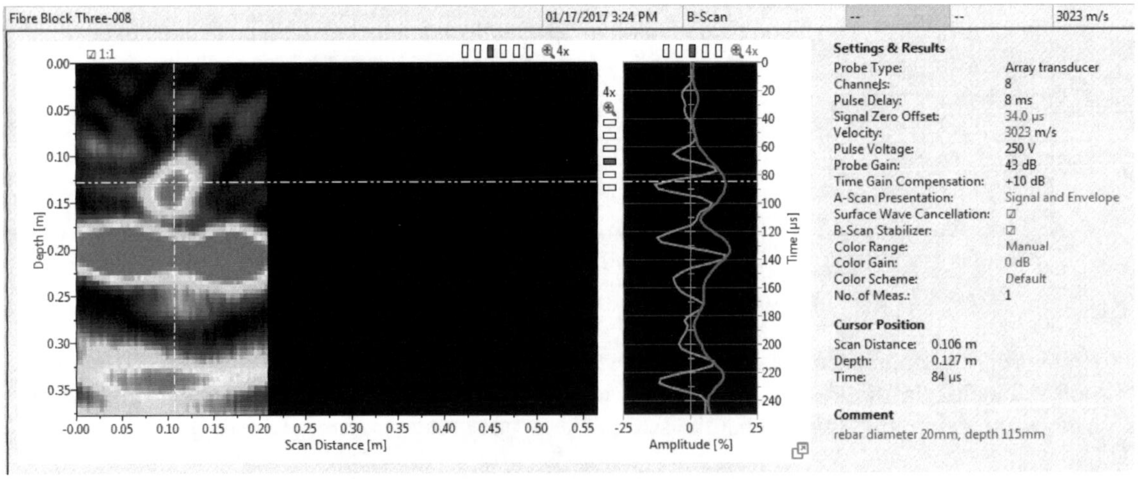

Figure 12. Rebar Depth Measurement

Here we can see that the rebar depth is estimated at 12.7 cm. The actual depth is 11.5 cm. This has to do with the measurement technique using the peak of the echo. While easier to detect than the onset of the received pulse, the effect of ringing of the signal is not always constant and this can lead to some errors. This is clearly something that needs to be refined if accurate depth information is required.

6. Conclusions

Traditional measurement techniques used for estimating cover depth use electromagnetic waves. The very nature of these technologies means that they cannot be used on concrete with steel fibre content. Ultrasonic pulse echo testing on the other hand has been shown to be an effective technique on this material. Rebars and pipes can be clearly detected and their relative cover can be determined. Accuracy of the cover measurements depends on accurate pulse velocity estimation which requires knowledge of the slab thickness. The technique to determine the onset of the received echo is in need of some improvement for really accurate cover estimation. Nevertheless, the technique looks very promising for this particular application.

References

[1] Bewehrungsüberdeckung in mit Stahlfasern bewehrtem Beton, Technischer Bericht, Projektarbeit FS 2016, Tanja Pfeiffer, 2016.

[2] Gebretsadik, Belayhun Tesfaye, "Ultrasonic Pulse Velocity Investigation of Steel Fiber Reinforced Self-Compacted Concrete" (2013). UNLV Theses, Dissertations, Professional Papers, and Capstones. 1828.

IOP Conf. Series: Materials Science and Engineering **246** (2017) 012017 doi:10.1088/1757-899X/246/1/012017

Fibre reinforced concrete in flexure and single fibre pull-out test: a correlation

M Manca[1], D Ciancio[1] and P Dight[1]
[1] The University of Western Australia, 35 Stirling Highway, Crawley 6009, WA, Australia

E-mail: mirko.manca@research.uwa.edu.au

Abstract. The aim of the present work is to assess whether a single fibre pull-out test can be related to the behaviour of multiple fibres in fibre reinforced concrete under bending condition. A simple model based on the stress block theory is described and compared with experimental results on three point bending tests with aligned fibres.

1. Introduction

Fibre reinforced concrete (FRC) is adopted in many civil applications such as tunnelling and flooring with the main benefit being improved toughness when compared to plain concrete. Such improvement has drastically reduced or eliminated the use of conventional steel reinforcement. In Australia, macro synthetic fibres are widely used in the mining sector as fibre-reinforced shotcrete (FRS). This offers a cheap and fast solution in ground control systems.

Despite the advantages in the use of fibres, high variability in the post-crack performance of FRC and FRS remains an issue which is still the object of many studies. Engineers often resort to empirical methods for design. The design of shotcrete underground linings, for example, relies on the Barton's Chart [1] and its modifications [2, 3] which relates the energy capacity of FRS based on empirical observations.

To fulfil this lack of understanding, there is a need to provide a simple tool to demystify the behaviour of FRC/FRS for design purposes that is accurate, can be replicated and can be easily implemented.

The concept of stress block is used for basic design in conventional flexure members in civil applications and offers a way to calculate the bending capacity of structural members. In order to model the tensile forces acting on cracked unreinforced concrete, a variety of stress profiles have been adopted in the literature, mostly using the fictitious crack model [4]. With the presence of fibres in the matrix, the choice of the stress profile dictates the accuracy of the model. Many researchers spread the effect of fibres throughout the entire cracked section.

A few studies have attempted to approach FRC with local forces induced by the bridging action of fibres, most notably [5] and [6]. Both researchers modelled the tensile forces imparted by the fibres with experimental results obtained from the pull-out test. To tackle the problem of fibres randomly distributed across the crack section, [5] assumed a pure random distribution of fibres (all orientations and embedment length have equal probability). In [6], the cracked section is divided in layers. For each layer an idealised most probable pull-out response was created such that it could describe the behaviour of all fibres in the beams for each embedment length

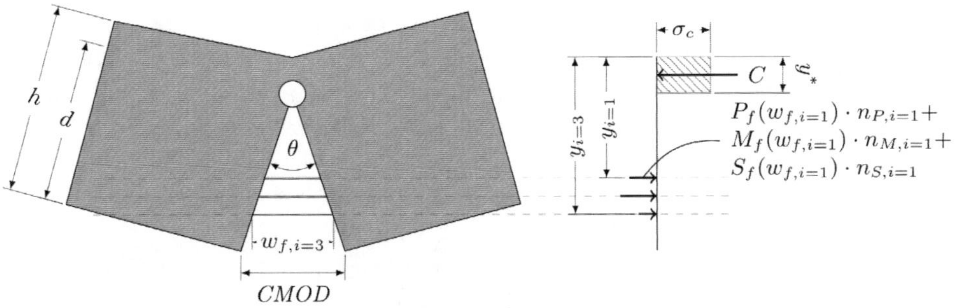

Figure 1. Schematic of the analytical model.

and orientation and is based on the most likely orientation and embedment length measured in each layer of a specific beam. The outcomes are different, from one side the first study concludes that this approach is valid and it also explains how the variability of the results of third point beams comes from the variability in the pull-out response. In [6], one of the conclusions is that the single fibre pull-out test may not be representative of the behaviour of fibres in FRC under bending conditions. In both cases, however, the randomness of fibres, and therefore high variability of the results, gives a complicated picture of the relationship between single fibre pull-out and fibre-reinforced concrete in bending.

The question that is yet to be answered and validated experimentally is whether a single fibre pull-out test can be effectively used to describe the complex behaviour of multiple fibres bridging the crack in flexure.

In light of recent developments based on beams with non-random aligned fibres [7], this paper aims at establishing a link between the single fibre pull-out test and bending capacity of FRC/FRS beams using the stress block theory. If the relationship is found on a simple case with aligned fibres, it can open the way to a more general formulation to model randomly dispersed fibres in FRC and FRS.

2. Model

In order to focus on the relationship between the single fibre pull-out test and the three point bending test, a semi-analytical model has been implemented. The model should be able to output the post-peak response of beams with aligned fibres based solely on the results obtained from the single fibre pull-out test as well as information on the location and number of fibres in the beam. Since the behaviour of a fibre pulling out of concrete strongly depends on how the fibre fails during the process, the mode of failure is an extra input of the model. For this particular combination of fibres and mortar used in this research three distinct modes of failure called *Pull-out mode*, *Mixed mode* and *Snap mode* have been identified. Further details can be found in section 4.

This semi-analytical model is based under the following assumptions:

- The problem can be simplified as a 2D problem (this assumption holds if the fibres are aligned).
- The cracked surface is plane and propagates at mid-span in the beam loaded in a three point bending configuration.
- The pin of the crack is located at a certain known height. It has been observed experimentally in [7] that cracks propagate some 90 % of the beam's height regardless of the number of fibres with only small variations during the test. The pin of the crack is where

the crack width is zero, the crack opens linearly to the bottom of the cross section where the crack mouth opening displacement is defined as *CMOD* (as shown in figure 1).

- The post-peak response is governed by the fibres, the compressive force at the top of the cross section equates the tensile forces coming from the fibres for equilibrium. Furthermore, the process zone (the area of cracked concrete that still offers some bridging action) is considered negligible whereas the compressive stresses are assumed constant.
- There is no interaction between fibres.
- The bridging action of each fibre can be obtained from a single fibre pull- out test and it is the same for all fibres with the same mode of failure.

The output of the model is the load P required to deflect the beam down to a certain amount of deflection δ that plays the role of independent variable. Other inputs are:

(i) Number of layers of fibres n_l in the cross section (see figure 1).

(ii) Number of fibres categorised by their mode of failure (*Pull-out mode*, *Mixed mode* and *Snap mode*) $n_{P,i}$, $n_{M,i}$ and $n_{S,i}$ for each layer i.

(iii) The functions $P_f(w_{f,i})$, $M_f(w_{f,i})$ and $S_f(w_{f,i})$ which relate the magnitude of the bridging force of each fibre to the crack width (w_f) and are obtained experimentally from the single fibre pull-out test. Each function describes a certain mode of failure. The crack width is calculated at each layer i.

(iv) $\delta - CMOD$ relationship. For simply supported beams, the following relationship can be used:

$$CMOD = \frac{4h}{L}\delta \tag{1}$$

where h is the height of the beam and L is the distance between the supports. The relationship has been experimentally validated in [7] and is valid for all beams regardless of the number of fibres.

With reference to figure 1, the following expressions hold:

$$\tan\theta/2 = \left(\frac{CMOD/2}{d}\right) \qquad CMOD \text{ found using Equation 1} \tag{2}$$

$$w_{f,i} = 2 \cdot (y_i - y^*) \cdot \tan\theta/2 \tag{3}$$

$$C = \sum_{i=1}^{n_l} [P_f(w_{f,i}) \cdot n_{P,i} + M_f(w_{f,i}) \cdot n_{M,i} + S_f(w_{f,i}) \cdot n_{S,i}] \qquad \text{Equilibrium} \tag{4}$$

$$M = \sum_{i=1}^{n_l} [P_f(w_{f,i}) \cdot n_{P,i} + M_f(w_{f,i}) \cdot n_{M,i} + S_f(w_{f,i}) \cdot n_{S,i}] \cdot (y_i - 0.5y^*) \qquad \text{Equilibrium about } C \tag{5}$$

$$P = \frac{4M}{L} \tag{6}$$

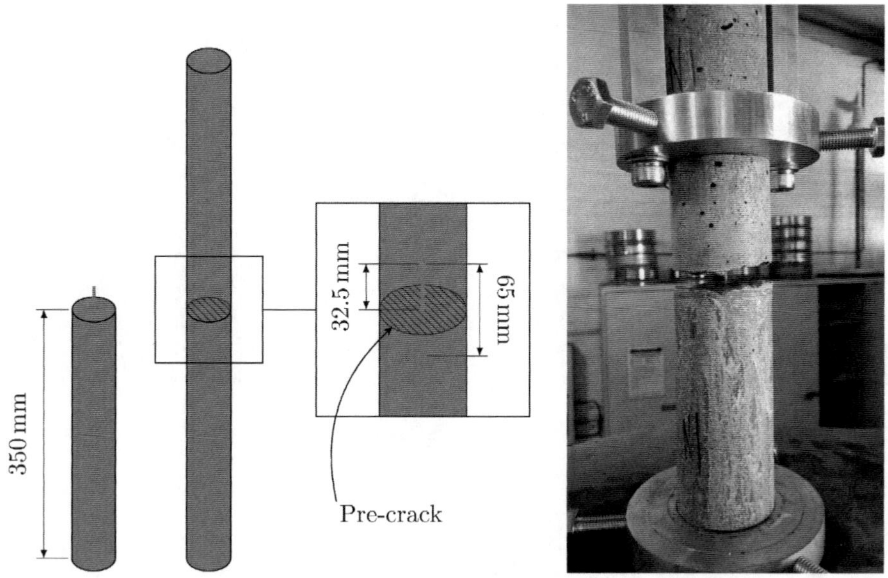

Figure 2. Schematic of the pull-out test specimen. On the left the lower part of the cylinder at the end of stage-one. Two days later the second half is cast such that the fibre results in the middle of the cylinder. On the right the set-up the the test.

where C is the resultant of the compressive stresses acting in the uncracked section of the beam which span from the top until y^* which is assumed, for simplicity, constant at 90 % of the height of the beam (i.e. $y^* = 0.1h$). The internal moment M is obtained from the equilibrium about C. Finally, the external load P (which is the main output of the model) is obtained equating the internal moment M with the external moment due to a load centrally located in a simply supported beam.

The model is admittedly very simple. Some assumptions (such as the position of the pin of the crack being constant) could be removed to improve the validity of the model. The aim of this preliminary study, however, is to investigate a possible relationship between the single fibre pull-out test and three point bending test with an approach as simple as possible. Further developments can take into account a more realistic model for improved accuracy.

3. Testing Procedure

3.1. Pull-out test

To complete the model described in section 2, one of the inputs is the load-displacement curve obtained from a fibre in a single pull-out test. For this purpose, a preliminary campaign with 10 specimens has been undertaken to quantify the variability of the results and to account for all possible modes of failures. The mortar and the fibres used in the pull-out tests are the same as reported in Reference [7].

Each specimen consist of a cylinder with a diameter of 50 mm and 700 mm in length made of mortar with a single fibre located in the middle of the cylinder. Each cylinder was prepared in two stages. In the first stage, the lower half of the cylinder (up to 350 mm) was cast and a fibre was placed according to the fibre configuration desired (figure 2). The second stage takes place two days later; the second half of the cylinder is cast to complete the specimen resulting in a single fibre fully embedded in mortar. Being the length the fibre 65 mm, 32.5 mm are embedded in the lower cylinder and the other half in the upper cylinder. This two-stage procedure ensures that the bond between the two halves of the cylinder is near zero in order to measure solely

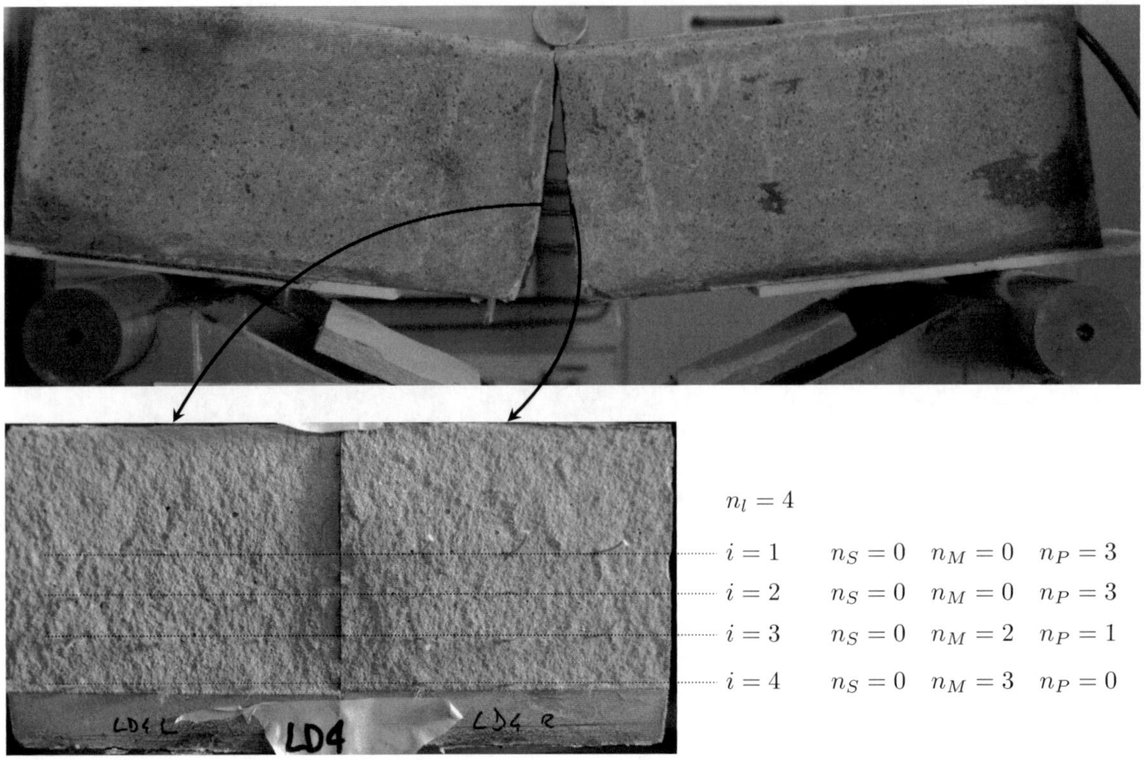

Figure 3. Cracked sections resulting from a three point bending test. The sections are analysed to visually identify the mode of failure for each fibre.

the bridging action of the fibre rather than the tensile capacity of the mortar. During the test, the lower half of the cylinder is clamped while the upper cylinder is connected to the mobile part of the testing apparatus. The test is set in displacement control at a speed of 1 mm/min to simulate a quasi-static crack opening conditions.

3.2. Three point bending test

As part of a wider study investigating the effects of the location and orientation of macro-synthetic fibres on the flexural response of FRC and FRS, a methodology has been developed to align and control the position of the fibres. A series of beams 600x150x150 mm have been tested in three point bending test with different number of fibres and different positions. In all beams, the fibres are aligned to the direction of the major stress and parallel to each other. The beams have been pre-notched at mid-span such that the crack propagates straight from the tip of the notch to the top of the cross section where the beam is loaded. The fibres have been centred in order to have an embedment length equal to half the length of the fibres (i.e. 32.5 mm). Further details and results on this experimental campaign can be found in [7]. After testing the beams, the cracked surface has been analysed to count the number of fibres and to assess their mode of failure. An example of this visual assessment is depicted in figure 3.

4. Results

The single fibre pull-out results showed, in general, low variability provided that the mode of failure and the embedment length are the same. The response of fibre does strongly depend on the way the fibre either breaks or pulls-out. At least three different modes of failure have been

IOP Conf. Series: Materials Science and Engineering **246** (2017) 012017 doi:10.1088/1757-899X/246/1/012017

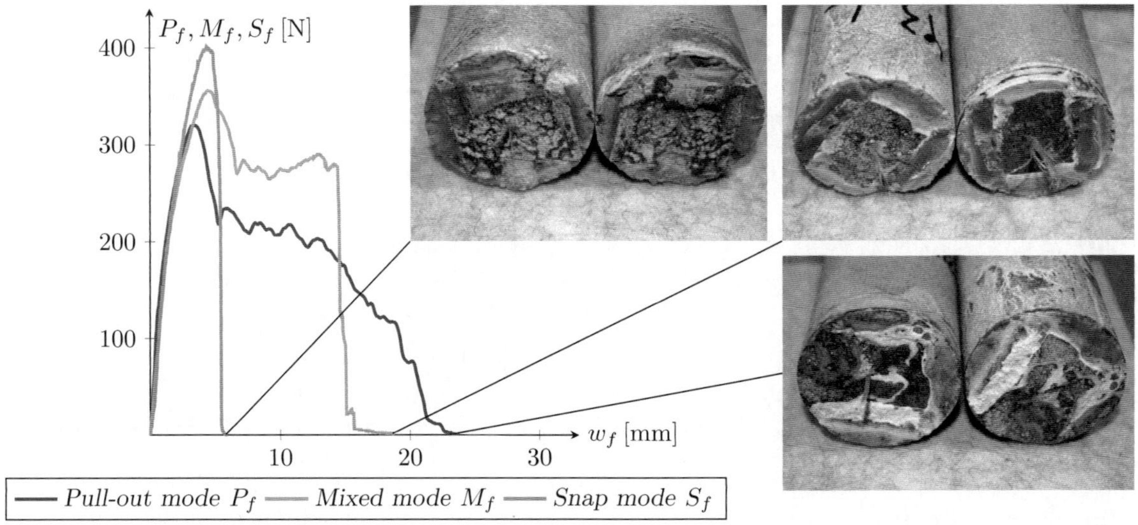

Figure 4. Pull-out test results for different modes of failure. The pictures show how the fibre looks after the test for each mode of failure.

identified:

(i) The fibre breaks in a brittle fashion (*Snap mode*)

(ii) The fibre remains intact and pulls-out or slips from one side (*Pull-out mode*)

(iii) The fibre breaks but only after it has partially slipped from one side; this mode of failure will be hereafter referred to as *Mixed mode*.

Example of these modes of failure are illustrated in figure 4.

After testing the beams in a three point bending configuration, both crack surfaces have been visually inspected to assess the mode of failure of the fibres as depicted in figure 3. The magnitude of the bridging force of each fibre described by the functions $P_f(w_f)$, $M_f(w_f)$ and $S_f(w_f)$ is taken directly from the pull-out results of figure 4 according to the mode of failure. The results of the model proposed in section 2 for various beams is plotted in figure 5 and 6.

5. Summary and Conclusions

There is, in general, good agreement between the model that links the pull-out behaviour and the experimental results obtained from the three bending point test. The model can reasonably describe all the beam series included in this study.

A possible source of error can result from the assumption that the crack surface is a plane which in turn affects the assumption that the pull-out length is exactly half the length of the fibre. The mortar was designed to minimise this effect and the beams have been pre-notched to promote the formation of the crack to be aligned with the point load. Nonetheless, in some specimens the pulled-out lengths of some fibres deviated slightly (within few millimetres) from the ideal figure of 32.5 mm. The accuracy of the model is also affected by the difficulty to distinguish fibres that snapped from those that failed in a mixed mode and the pull-out curve attributed to those fibres could be inappropriate.

The model, given its simplicity, works well in an ideal configuration where fibres are all aligned but cannot be applied yet to conventional FRC/FRS where the fibres are uncontrollably dispersed within the volume of the material. However, as a preliminary study on the behaviour of fibres in concrete and to answer the question of whether a single fibre pull-out test can be used

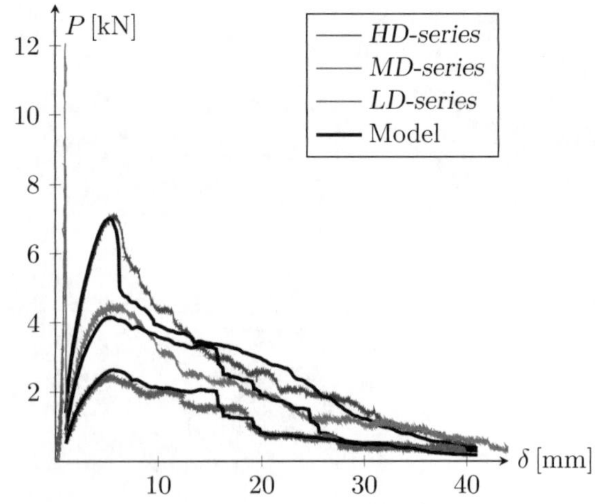

Figure 5. Comparison between the semi-analytical model and three point bending beams with different amount of fibres. *HD*, *MD* and *LD-series* stand for *High Density*, *Medium Density* and *Low Density* and are beams that contain 32, 20 and 12 fibres respectively.

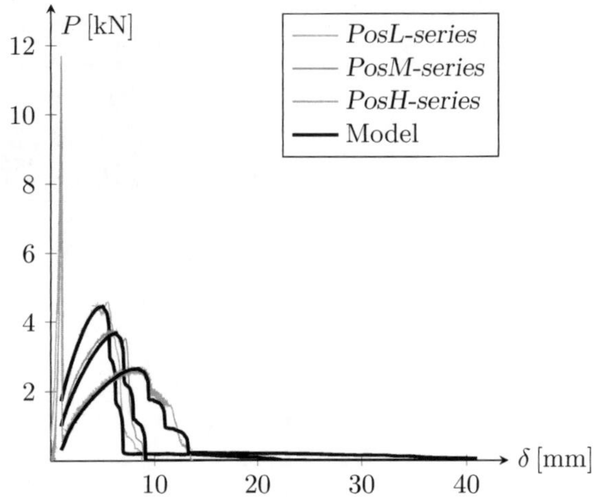

Figure 6. Comparison between the semi-analytical model and three point bending beams with fibres located at different heights. *PosL*, *PosM* and *PosH-series* stand for *Position Low*, *Position Medium* and *Position High* and refer to beams with the same amount of fibres (14) but all concentrated at the bottom of the crack (*PosL*), closer to the top (*PosH*) or in between (*PosM*).

to describe the behaviour of FRC/FRS, from the results in figure 5 and 6, it seems that pull-out test under direct tension represents the behaviour of multiple fibres embedded in mortar under bending conditions providing that the mode of failure of each fibre is assessed carefully.

Acknowledgement
The authors would like to thank Sika Australia Pty Ltd for the supply of synthetic fibres used in this experimental campaign. The first author acknowledges the financial contribution from Sika Australia Pty Ltd towards his fellowship.

References

[1] Grimstad E and Barton N 1993 Updating the Q system for NMT Report Norwegian Geotechnical Institute

[2] Grimstad E, Kankes K, Bhasin R, Magnussen A and Kaynia A 2002 Rock mass quality Q used in designing reinforced ribs of sprayed concrete and energy absorption Report Norwegian Geotechnical Institution

[3] Papworth F 2002 *Shotcrete Magazine* **1** 16–21

[4] Hillerborg A, Modeer M and Petersson P 1976 *Cement and Concrete Research* **6** 773–82

[5] Banthia H and Armelin N 1997 *Aci Materials Journal* **49**

[6] Jones P, Austin S and Robins P 2008 *Materials and Structures* **41** 449–63

[7] Manca M, Ciancio D and Dight P Effects of fibre location and density on the flexural toughness in FRC beams *Mechanics of Structures and Materials XXIV: Proceedings of the*

24th Australian Conference on the Mechanics of Structures and Materials vol 1 ed Hao H and Zhang C (Perth) pp 249–53

IOP Conf. Series: Materials Science and Engineering **246** (2017) 012018 doi:10.1088/1757-899X/246/1/012018

Micromechanical performance of interfacial transition zone in fiber-reinforced cement matrix

V Zacharda[1], J Němeček[1] and P Štemberk[1]

[1] Czech Technical University, Thákurova 7, 166 29 Prague 6, Czech Republic

E-mail: vojtech.zacharda@fsv.cvut.cz

Abstract. The paper investigates microstructure, chemical composition and micromechanical behavior of an interfacial transition zone (ITZ) in steel fiber reinforced cement matrix. For this goal, a combination of scanning electron microscopy (SEM), nanoindentation and elastic homogenization theory are used. The investigated sample of cement paste with dispersed reinforcement consists of cement CEM I 42,5R and a steel fiber TriTreg 50 mm. The microscopy revealed smaller portion of clinkers and larger porosity in the ITZ. Nanoindentation delivered decreased elastic modulus in comparison with cement bulk (67%) and the width of ITZ (~ 40 μm). The measured properties served as input parameters for a simple two-scale model for elastic properties of the composite. Although, no major influence of ITZ properties on the composite elastic behavior was found, the findings about the ITZ reduced properties and its size can serve as input to other microstructural fracture based models.

1. Introduction and motivation

Nowadays, the fiber reinforced concrete (FC) is a popular building material which is increasingly utilized for structural elements. Fiber concrete is characterized with randomly dispersed fibers that may have different length, thickness, composition and shape. The fibers are made from a wide range of materials like glass, steel, carbon or different kinds of synthetic materials. The most frequent material used for structural purposes is steel due to its beneficial role in enhancing composite tensile strength properties, permeability or impact resistance. For a good mechanical performance of FC the distribution and orientation of fibers in the composite volume plays a critical role.

The macroscopic FC composite behavior is also directly influenced by micromechanics of a fiber and interfacial zone that surrounds it. In the last decade, a lot of research on micromechanical characteristics of cementitious composite material has been conducted. With the development of advanced experimental techniques, like e.g. nanoindentors or scanning electron microscopes, it was possible to determine a variety of microscale material properties for cement paste [1,2,14], concrete [3], UHPC [4], mortars [5] and other materials. Microlevel elastic and viscoelastic properties, chemical composition or the microstructure of interfacial transition zone (ITZ) [3,5,6,7] have been revealed. So far, little research has been carried out for micromechanical properties of ITZ which is known for different morphology and chemical composition compared to the bulk cementitious matrix. Therefore investigations carried out in this paper are focused on micromechanics of a fiber and its influence on the overall composite behavior from the mechanical point of view.

Concrete is generally taken as a multiscale material consisting of several phases. At least two phases – aggregate and cement paste must be considered. But more correctly concrete is composed of at least three phases – aggregate, cement paste and interfacial transition zone that forms a thin layer surrounding the aggregate with different chemical and mechanical properties. If fibers are used in the concrete mixture they act as a separate phase surrounded by an additional ITZ with different mechanical properties [8,9,10,11] and microstructure [5,6,7] compared to the bulk matrix. In general, ITZ is located around each inclusion (sand, aggregate, steel fiber, reinforcement) in the concrete and the varying properties are caused by different reasons like locally different water/cement ratio, different confinement or compaction, different boundary conditions, etc.

Local differences of morphology and chemical composition can be sensed by microscopy whereas local mechanical differences can be well measured using nanoindentation. Nanoindentation has been already successfully used for characterization of a bulk cementitious matrix, The ITZ around aggregate [12] or reinforcement [8,9] or steel fiber in reinforcement in mortar [10], or various cementitious mixtures [12]. Again, the research conducted in this paper concentrates on ITZ around fibers in a normal FC (Fig. 1) and aims to quantify its local and overall influence on mechanical performance of the composite.

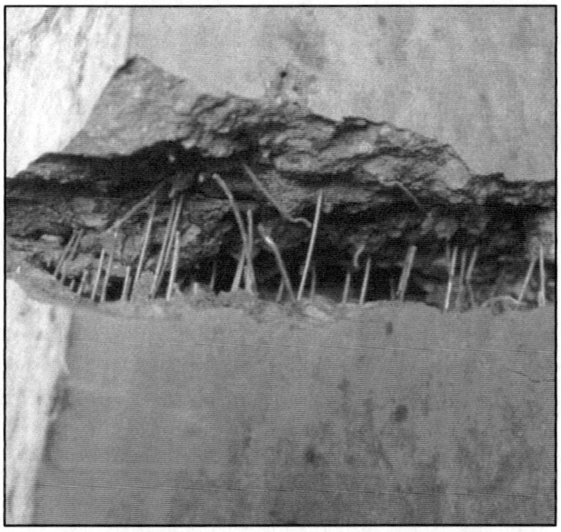

Figure 1. Tension fracture of a beam with pulled out fibers

2. Fiber concrete and the role of ITZ

2.1. Macroscopic mechanical role of fibers

The concrete with dispersed fiber reinforcement delivers the composite improved material characteristics and properties like higher ductility, toughness, tensile strength and impact resistance. Fibers can reduce shrinkage of material creep and reduce localized cracking by local bridging mechanism. Since the direction of a mechanical load may vary it is critically important to distribute homogeneously the fibers with fibers randomly oriented within the mixture or a cast element. Unfortunately, an addition of fibers increases mixture viscosity in a fresh state and makes mixing more difficult and energy expensive. Thus, lower volume ratios in the order of a few percent can be effectively used in reality.

The mechanical role of a fiber lies in bridging the cracks and transferring locally high tensile stresses that would otherwise lead to concrete cracking, to other composite locations. This leads to more spread or distributive fracture that is macroscopically manifested as increased ductility and the composite exhibits larger number of smaller cracks. Depending on a fiber-matrix bond, the deformation mechanism of a fiber involves either pulling out of fiber from the matrix or (less frequently) fiber rupture.

2.2. Microstructural effects of a fiber

As stated earlier, fiber is surrounded by ITZ zone which is chemically and mechanically different for the bulk. The scale of ITZ is in the order of tens of micrometres which is also a scale of hydration products within the cement matrix. There are five main hydration products in the hardened matrix, namely the calcium-silica or calcium-alumina hydrates (C-S-H and C-A-H gel), Portlandite (CH), unhydrated clinkers and pores.

Microstructure of ITZ around fibers is characterized with less portion of clinkers and more C-S-H gel and CH than in the bulk matrix. Microscopic studies show that ITZ zone contains smaller clinker or CH grains, has a higher local water/cement ratio, and thus a higher porosity [6,7]. It causes a weaker bond of the matrix to the fiber surface. The influence of the rigid obstacle formed by a fiber in the matrix is sometimes called the wall effect [7]. From the micromechanical point of view, the main difference between ITZ and the cement matrix bulk is still in the porosity which reduces the modulus of elasticity of ITZ by tens of percent. The thickness of ITZ is irregular and is related to the width of the fiber [12].

Micromechanical studies show that the average modulus of elasticity measured by nanoindentation for clinker is about ~ 120 GPa, low density C-S-H gel ~ 20 GPa, high density C-S-H gel ~ 31 GPa and CH ~ 40 GPa [13;14]. In the paper by Zhu and Bartos [9], different ITZ properties are found also for the bottom and top sides of the horizontal steel bar in cement. For instance, at the distance of 10 – 30 μm below the steel bar, the interfacial *E* modulus and hardness bottom side the steel bar was only 56% and 70% of the bulk, respectively [8]. The paper Hashin – Monteiro [11] says that the modulus of elasticity can be reduced in the ITZ to about 50% of its normal value so the ITZ is the weak link in concrete.

3. Material and microstructure

3.1. Tested material and sample preparation

The micromechanical performance of a fiber in cementitious matrix was studied on a sample of cement paste with a single embedded steel fiber. Cement paste was made from cement CEM I 42,5R with water to cement ratio equal to 0,4. As a reinforcement a straight steel fiber with Z-anchors (brand name TriTreg, Trinec, CZ) with length of 50 mm and thickness of 1,05 mm was used. The sample had a cylindrical shape with height of 65 mm and diameter of 30 mm. It was unmoulded three days after casting and stored in water for 28 days. After hardening, the samples were cut into slices (disks), the surface grinded and polished by a metallographic procedure [1] and prepared for testing by nanoindentation.

3.2. Characteristics of a cement matrix

Scanning electron microscopy and optical microscopy were employed for microstructural and chemical analysis of the matrix. Nanoindentation has been employed for characterization of individual hydration components of the matrix. The work has been conducted by the authors elsewhere [1] and only major conclusions are summarized here. Based on SEM imaging (12 images covering matrix area of 300 x 300 μm) and image analysis volume fractions have been assessed, Tab. 1.

Table 1. Volume ratio of single phases in cement CEM I 42,5R from SEM

Phase	C-S-H gel	CH	C-A-H	Clinker	Pores and cracks
Volume ratio [%]	67,69	17,46	5,54	9,01	0,29

As a complementary measurement, optical microscopy was performed using a digital camera (DinoCapture). The optical photo illustrates distribution of the observable phases in real colours (Fig. 3). The black and dark coloured spots in Fig. 3 belong to clinkers whereas lighter colours stand for main hydration products (C-S-H and CH). Capillary and smaller porosity cannot be revealed in the resolution of the optical picture.

Figure 2. SEM photo of cement matrix

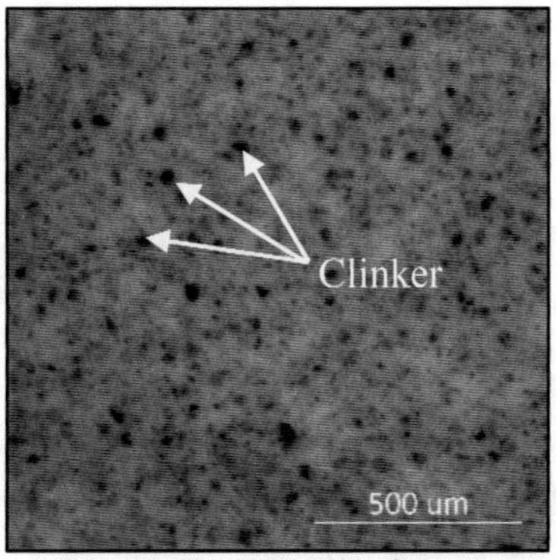

Figure 3. Optical microscopy

More precise porosity measurement was performed by mercury intrusion porosimetry (MIP) which gives overall porosity as well as pore distribution. A dried sample crushed from a larger piece with millimetre dimensions has been used for the analysis. The MIP technique gives pore sizes in about 0.001-100 μm. Figure 4 shows the distribution curve of pores in the tested sample.

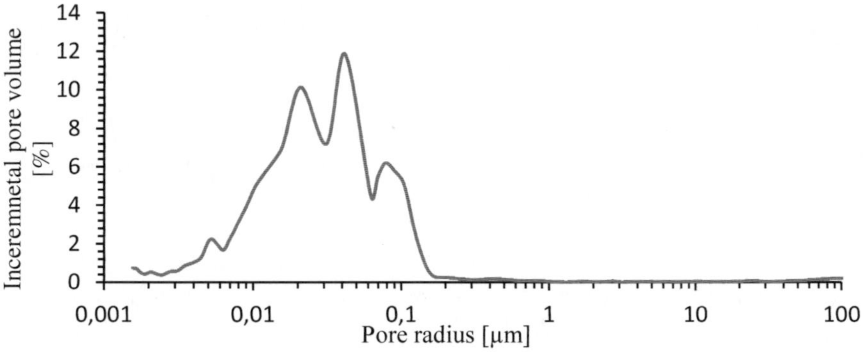

Figure 4. Distribution of pores

3.3. Microscopy of ITZ around fibers

Scanning electron microscopy was applied to show different distribution of phases in the microstructure. Six SEM photos were taken at random positions around the fiber. The scans cannot show all phases in the same contrast. The steel fiber with the highest atomic ratio is indicated by white colour whereas other phases are in dark grey colours. Unhydrous clinker particles can be detected as the lightest grey particles in the hydrated matrix (Fig. 5). The ITZ zone is characterized with less portion of the clinkers as a consequence of a locally higher water to cement ratio and more pronounced hydration of these phases. In contrast, higher w/c ratio causes increased porosity in the ITZ which can also be revealed from the images. Based on the different morphology, porosity and phase distribution an approximate width of ITZ is depicted in Fig. 6.

The chemical composition of ITZ was investigated by energy dispersive spectrometer. Output from EDS line scans made in perpendicular directions to the fiber are depicted in Fig. 7. Relative concentrations of the five main elements are shown in Fig. 7 as their relative weight ratios. Approximately equal elemental composition (with local fluctuations caused mainly by clinker inclusions) has been found which does not indicate any preferential or altered hydration in ITZ compared to bulk.

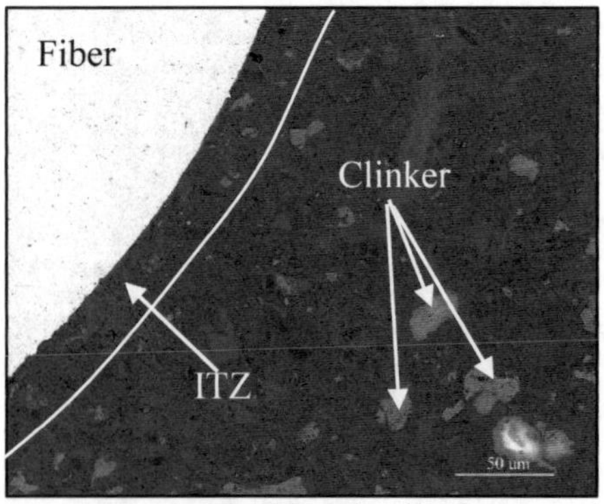

Figure 5. SEM detail of ITZ

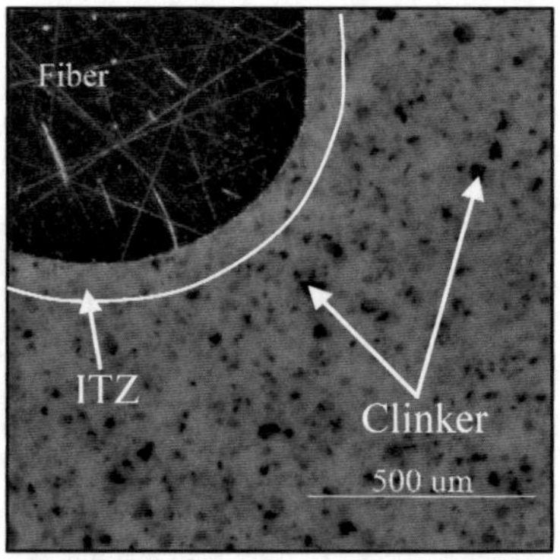

Figure 6. Light microscopy image of ITZ

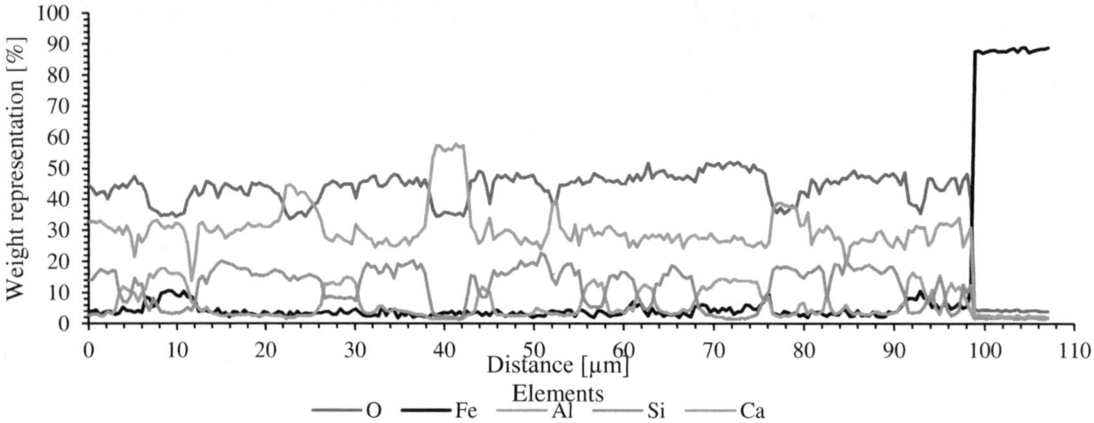

Figure 7. Relative weight ratios by 5 main elements on an EDS linescan

4. Micromechanical performance of ITZ around fiber

4.1. Nanoindentation parameters

As previously mentioned, the tested sample was prepared by saw cutting, grinding and polishing procedure prior to testing. Since roughness has direct influence to the quality of the measurement, it was checked before the experiment and kept within acceptable level which is in tens of nm.

Two positions on the sample were tested. First, a series of indents in the bulk matrix was conducted. Second, a series of indents in ITZ around the fiber was performed.

Nanohardness tester Hysitron TriboLab TI-700 equipped with Berkovich diamond tip was used. Load controlled test lasting for 26 seconds was prescribed to the maximum force of 2000 µN for each indent. The load function had a trapezoidal shape (Fig. 8). The first part of the function was a linear loading with speed of 40 mN/min lasting for 3 seconds. The second part was a constant loading for 20 seconds and the last part of the function was a linear unloading for 3 seconds.

A typical response (load-depth record) for various phases in the material is shown in Fig. 9.

The material properties were evaluated for each indent by the Oliver and Pharr [15] method. Subsequently, discrimination between the phases was done by statistical deconvolution in the bulk measurements [1,14].

IOP Conf. Series: Materials Science and Engineering **246** (2017) 012018 doi:10.1088/1757-899X/246/1/012018

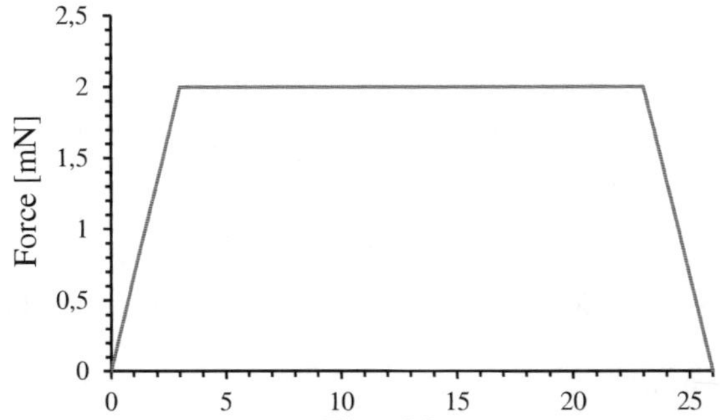

Figure 8. Load of function of indents

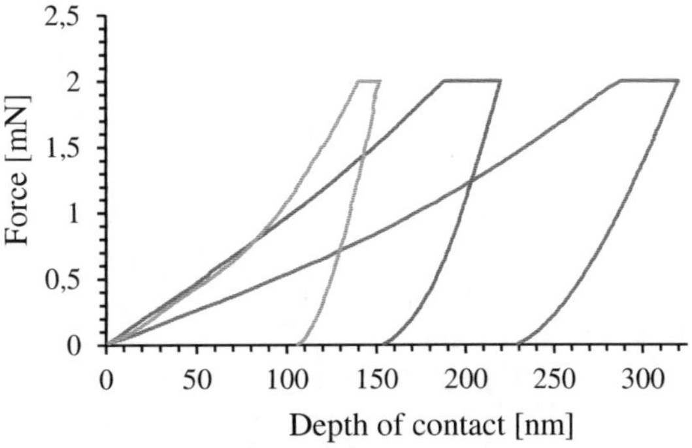

Figure 9. Indentation curves

4.2. Micromechanical response of the bulk cement paste

The results of the bulk cement paste were taken from authors' previous work [1] and is summarized here for readers convenience. The sample surface was covered with series of matrices with size of 10 x 10 = 100 indents with was 12,5 μm separation in between the indents. Overall, 850 indents were done. Elastic moduli were calculated by the theory of Oliver and Pharr [15]. Distributions of elastic moduli for individual micromaterial phases were determined using deconvolution method [1,14]. The resulting histogram of phase moduli is depicted in Fig. 10 an summarized in Tab. 2. The phase #1 corresponds to C-S-H gel, CAH and part of CH, #2 to mainly CH and small particle particle leftover of clinkers and #3 to large clinkers [1]. Note, that the number and volume fractions of the mechanical phases (Tab. 2) do not match exactly the chemically different phases found in the image analysis (Tab.1) although there is a certain overlap.

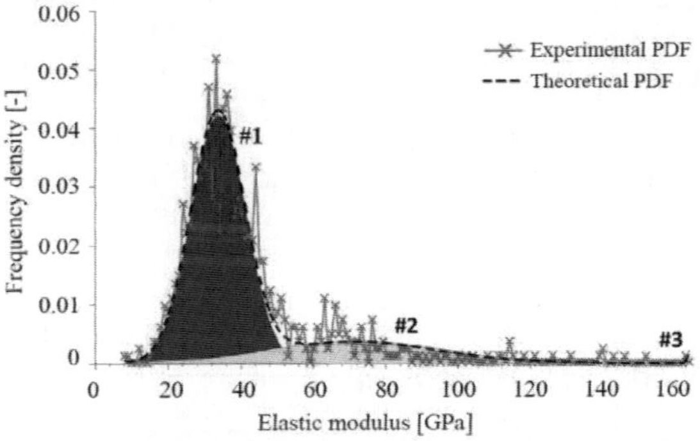

Figure 10. Histogram of theoretical and experimental probability density of elastic modulus in cement paste

Table 2. Values of modulus of elasticity for individual micromechanical phases in cement paste

	#1	#2	#3
Average value E [GPa]	33,58	70,51	164,59
Standard daviation [GPa]	7,3	22,83	0,55
Volume ratio [%]	77,72	22,03	0,25

4.3. Nanoindentation in ITZ

A series of indents was done in the ITZ around steel fiber in cement paste. Several rows leading in perpendicular direction to the fiber were performed. The separation of indents was lowered to 2 to 4 μm.

The Hysitron-700 system allows to make in-situ scan prior the indentation. The surface scan (similar to AFM scan) is shown in Fig. 11. The scan allows also precision positioning of the indents in the wanted direction. A row of prescribed 17 indents and resulting imprints scanned afterwards is depicted in Fig. 11.

The first two indents were prescribed on the surface of the steel fiber and other indents were positioned in ITZ and later in the bulk material coming from the fiber to the cement matrix. Some defective indentations caused by local porosity or roughness were eliminated from considerations. Overall, 27 rows in 10 regions around the steel fiber was done. In total, 336 indents were used for evaluation. The same methodology used for the bulk was also used here, i.e. the Oliver and Pharr [15] was used for obtaining modulus of elasticity for a single indent and profiles of the property were constructed for each row.

An example of individual row measurements is shown in Fig. 11 and 12 where microstructural position of rows and corresponding elastic moduli are shown.

Figure 13 shows average modulus of elasticity calculated from merged results from all respective positions with standard deviations. The graph shows a general trend of decreasing modulus as the position is closer to the fiber. A linear trend line was calculated and indicated in Fig. 13 which also shows an average bulk elastic modulus obtained from homogenization (calculated later in the paper) to which the ITZ modulus approaches.

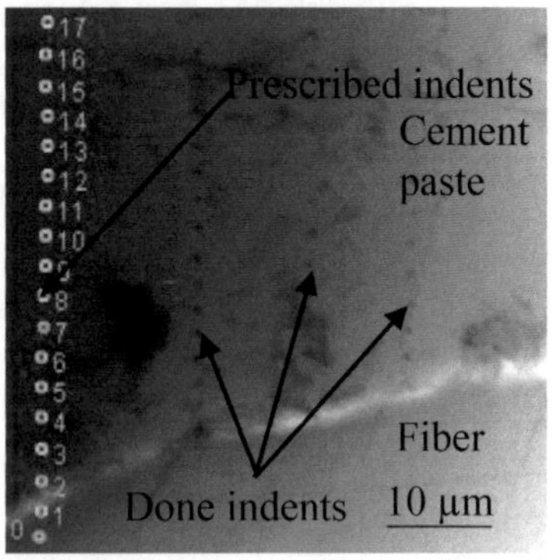

Figure 11. Scan of investigated region with marked and indented rows

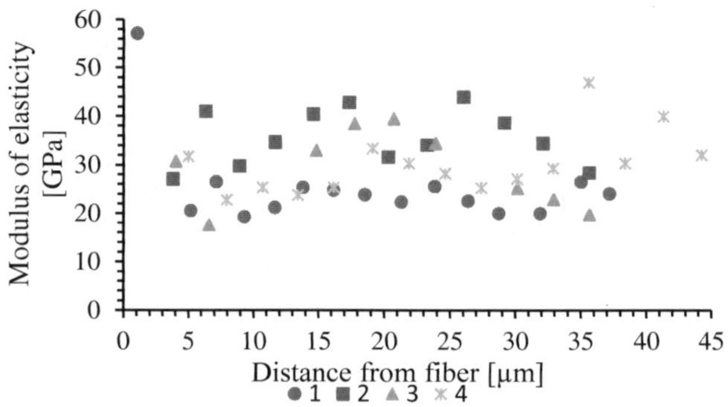

Figure 12. Profile of elastic moduli at individual positions (each colour indicates a group of indents belonging to a row)

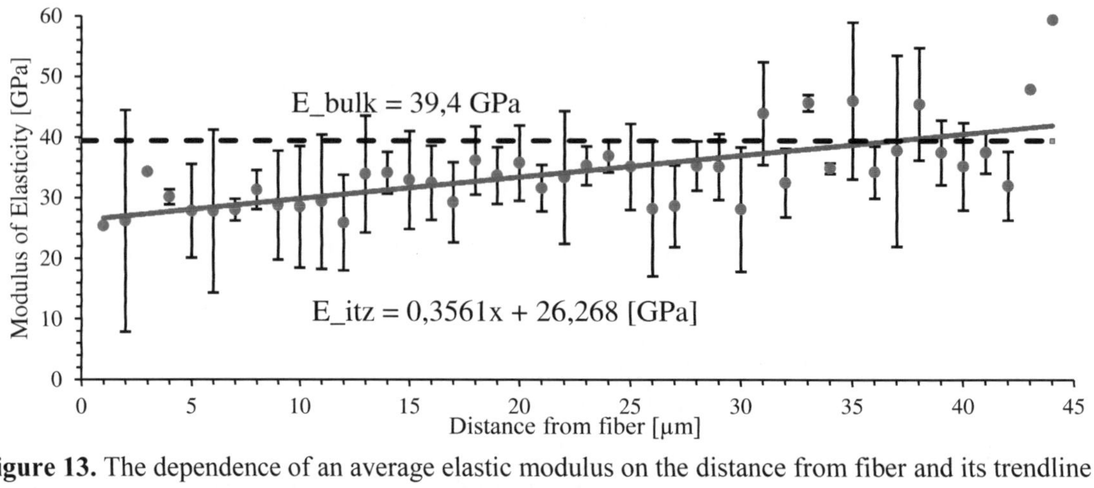

Figure 13. The dependence of an average elastic modulus on the distance from fiber and its trendline

5. A two-scale model for elastic modulus of fiber reinforced composite with ITZ

A simple two-scale model was created for demonstration of the influence ITZ on the overall elastic behaviour of the steel fiber reinforced cement paste. The first (lower) level of the model is composed of cement paste that further consists of four micromechanical phases obtained from nanoindentation

(Tab. 2). The second (upper) level is assumed to be composed of homogenized cement paste as matrix phase.

The study provided in this paper is focused on the mechanical influence considering the ITZ presence. Thus, simplified assumptions on the other parameters are therefore taken such as perfect bond of the phases and spherical shape of the inclusions. Other parameters like elastic moduli, volume fractions are taken from direct measurements (nanoindentation).

5.1. *Mori Tanaka elastic homogenization*

The Mori-Tanaka homogenization scheme assumes inclusions to be dispersed in a homogeneous matrix. The effective stiffness tensor L^{eff} is obtained as

$$L^{eff} = L^{(0)} + \sum_{r=1}^{n} c^{(r)} (L^{(r)} - L^{(0)}) A^{(r)} (c^{(0)} I + \sum_{r=1}^{n} c^{(r)} A^{(r)})^{-1} \tag{1}$$

where $L^{(0)}$ is the matrix stiffness tensor, $L^{(r)}$ is individual phase stiffness tensor, $c^{(0)}$ is the volume fraction of the matrix, $c^{(r)}$ is the volume fraction of individual phase, I is the fourth order unity tensor and $A^{(r)}$ is the concentration factor determined by Eshelby's inclusion problems solution as

$$A^{(r)} = \left[I + SM^{(0)} (L^{(r)} - L^{(0)}) \right]^{-1} \tag{2}$$

where $M^{(0)}$ is the matrix compliance tensor and S is the Eshelby tensor which depends only on the geometrical shape of an inclusion and material parameters of the matrix. For spherical inclusions embedded in an isotropic matrix the Eshelby tensor reduces to

$$S = \alpha^{(0)} I_v + \beta^{(0)} I_d \tag{3}$$

where I_v and I_d are volumetric and deviatoric parts of the unity tensor, respectively $\alpha^{(0)}$ and $\beta^{(0)}$ are

$$\alpha^{(0)} = \frac{3K^{(0)}}{3K^{(0)} + 4G^{(0)}}, \quad \beta^{(0)} = \frac{6(K^{(0)} + 2G^{(0)})}{5(3K^{(0)} + 4G^{(0)})} \tag{4}$$

where $K^{(0)}$ is the matrix bulk modulus and $G^{(0)}$ is the matrix shear modulus.

5.2. *Results from homogenization*

The input parameters for the first level homogenization are listed in Tab. 3. Homogenized elastic modulus obtained with the Mori-Tanaka scheme is 39.40 GPa. This value is sent to the second level homogenization as an input value for the matrix phase there. The second level homogenization input and output parameters are summarized in Tab. 4. A common steel fiber to matrix ratio of 1.5% is used for the present study. The assumption of a bilinear shape of the elastic modulus around fibers is taken (Fig. 13). Minimum elastic modulus estimated in ITZ (26.27 GPa) is considered as the ITZ modulus which is about 67% of the bulk E value. The width of ITZ is assumed to be defined by the point of intersection of the ITZ trend line and the bulk matrix elastic modulus. From Fig. 13, the width is assessed as 37 μm. Due to the uncertainty in the ITZ width estimate, a parametric study considering three ITZ widths (20, 40 and 60 μm) is performed. The volumetric content of the ITZ phase in a composite is given by the fiber dimensions and the ITZ width (for a single fiber the ITZ volume is given by the fiber surface area (165 mm^2) times the ITZ width 0.037 mm and is equal to 6.1 mm^3).

Negligible differences have been found for varying ITZ width in the given range and the homogenized composite elastic modulus was calculated as ~ 40.2 GPa (Tab. 4).

Table 3. Level 1- cement paste level homogenization

Phase	Vol. fraction	E [GPa]
#1 (matrix phase)	0.7772	33.58
#2	0.2203	70.51
#3	0.0025	164.59
Level 1-homogenized E (GPa)	1	39.40

Table 4. Level 2 - composite level homogenization

Phase	Vol. fraction	E [GPa]
Level 1 (matrix)	0.9829	39.40
Fibers	0.015	200
0. no ITZ, or	0	-
a. ITZ 20 μm, or	0.0011	26.27
b. ITZ 40 μm, or	0.0023	26.27
c. ITZ 60 μm	0.0034	26.27
Level 2-homogenized, case 0	1	40.25
Level 2-homogenized, case a	1	40.23
Level 2-homogenized, case b	1	40.21
Level 2-homogenized, case c	1	40.19

6. Discussion

Clearly, the ITZ width does not have a significant influence on the composite elastic modulus due to its negligible volumetric content in the mixture. This conclusion can be partially influenced by simplified assumptions (mainly the perfect bonding of the phases and spherical shape of inclusions). However, no major influence on the elastic modulus can be anticipated and the composite modulus does not vary much with or without considering ITZ in the homogenization. The reason is that the homogenized property depends mainly on the volumetric content of individual phases. In other words, volumetrically small phases have small influence on the resulting elastic modulus. Thus, neither small volumetric content of fibers (1.5%) nor ITZ content (0.1-0.3%) changes the cement paste elastic modulus much (pure cement paste E=39.40 GPa and composite E=40.2 GPa).

However, about 67% decrease in elastic modulus with respect to the cement bulk indicates also lower strength that is known to have the same trend as elastic modulus in concrete. It can be anticipated that the strength of a composite will be dictated by the weakest phase in the composite rather than its volumetric content. In this sense, the decreased ITZ properties and an approximate size of ITZ found in this paper can serve as input parameters to a fracture based model. Thus, future work should concentrate on the assessment of strength and fracture properties of ITZ and bulk and development of the composite strength model. Such model is, however, beyond the scope of the present paper.

7. Conclusion

The presented paper provides a micromechanical, microstructure and chemical study on ITZ between steel fiber and cement matrix. Microstructure of ITZ around fibers is characterized with less portion of clinkers than in a bulk matrix. Nanoindentation was used to deliver microscale elastic properties. An average modulus of elasticity of ITZ calculated from merged nanoindentation results showed a general trend of decreasing modulus as the position was closer to the fiber. A Minimum elastic modulus estimated in the ITZ (26.27 GPa) was considered as the ITZ modulus that is about 67% of the bulk E value. Homogenized elastic modulus of the bulk material obtained by the Mori-Tanaka scheme was calculated as 39.4 GPa. Taking fibers and ITZ into account, the second level homogenization exhibited composite elastic modulus of 40.2 GPa. The ITZ width does not have a significant influence on the composite modulus due to its negligible volumetric content in the mixture (0.1-0.3%).

However, it can be anticipated that the strength of composite will be defined by the weakest phase strength in the composite (which was proved to be ITZ) rather than its volumetric content.

Acknowledgement
Financial support of the Czech Science Foundation (project 17-05360S) is gratefully acknowledged. L. Polívka (Academy of Sciences, CZ) is acknowledged for his help with acquisition of SEM images.

References

[1] Němeček, J. *Mikromechanické vlastnosti cementových kompozitů*, Diplomová práce, Praha: ČVUT, Fakulta Stavební, 01/2017
[2] Němeček, J., Králík, V., Šmilauer, V., Polívka, L., Jäger, A. *Tensile strength of hydrated cement paste phases assessed by micro-bending tests and nanoindentation*, Cement and Concrete Composites 73 (2016), pp 164-173
[3] Diamond, S., Huang, J. *The ITZ in concrete – a different view based on image analysis and SEM observations*, Cement and Concrete Composites 23 (2001), pp 179-188
[4] Petráňová, V., Sajdlová, T., Němeček, J. *Micromechanical Homogenization of Ultra-High Performance Concrete*, Applied Mechanics and Materials Vol. 821 (2016), pp 518-525
[5] Nežerka, V., Němeček, J., Sližkova, Z., Tesárek, P. *Investigation of crushed brick-matrix interface in lime-based ancient mortar by microscopy and nanoindentation*, Cement & Concrete Composites 55 (2015), pp 122–128
[6] Scrivener, K. L., Crumbie, A. K., Laugesen, P. *The Interfacial Transition Zone (ITZ) Between Cement Paste and Aggregate in Concrete*, INTERFACE SCIENCE 12 (2004), pp 411–421
[7] Vargas, P., Restrepo-Baena, O., Tobón, J. I. *Microstructural analysis of interfacial transition zone (ITZ) and its impact on the compressive strength of lightweight concretes*, Construction and Building Materials 137 (2017), pp 381–389
[8] Allison, P.G., Moser, R.D., Weiss Jr., C.A., P.G. Malone, P.G., Morefield, S.W. *Nanomechanical and chemical characterization of the interface between concrete, glass–ceramic bonding enamel and reinforcing steel*, Construction and Building Materials 37 (2012), pp 638–644
[9] Zhu, W., Bartos, P.J.M. *Application of depth-sensing microindentation testing to study of interfacial transition zone in reinforced concrete*, Cement and Concrete Research 30 (2000), pp 1299-1304
[10] Wang, X. H., Jacobsen, S. He, J. Y., Zhang, Z. L., Lee, S. F., Lein, H. L. *Application of nanoindentation testing to study of the interfacial transition zone in steel fiber reinforced mortar*, Cement and Concrete Research 39 (2009), pp 701–715
[11] Hashin, Z., Monteiro, P.J.M. *An inverse method to determine the elastic properties of the interphase between the aggregate and the cement paste*, Cement and Concrete Research 32 (2002), pp 1291–1300
[12] Zhang J., Sun H., Wan J., Yi Z., *Study on microstructure and mechanical property of interfacial transition zone between limestone aggregate and Sialite paste*, Construction and Building Materials 23 (2009), pp 3393–3397
[13] Acker, P. *Swelling, shrinkage and creep: a mechanical approach to cement hydration* Material Structure, 37 (4) (2004), pp 237–243
[14] Němeček, J., Kralík, V., Vondřejc, J. *Micromechanical analysis of heterogeneous structural materials*, Cement & Concrete Composites 36 (2013), pp 85–92
[15] Oliver, W.C., Pharr, G.M. *An improved technique for determining hardness and elastic modulus using load and displacement sensing indentation experiments*, Journal of Material Research 7 (1992), pp 1564-1583

IOP Conf. Series: Materials Science and Engineering **246** (2017) 012019 doi:10.1088/1757-899X/246/1/012019

Vibrated and self-compacting fibre reinforced concrete: experimental investigation on the fibre orientation

A Conforti[1], G A Plizzari[1] and R Zerbino[2]

[1] Department of Civil, Environmental, Architectural Engineering and Mathematics, University of Brescia, Via Branze 43, Italy

[2] Department of Civil Engineering, La Plata National University. LEMIT, 52 entre 121 y 122, 1900 La Plata, Argentina

E-mail: antonio.conforti@unibs.it

Abstract. In addition to the fibre type and content, the residual properties of fibre reinforced concrete are influenced by fibre orientation. Consequently, the performance fibre reinforced concrete can be affected by its fresh properties (workability, flowing capacity) and by casting and compaction processes adopted. This paper focuses on the study of the orientation of steel or macro-synthetic fibres in two materials characterized by very different fresh properties: vibrated and self-compacting concrete. Four rectangular slabs 1800 mm long, 925 mm wide and 100 mm high were produced changing concrete and fibre type. From each slab, eighteen small prisms (550 mm long) were firstly cut either orthogonal or parallel to casting direction and, secondly, notched and tested in bending according to EN 14651. Experimental results showed that the toughness properties of a thin slab significantly varies both in vibrated and self-compacting concrete, even if in case of self-compacting concrete this variation resulted higher. Steel fibres led to greater variability of results compared to polymer one, underlining a different fibre orientation. A discussion on the relative residual capacity measured on the prisms sawn from the slabs and the parameters obtained from standard specimens is performed.

1. Introduction

It is well known that the residual properties of Fibre Reinforced Concrete (FRC) are directly related to the number of fibres crossing the fracture surface. Many studies confirm that fibres are oriented in concrete due to and different factors [1, 2, 3], being the casting and compaction processes [4], the geometry and type of the fibres and the wall effect some of the more relevant; very fluid mixtures more susceptible and in self-compacting FRC it was proved the existence of significant flow effect [5, 6, 7, 8, 9]. The influence of these factors increases if thin elements are considered [10, 11].

As a consequence, the *fib* Model Code 2010 takes into consideration the fibres orientation in the design of FRC elements [12]. In a similar way German standard DadStb [13] considers orientation factors which varying from 0.5 to 1.

This paper shows the results of an experimental study performed with the aim to analyse the orientation of fibres on 100 mm depth slabs and its influences on the residual properties of FRC. Vibrated and self-compacting FRC mixtures are compared; in addition the fibre type (steel or macro-synthetic) is also considered as a variable. A comparison of the relative residual capacity measured on the prisms sawn from the slabs and the parameters obtained from standard specimens is performed.

2. Experimental program

2.1. Materials

Four Fibre Reinforced Concretes (FRC) were produced incorporating a same volume (0.5 %) of steel (s) or polymer (p) fibres. Table 1 shows the characteristics of the selected fibres which have similar aspect ratio.

With each fibre, one FRC compacted by vibration (V) and one self-compacting FRC (SC) were prepared; the fibre dosages were 40 kg/m^3 and 4.6 kg/m^3 for steel and polymer fibres respectively.

Table 2 shows the mixture proportions, the identifications and fresh properties of each FRC.

Table 1. Characteristics of steel and macro-synthetic fibres.

Fibre designation	s	p
Type	Steel	Polypropylene
Shape	Hooked-end	Embossed
Length l [mm]	50	54
Diameter Ø [mm]	0.80	0.81
Aspect Ratio l/Ø	63	67
Tensile Strength [MPa]	> 1100	> 552
Elastic Modulus [GPa]	210	6
Relative density [kg/m^3]	7.85	0.91

Table 2. Fiber Reinforced Concretes.

Concrete identification	Vs	Vp	SCs	SCp
Concrete type	Vibrated (V)		Self-compacting (SC)	
Fibre type	s	p	s	p
Mixture proportions				
Portland cement CEM II/A-LL 42.5R [kg/m^3]	430		400	
Calcium carbonate filler [kg/m^3]	-		200	
Natural sand 0-4 mm [kg/m^3]	990		860	
Coarse aggregate 4-12 mm [kg/m^3]	725		725	
Water/cement ratio	0.46		0.44	
Superplasticizer (% of cement content)	0.2%		1.1%	
Fibre content [kg/m^3]	40	4.6	40	4.6
Fresh properties				
Slump [mm]	80	75		
Slump flow: diameter [mm] / flow time [s]			720 / 2.9	700 / 4.6

2.2. Test details

One 925 x 1800 x 100 mm slab, three 150 x 600 x 100 mm prisms as reference samples (RS) and four cubes 150 mm side were cast with each concrete. Slabs Vs and Vp were compacted by internal vibration, while the corresponding RS were compacted by external vibration. Slabs SCs and SCp were cast from the smaller side flowing along the 1800 mm length; the related RS were cast from its centre as it is indicated in the EN14651 standard [14].

All specimens were kept protected to avoid water evaporation during 1 day. After demoulded the specimens, moist curing was applied during 2 weeks and then they remains in outdoors conditions (relative humidity 70 ± 10 %, temperature 15 ± 10 °C).

With the aim of comparing the linear shrinkage of different mixtures, one RS of each concrete was introduced in a dry room (relative humidity 50 ± 3 %, temperature 20 ± 2 °C) at the age of 14 days; a mechanical length comparator with a precision of 0.00254 mm was used to perform the measurements.

Prisms of 100 x 150 x 550 mm were sawn from each slab in orthogonal (O) and parallel (P) directions of the flow one as it is indicated in Figure 1 (the arrow represents pouring point and casting direction). The samples were identified by a progressive number followed by cutting direction (O or P). These sawn prisms and the RS were notched and tested in bending following the guidelines of the EN 14651 Standard [14]. The specimens were turned on 90° for testing, thus the depth was 150 mm and the unique difference with the standard prisms was the beam width (100 mm instead of 150 mm). Bending tests were performed at 3 months. As results the limit of proportionality (f_L), and the residual flexural tensile strengths $f_{R,1}$, $f_{R,2}$, $f_{R,3}$ and $f_{R,4}$ corresponding to a Crack Mouth Opening Displacement (CMOD) value of 0.5, 1.5, 2.5 and 3.5 mm respectively were calculated. Figure 2 shows the instrumentation details of bending tests. At the end of bending tests, the number of fibres on the fracture surfaces was counted to determine the density of fibres.

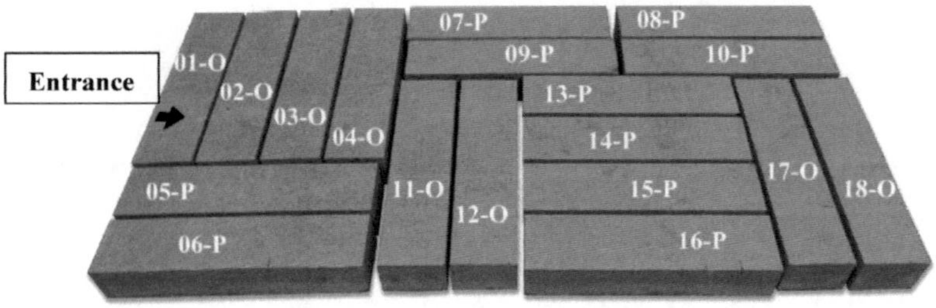

Figure 1. Schema of slabs and sample identification.

Figure 2. Bending tests.

3. Experimental results and discussion

As expected, no significant differences in drying shrinkage were found as the coarse aggregate volume was equal in all mixtures (see Figure 3).

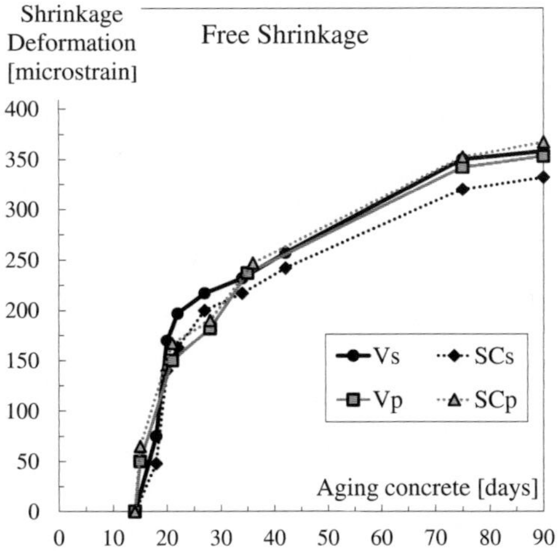

Figure 3. Free shrinkage deformation for vibrated (V) and self-compacting (SC) FRC.

Table 3 presents the mechanical properties of each FRC in terms of the mean values of cube compressive strength and flexural tensile properties measured on the reference specimens (RS), the limit of proportionality (f_L) and the residual strengths $f_{R,1}$ and $f_{R,3}$; the coefficients of variation (CV) are provided between brackets). In addition, and based on the bending properties, the FRC class according to the *fib* Model Code 2010 [12] were indicated. As it can be seen, vibrated FRC were characterized by a cube compressive strength near 61 MPa, while in the case of self-compacting FRC it was approximately 66 MPa. As expected, while the fibre type and content has no significant effect on compressive strength, the bending properties are strongly modified. However, for the same type and content of fibre, the post-cracking response is almost similar in vibrated and self-compacting FRC although the FRC had very different workability properties; when the FRC classes are considered no differences were found when V or SC FRC are compared.

Table 3. Mechanical properties of each concrete measured on standard specimens.

FRC	$f_{c,cube}$ [MPa]	f_L [MPa]	$f_{R,1}$ [MPa]	$f_{R,3}$ [MPa]	FRC class
Vs	62.7 (0.01)	5.74 (0.02)	6.71 (0.09)	6.16 (0.15)	6c
SCs	67.8 (0.06)	5.55 (0.04)	6.25 (0.19)	6.03 (0.20)	6c
Vp	59.5 (0.01)	5.64 (0.01)	1.94 (0.33)	2.76 (0.32)	1.5e
SCp	65.8 (0.03)	5.46 (0.07)	1.91 (0.14)	2.87 (0.23)	1.5e

As is well known the residual properties of FRC are directly related to the number of fibres crossing the fracture surface. With the aim of analysing the effect of fresh concrete properties on the orientation of fibres, Figure 4 compares the stress - CMOD curves obtained from the prisms sawn in orthogonal (O) and parallel (P) directions for the case of steel FRC (Vs and SCs). In a similar way, Figure 5 presents the stress - CMOD curves obtained on the prisms sawn from the slabs cast with macro-synthetic FRC (Vp and SCp). As a reference the stress - CMOD curves of the standard prisms (RS) are included in each plot.

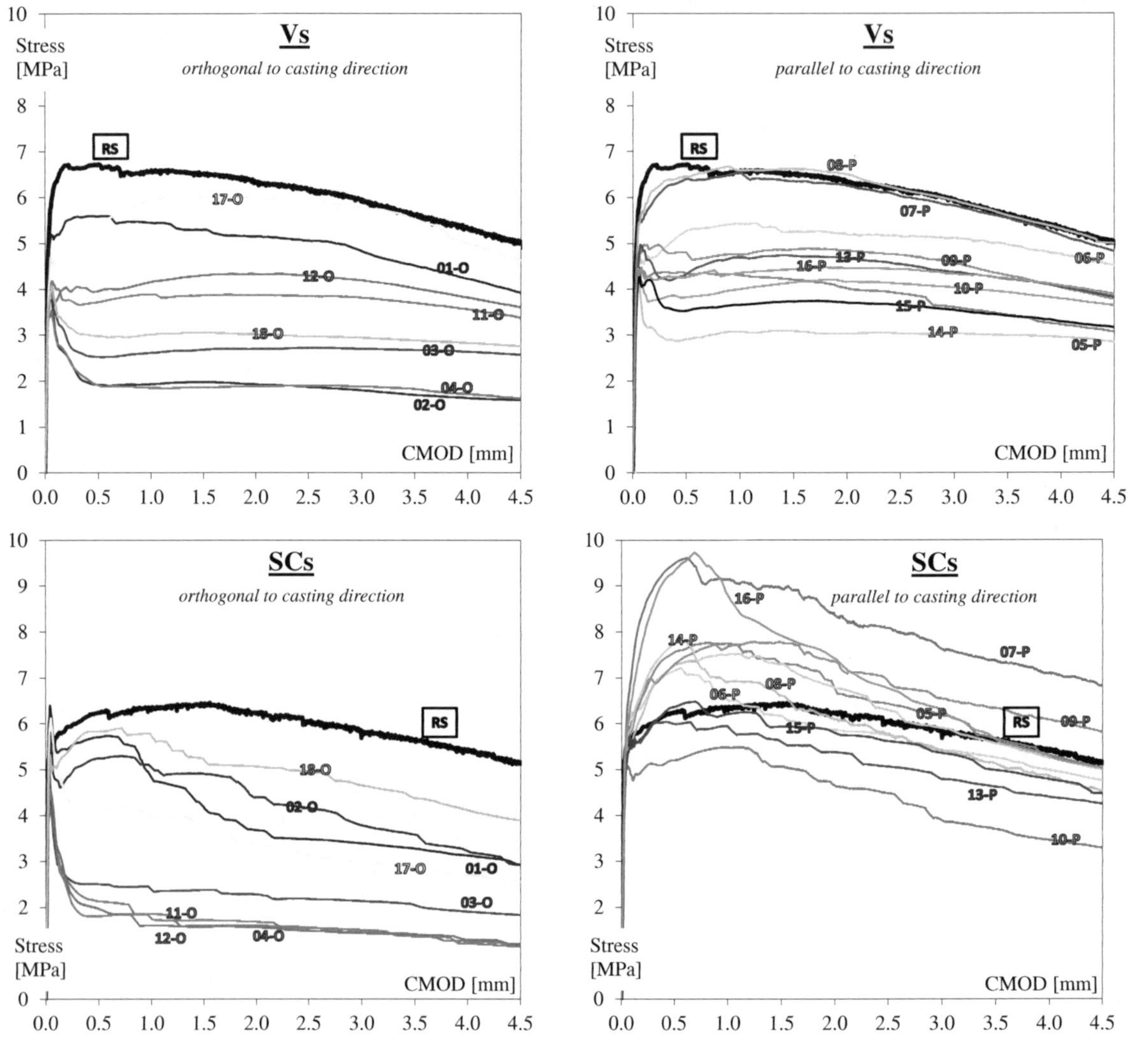

Figure 4. Nominal stress vs. CMOD curves for Vs and SCs.

It can be seen a great variability in the post-peak response of FRC, which is higher than the own variability usually found between standard specimens. In this case the orientation of fibres is strongly affected by wall and flow effects. Analyzing the results from steel FRC slabs, it appears that generally parallel (P) prisms show better residual properties than orthogonal (O) ones. In vibrated FRC all stress vs. CMOD curves are equal or lower than those corresponding to the RS indicating that the standard tests can be assumed as an optimal performance. The low performances can be associated to a lower content of fibres crossing the fracture surfaces. Although again in self-compacting FRC P prisms show better residual properties than O ones, the differences are significantly increased and in this case most parallel prims shows better performance than RS prisms; this fact clearly demonstrates the contribution of flow effects on fibre orientation. The curves of lower residual capacity correspond to O prisms obtained from the zone of higher flow rate or P prisms close to the end of the mould, where the wall effect of the end modifies the orientation of the fibres.

IOP Conf. Series: Materials Science and Engineering **246** (2017) 012019 doi:10.1088/1757-899X/246/1/012019

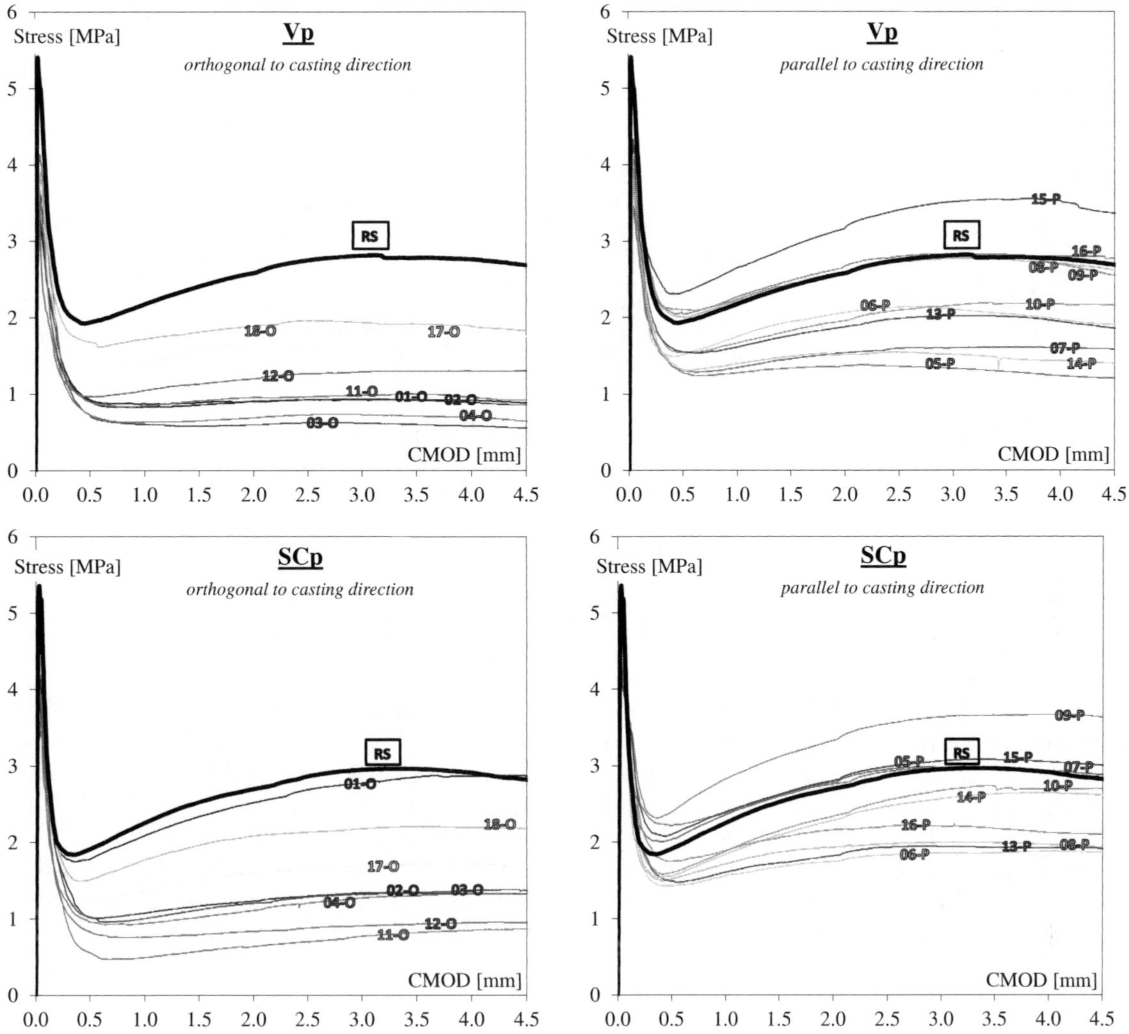

Figure 5. Nominal stress vs. CMOD curves for Vp and SCp.

Regarding polymer FRC, Figure 5 also shows a better post-peak performance of P-prisms than O-prisms. Nevertheless there are not strong differences when Vp and SCp curves are compared. In both cases the RS curves show the highest residual capacity in orthogonal specimens and near the middle post-peak behaviour in the case of parallel prisms.

Table 4 shows compares the values of the residual flexural tensile strength $f_{R,3}$ measured on the prisms sawn from the slabs with the results obtained from the reference specimens (RS); the mean values corresponding to orthogonal and parallel samples (O-prisms, P-prisms) are also given as relative values of RS. The CV are expressed between brackets. As it can be seen, the variability in sawn prisms results increases when comparing with RS results as wall and flow (mainly in the case of SC) effects modify the orientation of fibres in different sectors of the slabs. The lowest CV correspond to P-prisms and no significant differences in variability were found between the four FRC studied. Analysing the relative values it appears that the overall performance is near 70 % of the residual stresses measured on standard beams; O-prisms show $f_{R,3}$ near 50 % of RS and in the case of P-prisms it can be clearly seen that the residual capacity is higher in SC FRC than in V FRC making evident the flow effect mainly in the case of steel fibres.

Table 4. Mean values of the residual flexural tensile strength $f_{R,3}$ of samples obtained from the slabs.

Samples	$f_{R,3}$ [MPa]			
	Vs	SCs	Vp	SCp
RS	6.16 (0.15)	6.03 (0.20)	2.76 (0.32)	2.87 (0.23)
O-prisms	3.57 (0.41)	2.81 (0.47)	1.15 (0.47)	1.48 (0.44)
P-prisms	4.56 (0.21)	6.08 (0.16)	2.24 (0.30)	2.54 (0.22)
Mean (O, P)	4.12 (0.31)	4.62 (0.43)	1.75 (0.44)	2.07 (0.40)
	Relative values			
O/RS	0.58	0.47	0.42	0.52
P/RS	0.74	1.01	0.81	0.89
Mean/RS	0.67	0.76	0.63	0.72

As it is well known, the residual properties of FRC are directly related to the number of fibres crossing the fracture surface. Figure 6 shows the variation of $f_{R,1}$ and $f_{R,3}$ with the fibre density measured on the fracture surfaces of the prisms after bending tests, for steel FRC (Vs, SCs). It can be seen that there is a wide range of variation of fibre density between 0.3 and 1 fibres/cm^2. Both Vs and SCs specimens follow similar tendencies being the variability a bit higher in the case of SCs.

Analogous conclusions can be obtained from Figure 7, which shows the variation of $f_{R,1}$ and $f_{R,3}$ with the fibre density for polymer FRC (Vp, SCp); in this case the fibre densities varied between 0.2 and 0.8 fibres/cm^2, and no differences can be appreciated between vibrated or self-compacting FRC.

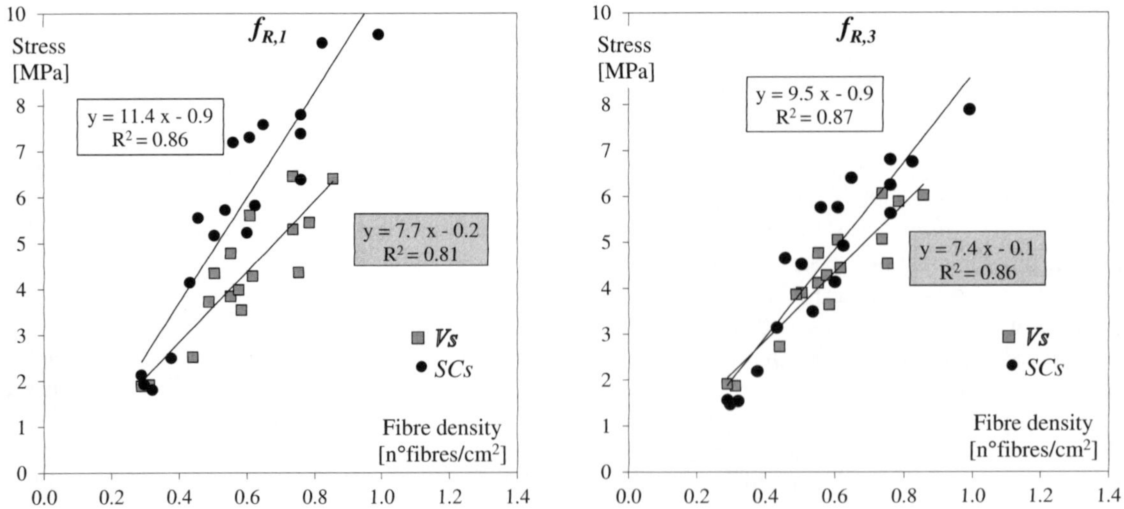

Figure 6. Residual strength $f_{R,1}$ and $f_{R,3}$ vs. fibre density on sample fracture surface for Vs and SCs.

IOP Conf. Series: Materials Science and Engineering **246** (2017) 012019 doi:10.1088/1757-899X/246/1/012019

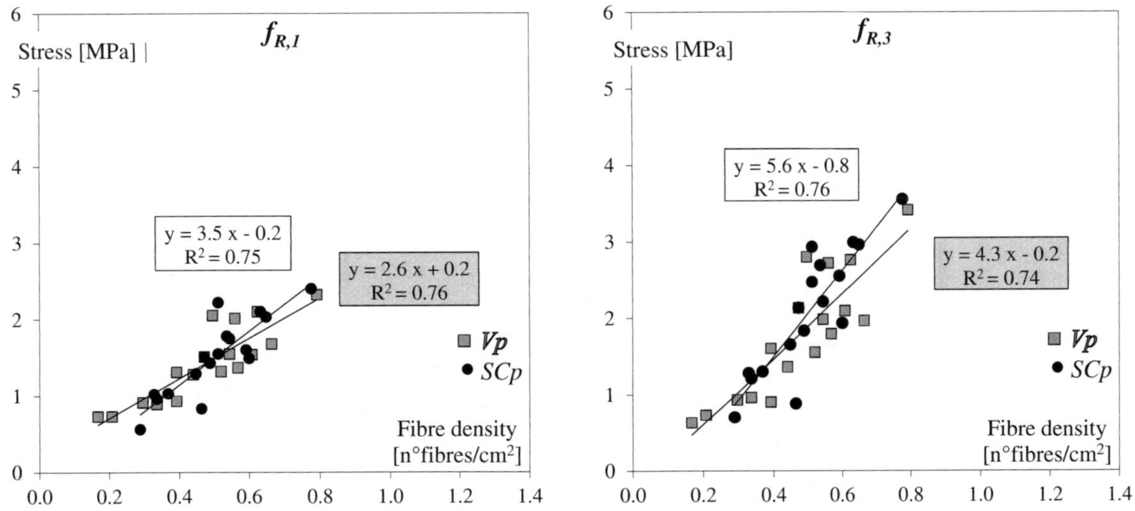

Figure 7. Residual strength $f_{R,1}$ and $f_{R,3}$ vs. fibre density on sample fracture surface for Vp and SCp.

Figure 8 and Figure 9 show the variation of $f_{R,3}$ for steel and polymer fibres in orthogonal ($f_{R,3y}$) and parallel directions ($f_{R,3x}$) respect to casting one. The residual mechanical properties of FRC slabs are different in the two considered directions, and these differences increase in SC-FRC and in the case of steel fibres. Wall effects always appeared in both directions (more evident in prisms orthogonally oriented to casting direction ($f_{R,3y}$). As expected, a flow effect appeared in SC-FRC, leading to higher performance of $f_{R,3x}$ compared to V-FRC, mainly in the case of steel fibres.

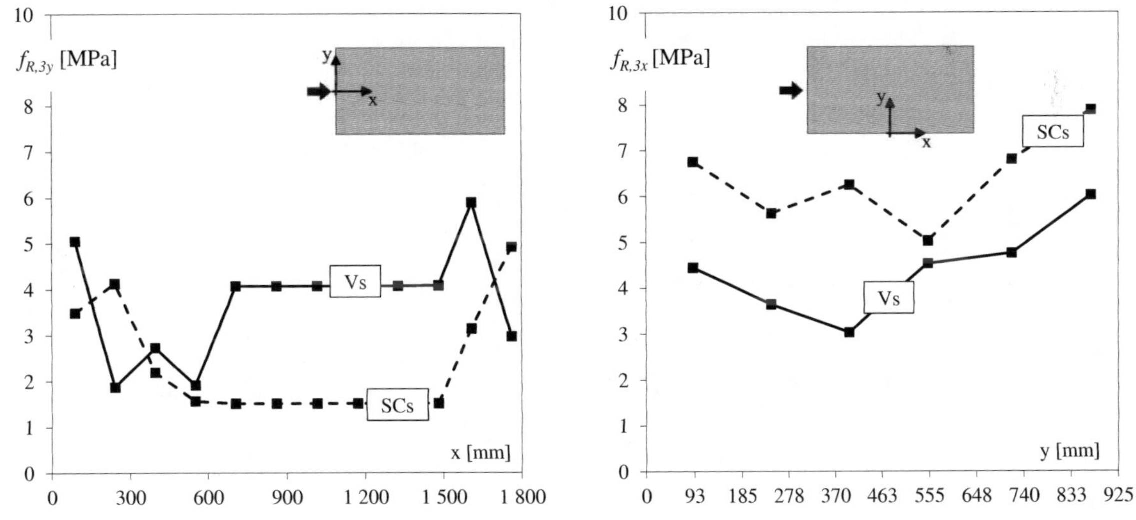

Figure 8. Variation of the residual flexural tensile strength $f_{R,3}$ in Vs and SCs slabs.

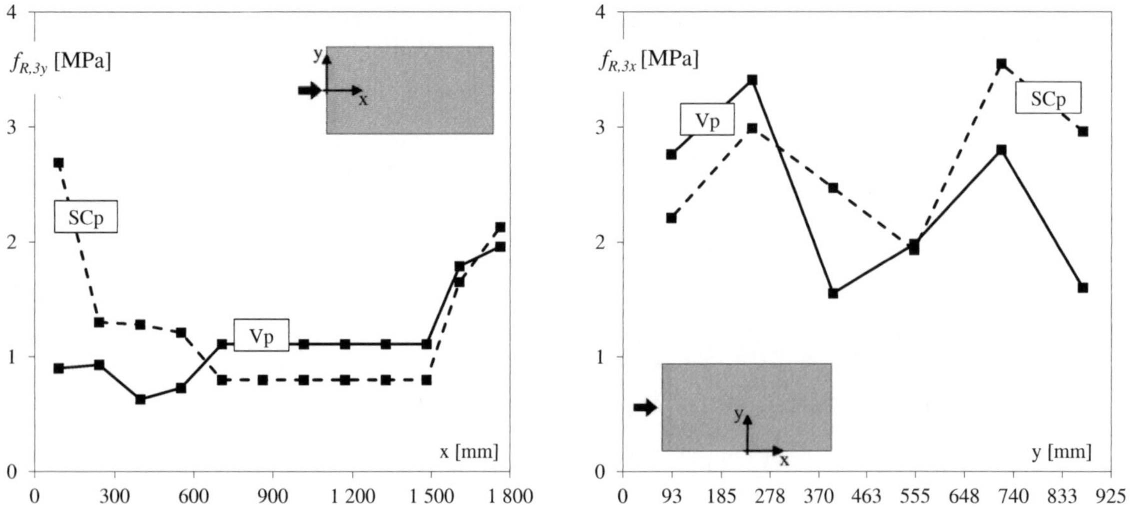

Figure 9. Variation of the residual flexural tensile strength $f_{R,3}$ in Vp and SCp slabs.

Finally, Figure 10 compares the residual bending parameters obtained from the sawn prisms with those obtained from the reference samples. In the case of steel FRC increases the variability in test results between V-FRC and SC-FRC, mainly in terms of the maximum values, which means that the flow effect is more pronounced with rigid fibres. On the contrary, both minimum and maximum values are similar in the case of macro-synthetic FRC for vibrated and self-compacting FRC.

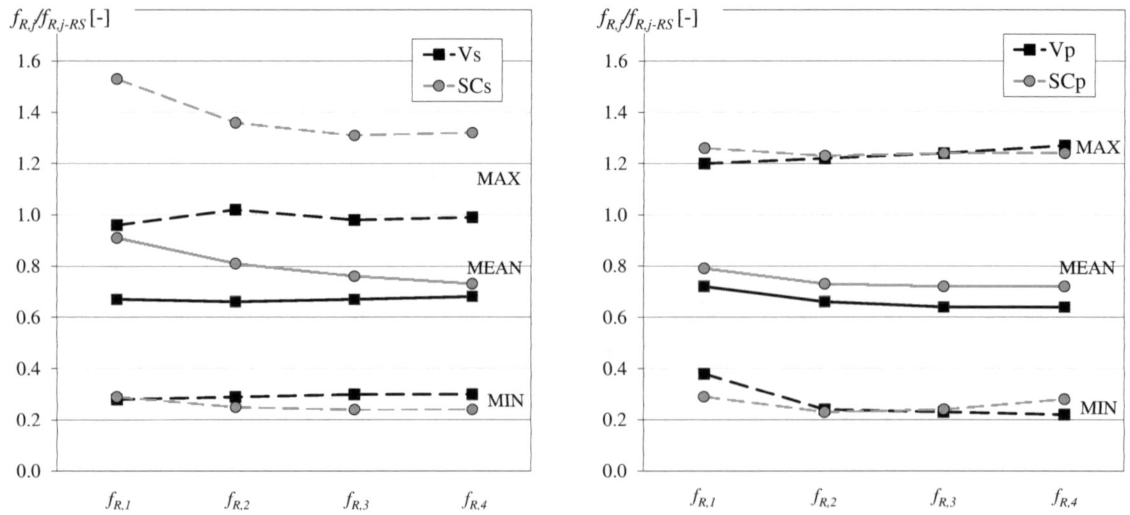

Figure 10. Comparison between residual properties of slabs and reference samples.

4. Concluding remarks

The orientation of fibres in 100 mm depth slabs cast with vibrated (V-FRC) and self-compacting (SC-FRC) steel and macro-synthetic FRC was studied. The following conclusions might be drawn:

- Wall and flow effects that modify the orientation of fibres along the slabs were verified. As a consequence, a great variability in the residual bending capacity in different directions and sectors of the slabs was found.
- Differences in fibre orientation increase in the case of SC-FRC, mainly if steel (more rigid) fibres are used.
- The mean values of the post-peak capacity found for the prisms sawn from the slabs were near 70% of the residual strengths measured on standard specimens.
- The residual properties were greater in the prisms parallel to the casting direction. In the prisms orthogonally oriented with respect to the casting direction the mean residual capacity decrease near 50% when compared to that obtained from standard specimens.

Acknowledgments
The Authors would like to give their appreciation to filler supplier NICEM s.r.l (Casazza, Italy) and to macro-synthetic and steel fibres suppliers: BASF Construction Chemicals Italy (Treviso, Italy) and Matassina s.r.l. (Vicenza, Italy), respectively. A special acknowledgement goes to Engineer Michele Vesentini, as well as to the technicians Luca Martinelli, Augusto Botturi, Domenico Caravaggi and Andrea Delbarba, for the assistance in performing the experimental program. The authors would like also to gratefully thank University of Brescia for the visiting professor program (year 2016).

References
[1] Sanal I, Ozyurt Zihnioglu N 2013 To what extent does the fiber orientation affect mechanical performance? *Constr and Build Mater* 44 p.671-681
[2] Ozyurt N, Mason TO, Shah SP 2007 Correlation of fiber dispersion, rheology and mechanical performance of FRCs *Cem and Concrete Comp* 29 p.70-79
[3] Stähli P, Custer R, van Mier JGM 2008 On flow properties, fibre distribution, fibre orientation and flexural behaviour of FRC *Mat and Struc* 41 p.189-196
[4] Tiberti G, Conforti A, Plizzari GA 2015 Precast segments under TBM hydraulic jacks: Experimental investigation on the local splitting behavior *Tunn Und Sp Tecn* 50 pp. 438-450
[5] Boulekbache B, Hamrat M, Chemrouk M, Amziane S 2010 Flowability of fibre reinforced concrete and its effect on the mechanical properties of the material *Constr and Build Mater* 24 (9) p. 1664-1671
[6] Torrijos MC, Barragán B, Zerbino R 2010 Placing conditions, mesostructural characteristics and post-cracking response of fibre reinforced self-compacting concretes *Constr and Build Mater* 24 p. 1078-1085
[7] Stähli P, van Mier JGM. 2007 Manufacturing, fibre anisotropy and fracture of hybrid fibre concrete *Eng Frac Mech* 74 p. 223-242
[8] Vandewalle L, Heirman G, Rickstal FV 2008 Fibre orientation in self-compacting fibre reinforced concrete *In: Gettu R, ed. 7th Int. RILEM Symp. on FRC* p. 719-728
[9] Cuenca E, Serna P 2010 Shear behavior of Self-Compacting concrete and Fiber-Reinforced concrete push-off specimens *Design, production and placement of self-consolidating concrete, RILEM Bookseries* 429-438
[10] Ferrara L, Meda A. 2006 Relationships between fibre distribution, workability and the mechanical properties of SFRC applied to precast roof elements *Mat and Struc* 39 p.411-420
[11] Zerbino R, Tobes JM, Bossio ME, Giaccio G 2012 On the orientation of fibres in structural members fabricated with self compacting fibre reinforced concrete *Cem and Concrete Comp* 34 p.191-200
[12] Fédération Internationale du Béton 2012 *fib Model Code for Concrete Structures* 2010 Vol 1 350 pages ISBN 978-2-88394-105-2; Vol 2 370 pages ISBN 978-2-88394-106-9
[13] Deutscher Ausschuss für Stahlbeton (DAfStb) *Guideline 2012 German Committee for Reinforced Concrete, Steel fibre reinforced concrete* p. 48
[14] EN 14651 2005 Test method for metallic fibre concrete *European Standard*

IOP Conf. Series: Materials Science and Engineering **246** (2017) 012020 doi:10.1088/1757-899X/246/1/012020

Reactive powder concrete reinforced with steel fibres exposed to high temperatures

T Kh Alrekabi [1], V M C F Cunha [2] and J A O Barros [2]

[1] Architecture Engineering Department, Engineering Faculty, Koya University, University Park, Danielle Mitterrand Boulevard, Koysinjaq KOY45, Iraq.
[2] ISISE, Department of Civil Engineering, School Eng., University of Minho, Campus de Azurém, 4800-058 Guimarães, Portugal.

E-mail: taghreed.khaleefa@koyauniversity.org

Abstract. An experimental investigation was carried out to assess the mechanical properties of reactive powder concrete (RPC) reinforced with steel fibres (2% in vol.) when exposed to high temperatures. The compressive, flexural and tensile strength, modulus of elasticity and post-cracking behaviour were assessed after specimens' exposure to different high temperatures ranging from 400 to 700ºC. The mechanical properties of the RPC were assessed for specimens dried for 24 hours at 60 ºC and 100 ºC. Partially dried specimens (60 ºC) exhibited explosive spalling at nearby 450 ºC, while fully dried RPC specimens (100 ºC) maintained their integrity after heating exposure. In general, the mechanical properties of RPC significantly decreased with the increase of the temperature exposure. The rate of decrease with temperature of the compressive, tensile and flexural strengths, as well the corresponding post-cracking residual stresses was higher for exposure temperatures above the 400 ºC.

1. Introduction

Since the appearance of the reactive powder concrete (RPC) in the early 90's, significant research has been carried out to develop and characterize this material. RPC is characterized by having very low water to binder ratio (W/B), and very dense microstructure formed by using a combination of powders including cement, silica fume, and ground quartz sand, as well by excluding the coarse aggregates from its composition. This combination endows RPC of ultra-high strength and high durability [1, 2]. When heat-treatment and/or pre-setting pressurization are used, RPC may attain compressive strengths ranging from 200 to 800 MPa [3, 4]. However, the high strength of RPC makes it extremely brittle; therefore, commonly fibres are added into the mixture (ranging from 1.5 to 3% in vol.) in an attempt of decreasing its brittleness [5].

RPC may be used on an assortment of applications, such as prestressed concrete structures, nuclear power plants, large-span arch roof structures and bridges and in structural rehabilitation. Within these structural applications, fire poses a tangible possible hazard. The potential risk of structural collapse due to fire resides in the highly brittle nature of RPC. Nevertheless, up to now, research on the fire / elevated temperatures exposure resistance of RPC is still scarce. Numerical investigations revealed that RPC is prone to explosive spalling under high temperature [6-8]. Moreover, experimental investigations [9] found that RPC spalled under high temperatures in a distinct manner than that of high-strength or high-performance concrete. Nonetheless, the spalling mechanisms of RPC still remain unknown. Liu and Huang [9] conducted a series of fire resistance tests and found that the residual compressive strength of RPC decreased with increasing fire duration.

This work presents and discusses the results of an experimental program to assess the mechanical properties of RPC reinforced with steel fibres (2% in vol.) when exposed to high temperatures. The compressive, flexural and tensile strength, modulus of elasticity and post-cracking behaviour were assessed after specimens' exposure to different high temperatures ranging from 400 to 700 ºC.

2. Experimental Program

2.1. Materials and concrete composition

The materials used in the production of the RPC were: cement, fine sand, silica fume, superplasticizer, tap water and steel fibres. Table 1 includes the final RPC composition per cubic metre. This mixture was selected after carrying out 9 preliminary trial batches. The guidelines given in [10] were adopted for fixing the quantity of cement, fine sand, and steel fibres. On the other hand, the quantities of silica fume, super-plasticizer and water were varied in order to obtain a mixture with proper rheological and mechanical properties, as well an adequate fibre distribution.

Table 1. RPC composition per m^3.

Cement [kg]	Fine sand [kg]	Silica fume [kg]	Superplasticizer [kg]	Water [kg]	Fibres [kg]
1000	1000	150	21.25	212.5	156

An ordinary Portland cement CEM I 42.5R produced by the Cement Company SECIL was adopted. Fine aggregates were used exclusively, namely, a natural river sand (non-siliceous), which was already clean and nearly free from impurities. The fine sand maximum nominal size was about 1 mm, which makes it suitable to be used in the production of RPC [11]. Figure 1 shows the grading of the adopted fine sand. Note that all the retained particles in the sieve with an opening of 1 mm were discarded, in spite of being very few.

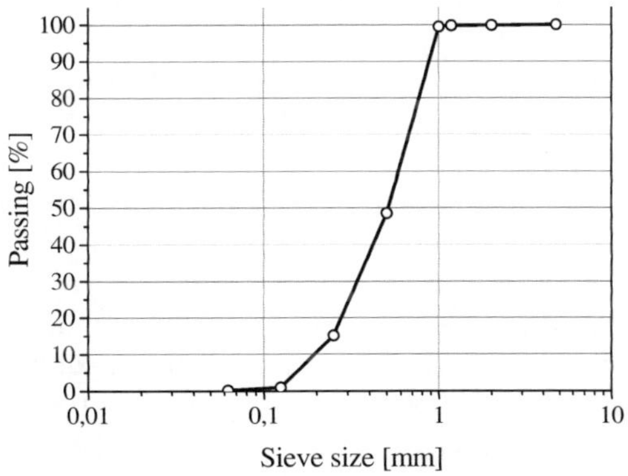

Figure 1. Grading of fine sand.

A greyish colored densified silica fume was used as an addition in the RPC mixes to enhance mainly the matrix densification. A superplasticizer commercially available, specifically Sika ViscoCrete® 20HE, was used as a high performance high water reducer admixture to produce the RPC. The solid content of this superplasticizer was 40% and its density was 1.08 kg/l. Tap water was used both for the mixture and curing of the specimens. Finally, high strength straight steel fibers with a copper coating and an aspect ratio (l_f/d) of 30 were used in all RPC mixtures. Table 2 includes the main properties of the used fibres.

Table 2. Geometric and mechanical properties of the steel fibres.

Length [mm]	Diameter [mm]	Tensile strength [MPa]	(l_f/d) [-]	Conformity standard
6	0.2	2800	30	ASTM-A820

2.2. Concrete mixture and curing procedures

The mixture procedure was performed using a rotary mixer with a capacity of 6 litters. Firstly, sand and cement were introduced into the mixer, respectively, and then mixed during 3 minutes. Afterwards, silica fume was added and mixed with the other constituents for more 3 minutes. Superplasticizer was added into the water jar and then stirred. This mixture was slowly added to the pre-mixed aggregates and binder into the mixer. After the addition of the total water and superplasticizer, the mixture procedure was carried out for more 3 minutes. The mixing process was stopped to remove and remix by hand any material bonded to the mixer's walls and flat beater, and then restarted for three additional minutes. This step was repeated until the mixture exhibited a good homogeneity. Finally, steel fibres were added slowly by hand during the mixing. The total mixing time was about 18 to 24 minutes.

The moulds were filled with the RPC mixture, and then vibrated during approximately 1 minute on a vibration table. After casting, all moulds were covered with a wet cloth for 48 hours to minimize the loss of moisture. Finally, after the 48 hours, the specimens were unmoulded and submerged under water in a climate chamber at a 20 °C temperature and 98% relative humidity until testing age, at about 28 days.

2.3 Exposure to high temperatures procedure

Prior to the exposure to the high temperatures procedure and mechanical testing, specimens were dried under two distinct moderate temperatures for about 24 hours, respectively, at 60 and 100 °C. This aimed to check the influence of the specimens' moisture degree on the eventual explosive spalling of the specimens under high temperatures. The moisture degree was only assessed in some prismatic specimens and revealed that at 100 °C the specimens were fully dried, whereas in specimens dried at 60 °C the core of the prismatic specimens was still wet. In the latter specimens, the average decrease of specimen's weight after dried at 100 °C was 7%.

An electric furnace with a maximum temperature of 1200 °C was used to subject the specimens to temperatures of, respectively, 400, 500, 600 and 700 °C. The heating and exposure procedures were similar for all the target temperatures. A heating rate of 10 °C/min up to the target temperature was maintained constant. After attaining the desired target temperature, the specimens were maintained at a constant temperature for one hour. Finally, specimens were allowed to gradually cool down over more than 24 hours. The cooling speed was slightly different for every target temperature. The temperature of all specimens was the same room temperature at the time of testing.

Table 3 includes the number of specimens that spalled when submitted to the high temperatures exposure for each drying condition. Additionally, it is included in round brackets the total number of specimens that have been exposed to each target temperature and drying conditions. Note that some specimens even though may not exhibited a visible failure pattern, most probably had severe micro-cracking due to damage induced by the thermal stresses gradient.

Table 3. Failure of specimens due to high temperatures exposure for distinct drying conditions.

Temp. Exposure		400 °C		500 °C		600 °C		700 °C	
Drying Temp.		60 °C	100 °C	60 °C	100 °C	60 °C	100 °C	60 °C	100 °C
Specimens	Compression	-	0 (4)	-	0 (4)	-	0 (4)	-	0 (2)
	Tension	0 (4)	-	2 (4)	-	4 (4)	-	-	-
	Flexure	0 (4)	-	4 (4)	-	3 (4)	-	-	0 (2)

The totality of specimens that have been exposed to 400 ºC for one hour did not suffered any spalling damage when dried at 60 ºC. Moreover, the specimens dried at 100 ºC also did not exhibit spalling, for 500, 600 and 700 ºC. The explosive spalling occurred for specimens dried at 60 ºC, consequently in the presence of moisture inside the specimen. The specimen failure was clearly audible in the laboratory due to the strong explosion of the specimens. This explosion occurred at nearby 450 ºC, which is in accordance with other research carried out for high strength concretes under high temperatures [12-13]. The latter occurrence maybe ascribed by the dehydration of the calcium hydroxide that occurs from 400 to 550 ºC, which could increase the build-up of high pore pressure [14]. Figure 2 depicts some examples of the observed specimen failure when submitted to high temperatures.

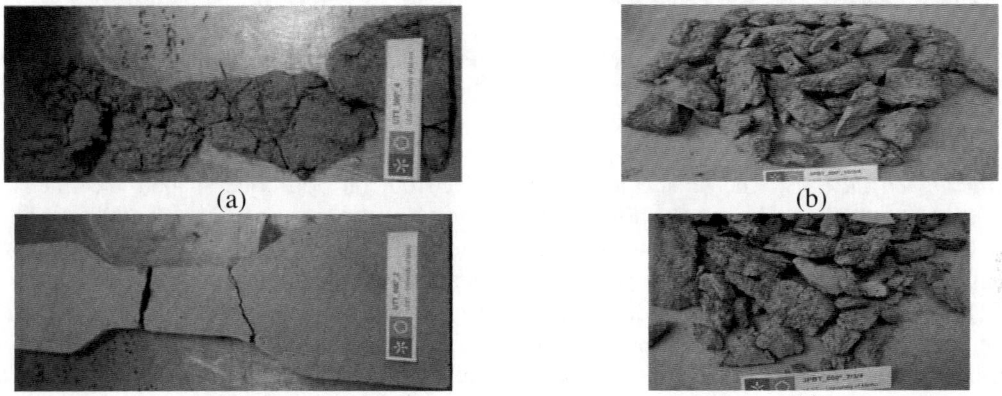

(a) (b) (c) (d)

Figure 2. Failure due to high temperature: (a) and (b) spalling at 500 ºC of direct tension and flexure specimens, respectively; (c) cracking of direct tension specimen at 600 ºC, and (d) spalling of flexure specimens at 600 ºC.

2.4. Test set-up and specimen geometry

A total of 52 specimens with distinct geometries were cast for performing compressive, flexural and uniaxial tensile tests. In general, four specimens per each series corresponding to a test typology and temperature exposure were used with the exception of the 700 ºC series. In the subsequent sections the specimen geometry and corresponding test set-up are described.

2.4.1. Compression tests. Cylindrical specimens with an approximate diameter of 44 mm and 105 mm height were employed for assessing the elasticity modulus and the compressive stress - strain relationship. Prior to testing, both top and bottom surfaces of all cylinders were carefully grinded to ensure plane and parallel surfaces. It was always guaranteed a height to diameter ratio higher than 2.

In a first stage, the test to determine the static modulus of elasticity was carried out under force control over a certain number of cycles. In the loading cycle, the stress was applied at a velocity of 0.2 MPa/s until a stress level of one third of the compressive strength is attained. The test ended when the strain difference, between two consecutive loading cycles, did not exceed 1×10^{-5}, according to [15]. Figure 3(a) shows the adopted set-up for the test of modulus of elasticity. On the other hand, the compressive stress-strain relationship of the developed RPC was obtained according to the test set-up depicted in figure 3(b). The latter test was carried out with closed-loop displacement control at a rate of 10 μm/s. The specimen's displacement was computed from the average readings of three LVDTs measuring the relative displacement the frame's rigid steel plates. Then, the strain was found by divided the average displacement on the length of the specimen.

IOP Conf. Series: Materials Science and Engineering **246** (2017) 012020 doi:10.1088/1757-899X/246/1/012020

(a) (b)

Figure 3. Compression tests set-up for determining the: (a) modulus of elasticity, (b) stress – strain relationship.

2.4.2 Uniaxial tensile tests. The tensile tests were carried out on dog-bone shaped specimens. The geometric details of the specimen are shown in the figure 4(a), whereas the test set-up is depicted in figure 4(b). The specimen deformation was obtained from the average readouts of two LVDTs with a gauge length of 120 mm. The tests were performed under closed-loop displacement control at a rate of 5 μm/s. The specimens were carefully fastened to the grips before testing, nevertheless after having been submitted to high temperatures, the specimens were slightly warped maybe due to the thermal stress gradient induced during the heating procedure. This may have introduced some damage in these specimens during their fixing process to the grips, which may have contributed to decrease their tensile strength.

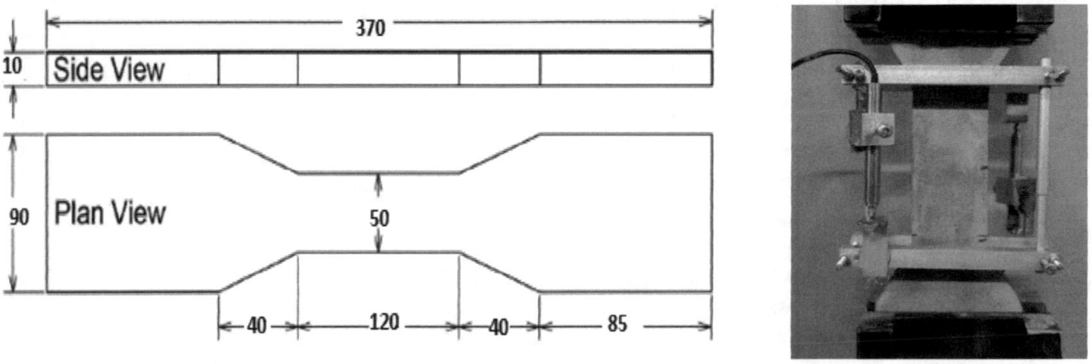

(a) (b)

Figure 4. (a) Dog-bone specimen (dimensions in mm), (b) uniaxial test set-up (dimensions in mm).

2.4.3 Three-point bending test. RPC prismatic specimens of 40×40×160 mm were prepared before testing. A notch at the mid-length of the prism with a depth of 1/4 times of the beam height (i.e. 10 mm) and a width of 1 mm was executed in the lateral surface all specimens, i.e. on a surface parallel to the casting direction. The loading span was 150 mm. Figure 5 shows the adopted test set-up. The tests were performed under closed-loop displacement control at a rate of 5 μm/s. One LVDT was used to assess the deflection at mid-span, figure 5(a), whereas the other one was used to obtain the crack mouth opening displacement, figure 5(b).

(a) (b)

Figure 5. Three-point bending test set-up: (a) view of full length of specimen, (b) crack mouth opening displacement measurement.

3. Experimental results

Table 4 resumes the main results obtained for the mechanical properties of the RPC tested under distinct high temperatures, namely, the average (Avg.) and coefficients of variation (CoV) for each.

Table 4. Main mechanical properties of the RPC exposed to high temperatures.

Temperature	Compressive strength [MPa]		Modulus of elasticity [GPa]		Tensile strength [MPa]		Flexural strength [MPa]	
	Avg.	CoV	Avg.	CoV	Avg.	CoV	Avg.	CoV
20 °C	137.6	4.8%	46.48	10.3%	4.8	28.8%	20.9	9.5%
400 °C	102.2	1.5%	30.9	5.8%	4.1	15.4%	15.4	14.0%
500 °C	84.4	1.2%	22.3	18.0%	3.6	-	-	-
600 °C	68.2	10.5%	17.5	7.2%	-	-	8.6	-
700 °C	65.4	5.2%	18.4	4.7%	-	-	1.0	13.9%

Figure 6 depicts the variation of the compressive strength, f_{cm}, modulus of elasticity, E_{cm}, flexural strength (normal stress under flexure), f_{fm}, and tensile strength, f_{tm}, with the high temperatures that specimens have been exposed.

Both the compressive strength and modulus of elasticity decreased approximately in a linear fashion up to a temperature of 600 °C and no significant variation was observed when increasing the temperature to 700 °C. The slight increase of the modulus of elasticity as well the slight decrease for the compressive strength, when increasing the temperature from 600 to 700 °C, is within the expected experimental scatter of this kind of material. Thus it can be concluded that the damage level of the matrix for those two temperatures would be similar. For an exposure temperature of 600 °C the reduction on the compressive strength and elasticity modulus when compared to the one obtained for room temperature was about 50 and 62%, respectively.

A significant decrease on the flexural strength with the increase of temperature was observed up to 700 °C. The variation was markedly nonlinear, and a steeper decrease of the flexural strength was observed after the 400 °C. The reduction of the flexural strength of specimens exposed to 600 and 700 °C when compared to the one obtained at room temperature was about 60 and 95%, respectively. On the other hand, the tensile strength revealed a slight decrease with temperature, however, since only it was possible to test specimens exposed up to 500 °C no definite conclusions can be withdraw.

3.1. Compressive stress-strain relationships

Figure 7 shows the average compressive stress-strain relationships of the RPC specimens exposed to the distinct high temperatures. It is clearly visible the decrease on the inclination of the pre-peak branch with the increase of temperature corresponding to a decrease of the matrix stiffness due to the thermal induced damage. Regarding the post-peak branch it is also noticeable a decrease on the residual strength with the increase of temperature. In particular, for temperatures above 600 °C, and for deformations higher than 0.01 the residual strength was practically null.

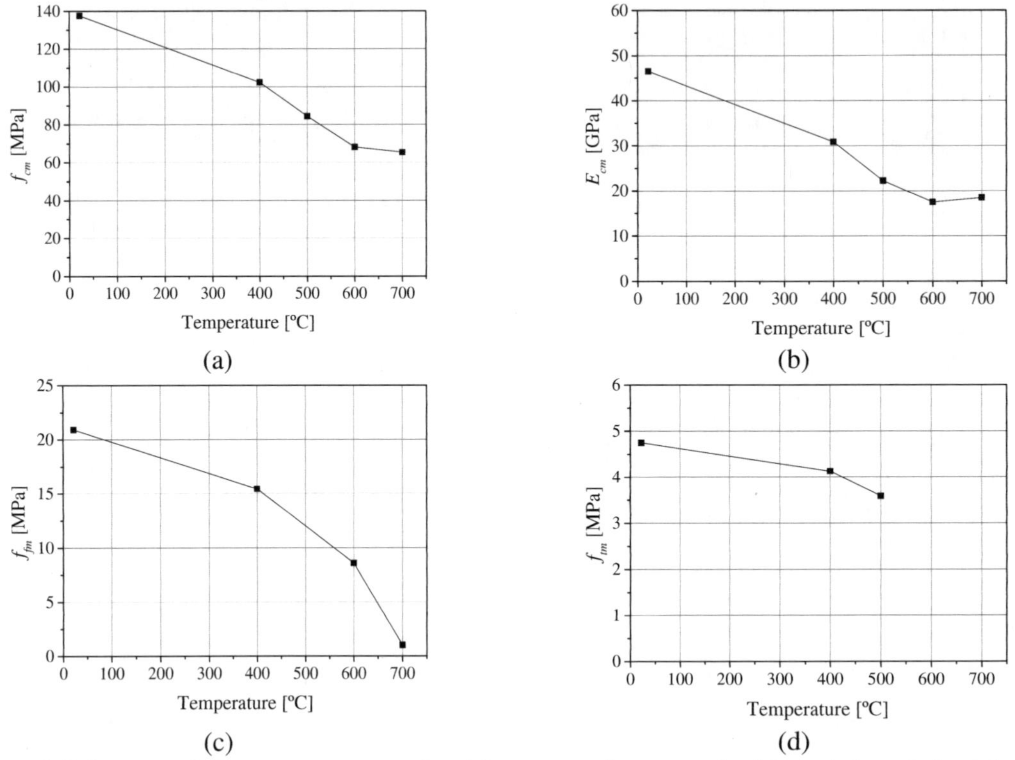

Figure 6. Influence of the temperature on the: (a) compressive strength, (b) modulus of elasticity, (c) flexural strength and (d) tensile strength.

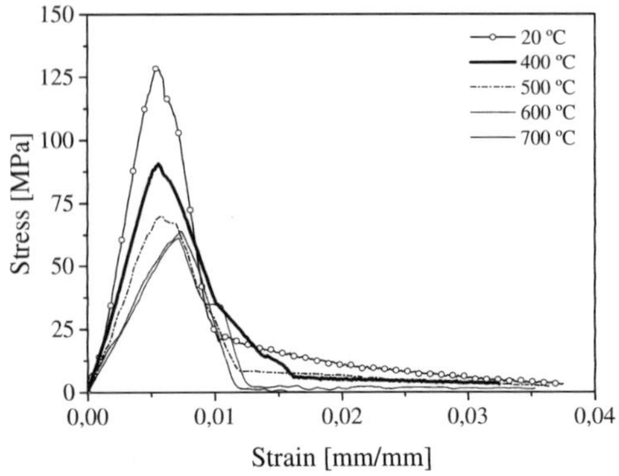

Figure 7. Compressive stress – strain relationships.

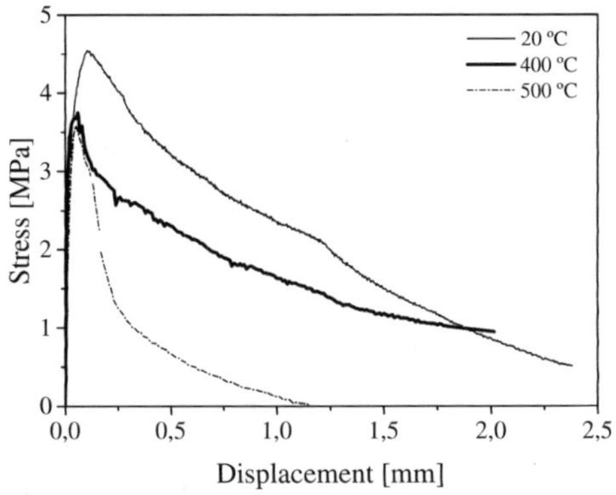

Figure 8. Tensile stress – displacement relationships.

3.2. Tensile stress – displacement relationships

Figure 8 depicts the average tensile stress – displacement curves for the series exposed to distinct temperatures. First of all, it should be noted that the displacement was measured over a gauge length of 120 mm, in the absence of a notch. Therefore, up to the localization of the macro-crack this measure will encompasses the specimen's elastic deformation in the aforementioned region. Since the appearance of multiple cracks was very limited, after the crack localization this displacement can be roughly regarded as the crack opening width, since the deformation up to this stage will be considerable small. The exposure to temperatures of 400 and 500 °C did not produced significant changes in the curve's branch up to the peak stress; the decreasing of stiffness in the pre- peak phase for these samples is similar to the stiffness decreasing for compression samples at the same phase. However, the post-peak branches were considerable distinct. In consequence of the significant reduction on the residual tensile strength with temperature, an increase of the slope of the softening branch immediately after tensile strength was also attained, representing the increase of brittleness of the RPC with the temperature. This was related to the thermal induced damage on the matrix that lead to a deterioration of fibre's bond strength and was more significant for the specimens submitted to a temperature of 500 °C. Moreover, for the latter series, specimens have failed for a very low crack opening width, of about 1.2 mm. On the other hand, the 20 and 400 °C series, at a deformation of 1.5 mm, still exhibited a considerable residual tensile strength, respectively, about 1.5 and 1.2 MPa.

3.3. Normal stress – deflection relationships under flexure
Figure 9 shows the relationship between the normal stress and deflection obtained in the three-point bending tests. For all specimens, in the pre-peak branch, it is observed a linear branch up to 80 to 90% of the maximum flexural stress followed by a nonlinear branch up to maximum stress. In the post-peak branch a gradual decrease of the residual stress was observed for the series non-exposed to high temperature. The series submitted to 400 and 600 °C revealed, besides a lower flexural strength, a steeper decrease of the residual flexural strength just after the peak stress was attained. The flexural and residual strength of the series subjected to 700 °C was very low, and after a deflection higher than 1 mm was practically null.

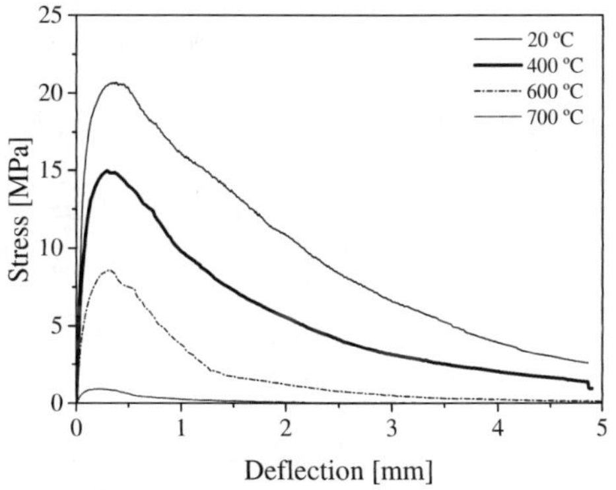

Figure 9. Normal stress – displacement curves under flexure

3.4. Post-cracking parameters

3.4.1. Uniaxial tensile test. The tensile strength, f_{ctm}, stress at crack opening width of 1 and 2 mm, respectively, σ_{1mm} and σ_{2mm}, were computed from the average stress – displacement relationships depicted in figure 8. Additionally, the dissipated energy up to a displacement of 1 and 2 mm, respectively, G_{1mm} and G_{2mm}, were also computed from the latter curves. Figure 10 plots the variation of those parameters with the increase of temperature. In general, the aforementioned parameters decreased with the increase of temperature being the unique exception the small increase of σ_{2mm} from 20 to 400 °C. Moreover, the decrease of the stresses and dissipated energy at a displacement of 1 and 2 mm were more significant for a 500 °C temperature. This may be ascribed to a higher pre-induced degree of micro-cracking of the matrix for temperatures higher than 400 °C, since the dehydration of the calcium hydroxide contributes to the increase of the pore pressure [14]. After this level of temperature, the reinforcement mechanisms provided by fibres are significantly affected detrimentally.

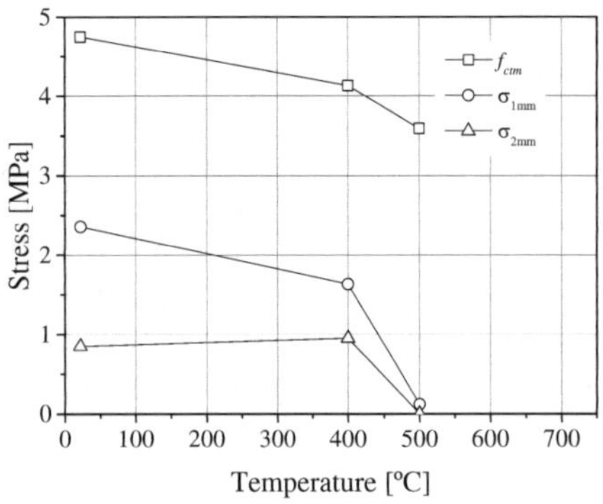

(a)

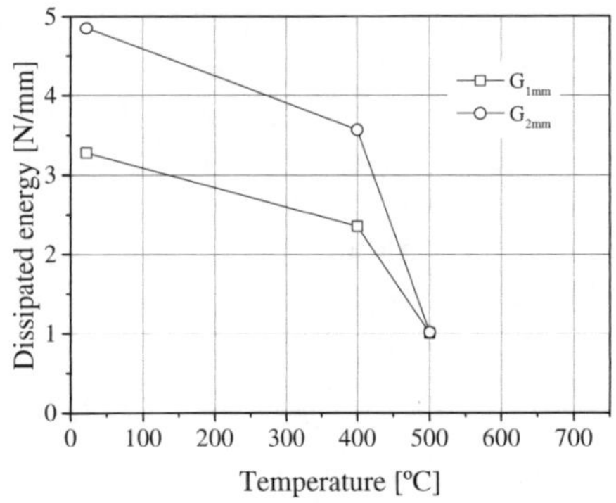

(b)

Figure 10. Influence of the temperature on the tensile post-cracking parameters: (a) f_{ctm}, σ_{1mm} and σ_{2mm}, and (b) G_{1mm} and G_{2mm}.

3.4.2. Three-point bending test. Table 5 includes the influence of temperature on the flexural tensile strengths, which were computed accordingly to the EN-14651 standard [16]. The residual strengths decreased with the increase of temperature, in particular for temperatures higher than 400 ºC. The residual strengths, f_{R1} and f_{R3} are of particular interest for service and ultimate limit state analysis, respectively. For an exposure temperature of 600 ºC, the f_{R1} and f_{R3} faced a reduction of about 63 and 90%, respectively. The smaller decrease of f_{R1} when compared to f_{R3}, may be ascribed to the mobilization of the fibres still being in an initial stage of the fibre pull-out, consequently leading to the mobilization of lower fibre pull-out loads. On the other hand, for f_{R3}, the fibre pull-out is on a much more advanced stage, taking into account that the matrix for this temperature (i.e. 600 ºC) has extensive thermal induced damage, the fibre / matrix bond strength is significantly reduced, therefore the fibres can easily be pulled out for lower loads resulting in a lower crack bridging stress.

Table 5. Flexural tensile strengths obtained from the three-point bending tests.

Temperature	f_{R1} [MPa]		f_{R2} [MPa]		f_{R3} [MPa]		f_{R4} [MPa]	
	Avg.	CoV	Avg.	CoV	Avg.	CoV	Avg.	CoV
20 ºC	20.3	10.1%	13.5	14.9%	8.4	16.5%	5.3	21.6%
400 ºC	14.0	16.9%	7.3	17.0%	4.1	17.2%	2.5	15.9%
600 ºC	7.5	0.0%	1.8	0.0%	0.8	0.0%	0.3	0.0%
700 ºC	0.8	16.2%	0.2	22.3%	0.1	40.5%	0.0	-

4. Conclusions

This work presented and discussed the results of an experimental program to investigate the mechanical behaviour of reactive powder concrete, RPC, reinforced with steel fibres when subjected to high temperatures ranging from 400 to 700 ºC. Based on the experimental results presented in this work, the following main conclusions can be pointed out:

1. RPC specimens could not be fully dried at 60 ºC for a period of 24 hours. For this drying condition, it was observed an explosive spalling failure at a temperature of approximately 450 ºC. On the other hand RPC specimens that have been dried at a temperature of 100 ºC for the same curing period did not exhibit explosive spalling failure, even at 700 ºC temperatures.
2. The compressive, flexural and tensile strength, as well as the elasticity modulus, decreased at a higher rate for temperatures higher than 400 ºC. For an exposure temperature of 600 ºC, the reduction on the compressive strength, elasticity modulus and flexural strength, when compared to the one obtained for room temperature, was about 50, 62 and 60%, respectively. On the other hand, the tensile strength revealed a slight decrease with temperature, however since only it was possible to test specimens exposed up to 500 ºC no definite conclusions can be withdraw.

3. In general, the residual uniaxial tensile strengths and the dissipated energy decreased with the increase of temperature, being the rate of decrease higher for temperatures above 400 °C.
4. The residual flexural tensile strengths also decreased with the increase of temperature, being also the rate of decrease higher for temperatures above 400 °C. Moreover, the decrease of f_{R3} with temperature was significantly higher than the one observed for f_{R1}, which could be ascribed to the fibre pull-out being on a much more advanced stage when f_{R3} is evaluated. The reduction of the fibre / matrix bond strength due to the extensive thermal induced micro-cracking enabled the fibres being pulled out for lower loads resulting in a lower crack bridging stresses.

Acknowledgements

The authors would like to acknowledge the Zhejiang Boen Company and MAPEI Company for providing gratuitously, respectively, the steel fibers and micro silica fume. The first author would also like to acknowledge the grant obtained under the scope of the Erasmus Mundus - Marhaba project. The third author wishes to acknowledge the grant SFRH/BSAB/114302/2016 provided by FCT.

References

[1] Cheyrezy M, Maret V, and Frouin L, 1995, Microstructural analysis of RPC (reactive powder concrete), J. Cem. &Con. Res. 25 1491–1500.
[2] Bonneau O, Poulin C, Dugat J, Richard P, and Aitcin P C , 1996, Reactive powder concretes: from theory to practice, J. Con. Inter. 18 47-9.
[3] Richard, P and Cheyrezy M H, 1994, Reactive powder concrete with high ductility and 200-800 MPa Compressive Strength, Concrete Technology Past, Present, and Future, ACI SP 144, editor; P. Kumar Mehta, S. Francisco, USA, 507-18.
[4] Shu-hua LIU, Li-hua LI and and Jian-wen FENG, 2012, Study on Mechanical Properties of Reactive Powder Concrete, J. Civ Eng. &Cons., 1:1. 6-11
[5] Richard P, and Cheyrezy M, 1995, Composition of reactive powder concrete", cement and concrete research, 25(7), 1501 –11.
[6] Schneider U, Diederichs U, Horvath J and et al., 2003, Verhalten von ultrahochfesten betonen (uhpc) unter randbeanspruchung (behaviour of ultra-high performance concrete (UHPC) under fire exposure), Beton- und Stahlbetonbau, 98(7), 408–17.
[7] Majorana C E and Pesavento F, 2000, Damage and spalling in HP and UHF concrete at high temperature," in Damage and Fracture Mechanics VI: Computer Aided Assessment and Control, vol. 6 of Structures and Materials, 105–17.
[8] Ratvio J, 2001, Ultralujan betonin kayttosovellukset Esitutkimus (preliminary study of ultra-strength concrete applications), VTT Tiedotteita, 2078, 3–45.
[9] Liu C T and Huang J S, 2009, Fire performance of highly flowable reactive powder concrete," J. Con. & Buil. Mat., 23(5), 2072–79.
[10] Ridha M M S, MohammadAli T Kh, Abbawi Z W, 2013, Behavior of axially loaded reactive powder concrete columns , J. Eng. & Dev., 17- 2, 193-209.
[11] Rahmatabadi M A D, 2015, Mechanical properties of reactive powder concrete under pre-setting pressure and different curing regimes, J. Str.& Civ. Eng. Res. 4(4), 354-58.
[12] Phan LT., Carino N J, 2001, Mechanical Properties of High Strength Concrete at Elevated Temperatures, Gaithersburg, Maryland: NISTIR 6726, Building and Fire Research Laboratory, National Institute of Standards and Technology,.
[13] Ali F, 2002, Is High Strength Concrete More Susceptible to Explosive Spalling than Normal Strength Concrete in Fire?, J. Fire Mater. 26 127-30.
[14] Zhang Q, Ye G, 2012, Dehydration kinetics of Portland cement paste at high temperature, J. Ther. Analy. & Cal. 110: 153-58.
[15] LNEC E397, 1993. Concretes – Determination of the compressive elasticity modulus. Technical report, National Laboratory for Civil Engineering, Lisboa, Portugal (in Portuguese).
[16] EN 14651, 2005. European Committee for Standardization, Brussels.

IOP Conf. Series: Materials Science and Engineering **246** (2017) 012021 doi:10.1088/1757-899X/246/1/012021

Flexural stiffness of the composite steel and fibre-reinforced concrete circular hollow section column

A Tretyakov [1], I Tkalenko [1], F Wald [1], J Novak [2], R Stefan [2] and A Kohoutková [2]

[1] Czech Technical University in Prague, Dept. of Steel and Timber Structures, Czech Republic

[2] Czech Technical University in Prague, Dept. of Concrete and Masonry Structures, Czech Republic

E-mail: alexey.tretyakov@fsv.cvut.cz

Abstract. The recent development in technology of production and transportation of steel fibre-reinforced concrete enables its utilization in composite steel-concrete structures. This work is a part of a project which focuses on development of mechanical behaviour of circular hollow section (CHS) composite steel and fibre-concrete (SFRC) columns at elevate temperature. Research includes two levels of accuracy/complexity, allowing simplified or advanced approach for design that follows upcoming changes in European standard for composite member design in fire EN1994-1-2 [1].

One part is dedicated to determination and description of flexural stiffness of the SFRC CHS columns. To determinate flexural stiffness were prepared series of pure bending tests at elevated and ambient temperature. Presented paper focuses on the results of the tests and determination of flexural stiffness at ambient temperature. Obtained outputs were compared to data of existing studies about concrete-filled tube members with plain concrete and values analytically calculated according to the existing European standard EN1994-1-1 [2].

1. Introduction

Concrete filled steel tube (CFST) members combine main advantages of both materials (concrete and steel) and provides better properties of the members in many aspects [3]. CFST gives more efficient construction process with no formwork needed rather than steel reinforced concrete composite. Infilling of the steel tube brings better mechanical properties of the members and increases fire resistance [4].

Numerous researches of CFST members' behaviour have been conducted in the last three decades. There is conventional set of questions in fire resistance assessment incorporated by their peculiar objects and characterized by principally similar ways of the solutions. Alberto, Romero and Espinos' latest researches were summarized in a proposal of a new method in EN1994-1-2 [1] for the fire design of concrete filled steel tubular columns [5]. This research is focused on investigating behaviour of CFST filled with plain and bar-reinforced concrete. Proposed approach aimed at developing European design solutions for CFST columns based on available test data, numerical and analytical simulations. However, EN1994-1-2:2005 [1] and most current proposition of standard development do not specify design solutions for composite steel and FRC columns in fire. Current practices of testing and modelling emphasize usage of higher strength materials [3], regardless of previous focuses on normal strength concrete infill. Since 1954 more than 150 circular hollow section steel to concrete composite (CFST) columns were tested in fire and less than 5% had SFRC infill.

Adding steel fibres to mixture increases ductility, compressive and tensile strength of concrete [6] and as a result flexural stiffness of the member, which is more efficient for subjecting to combined loads at ambient and elevated temperatures. Compared to the filling of CHS with plain concrete, SFRC's fiber plastic tension capacity to hold appearing cracks increases the volume of member's resistance to combined compression and bending [7]. Comparison of typical behaviour in tension of plain and SFR concrete is shown on *Figure 1.1*.

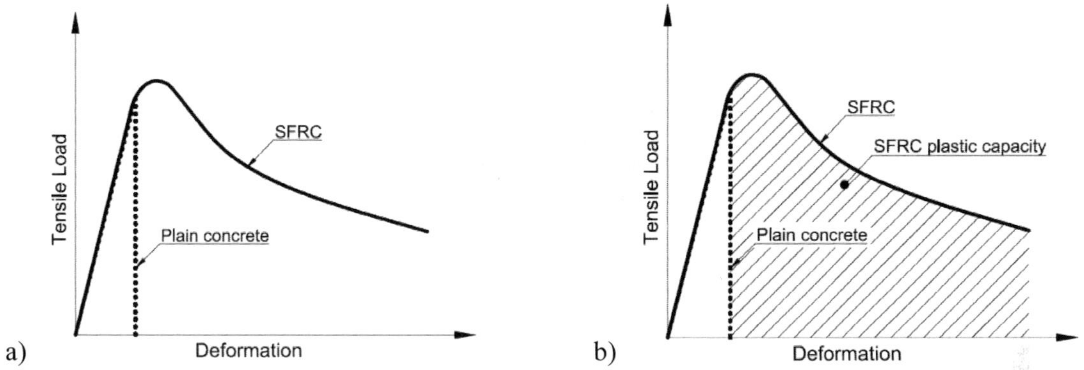

Figure 1.1 a) Behaviour in tension of plain and SFR concrete [7] **b)** Delineation of SFRC plastic capacity

From design rules of EN1994-1-1 [2] follows that improvement of flexural stiffness has negligible impact to buckling resistance. This demonstrates the importance of experimental and numerical investigation of flexural stiffness and determination of relevant design rules for CHS filled with SFRC.

Flexural behaviour of CHS sections filled with plain concrete was researched by many authors. Lin-Hai Han [8] [9] presented results from experimental studies on rectangular and circular CFST members and their comparison with different design approaches. Method used in his research was applied to compare flexural stiffness from current bending tests on CHS infilled by SFRC.

This paper shows the results obtained from the pure bending experiments at ambient temperature. Calculated values from tests are compared to EN1994-1-1 [2] design approach and to the selected results from Lin-Hai Han research [8] [9]. Received data is an inseparable part of further research focused on development of mechanical behaviour of circular hollow section (CHS) composite steel and fibre-concrete (SFRC) columns at elevated temperature.

2. Material properties

A welded circular hollow section with 323,9 mm diameter and 5 mm wall thickness made from steel S355, was chosen for the experiment.

The concrete composite is composed of easy available and widely used components which include Portland cement 42,5 R characterised by high early strength in accordance with CSN EN 197-1[10], water and siliceous aggregate. The workability of fresh concrete mass was maintained by using plasticizer Sika Visco Crete 1035 which also reduces the content of used mixing water. The chosen fibre reinforced concretes also contain two types of fibres. Single hook end steel fibres Dramix RC-80/60-BN with tensile strength 1225 MPa serve as reinforcement in the concrete composite. Whereas polypropylene fibres were added in concrete with an aim to reduce the risk of explosive concrete spalling which occurs at higher temperature at elevated temperature tests [11] [12].

IOP Conf. Series: Materials Science and Engineering **246** (2017) 012021 doi:10.1088/1757-899X/246/1/012021

3. Prue bending test

3.1. Specimens

For experiments were produced 3 m long specimens. To prove and prevent influence of friction between concrete and steel tube the specimens was produced in two types: with fully open ends (specimen B1) and with shear plate (specimen B2). List of the specimens is given in *Table 3.1.* Beams were stored for 28 days in the warmed-up warehouse. First test was carried out 29 days after specimens concreting.

Table 3.1. Specimen list

Specimen	Tube diameter, mm	Wall thickness, mm	Length, mm	End type	Quantity
B1	323,9	5	3000	Open ends	1
B2	323,9	5	3000	With shear plate	1

3.2. Test setup

Tests were performed at CTU in Prague laboratory. To reach a pure bending effect in specimen was chosen a four-point bending setup. During the test was measured deflection, tension and compression at steel tube, and displacement of the concrete beside the steel tube at the ends of the beam. Strain gauges were placed at the bottom and top surface of the steel tube at the distance 200 mm from the middle. Deflection was measured at 1/3 and 1/2 length of the beam. Setup of the test is shown on *Figure 3.1* and *Figure 3.2*.

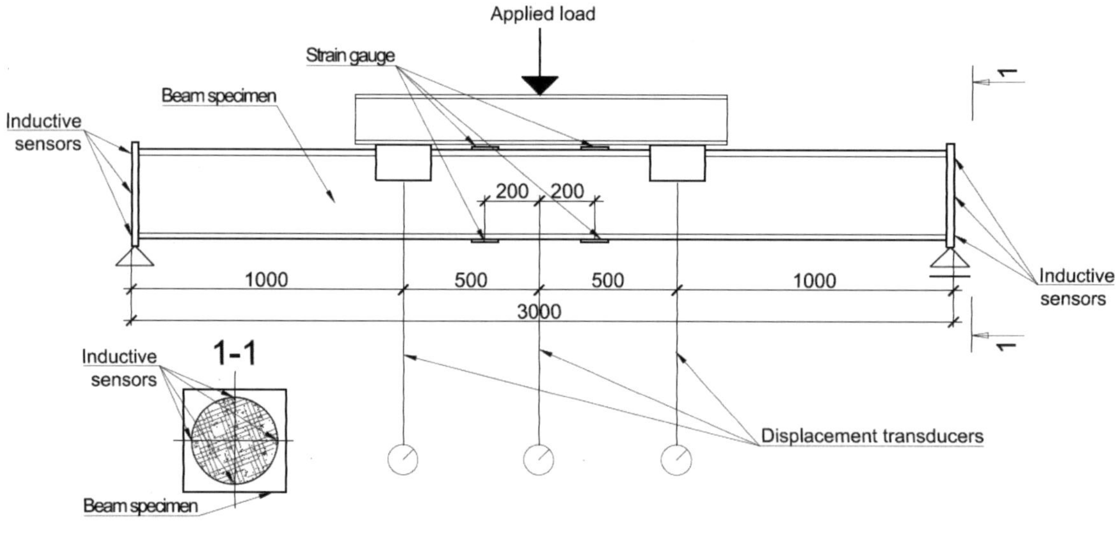

Figure 3.1 Bending test setup scheme.

a)

b)

Figure 3.2 Bending test setup photo a) Full view b) Inductive sensors setup.

3.3. Results

Dependence of vertical deflection in the middle of the specimen length curves is apparent on *Figure 3.3*. Behaviour of CFHS with SFRC member in bending can be divided into four stages: cracking linear, post-cracked linear, post-cracked non-linear and failure.

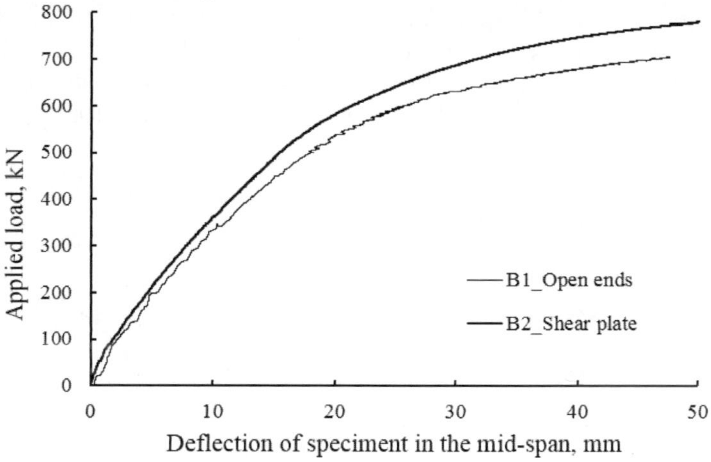

Figure 3.3 Vertical deflection in the 1/2 of specimen's length

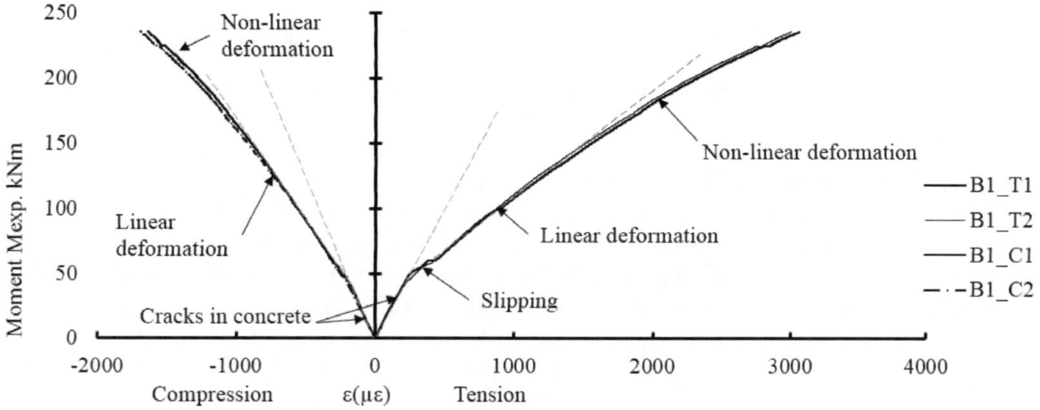

Figure 3.4 Moment-strain curves for specimen B1

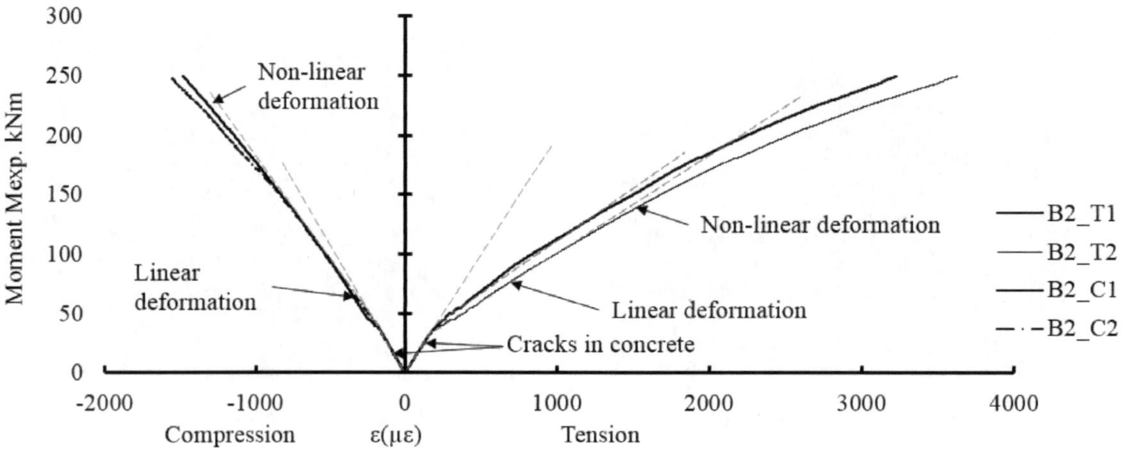

Figure 3.5 Moment-strain curves for specimen B2

Cracking linear stage ends with initial cracking of SFRC in tension area of composite section, post-cracked linear stage takes place until steel tube is yields, post-cracked non-linear stage is characterized by increasing of plastic deformation until failure. Typical moment-strain curve with determination of stages is shown on *Figure 3.4* and *Figure 3.5*. From the curve of specimen B1 on *Figure 3.4* could be seen the concrete core slipping influence, which was verified from the inductive sensors data.

From the moment-curvature diagram given on *Figure 3.6* it is apparent, that flexural stiffness decrease after cracks in concrete core growth. Flexural stiffness in first linear stage could be assumed as initial flexural stiffness of the section $EI_{exp.}$.

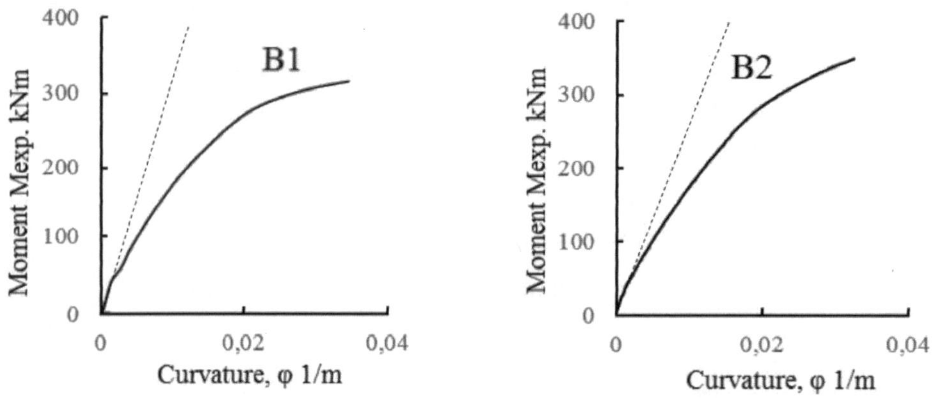

Figure 3.6 Moment-curvature diagrams

4. Determination of flexural stiffness and comparison

Determination of flexural stiffness for steel tubes filled with concrete without fibres is based on the effective member stiffness which could be assumed as addition of flexural stiffness of steel CHS and flexural stiffness of concrete core adjusted by correction factor [2]. According to EN1994-1-1[2] effective flexural stiffness of the member in comparison is defined as:

$$(EI)_{eff} = E_a I_a + 0{,}6 E_{cm} I_c \tag{4.1}$$

IOP Conf. Series: Materials Science and Engineering **246** (2017) 012021 doi:10.1088/1757-899X/246/1/012021

where E_a is the modulus of elasticity for the steel tube, E_{cm} is the modulus of elasticity for concrete, I_a and I_c are moments of inertia of the steel tube and concrete core without cracks. Modulus of elasticity for SFRC infill was taken from the experimental and numerical studies of mechanical properties and was adjusted as E_{cm}=36 GPa. Modulus of elasticity for the steel tube was adjusted as E_a=210 GPa. Values of flexural stiffness $(EI)_{eff}$ taken from the Lin-Hai Han research papers [8] [9] were changed according to adjusted E_a value. Results of the analytical calculation, together with selected values from Lin-Hai Han research, are given in *Table 4.1*. Comparison of the initial flexural stiffness $(EI)_{eff}$ / $EI_{exp.}$ ratio for members with SFRC and plain concrete infill [9] is shown on the *Figure 4.1*.

Table 4.1. Analytical flexural stiffness calculation results

Specimen	Dimension D x t x L	$(EI)_{eff}$ [kNm²]	$EI_{exp.}$ [kNm²]	$(EI)_{eff}$ / $EI_{exp.}$	Note
B1	323,9x5x3000	23700	31479	0,75	Current
B2	323,9x5x3000	23700	31555	0,75	
CVB-1	200x1,9x1400	2891	3414	0,85	Han *et al* (2006)
CVB-2	200x1,9x1400	2891	3216	0,90	
CB5-1	180x3x1800	2305	2383	0,97	
CB6-1	180x3x1800	2305	2371	0,97	
CB6-2	180x3x1800	2305	2321	0,99	

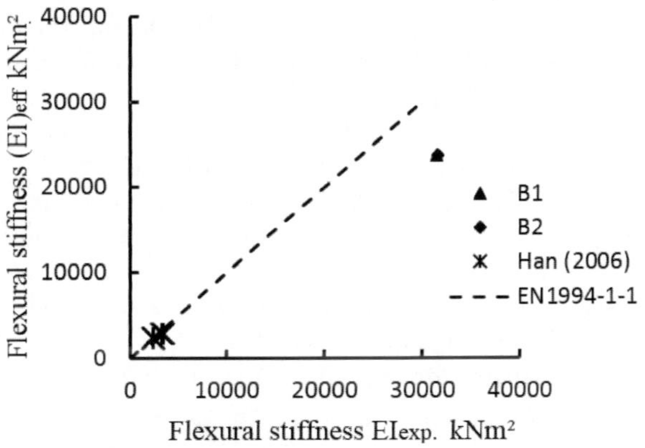

Figure 4.1 Comparison of the initial flexural stiffness values

Value of the flexural stiffness $(EI)_{eff}$ for all examined members is lower than obtained from the tests. From the presented comparison could be seen that, current specimens with SFRC infill have lower $(EI)_{eff}$ / $EI_{exp.}$ initial stiffness ratio than that with plain concrete core.

5. Conclusion

Two pure bending tests at ambient temperature were performed. Outputs from the experimental studies were analysed and presented. Experimental and design initial flexural stiffness of CHS with SFRC core were determined and compared to that for CHS with plain concrete infill. From the comparison is apparent that compared to analytical method in EN1994-1-1[2], initial stiffness for members with plain concrete infill is lower. Stiffness ratio $(EI)_{eff}$ / $EI_{exp.}$ for specimens with plain concrete core was determined in range of 0,85-0,99. Flexural stiffness ratio for current specimens B1 and B2 was determined as 0,75. Therefore experimental initial flexural stiffness from current tests is 25% higher than at EN1994-1-1[2] design approach and up to 24% higher compared to that for CHS with plain concrete infill. For the both current specimens, effective flexural stiffness obtained from EN1994-1-1[2] design approach could be assumed as conservative.

Given conclusion leads to the further numerical and experimental analyses. As a next step for the studies of CHS with SFRC core at ambient temperature, could be specified influence of the shear plates, determined and compared moment capacity and serviceability-level of section flexural stiffness [8][9].

Obtained data would be used for the research of steel CHS with SFRC infill behaviour at elevated temperature.

Acknowledgment
The authors would like to express their sincere gratitude to the Czech Science Foundation for the help provided through the project "Models of steel and fibre concrete composite columns exposed to fire", carried out with a financial grant GAČR 15-19073S.

References

[1] CEN. EN 1994-1-2, *Eurocode 4: Design of composite steel and concrete structures. Part 1-2: General rules – Structural fire design*. 2005,Brussels, CEN

[2] CEN. EN 1994-1-1, *Eurocode 4: Design of composite steel and concrete structures. Part 1-1: General rules and rules for buildings*. 2004, Brussels, CEN

[3] Liew J.Y.R., Xiong M.X., Xiong D.X. 2015 *Design guide for concrete filled tubular members with high strength materials to Eurocode 4*, Research Publishing, Singapore

[4] Zhao X., Han H. and Lu H. 2010 *Concrete-filled tubular members and connections*. New York: Spon Press

[5] Albero V., Espinos A., Romero M.L., Hospitaler A., Bihina G., Renaud C. 2016 *Proposal of a new method in EN1994-1-2 for the fire design of concrete filled steel tubular columns*. Engineering Structures vol. 128, pp. 237–255

[6] ACI Committee 544, 1988 *Design Considerations for Steel Fiber Reinforced Concrete,* ACI 544,4 R American Concrete Institute, Farmington Hills, MI

[7] Adhikari S., Patnaik A. 2012 *Potential applications of steel fibre reinforced concrete to improve seismic response of frame structures* NED University Journal of Research, pp. 113-128

[8] HAN, L.-H. 2004 *Flexural behaviour of concrete-filled steel tubes* Journal of Constructional Steel Research vol. 60(2), pp. 313-337

[9] HAN L.-H., Hui LU, Guo-Huang YAO and Fei-Yu LIAO 2006 *Further study on the flexural behaviour of concrete-filled steel tubes* Journal of Constructional Steel Research 62(6) pp.554-565

[10] The Czech Standard ČSN EN 197-1 *Cement - Part 1: Composition, specifications and conformity criteria for common cements,* ČNI, 2002, (in Czech)

[11] Novák J. and Kohoutková A. 2017 *Fire response of Hybrid Fiber Reinforced Concrete to High Temperature. In: Procedia Engineering* Lausanne: Elsevier, Modern Building Materials, Structures and Techniques, vol. 172, pp. 784-790

[12] Goremikins, V., Blesak, L., Novak, J. & Wald, F. 2016 *Experimental method on investigation of fibre reinforced concrete at elevated temperatures*, Acta Polytechnica, vol. 56, no. 4, pp. 258-264

IOP Conf. Series: Materials Science and Engineering **246** (2017) 012022 doi:10.1088/1757-899X/246/1/012022

Multi-scale Pull-Out Resistance of Steel Reinforcing Bar embedded in Hybrid Fiber Reinforced Concrete (HyFRC)

A Lin[1] and C P Ostertag[1]

[1]Department of Civil and Environmental Engineering, University of California, Berkeley, CA 94720, USA

E-mail: alin1988@berkeley.edu

Abstract. This paper investigates the pull-out resistance of a steel reinforcing bar embedded in a strain hardening hybrid fiber reinforced concrete that utilizes both polyvinyl alcohol (PVA) microfibers and hooked-end steel macrofibers. Both unconfined and specimens confined with transverse steel spiral reinforcements were investigated. Pull-out tests on rebars embedded in unconfined concrete revealed brittle splitting failure whereas rebars embedded in unconfined HyFRC exhibited ductile frictional pull-out behavior. The high early rebar pull-out resistance in HyFRC is linked to an improved pull-out resistance of the steel macrofibers due to the micro/macrofiber synergy. Steel macrofiber pull-out tests revealed that the PVA microfibers are very effective in enhancing the pull-out resistance of the steel macrofibers which then lead to an overall performance enhancement of the rebar pull-out resistance.

1. Introduction

A Hybrid Fiber Reinforced Concrete (HyFRC) was developed combining polyvinyl alcohol (PVA) microfibers and hooked-end steel macrofibers to obtain strain hardening behavior at relatively low fiber volume fraction [1]. When damage is induced in HyFRC, the PVA microfibers and steel macrofibers are effective in bridging the micro and macrocracks respectively. By providing such multi-scale crack resistance, the HyFRC can achieve both a strain hardening and deflection hardening behavior despite a fiber content as low as 1.5% [1]. The superior mechanical properties and improved workability due to the low fiber content make HyFRC a practical material to enhance the performance of reinforced concrete structures.

This paper aims to investigate the pull-out resistance of a steel reinforcing bar embedded in a HyFRC matrix. The pull-out resistance is a function of the rebar/matrix bond which strongly affects the performance of reinforced concrete members. When the HyFRC is combined with longitudinal steel reinforcing bars and tested in uniaxial tension, significant tension stiffening was observed due to the rebar/matrix bond enhancement and the continuous load carrying capacity of the matrix [2].

This paper reports on results obtained from rebar pull-out tests that were conducted to provide insights into the mechanisms responsible for the bond enhancement. Pull-out tests have been performed on rebars embedded in high performance fiber reinforced composites that utilize one single type of fiber [3] but not yet in a HyFRC matrix. In addition, hooked-end steel macrofibers are being pulled out of mortar matrices reinforced with PVA microfibers to investigate the macro/microfiber synergy and its effect on the pull-out performance of rebars embedded in a HyFRC matrix.

2. Experimental program

2.1. Rebar pull-out test

2.1.1. Materials. Rebar pull-out tests were performed with specimens that either utilize a HyFRC or a concrete control matrix without fibers (denoted here as CC). The HyFRC mix design is given in table 1. Coarse aggregates with maximum size of 3/8 inch (9.52 mm) and Vulcan sand with 2.9 fineness modulus were used for both the HyFRC and concrete control specimens. Three types of fibers were utilized in HyFRC: PVA microfibers and two types of hooked-end steel macrofibers, SF-1 (DRAMIX RC-80/60-BN fiber) and SF-2 (DRAMIX ZP305 (Bekaert) fiber). The dimensions and mechanical properties of the fibers are given in table 2.

Table 1. HyFRC mix design (per m^3)

Mixture	Cement (kg)	Fine aggregate (kg)	Coarse Aggregate (kg)	Water (kg)	SF-1 (% vol.)	SF-2 (% vol.)	PVA (% vol.)	Water-to-cement ratio
HyFRC	423	852	774	228	0.8	0.5	0.2	0.54

Table 2. Fiber dimensions and mechanical properties

Fiber	Material	Length (mm)	Diameter (mm)	Strength (MPa)	Stiffness (GPa)
SF-1	Steel, hook end	60	0.75	1225	210
SF-2	Steel, hook end	30	0.55	1350	210
PVA	Polyvinyl alcohol	8	0.038	1600	40

2.1.2. Test specimens. Pull-out tests were performed on unconfined and confined HyFRC and unconfined and confined concrete control specimens. All confined specimens were reinforced with transverse steel spiral reinforcements. The dimensions of the rebar pull-out specimens are shown in figure 1. The geometry of the specimens are similar to the ones described in [4]. All rebar pull-out specimens consist of a ribbed No. 8 steel rebar (with dimensions given in table 3) centrally embedded in a cylindrical concrete matrix (152.4 mm of diameter and 304.8 mm of height). The ratio between concrete cover thickness and rebar diameter (c/d_b) is 2.5. The rebar at the top and the bottom of the specimen was debonded by PVC (Polyvinyl Chloride) pipes which are shown as dashed lines in figure 1. The bond length (L_e) of the rebar was 3.5 times the rebar diameter (d_b) and included approximately 5 ribs. The confined specimens that contain transverse steel spiral reinforcements are shown in figure 2. The spiral wire is an ASTM A641, Zinc–Coated (Galvanized) Carbon Steel with a tensile strength of 483 MPa (70 ksi) and a diameter of 3.43mm. Transverse reinforcements of 0.75% and 1.5% were provided by 25.4mm and 12.7 mm of spiral spacing respectively. Information on all rebar pull-out specimens are given in table 4. Specimens that were not confined by spirals are denoted as NC. Specimens confined by spirals are denoted as C with the % of transverse reinforcement given in parenthesis. For example CC-C(1.5) indicates that the concrete control specimen is confined with 1.5% transverse spiral reinforcement.

An external vibrator was used during casting to achieve good compaction. After casting the specimens were moist cured for 28 days in their plastic molds with their top faces covered by wet fabrics and plastic sheets. Both cylindrical HyFRC and CC specimens (102x203mm) were cast and tested in compression to determine their compressive strength.

Table 3. Rebar characteristics

Rebar size	Diameter (mm)	Height of rib (mm)	Rib spacing (mm)	Rib height/rib spacing
No. 8	25.4	2.845	18.29	0.156

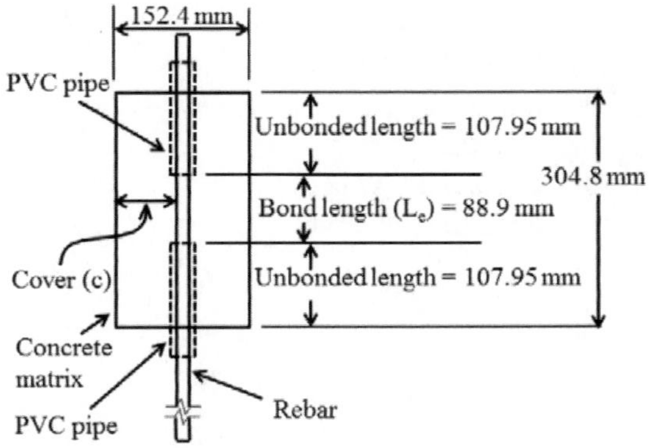

Figure 1. Dimensions for rebar pull-out specimens

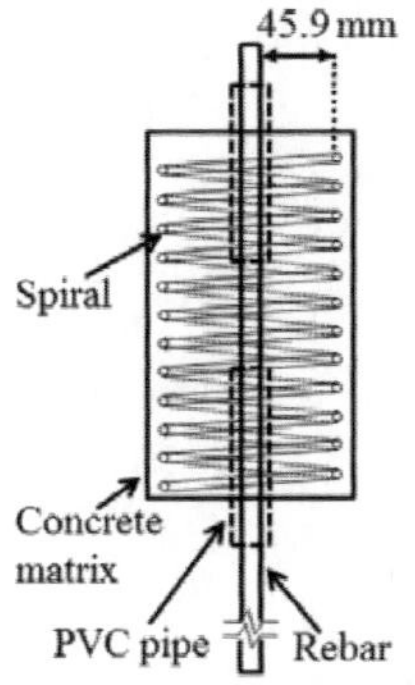

Figure 2. Spiral configuration for rebar pull-out specimens

Table 4. Rebar pull-out Specimens

Specimen name	Presence of spiral	Spiral Reinforcement	Spiral spacing (mm)	Number of specimens
CC-NC	No	0	-	3
CC-C(0.75)	Yes	0.75%	25.4	3
CC-C(1.5)	Yes	1.5%	12.7	3
HyFRC-NC	No	0	-	3
HyFRC-C(0.75)	Yes	0.75%	25.4	3

2.1.3. Test set-up and procedure. Monotonic rebar pull-out tests were conducted on a 534 kN (120 kip) capacity universal testing machine. The test set-up is shown in figure 3. The matrix surrounding the rebar was supported by a bearing plate on the upper machine head and the rebar was gripped by the lower machine head. The pull-out experiment was performed by moving the upper head upwards at a rate of 2.54 mm/min. The debonded region of the rebar at the bottom of the specimen kept the bond region away from the bearing plate to prevent local confinement. The load applied to the rebar was recorded by a load cell and the slippage of the rebar was measured with potentiometer 1 shown in figure 3.

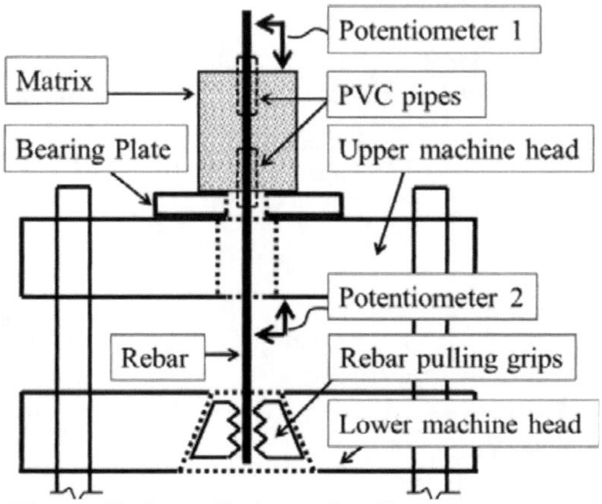

Figure 3. Overall view of pull-out test set-up

2.2. *Fiber pull-out test*

2.2.1. Materials. Mortar mixtures for the hooked-end steel macrofiber pull-out tests are given in table 5. These mortar mixtures contain the same fine aggregates and PVA microfibers as the rebar pull-out specimens. The mortar mixture of the fiber pull-out specimens was designed to be identical to the mortar proportions of the HyFRC mixture. Hence, the 0.29vol% of PVA microfibers used in the mortar matrix corresponds to the 0.2 vol% of PVA microfibers in the HyFRC specimens.

Table 5. Mixtures for fiber pull-out specimens (per m^3)

Mixture name	Cement (kg)	Fine aggregate (kg)	Water (kg)	PVA fiber (Vol. %)	W/C ratio
M-0%PVA	611.4	1232.9	330.2	0	0.54
M-0.29%PVA	609.6	1229.3	329.2	0.29	0.54

2.2.2. Specimens and testing. Figure 4 shows the dimensions of the fiber pull-out specimens. The steel macrofiber SF-1 (see table 2 for dimensions) is centrally embedded in a cylindrical mortar specimen of 38 mm in height and 38 mm in diameter. The embedment length of the steel fiber was kept constant for all specimens. All mortar specimens were vibrated by a table vibrator during casting to achieve good compaction and cured in a fog room for 28 days.

The fiber pull-out tests were conducted on a 267 kN (60 kip) capacity universal testing machine and the test set-up is shown in figure 5. The specimens were glued to the bottom plates by epoxy to induce a matrix stress field similar to the stress field around a fiber when the fiber is bridging a crack [5]. The grip with a set-screw clamps onto the fiber and the fiber is being pulled out with the upward movement of the upper machine head. The applied axial load and the fiber displacement were measured by a load cell and a potentiometer, respectively.

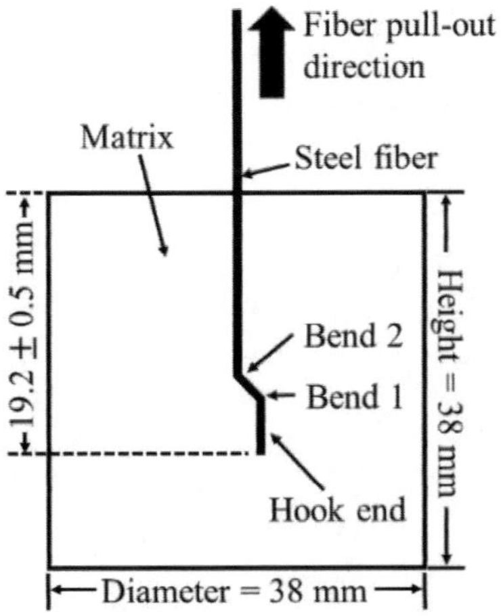

Figure 4. Dimensions for fiber pull-out specimens

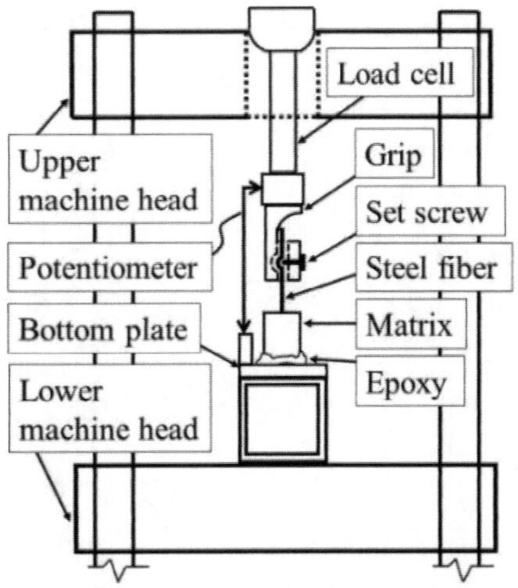

Figure 5. Set-up for fiber pull-out test

3. Test results

3.1. Rebar pull-out test

From the test results, the averaged bond stress along the bond length is calculated by the following equation:

$$\tau = \frac{F}{\pi\, d_b\, l_e} \qquad (1)$$

where τ is the bond stress, F the axially applied load, d_b the rebar diameter, and l_e the bond length. In equation (2) the bond stresses are normalized according to [4] to allow for direct comparison independent of f_c'.

$$\tau_{normalized} = \tau \left(\frac{36.8}{f_c'}\right)^{1/2} \qquad (2)$$

where $\tau_{normalized}$ is the normalized bond stress, τ the experimentally determined bond stress determined from equation (1), and f_c' the measured concrete compressive strength. The normalized bond stresses provide information on the rebar pull-out resistance. The results of the rebar pull-out tests are given in table 6. The normalized bond stresses are plotted against rebar slippage in figure 6. Figure 6a represents the total slippage of the rebar and figure 6b highlights the initial slippage of the rebar up to peak stress.

As shown in figure 6 the unconfined CC not only exhibits the lowest peak stress but also reveals abrupt failure after reaching the peak stress. Only confinement provided by transverse spiral reinforcements could prevent such brittle failure in CC specimens. On the other hand, the rebar embedded in HyFRC exhibits not only a higher peak bond stress but also a gradual post-peak softening behavior even without transverse confinement. In addition, only specimens with HyFRC matrices exhibit significant increase in pull-out resistance in the ascending branch prior to reaching the peak bond stress as evident from figure 6b.

Table 6. Results for rebar pull-out test program

Rebar pull-out specimen name	Averaged compressive strength (MPa)	Peak of normalized bond stress (MPa)		Slip at peak bond stress (mm)		Failure mode
		Individual value	Average	Individual value	Average	
CC-NC-1	41.93	14.74	15.16	0.965	0.851	Splitting failure
CC-NC-2		15.64		0.856		
CC-NC-3		14.95		0.729		
CC-C(0.75)-1		16.26	15.64	1.483	1.077	Frictional pull-out
CC-C(0.75)-2		15.43		1.059		
CC-C(0.75)-3		15.23		0.691		
CC-C(1.5)-1	36.50	18.53	18.05	1.361	1.674	Frictional pull-out
CC-C(1.5)-2		17.78		2.189		
CC-C(1.5)-3		17.91		1.468		
HyFRC-NC-1	40.04	20.19	20.39	0.808	1.021	Frictional pull-out
HyFRC-NC-2		20.33		1.349		
HyFRC-NC-3		20.60		0.907		
HyFRC-C(0.75)-1		25.70	23.98	1.808	1.783	Frictional pull-out
HyFRC-C(0.75)-2		22.67		1.544		
HyFRC-C(0.75)-3		23.49		1.999		

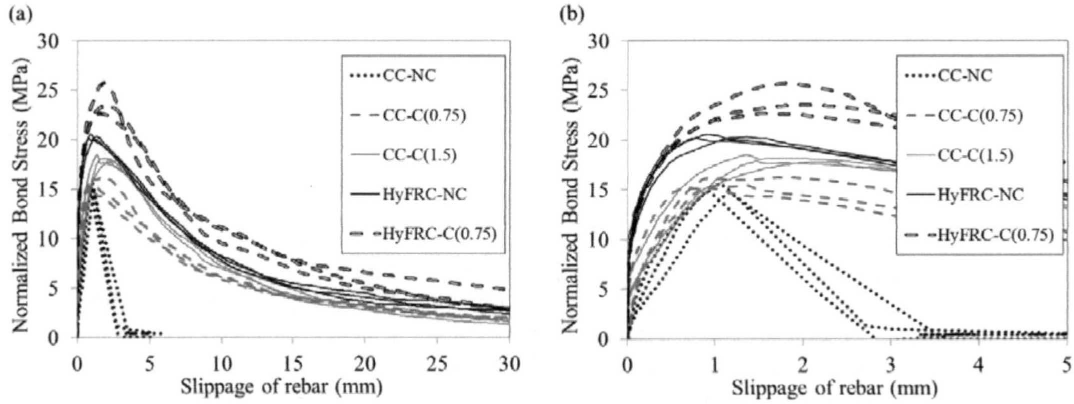

Figure 6. Rebar pull-out responses; (a) total slippage of rebar; (b) slippage of the rebar in the ascending portion up to the peak bond stress.

Two to four fully developed splitting cracks formed in the unconfined CC specimens at the peak bond stress. These spitting cracks then propagate in an unstable manner and lead to the observed brittle splitting failure shown in figure 7. On the other hand, by either adding transverse spiral reinforcements to the CC matrix or replacing the CC matrix with HyFRC the crack pattern changes and the brittle splitting failure can be prevented. Instead of a few dominant splitting cracks a crack pattern with multiple hair-line splitting cracks develops in the radial direction. In figure 8 which represents a typical saw-cut cross section of unconfined HyFRC specimens these multiple fine splitting cracks are indicated by dashed lines.

IOP Conf. Series: Materials Science and Engineering **246** (2017) 012022 doi:10.1088/1757-899X/246/1/012022

In both unconfined HyFRC specimens and confined CC specimens, frictional pull-out failure was observed instead of the brittle splitting failure which dominated the CC specimens.

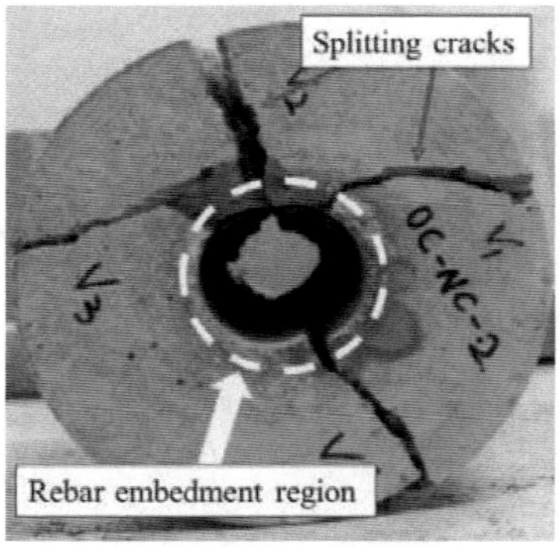

Figure 7. Splitting failure for unconfined CC specimens

Figure 8. Saw-cut cross section for HyFRC-NC after frictional pull-out failure. The dashed lines represent the multiple fine splitting cracks

3.2. *Fiber pull-out test*

The results of the steel fiber pull-out test conducted with unreinforced and PVA microfiber reinforced mortar matrices are given in table 7. The fiber displacement includes both the elastic elongation and the slippage of the fiber.

Table 7. Results of fiber pull-out test program

Fiber pull-out specimen name	Peak load (N)		Fiber displacement at peak load (mm)	
	Individual value	Average	Individual value	Average
M-0%PVA-1	330.3	334.6	0.89	0.89
M-0%PVA-2	340.2		0.86	
M-0%PVA-3	338.4		0.95	
M-0%PVA-4	331.9		0.83	
M-0%PVA-5	325.5		0.88	
M-0%PVA-6	341.5		0.91	
M-0.29%PVA-1	329.5	339.2	0.79	0.87
M-0.29%PVA-2	324.3		0.87	
M-0.29%PVA-3	325.4		0.89	
M-0.29%PVA-4	380.1		0.97	
M-0.29%PVA-5	333.5		0.84	
M-0.29%PVA-6	342.7		0.84	

The average load-fiber displacement curve for each type of mortar is plotted in figure 9 with the total displacement of the fibers given in figure 9a and the ascending portion up to peak load highlighted in figure 9b. The pull-out response of the steel macrofiber embedded in either the unreinforced or PVA microfiber reinforced mortar matrices are similar in shape exhibiting two peaks, P1 and P2, as shown in figure 9a. These peaks are associated with the plastic deformation of the hooked ends of the steel fiber at two locations [5] indicated in figure 4 by bend 1 and bend 2.

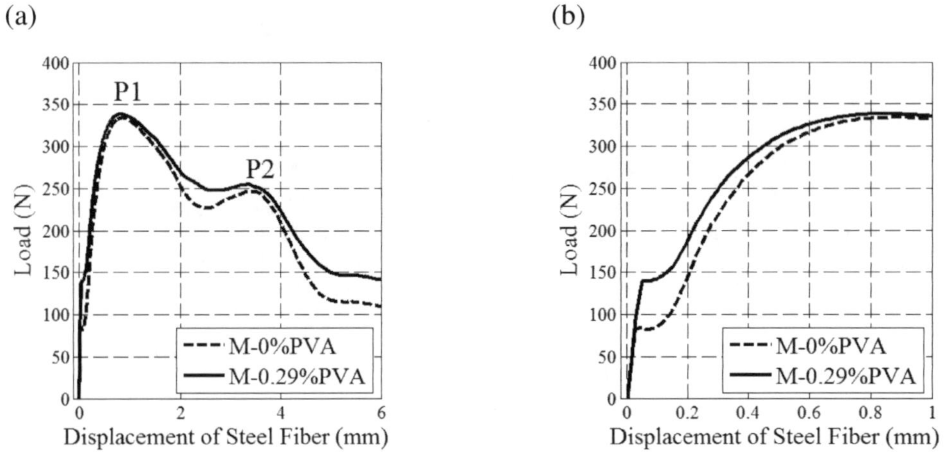

Figure 9. Averaged fiber pull-out responses in (a) high displacement range (b) low displacement range

The difference in fiber pull-out behavior between unreinforced and PVA fiber reinforced mortars is more apparent in the ascending portion of the curves as shown in figure 9b. The curves exhibit a linear ascending branch followed by a plateau like region. The linear portion of the curve is being extended from 82.8N in unreinforced to 139.1N in PVA microfiber reinforced mortar specimens. The linear ascending branches in figure 9b are due to the elastic elongations of the steel fibers. After the plateau region fiber slippage becomes the dominant mechanism.

4. Discussion

4.1. Pull-out resistance between rebar and concrete matrix

The mechanical interlock between rebar ribs and the surrounding concrete matrix provides most of the pull-out resistance. When the rebar slips against the matrix, the radial components of the inclined bearing force between rebar ribs and the matrix induce a tensile stress field in the matrix as schematically shown in figure 10. The tensile stresses are responsible for the formation of splitting cracks. The brittle failure observed in the unconfined CC matrix was caused by the unstable propagation of such splitting cracks. In HyFRC on the other hand, the propagation of splitting cracks is resisted by both micro and macrofibers. The PVA microfibers are effective in bridging and arresting such cracks at onset while the steel macrofibers resist their widening and propagation. Both, the bridging of PVA microfibers and steel macrofibers induce intense crack closure stresses onto the crack surfaces as shown in figure 11. These crack closure stresses which result in normal stresses acting on the rebar enhance the rebar/matrix bond and increase the pull-out resistance of the rebar and ultimately change the failure mode to a more ductile frictional pull-out behavior.

While spiral reinforcements also induce normal stresses on the rebar during rebar pull-out the unconfined HyFRC provides higher pull-out resistance compared to the CC matrix reinforced with 0.75% spirals. At least 1.5% of spiral reinforcement was required to confine the CC matrix to achieve similar peak stress and post-peak behavior as unconfined HyFRC. Adding spiral reinforcement to HyFRC further increases the pull-out resistance due to the additional normal stresses acting on the rebar.

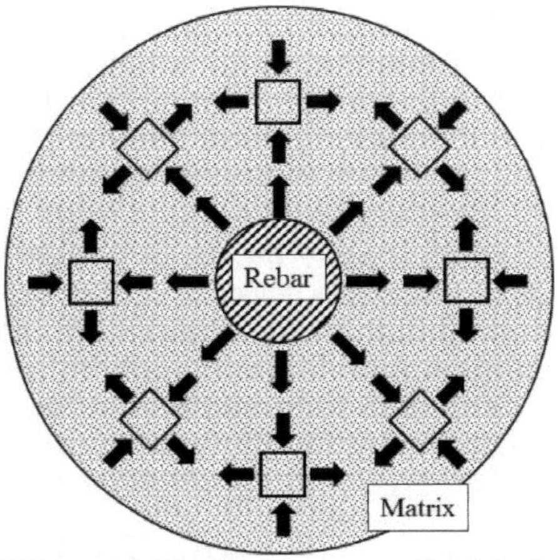

Figure 10. Tension ring stress field during rebar pull-out

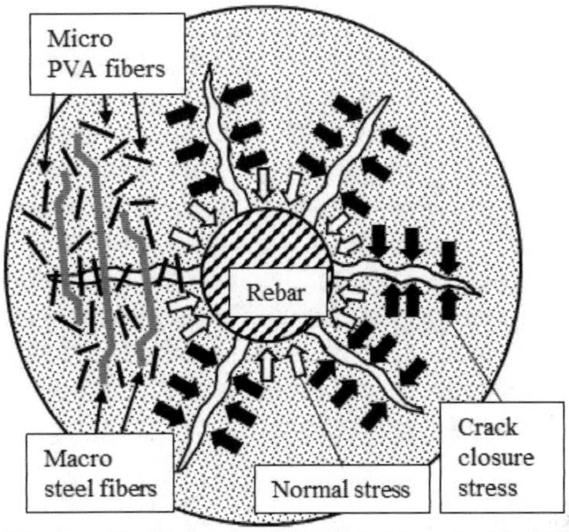

Figure 11. Rebar bond mechanism in HyFRC

Besides the higher peak and post peak bond stress observed in unreinforced HyFRC compared to unreinforced and spiral reinforced CC the ascending curves (shown in figure 6b) are far steeper in unreinforced HyFRC. The steeper ascending curves reveal a rather significant slip stiffening behavior up to the peak stress, while the addition of spiral reinforcements in both CC and HyFRC had only a minimal effect on the ascending curve. These differences can be explained by the effectiveness of fibers and the lack of spiral reinforcement in resisting splitting crack formation at onset. The fibers are located close to the rebar and hence are able to bridge the splitting cracks and prevent their opening during the early crack initiation phase. On the other hand, the transverse spiral reinforcement was not mobilized until the splitting cracks have already reached a certain length and crack width. Hence, the minimal effect of the spiral reinforcement on slip stiffening can be explained by the lack of crack closure stresses acting on the splitting cracks during the crack initiation phase.

4.2. Micro/macro fiber synergy in a HyFRC-rebar system

When steel macrofibers bridge the splitting cracks that form during rebar pull-out, the fibers initially stretch elastically and then eventually start to pull out. Once fiber pull-out takes place both the crack closure stresses acting along the splitting cracks and the normal stress acting on the rebar are being reduced which changes the pull-out response of the rebar. However, in HyFRC the pull out of the steel macrofibers is delayed up to higher load levels due to the presence of the PVA microfibers. When hooked-end steel macrofibers pull out of the matrix, microcracks form around these steel fibers. These microcracks, if not bridged by PVA microfibers, weaken the bond between the matrix and the steel fibers which reduce the pull-out resistance of the steel fibers.

The observed increase in pull-out resistance of the rebar embedded in HyFRC can be explained by a process that occurs on multiple scales. On the smaller scale, the pull-out resistance of the steel fibers is enhanced by PVA microfibers bridging the microcracks around the steel fiber which enhances the fiber/matrix bond. On the larger scale, the pull-out resistance of the rebar is enhanced by the steel macrofibers bridging the splitting cracks and their ability to maintain higher closure stresses across the splitting crack surfaces due to the presence of PVA microfibers. The higher normal stresses acting on the rebar due to the higher closure stresses enhance the rebar/matrix bond. This micro/macrofiber synergy provides the basis for the enhanced pull-out resistance of the rebar when pulled out from a HyFRC matrix.

5. Conclusion

A rebar embedded in a HyFRC matrix exhibits a far higher pull-out resistance compared to the same rebar being pulled out from a concrete matrix. In addition, even an unconfined HyFRC matrix is able to change the failure mode from a brittle splitting failure to a more ductile frictional pull-out behavior. Although transverse spiral reinforcements also provide a frictional pull-out failure in confined concrete control specimens, a spiral reinforcement as high as 1.5% is required to achieve peak/post-peak rebar pull-out resistance comparable to the resistance provided by unconfined HyFRC. While HyFRC enhances the pull-out resistance prior to the peak bond stress only a small enhancement could be observed with transverse spiral reinforcements. The experimental results reveal that this enhancement in HyFRC was partially due to a steel fiber bond enhancement induced by PVA microfibers. In a HyFRC member reinforced by a steel reinforcing bar, such multi-scale pull-out resistance can provide superior composite behavior between rebar and matrix.

References

[1] Blunt JD and Ostertag CP 2009 Deflection hardening and workability of hybrid fiber composites *ACI Mater. J.* **106** p. 265-72.

[2] Moreno DM, Trono W, Jen G, Ostertag C, and Billington SL 2014 Tension stiffening in reinforced high performance fiber reinforced cement-based composites *Cem. Concr. Compos.* **50** p. 36-46.

[3] Krstulovic-Opara N, Watson KA, and LaFave JM 1994 Effect of increased tensile strength and toughness on reinforcing-bar bond behavior *Cem. Concr. Compos.* **16** p. 129-41

Hota S and Naaman AE 1997 Bond stress-slip response of reinforcing bars embedded in FRC matrices under monotonic and cyclic loading *ACI Struct. J.* **94** p. 525-36

Chao S-H, Naaman AE and Parra-Montesinos GJ 2009 Bond behavior of reinforcing bars in tensile strain-hardening fiber-reinforced cement composites *ACI Struct. J.* **106** p. 897-906.

[4] Eligehausen R, Popov EP and Bertero VV 1983 Local bond stress-slip relationships of deformed bars under generalized excitations *Earthquake Engineering Research Center Report* No. UCB/EERC-83/23, University of California, Berkeley, CA

[5] Markovich I, van Mier JGM, and Walraven JC 2001 Single fiber pullout from hybrid fiber reinforced concrete *HERON* **46** p. 191-200.

IOP Conf. Series: Materials Science and Engineering **246** (2017) 012023 doi:10.1088/1757-899X/246/1/012023

Application of UHPC thin-walled elements for multi-purpose use table

O Slabý[1], J Vašková[1] and V Veselý[2]

[1] Department of Concrete and Masonry Structures, Faculty of Civil Engineering, CTU in Prague, Thákurova 7, 166 29 Prague 6, Czech Republic
[2] Betotech, s.r.o., Beroun 660, 266 01 Beroun, Czech Republic

E-mail: ondrej.slaby@fsv.cvut.cz

Abstract. This paper deals with thin-walled elements made of high performance concrete and the application of these elements in practice. This material is nowadays used more often in civil engineering and also in architecture, it enables realization of lighter and thinner products. The paper presents design and manufacturing of a ping-pong table for multi-purpose use, which is assembled from thin-walled UHPC elements. The intention was to utilize the advantages of the material and design and construct a demountable table, which could be easily transported; hence the weight of each element will be lower than 100 kg. The thickness of the concrete board of the table was only 15 mm. The paper describes the whole development of the product – from the design and the software verification, the development of casting technologies and the development of suitable concrete composite to the final realization and testing of the manufactured table.

1. Introduction

Nowadays progress and research in the area of high performance concrete and concrete technology allows to produce more interesting and thin-walled structures and concrete products. The research of UHPC is mainly focused on increasing the load-bearing capacity, ductility, durability and other required properties. The application of various types of fibres in the composite is significant in the development of UHPC. The usage of suitable additives and admixtures is also necessary. Nowadays concrete is used not only as a structural material, but it is used more often also as a material for furniture, for construction of machines, for façades tiles or other complex shape products. High performance fibre reinforced concrete compared to the composite without fibre material has significantly higher tensile strength and toughness and can be successfully used for thin-walled components with thicknesses of 10-15 mm without traditional reinforcement.

The aim of the research was to exploit the advantages of the UHPC material and make a ping pong table for multi-purpose use. Common existing tables are mostly represented by green, massive, heavy and relatively unattractive products. The main idea was to design a table from separate elements that can be easily transported without lifting machine, only by two persons. So the maximum weight of each element could be 100 kg. It was necessary to optimize the dimensions of each element. The aim was also to design simple shape of particular elements that could be simply joined together.

The concreting of the samples, the testing and the final production of the pilot table were realized in cooperation with Betotech, s.r.o. in Králův Dvůr.

The realization of the project was divided in stages:
- Design of elements and their optimization
- Numerical analysis of load cases
- Design of the suitable concrete mixture and its testing
- Technology of concreting and production of pilot samples
- Manufacturing of the table and its testing

2. Product design

The table consists of two halves – two separate tables. Each table is composed of two basic elements – the table board and the table legs. The advantage of the proposed table is that each half can be used separately – for example as a dining table for outdoor or indoor use.

Figure 1. UHPC table from thin-walled elements

The table board is rectangular with dimensions of 1525 x 1370 mm and its weight is almost 100 kg. The thickness of the board is only 15 mm. The board is stiffened by the ribs – edge rib around the periphery and two lengthways ribs. The ribs provide better load distribution. Holes in the lengthways ribs are used for connection the board with the legs.

The table leg is a closed frame with cross sectional dimensions of 120 x 60 mm. The total size of the table leg is 1100 x 745 mm and the weight is 60 kg. There is a pair of notches in the upper part of leg used as a supporting of the table board.

The table composed of two boards and four table legs weighs 440 kg, that is one half 220 kg.

3. Load cases and numerical analysis

In the next step the designed table was analyzed by Finite Element Method in the Software Dlubal RFEM. Various load cases and load combinations, which may with high probability occur during use, manufacturing or transport, were analyzed. There are no prescribed load values and load combinations in standards for this type of product. Therefore seven load cases were proposed that are likely to happen with high probability. As an example, load cases that are simulating people sitting on the edge of the table, uniformly distributed load on the table board or load case that is simulating the transport the assembled table, are mentioned here. One of the load cases was horizontally applied force, simulating pushing the table by a table tennis player. This load case was important for the design of the connection between the board and the leg. The stability and straining of the joints were also checked. All the results were compared with the strength characteristics of the designed concrete. The maximum tensile stress from all the load cases was 14,9 MPa. The table was designed assuming linear behavior. The tension stresses calculated in numerical analysis for all loading cases were lower than tensile strength measured in laboratory tests, the cracks in concrete would not occur.

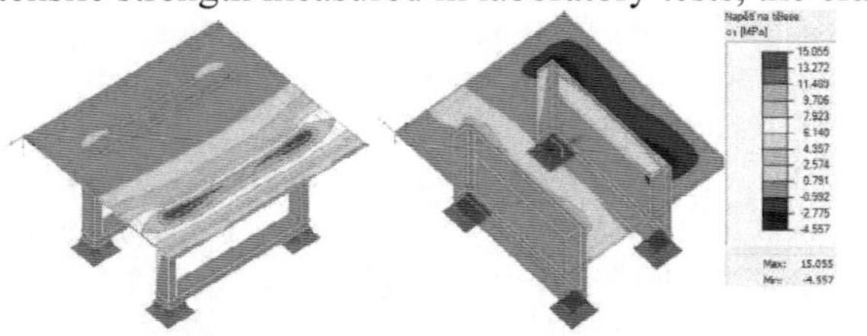

Figure 2. Load case – line load on the edge of the board 1,5 kN/m′ and numerical analysis of the load case

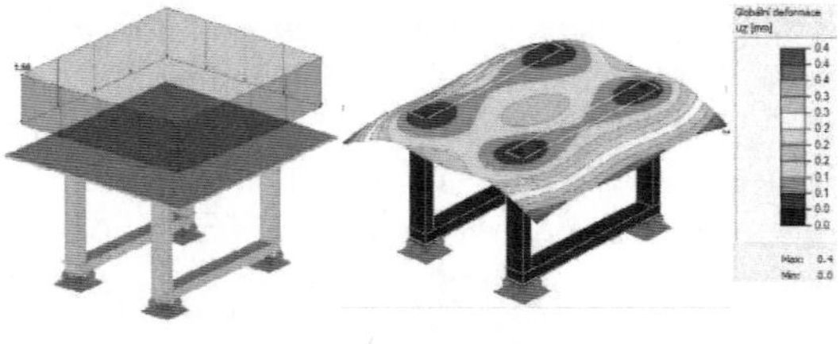

Figure 3. Load case – uniformly distributed load 1,5 kN/m2 and the result of board deformation

The results of the numerical analysis pointed out that the peaks of tension stress in area of contact and connection of the board and the leg must be eliminated. That was realized simply by an intermediate 4 mm thick layer made of special rubber. The rubber elements provide better contact of both elements (board and leg) and also compensate the irregularities and deviations generated during manufacturing. The rubber layer was also considered in the numerical analysis.

Connection of the legs and board is provided by eight steel connections – two in each leg. There is a internal thread glued in the lengthway rib of the board and a hole for the screw in the leg. The joint is then simply realized by a wrench for tightening the screws. The screw passes through the hole in the leg and is anchored into the internal thread in the board. The designed connection ensures perfect connection of elements, the overall stability and also enables fast and easy installation.

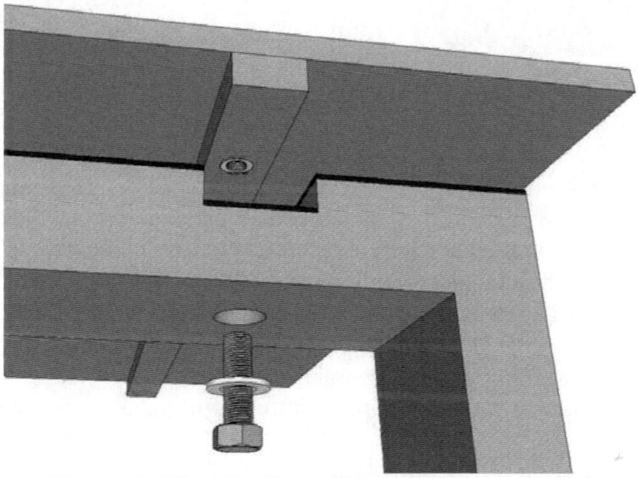

Figure 4. Visualization of board and table leg joint

4. Ultra-high performance fibre reinforced concrete

It was necessary to design completely new mixture of concrete for the designed table that will meet all the requirements – especially the high tensile strength (there is no other additional reinforcement in addition to fibres in the concrete mixture) and very good workability (because of the thickness of the board is only 15 mm). For the final mixture, Portland cement CEM I 52,5 was used, aggregate size up to 2 mm, microsilica, slag, superplasticizer and high strength steel fibres 10 mm long. The water-cement ratio was less than 0,27.

Several testing samples were realized for the verification of the designed concrete mixture. Tensile strength was determined in a three-point bending tests of prisms 40x40x160 mm. The tensile strength was 22 MPa after 28 days. A board with dimensions 700x150x15 mm was concreted to verify the

IOP Conf. Series: Materials Science and Engineering **246** (2017) 012023 doi:10.1088/1757-899X/246/1/012023

surface quality and edge quality. Smaller samples 150x40x15 mm were prepared by cutting of the 700x150x15 mm sample and they were tested in three-point bending test. These samples were important, because the thickness of the sample was same as the thickness of the final board of the table. The strength of these samples (influenced by "size effect" of the smaller thickness) was 15,5 MPa. The compressive strength tested on cubes 100x100x100 mm was 152 MPa. All the strength values satisfy all considered calculated load cases. After strength tests it was found that the fracture surface of the specimens contained an amount of air pores. This problem was solved by modifying the aggregate fractions and by modifying the amount of superplasticizer.

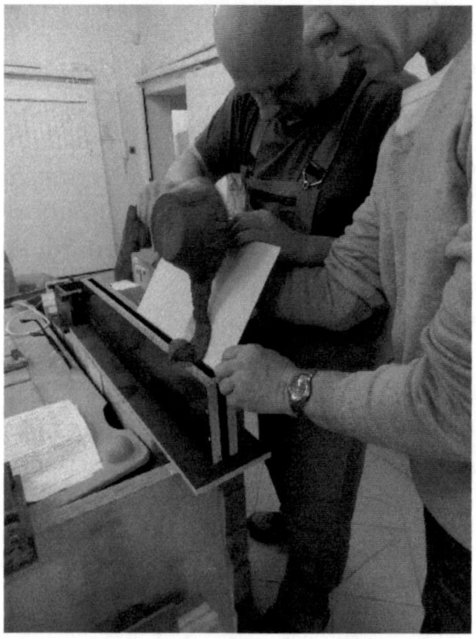

Figure 5. Concreting of the board sample 700x150x15 mm in vertical position

Figure 6. Detail of the fracture surface

There were manufactured more pilot products after all necessary strength tests. The aim of these pilot elements was to verify the possibility of concreting such thin and large elements, namely of the board. Pilot tests showed some problems that had to be solved and considered before the producing of final functional pilot table. The changes were made in the technology of concreting above all – originally it was supposed that the board will be concreted in the vertical position but after the pilot concreting it was necessary to change the technology and concrete the board in horizontal position. Pilot products also verified the in terms of appearance; the smooth surface of the concrete board and the edges. The good formwork was the most important thing for the next production of the table.

5. Manufacture and installation of the pilot table

The formwork of the board comprised of two parts and the concreting was realized in two phases. In the first phase, precise dose of concrete corresponding exactly with the thickness of 15 mm was concreted in the flat formwork. After the first phase the formwork was covered from above with the "negative" form and all the ribs were concreted. The formwork had to be very stiff and accurate because of the precise dimensions of each element. The formwork of the legs was more simple; it was comprised of a "closed frame".

Figure 7. Concreting of the board in horizontal position

Figure 8. Concreting of the ribs

After the maturity – 28 days it was possible to complete the table. It was necessary to drill the holes in the legs for the connecting elements, glue the threaded fitting into the holes in the ribs of the board and to put the rubber inserts. The first step of the installation was the placing of the legs which were provisionally fixed, then the boards were placed on the legs and connected with the connecting joints. The table was finally seated and leveled.

There were examined some of load cases considered in the numerical analyses after assembling the table. Perhaps most interesting was loading of the table board by 6 persons. In all load conditions the table met the expectations and safely comply with the requirements. The table was also tested by professional table tennis players J. Panský and M. Orlowski in an exhibition and the properties are convenient for the table tennis game.

Figure 9. The load test by 6 persons

Figure 10. Exhibition by table tennis players J. Panský and M. Orlowski

6. Conclusion

The aim of the research was to show the excellent properties of UHPC in reality and manufacturing of very thin and elegant elements of UHPC for the table for multipurpose use. All the elements had maximum weight of 100 kg, smooth surface and good durability in comparison with common tables. The advantage is also the simple and quick installation.

The technology of concreting and manufacturing of thin UHPC elements was verified by experiments and pilot products. The realization of this table enabled also testing the theory in practice and improving some manufacturing processes. It was verified that it is possible to produce such a thin and large elements that will withstand high levels of load. It should be mentioned that the price of this manufactured table is higher than tables from common materials, but the look, multipurpose use and the durability in exterior are significant advantages.

The article points out one of the possible usage of UHPC material with a great potential.

Acknowledgement

The work was supported by the grant No. SGS17/049/OHK1/1T/11 "Fibre Reinforced Concrete with Specific Structures - Behavioural Analysis and Applications". The tests and the final production were realized in the laboratory of Betotech s.r.o. The formwork was developed and realized in cooperation with PERI, spol. s.r.o.

References

[1] Slabý O 2015 Members from ultra-high performance concrete – design and applications, Bachelor´s thesis, Faculty of Civil Engineering, CTU in Prague

[2] Collepardi M 2009 *The New Concrete* Prague ČKAIT ISBN 978-80-87093-75-7

[3] Table tennis. Table tennis tables, functional and safety requirements, test methods, ČSN EN 14468-1, Czech office for standards, metrology and testing, Prague 2015

[4] Bartos P JM 2012 Fiber reinforced concrete *Concrete construction of the 21st century - Concrete with added value* 12 (separate annex of magazine) p 69-77 ISSN 1213-3116

[5] Zelenka T, Kohoutková A, Vodička J, Hrdoušek V, Broukalová I 2015 Design of pedestrian bridge from textile reinforced concrete *8th International Conference Fibre concrete 2015 - Technology, Design, Application* CTU in Prague pp. 1-9 ISBN 978-80-01-05683-7

[6] Tipka M, Vašková J 2015 Dependence of Crack Opening on Deflection During Bending Test of Fibre Reinforced Concrete *8th International Conference Fibre concrete 2015 - Technology, Design, Application* CTU in Prague pp. 1-8 ISBN 978-80-01-05683-7

IOP Conf. Series: Materials Science and Engineering **246** (2017) 012024 doi:10.1088/1757-899X/246/1/012024

Quantification of cracking localization in fibre-reinforced concrete beams

S Werzbrger[1], Y S Karinski[2] and A N Dancygier[1,2]

[1] Faculty of Civil and Environmental Engineering, Technion – Israel Institute of Technology, Technion City, Haifa 32000, Israel
[2] National Building Research Institute, Technion - Israel Institute of Technology, Technion City, Haifa 32000, Israel

E-mail: karinski@technion.ac.il

Abstract. The paper presents an analysis of cracking localization in beams made of fibre-reinforced concrete with conventional reinforcement. This phenomenon has been quantified by the ratio **m/n** between the number of the significantly wide cracks and the total number of flexural cracks that developed within the constant moment zone of the beams. It is shown that beams with larger reinforcement ratio had larger values of **m/n**, that is, lower cracking localization. For the given concrete mix, when the reinforcement ratios ρ were larger than ~2% cracking localization was diminished. Furthermore, for low values of ρ, **m/n** ratios of the specimens with lower fibre contents are somewhat larger than those of the specimens with larger content. It is also shown that as the reinforcement ratio decreases below ~0.5% the normalized mid-span ultimate deflections decreases as well, which corresponds with the observation of the cracking localization phenomenon for low reinforcement ratios. For larger reinforcement ratio, when **m/n** increases above ~0.6, there is no effect of cracking localization on the ultimate deflection. Moreover, in this range of r the values of d are ≳ 1.0, as commonly expected for RFC structural elements. It is noted that the ultimate deflection of the beams corresponds to their flexural ductility.

1. Introduction

Concrete toughness may be increased by addition of fibres [1,2], which are expected to influence the cracking process and reduce crack width and spacing [3]. Hence, it is also expected that beams made of fibre-reinforced concrete with conventional reinforcement (RFC) will exhibit improved crack control, as well as increased structural ductility. However, there are studies that indicate that RFC beams showed reduced ductility when they included relatively low amounts of conventional reinforcement [4-6]. This phenomenon is caused by cracking localization, when under the ultimate load, one or only few cracks widen more than the other flexural cracks do. As a result of this widening, the part of the rebars, bridging the crack, is significantly strained and this leads to a relatively short plastic hinge in these beams.

A similar phenomenon of cracking localization was observed in experimental studies of RFC bars under axial tension for both normal [7, 8] and high [7, 9,10] strength concretes. The experiments show that the localization caused by a given amount of fibers is more pronounced for low conventional reinforcement ratios [8]. Furthermore, the results show that for a given conventional reinforcement ratio, cracking localization is more significant for larger amounts of fiber content (and vice versa) [11, 12]. Note that the specimens in these experiments had a single, centrally-located rebar.

In order to explain this phenomenon a probabilistic model for tensile RFC bars has been introduced [13] and verified [8, 14] by the authors. Findings of these experimental and theoretical studies, which refer to axially loaded elements, can be applied in flexural elements, by considering this element as the tensile zone of a beam. However, this application cannot be done directly because of the differences between the tensile bar and tensile zone of a flexural element, such as the non-uniform strain distribution along the cross-section and the non-central location of the rebars.

This paper presents an additional analysis of the results [5], with regards to quantification of cracking localization.

2. General

The experimental program [5] included twenty five, four-point bending tests of fiber-reinforced concrete beam specimens with conventional deformed rebars. The specimens were normal strength concrete beams with cross-sections that were 240-mm width by 300-mm height and different amounts of conventional reinforcement and steel fibers. Each beam had a span of 3.2 m with 1.5 m between the two equal loads acting on it (the constant moment zone) and two 0.85-m shear spans. Figure 1 shows a scheme of the test setup. The tests were performed under manual displacement control till failure. Cracking was monitored during the test, before and after yielding of the longitudinal reinforcement.

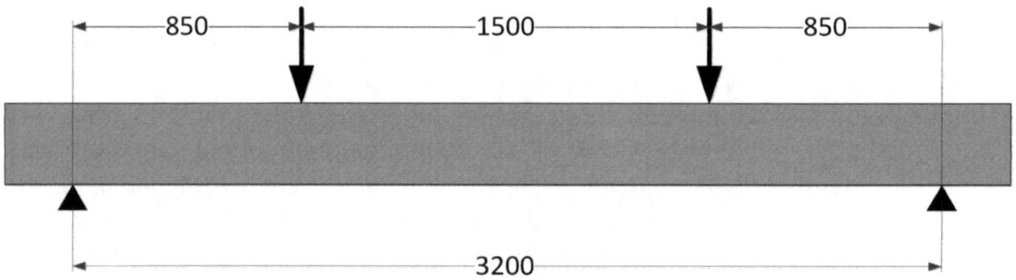

Figure 1. Scheme of the test setup (All dimensions are in mm).

Table 1. Details of the specimens.

Specimen type	No. of specimens	d_s (mm)	ρ_f (kg/m^3)	Longit. rebars	ρ (%)	f_c {1} (MPa)
NF-0-015	2		0			29.8 (1.7)
40-015	2	27	40	2ϕ8	0.15	30.9 (0.3)
NF-1-015	3		60			25.8 (1.0)
0-039	1		0			31.8 (0.6)
40-039	2	30	40	2ϕ8 + 2ϕ10	0.40	30.4 (0.4)
60-039	2		60			30.6 (0.2)
NF-0-063	1	33	0	2ϕ16	0.63	32.5 (0.4)
NF-1-063	3		60			25.4 (0.5)
0-094	1	33	0	3ϕ16	0.94	32.5 (0.4)
60-094	2		60			33.4 (0.5)
NF-0-126	1	33	0	4ϕ16	1.26	29.7 (1.0)
NF-1-126	2		60			26.7 (1.4)
NF-0-330	1	58	0	5ϕ22	3.27	26.5 (1.0)
NF-1-330	2		60			24.3 (0.9)

{1} Values in brackets denote standard deviations

The concrete mix was made of Portland cement type CEM I-52.5N with dolomite coarse aggregates (maximum size 19 mm), and natural quartz sand. The fibrous mixes included 35-mm long, hooked-end steel fibers with an aspect ratio of 64. The fibers content ρ_f was 40 and 60 kg/m^3 (0.50 and 0.76% in volume). The longitudinal conventional reinforcement consisted of deformed rebars with different reinforcement ratios ρ, which varied from 0.15 to 3.3%. In addition, control specimens of each reinforcement ratio but without fibers, were also tested. Details of the specimen notations, their reinforcement and the effective cover d_s (distance from the bottom of the cross-section to the center of the longitudinal reinforcement) are given in Table 1. In addition, the table includes the average concrete uniaxial compressive strength f_c that were taken by 150-mm cubes (28 days wet cured).

3. Results – cracking localization

Load-deflection curves of the beams, which are presented in [5] show significant reduction of flexural ductility of RFC specimens with low amounts of conventional reinforcement. This phenomenon is more pronounced for larger amounts of fibers, as illustrated in Figure 2 for ρ = 0.39%.

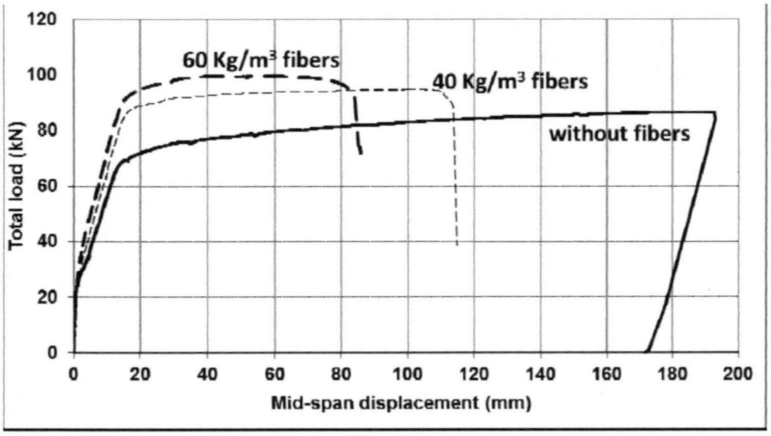

Figure 2. Reduction of flexural ductility in beam specimens with ρ = 0.39%.

In order to quantify the cracking localization that leads to this phenomenon, the following further analysis of the test results has been performed.

The flexural cracks that developed within the constant moment zone (i.e., between the loading points, see Figure 1) during the tests, were counted for each of the specimens (Table 1). The average number **n** for each type of specimen was taken into account in the analysis.

Post-test observation of the crack pattern shows that in the RFC beams with the lowest amount of conventional reinforcement (NF-1-015 and 40-015) only a single crack widened significantly compared with the other cracks. When ρ was increased, the number of significantly-wide cracks increased as well and for large reinforcement ratios no cracking localization was observed. Examples of cracked beams are shown in Figure 3.

For each specimen, the number of the significantly wide cracks was recorded after the test, and averaged for each type of specimen (these average numbers are denoted here as **m**). The ratio **m/n** ≤ 1.0 is a parameter that quantifies the level of cracking localization. Smaller values of **m/n** indicate more intensive cracking localization.

The results of the crack count in the RFC specimens, as well as the ratios **m/n** are listed in Table 2 and plotted versus the conventional reinforcement ratio ρ in Figure 4.

(a) Cracking localization (single wide crack) in specimen NF-1-015 (ρ = 0.15%)

(b) No cracking localization in specimen NF-1-126 (ρ = 1.26%)

Figure 3. Examples of post-test cracked RFC beams

Table 2. Results of the cracks count in the RFC specimens.

Specimen type	No. of specimens	n	m	m/n
40-015	2	11	1.5	0.14
NF-1-015	3	9.3	1	0.11
40-039	2	14	3.5	0.25
60-039	2	15.5	3.3	0.21
NF-1-063	3	12	7	0.58
60-094	2	12.5	8	0.64
NF-1-126	2	14.5	10	0.69
NF-1-330	2	12.8	12.5	0.98

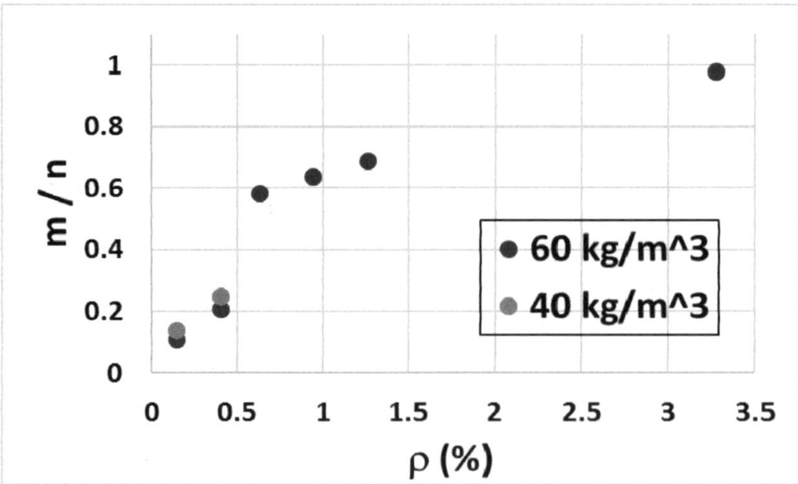

Figure 4. Cracking localization in RFC beams

It is evident from the figure that beams with larger reinforcement ratio had larger values of **m/n**, that is, lower cracking localization. For the concrete mix with 60 kg/m^3 that was tested, when the reinforcement ratios were larger than ~2% cracking localization was diminished. Furthermore, for low values of ρ, **m/n** ratios of the 40-kg/m^3 specimens are somewhat larger than those of the 60-kg/m^3 specimens. This result conforms to the trend, according to which no cracking localization is expected in reinforced concrete beams without fibers.

Figure 5 shows the mid-span ultimate deflections δ of the RFC beams, normalized with the ultimate deflections of the corresponding beams without fibers (the values of the deflections are reported in [5]).

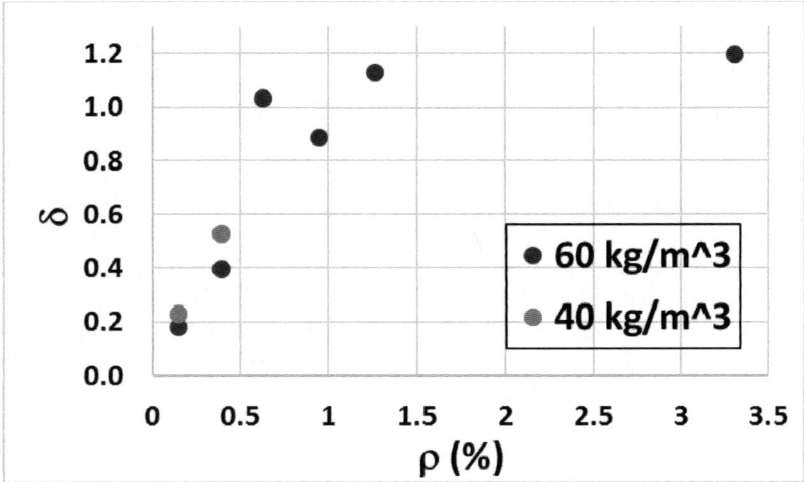

Figure 5. Normalized mid-span ultimate deflections δ of the RFC beams

It can be seen that as the reinforcement ratio decreases below ~0.5% the normalized mid-span ultimate deflections decreases as well. This result corresponds with the above observation of the cracking localization for low reinforcement ratios (see Figure 4). For larger reinforcement ratio, when **m/n** increases above ~0.6, there is no effect of cracking localization on the ultimate deflection. Moreover, in this range of ρ the values of δ are $\gtrsim$ 1.0, as commonly expected for RFC structural elements. It is noted that the ultimate deflection of the beams corresponds to their flexural ductility (e.g., [15]).

4. Conclusions

The paper presents an analysis of cracking localization in RFC beams, based on previously published experimental results. This experimental program included twenty five, four-point bending tests of fiber-reinforced concrete beam specimens with conventional deformed rebars. The cracking localization phenomenon has been quantified by the ratio **m/n** between the number of the significantly wide cracks and the total number of flexural cracks that developed within the constant moment zone of the beams.

It is shown that beams with larger reinforcement ratio had larger values of **m/n**, that is, lower cracking localization. For the concrete mix with 60 kg/m^3 that was tested, when the reinforcement ratios were larger than ~2% cracking localization was diminished. Furthermore, for low values of ρ, **m/n** ratios of the 40-kg/m^3 specimens are somewhat larger than those of the 60-kg/m^3 specimens. This result conforms to the trend, according to which no cracking localization is expected in reinforced concrete beams without fibers.

It is also shown that as the reinforcement ratio decreases below ~0.5% the normalized mid-span ultimate deflections decreases as well, which corresponds with the observation of the cracking localization phenomenon for low reinforcement ratios. For larger reinforcement ratio, when **m/n** increases above ~0.6, there is no effect of cracking localization on the ultimate deflection. Moreover, in this range of ρ the values of δ are ≳ 1.0, as commonly expected for RFC structural elements. It is noted that the ultimate deflection of the beams corresponds to their flexural ductility.

Acknowledgment
This work is partly supported by the Israeli Ministry of Immigrant Absorption (through a joint grant from the Centre for Absorption in Science and the Committee for Planning and Budgeting of the Council for Higher Education under the framework of the KAMEA Program). The research grant is greatly appreciated.

References
[1] Bentur A and Mindess S 2007 *Fibre Reinforced Cementitious Composites, 2ND Edition* (New Yory: Taylor & Francis)
[2] International Federation for Structural Concrete (fib) SAG 5 2013 *Model Code for concrete structures 2010* (Berlin: Wilhelm Ernst & Shon)
[3] Fantilli A P, Ferretti D, Iori I and Vallini P 1999 Behaviour of R/C elements in bending and tension: the problem of minimum reinforcement ratio *Minimum Reinforcement in Concrete Members*, ed Carpinteri A (Amsterdam: Elsevier) pp 99-125
[4] Schumacher P, Walraven J C, den Uijl J A and Bigaj-van Vliet A 2009 Rotation capacity of self-compacting steel fibre reinforced concrete beams *HERON* **54** pp 127-61
[5] Dancygier A N and Berkover E 2016 Cracking localization and reduced ductility of fiber-reinforced concrete beams with low reinforcement ratios *Eng. Struct.* **111** pp 411-424
[6] Dancygier A N and Savir Z 2006 Flexural behavior of HSFRC with low reinforcement ratios *Eng. Struct.* **28**(11) pp 1503-1512
[7] Deluce J R and Vecchio F J 2013 Cracking behavior of steel fiber-Reinforced concrete members containing conventional reinforcement. *ACI Str. J.* **110**(3) pp 481-90
[8] Dancygier A N, Karinski Y S and Navon Z 2016 Experimental quantification of the cracking localization phenomenon in tensile R/FRC bars. *Proc fib Symposium "Performance-based approaches for concrete structures, Materials technology, structural design, analytical modelling, conformity assessment and testing"*, Cape Town, South Africa
[9] Redaelli D and Muttoni A 2007 Tensile behaviour of reinforced ultra-high performance fiber reinforced concrete element. *Proc fib Symposium: Concrete structures: stimulators of development.* Dubrovnik, Croatia
[10] Redaelli D 2006 Testing of reinforced high performance fibre concrete members in tension. *Proc fib 6th Int PhD Symposium in Civil Engineering*, Zurich, Switzerland pp 122-123.
[11] Yang Y, Walraven J C and den Uijl J A 2009 Combined effect of fibers and steel rebars in high performance concrete *HERON* **54**(2/3) pp 205-224
[12] Dancygier A N, Karinski Y S and Navon Z Cracking localization in tensile conventionally reinforced fibrous concrete bars (submitted for publication)
[13] Dancygier A N and Karinski Y S 2014 Probabilistic model of the cracking localization in axially loaded fibrous reinforced concrete bars *Eng. Struct.* **79** pp 417-426
[14] Karinski Y S, Dancygier A N and Navon Z 2017 Experimental verification for a probabilistic model of fibers distribution along a reinforced concrete bar *Mater. Struct.* (In press)
[15] Park R and Paulay T 1975 *Reinforced Concrete Structures*, (New York : John Wiley & Sons)

IOP Conf. Series: Materials Science and Engineering **246** (2017) 012025 doi:10.1088/1757-899X/246/1/012025

Punching Shear in Steel Fibre Reinforced Concrete Slabs without Traditional Reinforcement

K H Tan[1] and A Venkateshwaran[2]

[1,2] Department of Civil and Environmental Engineering, National University of Singapore, Singapore

E-mail: tankh@nus.edu.sg

Abstract. The punching shear capacity of steel fibre reinforced concrete (SFRC) slabs without traditional steel bar reinforcement was investigated by conducting central point-load tests on twelve square slabs. The test parameters covered fibres with different multi-hook ends, concrete compressive strength, reinforcing index and slab thickness. The statistical performance of two existing models for the prediction of punching shear capacity of SFRC slabs without traditional reinforcement was examined. The load carrying capacity of these slabs were also assessed using the yield line theory. It is noted that the slabs failed primarily in flexure and the yield line theory predicted the load carrying capacities of the slabs most accurately. The reason for a flexural failure in SFRC slabs without steel bars is attributed to the lesser energy required in the propagation of an existing flexural cracks than in the creation of a new circumferential cracks around the column face.

1. Introduction

Punching shear is undesirable in structural concrete flat slab systems owing to its brittle nature, which may lead to progressive collapse of a building [1]. Several methods have been implemented to enhance the punching shear resistance of flat slabs and these include the use of traditional shear stirrups, shear studs, helix reinforcement and lattice reinforcement, apart from the provision of drop panels [2-6]. With test results affirming the success of steel fibre reinforcement in enhancing the punching shear resistance of slabs, several researchers and committees have considered its application in combination with traditional steel bar reinforcement [7-13].

Recent experimental studies reveal the possibility of constructing steel fibre reinforced concrete (SFRC) flat slabs devoid of traditional steel bars) [14-17]. Researchers have successfully used the yield line theory to analyse the flexural behaviour and predict the load-carrying capacity of such flat slabs [2, 16-17]. However, recent tests showed that SFRC flat slabs without traditional reinforcement could also experience a punching shear mode of failure under highly concentrated loads [17]. Considering the results obtained on different slab panels [17], there is uncertainty as to whether a pure flexural failure or a flexural failure leading to punching could occur under a concentrated load.

In this study, a test programme consisting of central point-load tests on twelve simply supported, square SFRC slabs without traditional reinforcement was carried out to examine the effect of volume fraction of steel fibres, slab thickness, concrete compressive strength and fibre hook-end configuration on the punching shear capacity. The performance of the yield line theory and two punching shear prediction models applicable to SFRC slabs without traditional reinforcement were evaluated and the failure mode further clarified.

2. Theoretical Considerations

In general, the punching shear behaviour of a slab can be examined by testing a slab simply-supported on all four edges under a concentrated load at the center. The edges simulate the lines of contra-flexure in a flat slab system. Under such a loading system, the slab would develop cracks in the radial direction at its bottom face. This could either lead directly to a punching shear failure, or to extensive cracking that might result in a flexural failure mode.

2.1. Existing Prediction Models for Punching Shear Capacity

The punching shear capacity of a SFRC slab, V_R, can be considered to comprise two components, that is, the contribution from "concrete", $V_{R,c}$, and the contribution from "steel fibers", $V_{R,f}$. Several international standards [3, 12, 18] have provided expressions to predict the "concrete" contribution. In general, the value of $V_{R,c}$ is a function of the tensile strength of concrete (expressed in terms of cylinder compressive strength of concrete, f_c') and effective depth, d, to the tensile steel reinforcement. EN 1992: 2004 [18] and fib Model Code 2010 [12] also consider $V_{R,c}$ to be governed by the reinforcing steel ratio and yield strength of steel reinforcement respectively. Also, several investigators on punching shear in SFRC slabs have proposed models to predict both the "concrete" and the "steel fiber" contribution to the punching shear capacity [7, 10, 11-13 19-21]. The "steel fiber" contribution, $V_{R,f}$, is primarily expressed as a function of the fiber content and thereby, the residual tensile strength of SFRC.

Among the several analytical and semi-analytical punching shear capacity prediction models, only those proposed by Choi et al. [11] and Grimaldi et al. [21] are applicable for SFRC slabs without traditional reinforcement. In other prediction models, the predicted load-carrying capacity reduces to zero when the percentage of steel reinforcement is equal to zero.

2.2. Flexural Capacity of SFRC Slabs Under Concentrated Load

The yield line theory has been used with reasonable accuracy to predict the flexural capacity of steel fiber reinforced concrete slabs [16-17]. Figure 1 indicates two possible yield line patterns for a square simply supported slab under a central point load.

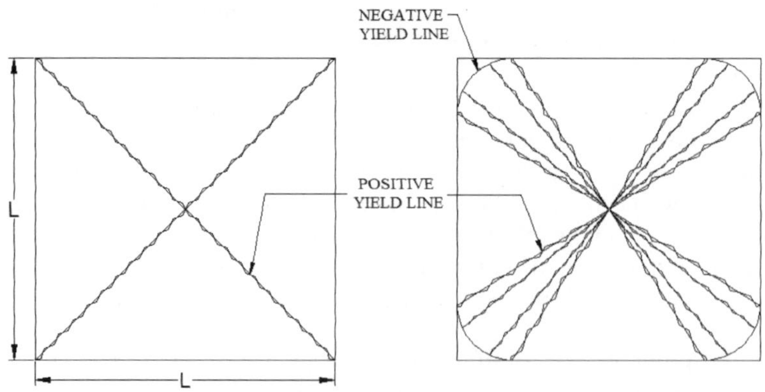

(a)Basic Pattern (b) Part-Fan Pattern

Figure 1. Yield line patterns

IOP Conf. Series: Materials Science and Engineering **246** (2017) 012025 doi:10.1088/1757-899X/246/1/012025

The basic yield line pattern (Figure 1a) gives a failure load, P_B, equal to $8M_B$, where M_B is the moment of resistance of the slab section in kNm/m while the part-fan yield line pattern (Figure 1b) predicts a failure load equal to $2\pi M_B$,. The actual yield line pattern observed in tests is however a mix of these two patterns. Since the yield line theory gives an upper bound solution, the lower of the two failure loads, that is, $2\pi M_B$ shall be used to predict the ultimate load carrying capacity in flexure of centrally loaded square slabs.

Based on the idealized stress-strain relations for SFRC, the strain and stress distribution across a SFRC section without traditional steel bars at ultimate limit state can be obtained as shown in Figure 2. A rigid-plastic stress block in tension is assumed with residual tensile strength equal to $\eta_1\eta_2 f_{Ftu}$, where f_{Ftu} = residual tensile strength of SFRC corresponding to the Model Code 2010 rigid-plastic model [12] (= $f_{R3}/3$) (of which test values are given in Table 1); and f_{R3} = residual flexural strength of SFRC determined from three-point bend tests on notched prisms in accordance with the BS EN 14651 [18]. Also, η_1 = reduction factor accounting for the effect of slab thickness on the fiber orientation (taken as $e^{0.0024(125-H)}$ [22]) with H being the thickness of the slab; and η_2 = reduction factor to account for the low fiber content along the lines of crack propagation in flexure (= 0.82 [23]).

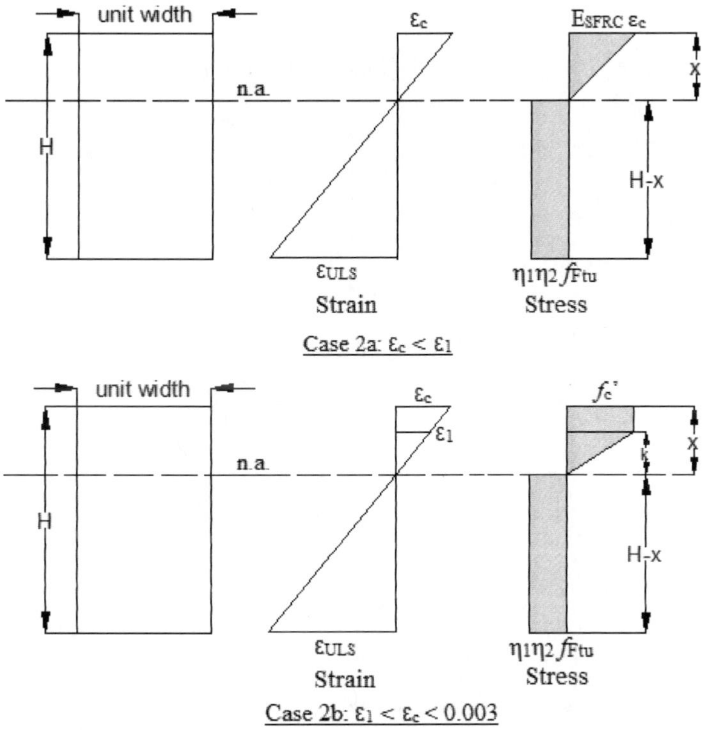

Figure 2. Strain and stress distributions at an SFRC section without traditional reinforcement at ultimate limit state.

In SFRC slabs without traditional bar reinforcement, the ultimate limit state for flexure occurs when the extreme tensile fibre strain, ε_t, reaches 0.02 or $2.5/H$, whichever is earlier [12]. The neutral axis, x, is determined by considering force equilibrium on the section, and the moment of resistance, M_B, is then calculated by taking moments of internal forces about the neutral axis.

IOP Conf. Series: Materials Science and Engineering **246** (2017) 012025 doi:10.1088/1757-899X/246/1/012025

3. SFRC Slabs Without Traditional Reinforcement

3.1. Specimen Details and Test Setup

Twelve square SFRC slabs without traditional reinforcement were tested under central point–loading. The slabs measured 700 mm by 700 mm in plan and had thickness varying from 90 mm to 150 mm. Details of the specimens are given in Table 1. Two types of hooked–end fibers, namely, 4D- and 5D-fibers having 1.5- and double-hook ends (Figure 3), respectively, were used in the concrete mixture.

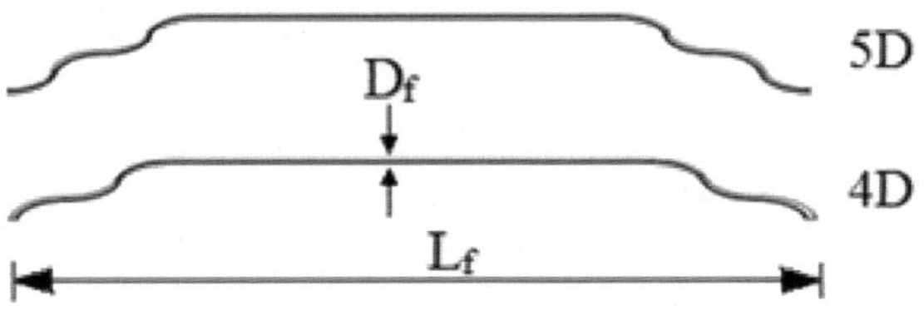

Figure 3. Steel fibers with multiple hook-ends.

Table 1. Details of tested SFRC specimens without traditional reinforcement.

Specimen Label*	Concrete Class	f_c' (MPa)	f_{Ftu} (MPa)	Fiber Type	Fiber volume (kg/m3)	Slab thickness (mm)	Remarks
M42	C30/37	36.0	1.66	4D	20	120	Variation of fiber volume
M44		36.0	2.21		40		
M46		31.4	2.70		60		
M48		36.1	3.30		80		
M52		37.4	1.67	5D	20		Variation of fiber volume
M54		36.5	2.21		40		
M56		40.8	2.80		60		
M58		42.2	3.36		80		
M54A		34.9	2.19		40	90	Variation of slab thickness
M54B		34.9	2.19		40	150	
L54	C16/20	21.1	2.03		40	120	Variation of concrete strength
H54	C50/60	51.8	2.36		40		

Note: All fibres were 60 mm long and 0.9 mm in diameter.

* Prefix L, M and H indicate low, medium and high strength concrete respectively. Suffix A and B indicate slab thickness of 90 and 150 mm, respectively.

Aggregates, sand, cement and water were added into an 80-litre pan mixer in sequence. To avoid balling, the steel fibers were separated by hand and added into the mixture in the last step to simulate the addition of fibers into a mix truck as highlighted in the guidelines by ACI 544.3R-93 [24].

The specimens were consolidated via internal vibration. After the concrete was allowed to harden for a day, the wooden moulds were removed and the slabs were left to cure under wet gunny sacks for the next 28 days. Concrete cubes were cured under the same conditions so as to ensure strength readings are representative of the slab.

The tests were conducted using the 100 ton MTS Actuator System. As shown in Figure 4, the load was applied over a patch measuring 100 mm by 100 mm at the centre of the slab, at a displacement rate of 0.1 mm/min. The slab was simply supported on a square steel frame with supporting lines measuring 600 mm on each side of the slab. At the bottom of the slab, a displacement transducer was placed at the centre to measure the slab deflection. The slabs were loaded monotonically to failure. After the peak load was reached, the loading rate was gradually increased to 0.3 mm/min and the slab was further loaded to obtain the descending portion of the load-deflection curve. The test was terminated when the load dropped to about 50 – 60% of the peak load, to ensure that no further load rebound would occur.

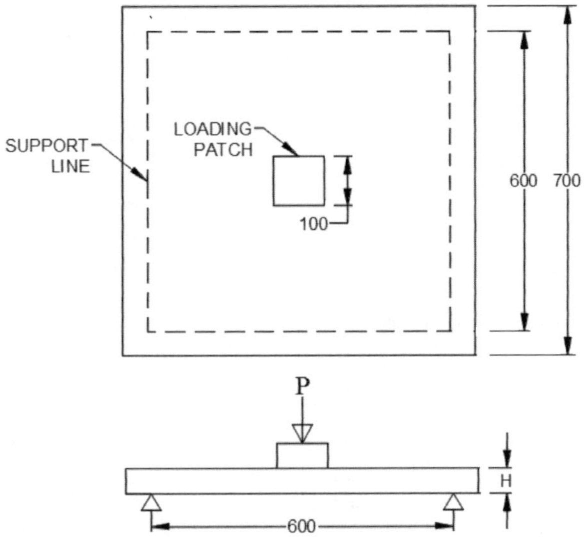

Figure 4. Central point load tests on SFRC slabs without traditional reinforcement.

3.2. Test Results

The typical crack pattern on the underside of the slabs is shown in Figure 5. The formation of yield lines is clearly observed indicating a flexural mode of failure. The load–deflection curves for the tested slabs are shown in Figure 6.

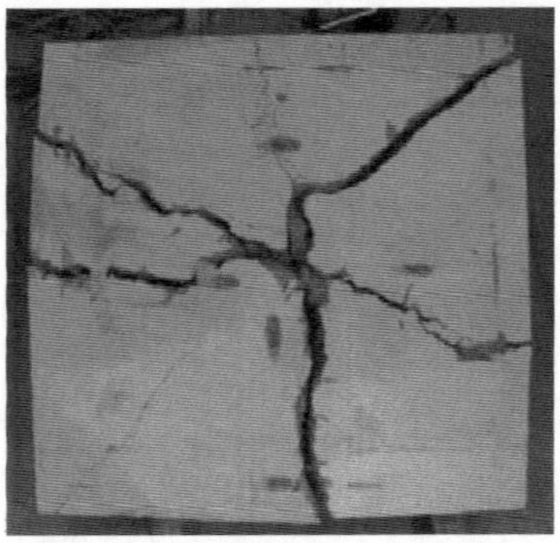

Figure 5. Crack pattern on the underside of slab M44.

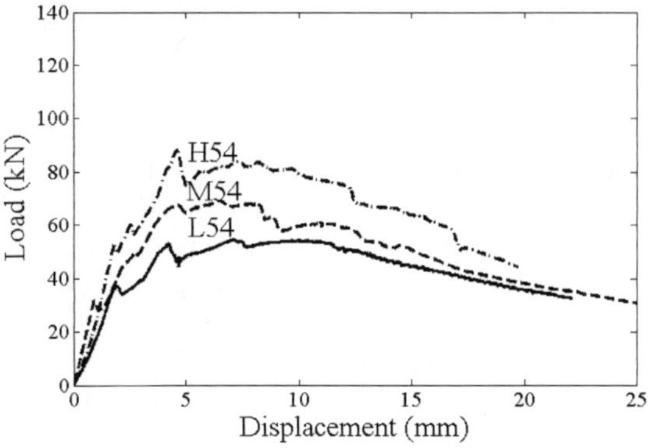

(a) Effect of concrete compressive strength

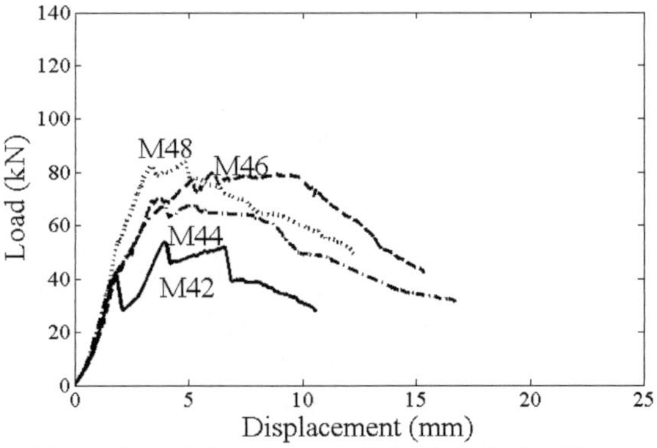

(b) Effect of fiber content (SFRC with 4D fibres)

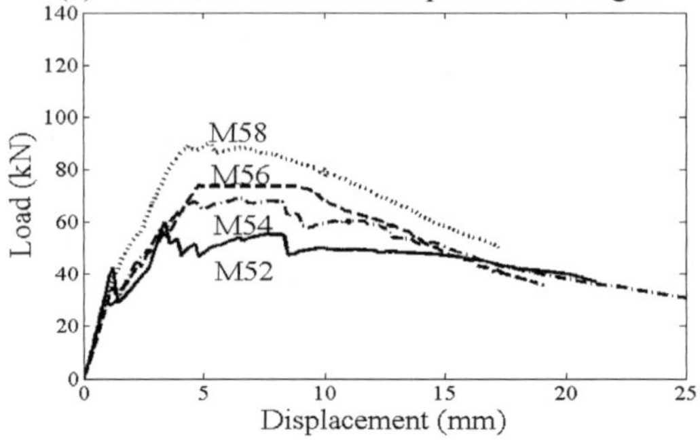

(c) Effect of fibre content (SFRC with 5D fibres)

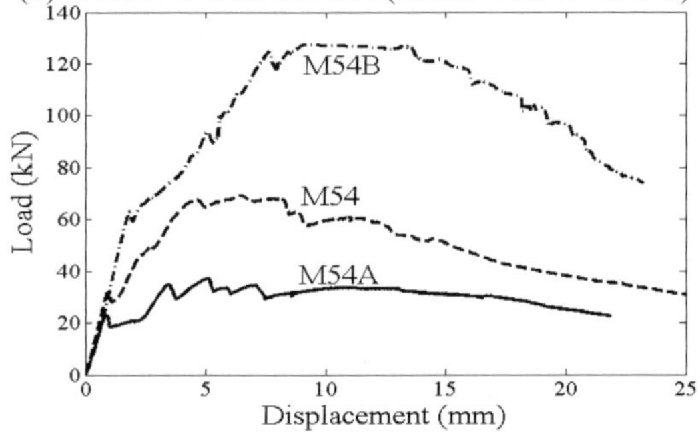

(d) Effect of slab thickness

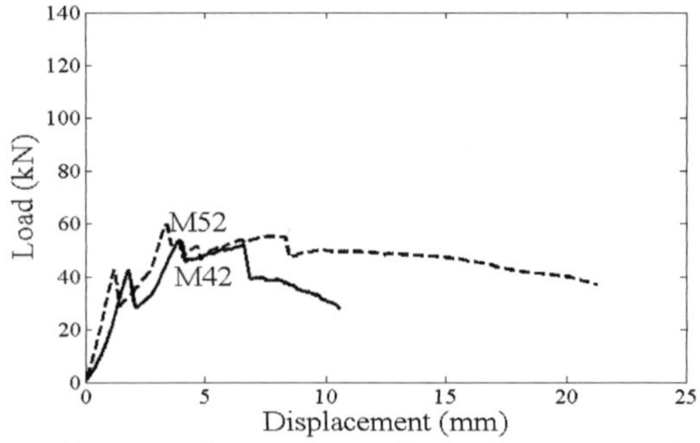

(e) Effect of end anchorage (Fibre content = 20 kg/m^3)

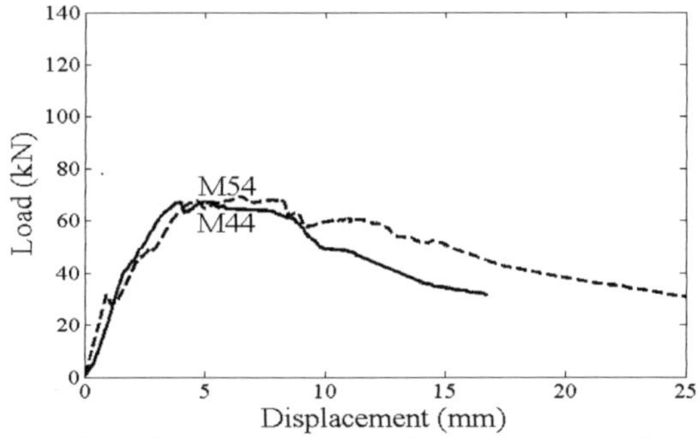

(f) Effect of end anchorage (Fibre content = 40 kg/m^3)

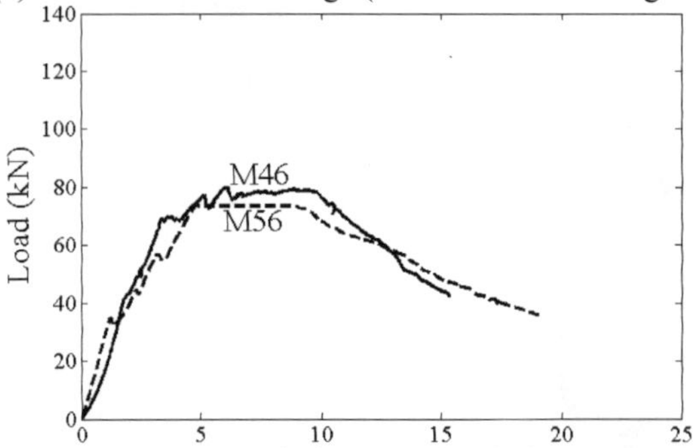

(g) Effect of end anchorage (Fibre content = 60 kg/m^3)

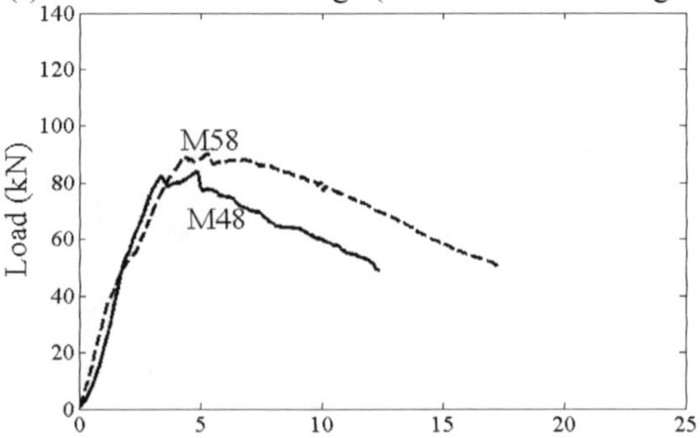

(h) Effect of end anchorage (Fibre content = 80 kg/m^3)

Figure 6. Load – Displacement curves for SFRC slabs without traditional reinforcement.

The curves display ductile behaviour from a deflection of about 3 to 5 mm to a deflection of around 8 to 10 mm in many slabs after which, the load drops drastically. The slab thickness affected the ultimate load-carrying capacity most [Figure 6(d)], followed by the reinforcing index [Figures 6(b) and (c)] and concrete compressive strength [Figure 6(a)]. The end anchorage of hooked fibers has a marginal effect on the peak load. Figures 6(e) to (h) indicate that specimens with 5D fibres displayed a slightly higher peak load compared to those with 4D fibres with the exception of M56. Specimens with 5D fibres also exhibited more ductile behaviiour than those with 4D fibres and sustained larger deflections [Figures 6(e) to (h)].

3.3. Predictions by Yield Line Theory and Punching Shear Prediction Models

Table 2 shows the statistical performance of the yield line theory and the two punching shear prediction models by Grimaldi et al. [21] and Choi et al. [11], comparing with the test results.

Table 2. Statistical Performance of Prediction Models.

Test/Predicted Values	Yield Line Theory	Grimaldi et al. [21]			Choi et al. [11]
		Assuming Truncated Cone Failure Surface, with		Assuming Truncated Pyramid Failure Surface, with	
		$V_{R,c}$ by EN 1992, 2004	$V_{R,c}$ by ACI 318-14	$V_{R,c}$ by ACI 318-14	
AVERAGE	0.99	0.19	0.43	0.44	1.26
STDEV	0.19	0.03	0.06	0.07	0.58
COV	0.19	0.16	0.15	0.15	0.46
MIN	0.81	0.16	0.32	0.32	0.72
QUARTILE 1	0.83	0.17	0.40	0.41	0.93
MEDIAN	0.96	0.17	0.42	0.43	1.15
QUARTILE 3	1.06	0.19	0.47	0.49	1.33
MAX	1.43	0.26	0.53	0.55	2.82

Since the model by Grimaldi et al. [21] considers $V_{R,c}$ to be constant for a given concrete compressive strength, it consequently over predicts the load-carrying capacity of the SFRC slabs without traditional reinforcement by a big margin (Table 2). Choi et al.'s model [11] considers the neutral axis depth in the prediction of punching shear capacity of SFRC slabs. However, owing to the conservative predictions of contribution by steel fibres, $V_{R,f}$, to the punching shear capacity, the model under predicts the mean load-carrying capacity of the test slabs by 26% with a large standard deviation of 0.58.

The yield line theory accurately predicts the average load-carrying capacity of SFRC slabs without traditional reinforcement within 1% of the observed values with a standard deviation of 0.19. Thus, it is deduced that a flexural failure mode occurs in SFRC slabs under a concentrated load. This is probably due to the lesser energy required to propagate existing yield lines than otherwise needed to form new circumferential cracks around the column face typical of punching failure.

4. Conclusion

In this study, an investigation on the punching shear capacity of SFRC slabs without traditional reinforcement was carried out. Predictions from two punching shear prediction models and the yield line theory were compared with results from twelve SFRC slabs. It was observed that the yield line theory provides accurate predictions of the load carrying capacity of the SFRC slabs while the punching shear prediction models fail to predict the observed values accurately. It is noted that all tested slabs experienced a flexural mode of failure rather than a punching shear failure.

Acknowledgement
The authors acknowledge the contribution of Mr. Lee Y C S in carrying out the tests as part of his final year undergraduate dissertation. The support of Bekaert Singapore Pte Ltd in providing the steel fibres is gratefully acknowledged. The authors also thank the staff and technicians in the Structural Engineering Laboratory, National University of Singapore for their assistance in executing the test programme.

References
[1] Gardner, N., Huh, J. and Chung, L., 2002. Lessons from the Sampoong department store collapse. *Cement and Concrete Composites,* 24(6), pp. 523-529.
[2] ACI 544.6R, 2015. *Report on design and construction of steel fiber-reinforced concrete elevated slabs,* Farmington Hills, MI: American Concrete Institute (ACI).
[3] ACI 318, 2014. *Building code requirements for structural concrete and commentary,* Farmington Hills, MI: American Concrete Institute (ACI).
[4] Corely, W. G. and Hawkins, N. M., 1968. Shear head reinforcement for slabs. *ACI Journal Proceedings,* Volume 65, pp. 811-824.
[5] Kang, T. and Wallace, J., 2006. Punching of reinforced and post-tensioned concrete slab-column connections. *ACI Structural Journal,* Volume 103, pp. 531-540.
[6] Park, H. et al., 2007. Lattice shear reinforcement for slab-column connections. *ACI Structural Journal,* Volume 104, pp. 294-303.Corely, W. G. and Hawkins, N. M., 1968. Shear head reinforcement for slabs. *ACI Journal Proceedings,* Volume 65, pp. 811-824.
[7] Narayanan, R. and Darwish, I., 1987. Punching shear tests on steel-fiber-reinforced micro-concrete slabs. *Magazine of Concrete Research,* 39(138), pp. 42-50.
[8] fib Model Code, 1990. *fib Model Code for Concrete Structures,* Lausanne, Switzerland: fib – International Federation for Structural Concrete, Verlag, Ernst and Sohn, 1991.
[9] Tan, K.H and Paramasivam, P., 1994. Punching shear of steel fiber reinforced concrete slabs. *Journal of Materials in Civil Engineering, ASCE,* 6(2), pp. 240-253.
[10] Harajli, M., Maalouf, D. and Khatib, H., 1995. Effect of fibers on the punching shear strength of slab-column connections. *Cement and Concrete Composites,* Volume 17, pp. 161-170.
[11] Choi, K., Taha, M., Park, H. and Maji, A., 2007. Punching shear strength of interior concrete slab-column connections reinforced with steel fibers. *Cement and Concrete Composites,* 29(5), pp. 409-420.
[12] fib Model Code, 2010. *fib Model Code for Concrete Structures,* Lausanne, Switzerland: fib - International Federation for Structural Concrete, Verlag, Ernst and Sohn, 2013.
[13] Neto, M., Barros, A. and Melo, S., 2014. Model to simulate the contribution of fiber reinforcement for the punching resistance of RC slabs. *Journal of Materials in Civil Engineering, ASCE,* 26(7), pp. 04014200,1-10.
[14] Destree, X., 2001. Steel fiber reinforcement for suspended slabs. *Concrete,* Volume Sept., pp. 58-59.
[15] Destree, X. and Mandl, J., 2008. *Steel fiber only reinforced concrete in free suspended elevated slabs: case studies, Design assisted by testing route, comparison to the latest SFRC standard documents.* Amsterdam, s.n., pp. 437-443.
[16] Ward, P., 2010. SFRC suspended slabs in flexure. *The Structural Engineer,* 88(1), pp. 16-19.
[17] Hedebratt, J. and Silfwerbrand, J., 2014. Full-scale test of a pile supported steel fibre concrete slab. *Materials and Structures,* 47(4), pp. 647-666.
[18] BS EN 1992, 2004. *Design of concrete structures. Part 1-1: General rules and rules for buildings.* Brussels: European Committee for Standardization (CEN), Eurocode 2.
[19] Long, N.-M., Maian, R. and Toan, T. Q., 2012. Punching shear capacity of internal SFRC slab-column connections. *Journal of Structural Engineering, ASCE,* Volume 138, pp. 613-624.

[20] Shaaban, A. and Gesund, H., 1994. Punching shear strength of steel fiber reinforced concrete flat plates. *ACI Structural Journal,* 91(4), pp. 406-414.

[21] Grimaldi, A., Alberto, M. and Rinaldi, Z., 2013. Experimental behaviour of fiber reinforced concrete bridge decks subjected to punching shear. *Composites: Part B,* Volume 45, pp. 811-820.

[22] Michels, J., Waldmann, D., Maas, S. and Zurbes, A., 2012. Steel fibers as only reinforcement for flat slab construction - Experimental investigation and design. *Construction Buildings and Materials,* 26(1), pp. 145-155.

[23] Amin A., Foster S.J. and Muttoni A. 2015. Derivation of the σ-w relationship for SFRC from prism bending tests, *Structural Concrete*, 16(1), pp. 93-105

[24] ACI 544.3R-93, Reappr. 1998. *Guide for specifying, proportioning, mixing, placing and finishing steel fiber reinforced concrete,* Farmington Hills, MI: American Concrete Institute.

IOP Conf. Series: Materials Science and Engineering **246** (2017) 012026 doi:10.1088/1757-899X/246/1/012026

Seismic behaviour of synthetic-FRC columns

P Paultre[1] and R Eid[2]

[1] Department of Civil Engineering, University of Sherbrooke, Sherbrooke, Quebec, J1K 2R1, Canada
[2] Civil Engineering Department, SCE - Shamoon College of Engineering, Beer Sheva 84100, Israel

E-mail: Patrick.Paultre@USherbrooke.ca

Abstract. Inclusion of short discrete fibres into the concrete mixture can increase the compressive strength and ductility of normal-strength concrete (NSC) and high-strength concrete (HSC) column specimens under compressive loading as already has been shown by several studies. Concrete design codes ensure ductile behaviour of columns by setting a requirement for a minimum amount of transverse steel reinforcement. Therefore, the inclusion of discrete short fibres into the concrete mixture combined with a reduced amount of lateral reinforcement can be an alternative to the latter full amount required by the codes. This paper presents tests that were performed on large-scale fibre-reinforced NSC circular columns under cyclic flexure and constant axial load simulating earthquake loading. The aim of this test program is to examine the combined confinement effect of steel or synthetic fibres and the transverse steel reinforcement type (spirals or hoops) on the structural performance of RC columns. The results show that in terms of ductility and energy dissipation, the behaviour of the fibre-reinforced concrete (FRC) specimens is improved compared to the nonfibrous ones. This behaviour is also predicted by the proposed confinement model which takes into account the mechanical and the geometrical properties of the concrete and the reinforcement as well as those of the fibres.

1. Introduction

In the last few decades, the evolution in technology has allowed the development and the use of fibers as a component into the concrete mix. This composite material, which is known as fiber-reinforced concrete (FRC), is basically concrete with discrete fibers distributed in a random way. The presence of fibers into the concrete matrix offers several advantages like the reduction of shrinkage (cracking control), the enhancement in toughness and in tensile strength, as well as the improvement of strain characteristics while cracking occurs. Furthermore, it is well known that concrete is a brittle material and the steel inside the concrete has the role of taking the tensile stresses. In the long-term behavior, however, the structure may suffer deterioration caused by corrosion of this reinforcing steel, leading to the need of essential repairs. The FRC has been found to be a more impermeable material, so this corrosion phenomenon could be minimized as well as the required maintenance. Basically, the application fields of FRC concern nonstructural and non-principal structural elements like slabs on grades, pools, and stabilization of ground slopes. However, the advantages of FRC already observed keep the attention of the research community for the possibility of using this material in the structural field.

Despite a large variety of fibers developed for FRC (steel, glass, natural and synthetic fibers), research studies for the structural application fields have been more interested in steel and synthetic fibers. The ACI Committee 544 [1] reported design considerations for steel fiber reinforced concrete (SFRC) limited to some structural applications. On the other hand, there are few published research studies (e.g., [2-4]) that indicate positive field experiences for the synthetic fiber reinforced concrete (SNFRC) and show several advantages over the SFRC. For example, the ability of synthetic fibers to melt during a fire allows the steam pressure inside the concrete to dissipate through the pathways left by the melted fibers, and this in turn helps to reduce severe spalling [4]. This is not observed when using SFRC. Also, SNFRC elements placed in aggressive environments will have better appearance than those made of SFRC because the surface of elements containing steel fibers will corrode with time [5]. Thus, there is an interest to continue examining the structural applications of SNFRC composite material. Such research will enable the development of analytical models and appropriate design procedures. Concrete design codes ensure ductile behavior of columns by setting a requirement for a minimum amount of transverse steel reinforcement. Studies have shown, however, that the use of FRC can enhance the post-peak behavior and hence, the ductility and the energy dissipation ability of concrete columns subjected to axial loading. The higher resistance to crack growth and the excellent durability of FRC over nonfibrous concrete can result in higher cost-effective value. This paper presents tests on large-scale circular SNFRC columns subjected to a combined constant axial load and reversed cyclic flexure simulating earthquake loading. The main objective of the research program is to study the seismic behavior of SNFRC circular columns subjected to constant axial load while relating the amount of transverse reinforcement and the volumetric synthetic fiber ratio to the gain in ductility, energy dissipation, resistance and damage.

2. Experimental program

Figures 1 and 2 depict the reinforcement details and the experimental setup of the full size synthetic fibre-reinforced NSC (SNFRC) circular specimens (305 mm diameter and 2000 mm height) that were tested under cyclic flexure and constant axial loading. To examine the influence of adding synthetic fibres to the concrete mixture on the cyclic behavior of NSC columns, three NSC and three SNFRC column specimens were tested. The synthetic fibres have a length of 50 mm and an aspect ratio of 74 (see figure 3 - for more details see Osorio et al. [6] and Rivera [7]). The specimens had two types of transverse reinforcement – hoops and spirals. Figures 4-5 show the lateral force versus tip displacement of two specimens: C30S100FP0 and C30S100FP1 with hoops reinforcement (targeted concrete strength of 30 MPa, tie spacing of 100 mm, targeted axial load of 25% of , and FP0 and FP1.0 indicate fiber volumetric ratio of 0.0% and 1.0%, respectively). These two specimens had a longitudinal reinforcement ratio of 0.025 and a volumetric transverse steel reinforcement ratio of 0.015 for a yield strength of around 440 MPa. Moreover, figures 6-7 show the lateral force versus tip displacement of two similar specimens with spiral reinforcement. The figures illustrate that, in terms of ductility and energy dissipation, the SNFRC specimens exhibit improved behaviour compared to the NSC columns.

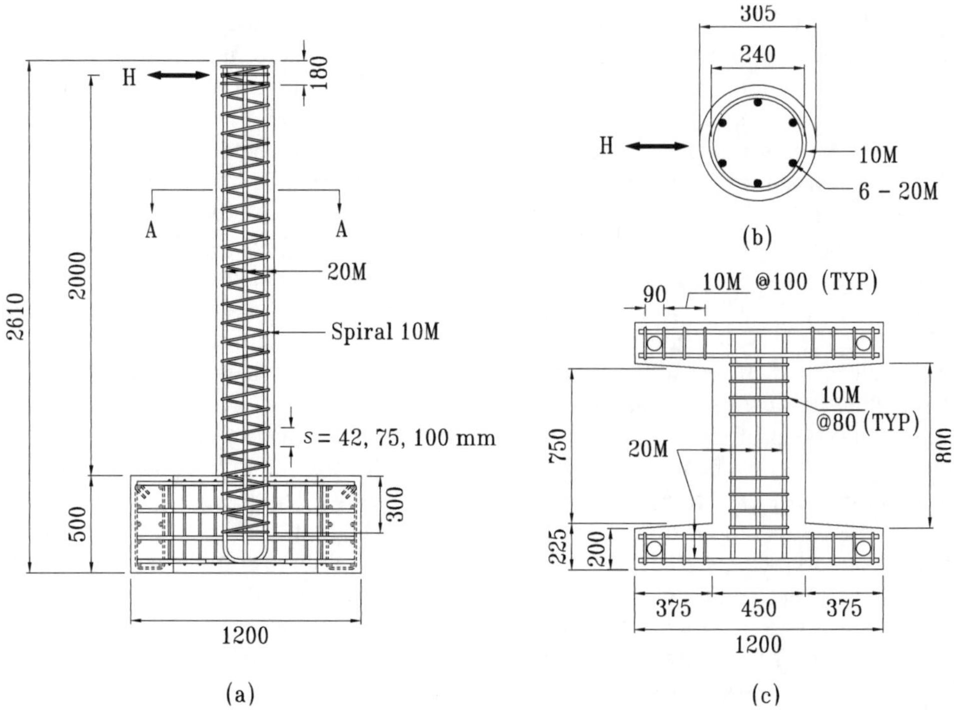

Figure 1. Reinforcing cage and instrumentation details for specimens with spiral reinforcement (dimensions in mm).

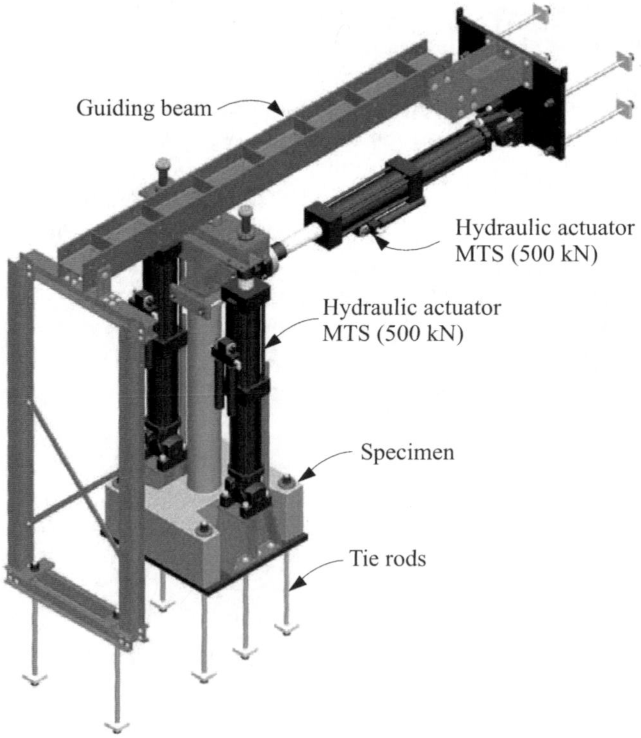

Figure 2. Experimental setup.

IOP Conf. Series: Materials Science and Engineering **246** (2017) 012026 doi:10.1088/1757-899X/246/1/012026

Figure 3. Synthetic fibers used in the research study

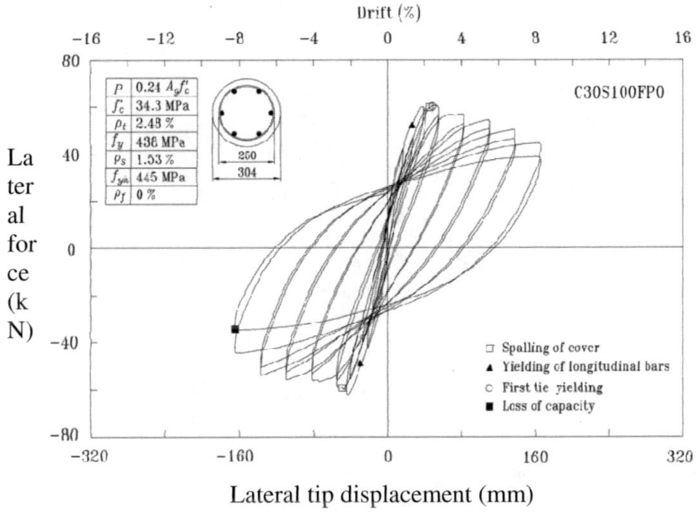

Figure 4. Experimental lateral force versus lateral tip displacement for nonfibrous RC specimen C30S100FP0 (with hoops transverse reinforcement)

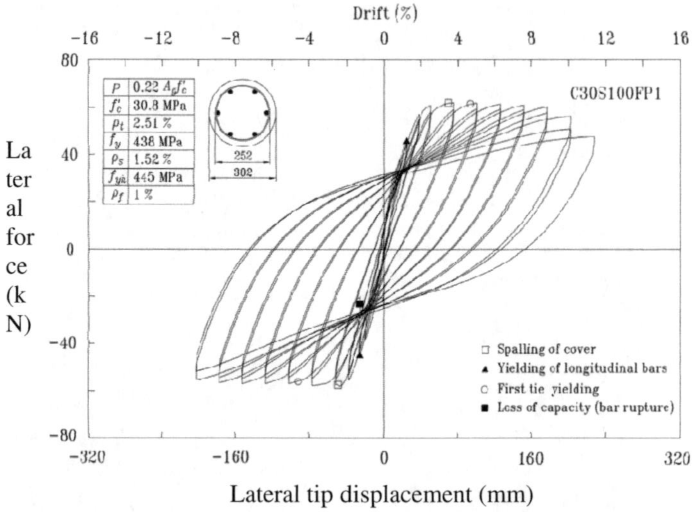

Figure 5. Experimental lateral force versus lateral tip displacement for SNFRC specimen C30S100FP1 (with hoops transverse reinforcement)

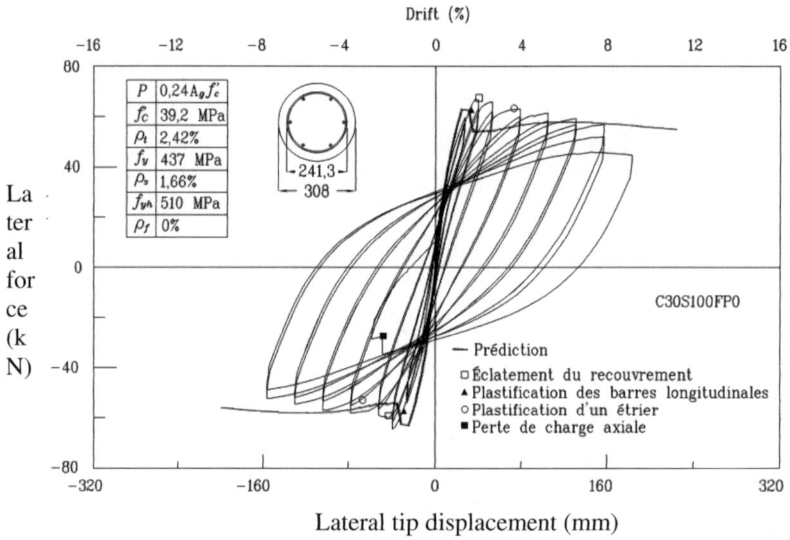

Figure 6. Experimental and analytical (envelop) lateral force versus lateral tip displacement for nonfibrous RC specimen C30S100FP0 (with spiral reinforcement)

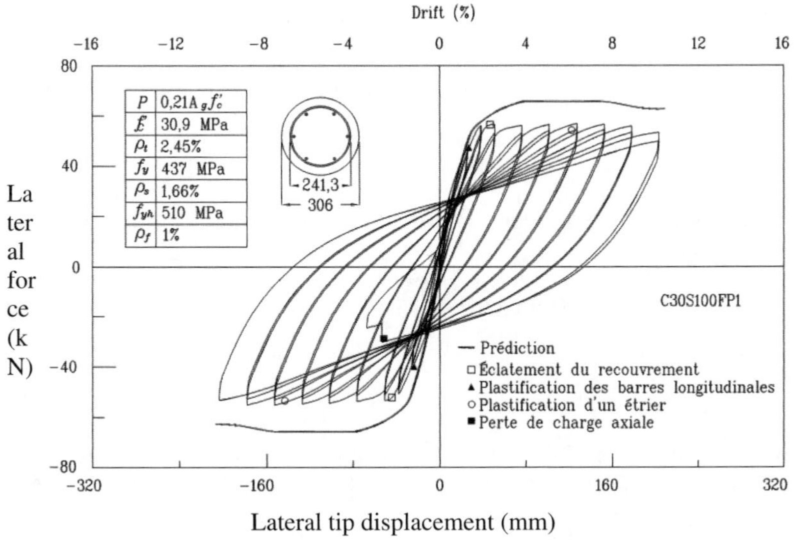

Figure 7. Experimental and analytical (envelop) lateral force versus lateral tip displacement for SNFRC specimen C30S100FP1 (with spiral reinforcement)

3. Analytical stress-strain curve

Several models have been proposed in the literature to predict the axial stress-strain behaviour of NSC and HSC confined by transverse steel reinforcement (e.g., [8-10]). Cusson and Paultre [8] and Légeron and Paultre [10] based their models on the effective confining lateral uniform pressure, f_{le}, which is a function of the lateral steel stress at concrete peak stress, and an effective confinement coefficient, K_e [11-12]. This coefficient is based on the assumption of an arch action between the steel ties and on empirical results [11].

The expressions for the confined concrete compressive peak stress, f'_{cc}, and its corresponding strain, ε'_{cc}, were extracted from experimental results by regression analysis [8,10] as a function of the effective confinement pressure at concrete peak stress f_{le}:

$$\frac{f'_{cc}}{f'_c} = 1 + 2.4(f_{le}/f'_c)^{0.7}, \quad \varepsilon'_{cc} = \varepsilon'_c + 0.21(f_{le}/f'_c)^{1.7} \tag{1}$$

where f'_c is the unconfined concrete strength and f_{le} is the effective confinement pressure at concrete peak stress:

$$f_{le} = K_e f_{ls} + f_{lfb} \tag{2}$$

The lateral pressure at peak concrete stress due to the action of the transverse reinforcement, f_{ls}, is derived from the force equilibrium of the half cross-section of the confined column:

$$f_{ls} = \frac{A_{sh} f_{hc}}{sc} \tag{3}$$

where c is the concrete core size, s is the steel tie spacing, f_{hc} is the stress in the lateral steel at peak concrete stress, and A_{sh} is the total cross-section area of the transverse reinforcement.

Generally, the production of fibre-reinforced concrete is made with short fibres having aspect ratio in the range of $l_f/d_f = 50-100$ where l_f is the fibre length and d_f is the fibre diameter or equivalent diameter. The tension failure of such composites in the post-cracking zone occurs primarily by fibre pull-out [13]. Assuming a constant frictional bond strength, τ_{fu}, and an average fibre embedded length equal to l_f/d_f, the confinement pressure developed due to the action of the fibres in an axially loaded fibre-reinforced concrete column is:

$$f_{lfb} = \eta_\theta \tau_{fu} \upsilon_f \, l_f/d_f \tag{4}$$

where η_θ is the fibre orientation efficiency factor. Aveston et al. [14] suggested that this efficiency factor can be taken as $\eta_\theta = 0.5$.

The prepeak branch of the stress-strain curve is based on a relationship originally proposed by Popovics [15]:

$$f_c = f'_{cc} \frac{k(\varepsilon_c/\varepsilon'_{cc})}{k-1+(\varepsilon_c/\varepsilon'_{cc})^k}, \quad \varepsilon_c \le \varepsilon'_{cc} \tag{5}$$

where k is a parameter controlling the slope of the ascending branch. The post-peak branch of FRC stress-strain curve is a modification of an expression proposed by Fafitis and Shah [16]:

$$f_c = f'_{cc} \exp\left[k_1(\varepsilon_c - \varepsilon'_{cc})^{k_2}\right], \quad \varepsilon_c > \varepsilon'_{cc} \tag{6}$$

where k_1 and k_2 are parameters controlling the shape of the post-peak branch [8]. Figures 6-7 give the analytical predictions of the proposed model with experimental results. The figures show good agreement between the overall behaviour predicted by the model and the corresponding experimental results.

4. Conclusions

This paper presents a test program with the aim of examining the combined confinement effect of synthetic fibres and transverse steel reinforcement on the structural performance of NSC columns. The results show that in terms of ductility and energy dissipation, the behaviour of the FRC specimens is improved compared to the nonfibrous columns. This behaviour is also predicted by the proposed confinement model which takes into account the mechanical and the geometrical properties of the concrete and the reinforcement as well as those of the fibres. Predictions were found to be in good agreement with experimental results.

References

[1] ACI Committee 544.4R 1999 Design considerations for steel fiber reinforced concrete *American Concrete Institute* (Farmington Hills/Michigan)

[2] Djumbong A, Cusson D and Paultre P 2008 Experimental study of influence of synthetic fibres on behavior of cylindrical high-strength concrete columns *the 37th Annual Conference of the Canadian Society for Civil Engineering*, (Quebec City/Canada) pp 10

[3] Nguyen H 2006 Shear Behavior of polypropylene fiber reinforced concrete beams *MASc Department of Civil Engineering University of Sherbrooke* (Sherbrooke/Canada) pp 98 (in French)

[4] Han C, Hwang Y, Yang S and Gowripalan N 2005 Performance of spalling resistance of high performance concrete with polypropylene fiber contents and lateral confinement *Cement and Concrete Research* **35** pp 1747-1753

[5] Chernov V, Zlotnikov H and Shandalov M 2006 Structural synthetic fiber-reinforced concrete-experience with marine applications *Concrete International* **28** 8 pp 56-61

[6] Osorio L I, Paultre P, Eid R and Proulx J 2014 Seismic behavior of synthetic-fiber-reinforced circular columns *ACI Structural Journal* **111** 1 pp 189-200

[7] Rivera E J 2009 Flexure-compression behaviour of synthetic fibre-reinforced concrete circular columns *MASc Department of Civil Engineering University of Sherbrooke*

IOP Conf. Series: Materials Science and Engineering **246** (2017) 012026 doi:10.1088/1757-899X/246/1/012026

(Sherbrooke/Canada) pp 135 (in French)

[8] Cusson D and Paultre P 1995 Stress-strain model for confined high-strength concrete *Journal of* Structural *Engineering* **121** 3 pp 468-477

[9] Razvi S R and Saatcioglu M 1999 Confinement model for high-strength concrete *Journal of* Structural *Engineering* **125** 3 pp 281-289

[10] Légeron F and Paultre P 2003 Uniaxial confinement model for normal- and high-strength concrete columns *Journal of Structural Engineering* **129** 2 pp 241-252

[11] Sheikh S A and Uzumeri S M 1982 Analytical model for concrete confinement in tied columns *Journal of the Structural Division* **108** 12 pp 2703-2722

[12] Mander J B, Priestley M J N and Park R 1988 Theoretical stress-strain model for confined concrete *Journal of Structural Engineering* **114** 8 pp 1804-1826

[13] Bentur A and Mindess S 1990 Fiber reinforced cementitious composites *Elsevier Applied Science* (New York)

[14] Aveston J, Mercer R A and Sillwood J M 1974 Fiber reinforced cements - scientific foundation for specification *composites, standards testing and design, proceedings of the national physical laboratory conference* (UK) pp 93-103

[15] Popovics S 1973 A numerical approach to the complete stress-strain curves for concrete *Cement and Concrete Research* **3** 5 pp 583-599

[16] Fafitis A and Shah S 1985 Predictions of ultimate behavior of confined columns subjected to large deformations *ACI Structural Journal* **82** 4 pp 423-433

IOP Conf. Series: Materials Science and Engineering **246** (2017) 012027 doi:10.1088/1757-899X/246/1/012027

The effect of the length of macro synthetic fibres on their performance in concrete

K P Juhász[1] and V Kis[2]

[1] Department of Mechanics, Materials & Structures, Budapest University of Technology, Műegyetem rkp. 3, 1111 Budapest, Hungary
[2] Fehérvári út 168-178./B, 1116 Budapest, Hungary

E-mail: office@jkp.hu

Abstract. Nowadays macro synthetic fibres are able to compete with steel fibres despite their low Youngs Modulus. This is due to their different pull-out mechanism and a larger number of individual fibres per kilo compared to steel fibres. Macro synthetic fibres bond to the concrete along their full length, usually with an embossed surface, while steel fibres are mostly anchored by their hooked ends. If the bond is defined by the length of the embossed surface, logically the longer the synthetic fibre the higher post-crack capacity. In this paper the same type of macro synthetic fibre was researched with different lengths but at the same dosage. The consistency of the fresh concrete together with the quality of the distribution of the fibres have been analysed and compared with the residual strength. After analysing these data the optimum fibre length was able to be determined.

1. Introduction

Fibre reinforcement generally increases the ductility of the concrete by increasing the fracture energy of the quasi-brittle concrete. By adding fibre the consistency and workability of the concrete will also change, and has to be managed by using a proper concrete mix design. The effect of the fibre can be characterized by the increase of the fracture energy, dividing the fracture energy of the fibre reinforced concrete into two parts: fracture energy of the plain concrete (G_f) and the added fracture energy (G_{ff}). This added fracture energy is dependent on the fibre type, dosage and the concrete matrix [1].

Macro synthetic fibres generally pull-out from the concrete matrix first, the fibre will only rupture at bigger crack widths. The length of the fibre will increase the energy absorption, but after a certain fibre length it could also impair the consistency, workability and the quality of the concrete mix, which could lead to lower strength values.

Furthermore the orientation of the fibres also depend on their length. The added fracture energy is a material constant if there is no wall effect. In case of formwork structures there is a wall effect which is nominally a distance of half the length of a fibre, where the orientation of the fibres can be changed. This wall effect cannot be neglected during the evaluation and has to be calculated.

The fibre length for steel fibre also has an effect on the mixing properties of the aggregates [2, 3], although this effect is much less than for synthetic fibre [4]. Because of the flexibility of the individual fibres, macro synthetic fibres can bend around the aggregate and thus do not change the distance between the aggregate.

IOP Conf. Series: Materials Science and Engineering **246** (2017) 012027 doi:10.1088/1757-899X/246/1/012027

The optimal length of macro synthetic fibre was researched and an exact value of 54 mm was given by MacDonald [5]. Other researchers developed a testing method to specify the optimal length based on the characteristic length, this length is where the fibre will pull out from the matrix before rupture [6].

Our research will investigate the effect of the length of macro synthetic fibres in a direct way from beam test results where fibres with a different lengths but with the same dosage were mixed in similar concrete.

2. Problem statement

Manufacturers produce macro synthetic fibres with different lengths. The macro fibres can increase the ductility of the concrete by increasing the fracture energy. Research has identified methods to determine the characteristic length (l_c) of steel fibres [6] determining where the fibre will pull-out from the concrete matrix before rupture. Fibre with an anchorage length longer than this characteristic length will rupture instead of pull-out, which will lead to a lower added fracture energy.

In the case of macro synthetic fibres this length is not necessarily the optimal length. Firstly, using standard test methods the crack opening is measured to CMOD=3.5 mm (Crack Mouth Opening Displacement), which is approximately a crack width of 2.5-3 mm. If the fibre ruptures at a higher value this won't be seen in the test result. Secondly, even if these fibres rupture during the test this would only be seen at the higher values. Thirdly, if the fibres are randomly but perfectly distributed and orientated, then fibres only in the perfect position and orientation (anchorage in the middle point and perpendicular to the cracked section) would rupture. Fibres with a shorter anchorage length or with a smaller angle will probably pull-out from the matrix.

Based on the above determining the characteristic length of the fibre could lead to an improper optimal length. In our research the effect of the length of the fibres is investigated by a direct test method: instead of a pull-out test using a concrete beam tests.

Increasing the length of macro synthetic fibres increases its anchorage in the concrete and also its pull-out force in order to reach an upper limit where the fibre will rupture before being pulled-out. Based on this the longer fibre will lead to an increase in added fracture energy up to a certain length. However, because of practical reasons the fibre's length cannot be increased indefinitely. The results of previous research and practical experience also shows that during mixing fibre-balls can be formed. These balls can grow during mixing as other fibres join to the ball. These fibre-balls degrade the quality and performance of the fibre reinforced concrete and mostly must be taken out during the mixing. This leads to less fibre in the concrete and obviously a decrease in its added fracture energy. To some extent the formation of these fibre-balls depends on the length of the fibre: as longer fibres are easier to lock to each other than the smaller ones.

Fibre orientation is random and although the formwork has a wall-effect on the fibres which restricts their orientation. The wall effect is only evident on their orientation if the fibres make contact with the wall. For the same beam the wall-effect is bigger with longer fibres, which will also be analysed here.

3. Laboratory tests

3.1. Concrete mix design

The concrete mix design used was a typical industrial flooring concrete, with a concrete strength class C30/37. The water/cement ratio was 0.4. The actual mix design can be seen in Table 1.

Table 1. Concrete mix design

Name	Type	Dosage [kg/m^3]
Cement	CEM I 42.5 R	400
Water		160
Sand	0/4	735
Fine Aggregate	4/8	294
Course Aggregate	8/16	808
Superplasticizer	Mapei Dynamon NRG 1012	3.2
Fibres	BarChip 30, 40, 48, 55, 70 mm	4.0

3.2. Fibre types
All the fibres used were the same synthetic macro fibre type: BarChip (Elasto Plastic Concrete), but with different lengths: 30 mm, 40 mm, 48 mm, 55 mm and 70 mm. The dosage was the same: 4 kg/m^3. The fibres are made from modified polyolefin, with an effective diameter of 0.72mm and with an embossed surface.

3.3. Test matrix
There were 7 beams for every type of concrete. The test matrix can be seen in Table 2.

Table 2. Test matrix

Sign	Concrete	Fibre type	Dosage [kg/m^3]	Dosage [number of pieces]
BC30 1-7	C30/37	BarChip 30	4.0	93240
BC40 1-7	C30/37	BarChip 40	4.0	69930
BC48 1-7	C30/37	BarChip 48	4.0	58275
BC55 1-7	C30/37	BarChip 55	4.0	50858
BC70 1-7	C30/37	BarChip 70	4.0	39960

3.4. Specimens and testing method
The concrete was made using a Collomatic XM2 forced-action mixer, which represents well the industrial environment and provides perfect mixing. At first the plain concrete was mixed for 1 minute, then the fibre was added and mixed for a further 1 minute. The fibres mixed well and the fresh concrete consistency was F2. The concrete was poured into the steel formwork and was vibrated for 15 seconds. The upper surface of the beam were smoothed with a steel tool. The specimens were stored in air for 15 days. The size of the beams were 150 mm x 150 mm x 550 mm and tested at 500 mm span. In the middle of the beam there was a notch 2 mm wide by 25 mm deep for the CMOD measurement.

The testing method was a 3 point bending beam test according to RILEM TC162 [7]. The beams were loaded in the middle, above the notch. The testing speed was 0.2 mm/min. The tests were done in the Laboratory of Czako Adolf, Department of Mechanics, Materials and Structures, Budapest University of Technology and Economics. The test was carried out using a universal testing machine: Zwick-Roell Z150, with a capacity of maximum 150 kN.

3.5. Evaluated results
The loading force, F in N and the CMOD in mm were measured during the test as an output data. After the bending test the beam was divided into two parts using the loading machine to break the beam completely to be able to see the cracked surface clearly.

The location of the fibres were analysed by firstly dividing each cracked face into 6 strips each with a thickness of 25 mm. The 6th strip was the notch as can be seen in figure 1a. Secondly, the effect of the formwork was analysed by counting the fibres in the zones where the wall-effect occurs. Fibres were counted near the surface at a distance of a half length of fibre (l_f), as can be seen in figure 1b. During counting both the fibres that were pulled-out and the ruptured fibres were recorded.

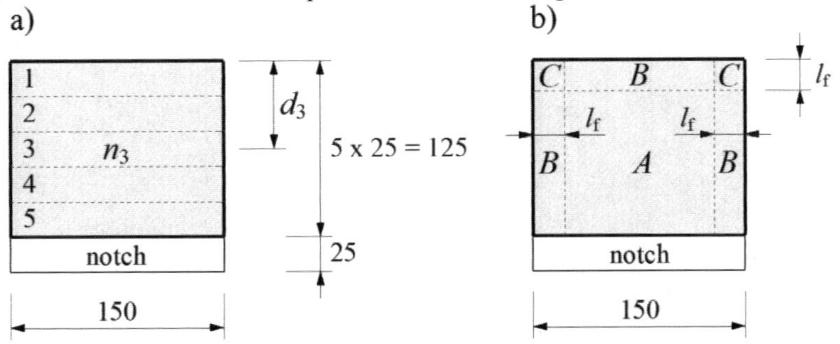

Figure 1. Face of the cracked surface for fibre counting

4. Test results

4.1. Distribution and amount of the fibres in the cracked section

The distribution and position of the fibres on the cracked surface had a significant effect on the results so this must be taken into account during the evaluation. The point where the fibres crossed the cracked section were measured from the top of the beam which was the edge of the beam that was loaded. This is taken the compressed zone as an approximation to zero thickness. Using this method the *fibre-moment* (S_f) can be determined. Assuming perfect mixing the fibres crossing the crack per unit area can be determined. This is based on the work of many researchers [8, 9, 10] and which could lead to the determination of the *ideal fibre-moment* ($S_{f,i}$). By comparing the actual and ideal fibre-moments the ideal *fibre-work* (W_f, area under the load-CMOD diagram up to 3.5 mm) can be calculated. This method is explained in detail in [11].

4.2. Wall-effect

The fibres have free orientation in the concrete if unaffected by the aggregates. To calculate the number of fibres crossing the cracked section the average length of the fibres in perpendicular direction to the cracked section need to be calculated. If this value is known the specific number of fibres crossing the cracked section can be determined according to equation (1) which was derived by Krenchel [12].

$$n_a = \frac{2\, l_{f,x,m}\, N}{V}$$

where n_a number of fibers crossing the section, pieces/mm^2
$l_{f,x,m}$ average fiber length in the undistributed zone, mm
N number of fibers in the volume
V volume, mm^3

It can be seen that if the dosage is the same the number of the fibres crossing the section are independent of the fibre length. In this test series this means that in the undisturbed zone of the section (where there is no wall-effect, zone A at figure 1/b) the number of fibres crossing the cracked section is the same.

The average fibre length was first calculated by Romualdi and Mandel [8]. They used polar coordinates to define the orientation of the fibre, but this had a densification in the z axis if the input parameters were chosen randomly (see figure 2a). In order to eliminate this, cylinder coordinates must be used. First the angle between the x axis, and the distance from the xy plane: t. By defining the orientation of the fibre with $\sphericalangle$ and t, and choosing the parameters randomly, the distribution is equable (see figure 2b).

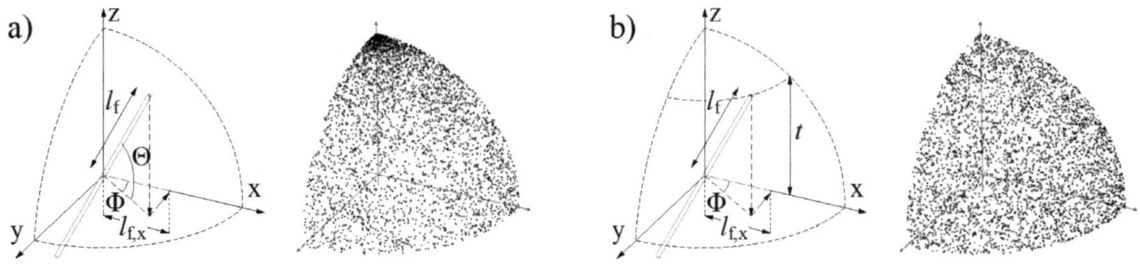

Figure 2. Definition of the fibre in the coordinate system and the dispersion at random input parameters a) ∢-Θ b) ∢-t

The average fibre length in direction x is defined in equation (1):

$$l_{f,x,m} = \frac{\int_0^{\frac{\pi}{2}}\int_0^{l_f} l_f \cos\Phi \cos\left(\arcsin\frac{t}{l_f}\right)\Delta\Phi\,\Delta t}{\frac{\pi}{2} l_f} = 0.5\,l_f \tag{1}$$

Including the wall effect the equation will change as (2):

$$l_{f,x,m} = \frac{\int_0^{\frac{\pi}{2}}\int_0^{T} l_f \cos\Phi \cos\left(\arcsin\frac{t}{l_f}\right)\Delta t\,\Delta\Phi + \int_0^{\frac{\pi}{2}}\int_T^{l_f} l_f \cos\Phi \cos\left(\arcsin\frac{T}{l_f}\right)\Delta t\,\Delta\Phi}{\frac{\pi}{2} l_f} \tag{2}$$

where T is the wall distance from the middle point of the fibre.

This average fibre length is valid only for steel fibres (rigid material). Assuming that fibres that make contact with the mould are deflected (flexible material) the following model in figure 3 can be used. The average length of the fibres in the case of steel (rigid material) and synthetic (flexible material) can be seen in figure 4 and 5, respectively. It can be seen that the effect of the formwork is lower in case of synthetic fibre, therefore less fibre will cross the cracked section compared to steel fibre

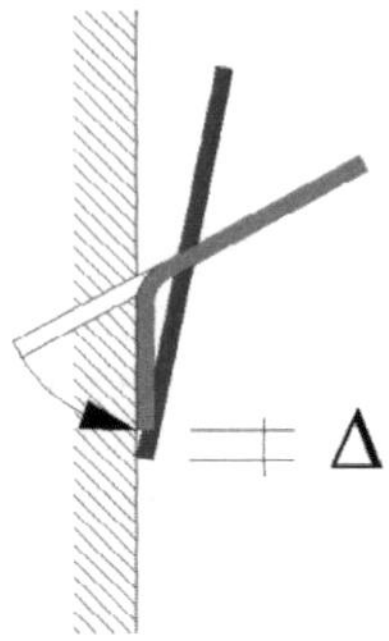

Figure 3. rigid (steel) and flexible (synthetic) model

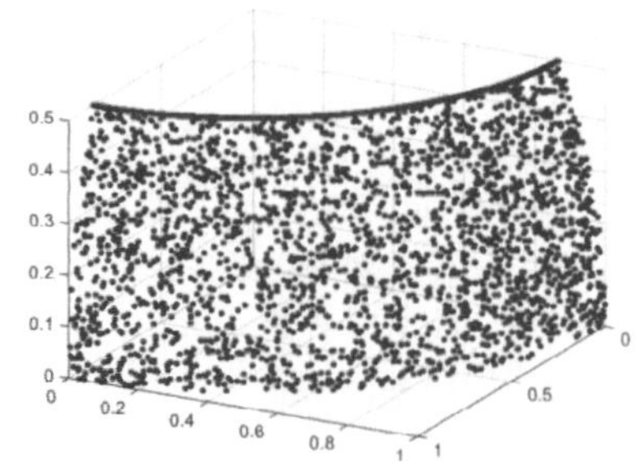

Figure 4. Distribution of the steel fibres at T=0.5 l_f

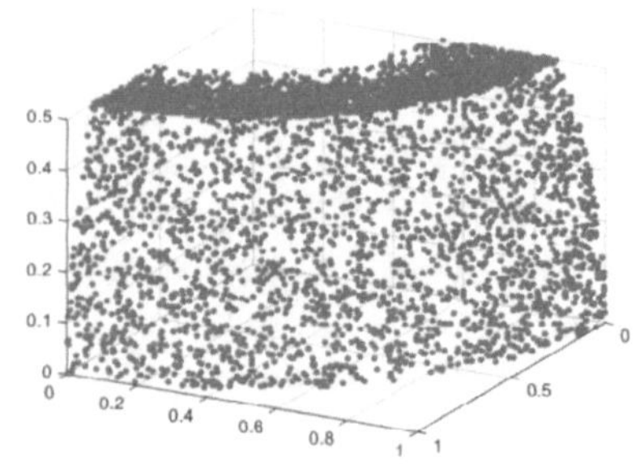

Figure 5. Distribution of the synthetic fibres at T=0.5 l_f

IOP Conf. Series: Materials Science and Engineering **246** (2017) 012027 doi:10.1088/1757-899X/246/1/012027

The numbers of fibres according to both the theoretical and actual test results can be seen in figure 6. According to the theoretical calculation the number of fibres crossing the cracked section is the same if the same dosage was used. The theoretical model in most cases gave a higher result, which could be because of the difficulty of counting the number of fibres. Fibres on the surface could be pulled-out (figure 7a) or ruptured (figure 7b and 7c). The pulled-out fibres are visible only one side of the cracked surface. The point of fibre's rupture could be either seen outside of the concrete (ruptured fibres, figure 7b) or could have occurred in the concrete (ruptured-in, figure 7c). In this second case only half of the fibres are counted, because the part which has ruptured in the concrete is not visible. Because of this the number of fibres that were actually counted could be less than were in the concrete.

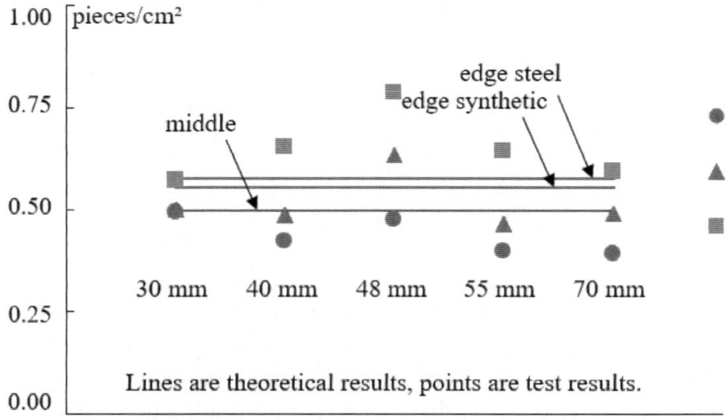

Figure 6. Comparing the theoretical and test results

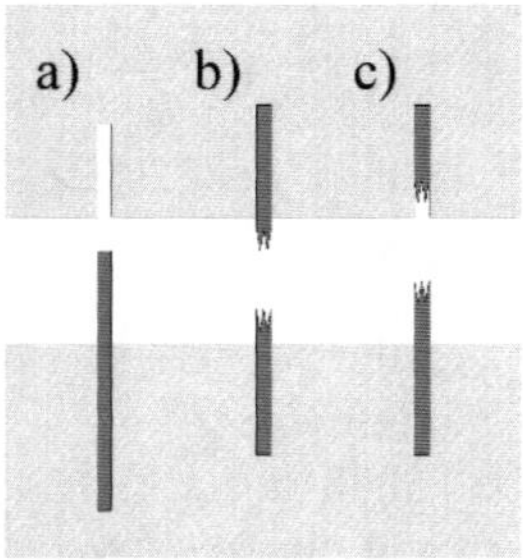

Figure 7. Explanation of the pulled-out (a), ruptured (b) and ruptured-in (c) fibres

4.3. Effect of the length of the fibres to the fibre-work

According to previous research fibre has pulled-out from the concrete at 15 mm anchorage length, but has ruptured at 22.5 mm. Because of this the characteristic length (l_c) is between 15 mm and 22.5 mm, so the optimal length of the fibre is between 30 mm and 45 mm.

Plotting the length of the fibre and its *fibre-work* (W_f) in a diagram shows a different correlation than was expected which can be seen in figure 8. The length of fibre which has the highest *fibre-work* is the fibre with 55 mm length, after this length a significant decrease in performance can be seen at 70 mm. Assuming zero *fibre-work* at zero fibre length a line can be fitted on the points up to 55 mm from the origin. This would represent a linear relationship between the length of the fibre and its effect on the *fibre-work*. After this point the *fibre-work* decreased, although the flexural strength of the concrete did not change. This meant that the length of the fibre did not have any influence on the matrix strength, so the fall in the *fibre-work* is because of the changes in the fibre's energy absorption mode.

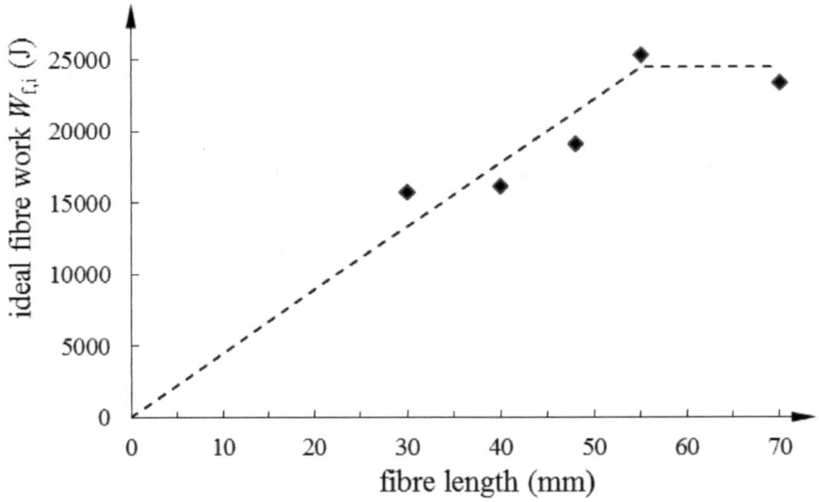

Figure 8. Fibre length and *fibre-work* correlation

5. Conclusion

The length of the fibre has a major effect on its performance in fibre reinforced concrete, both with steel and macro synthetic fibres. With steel fibres the length will change the mixing properties of the fresh concrete as the longer fibres will generate more space between the aggregates, while the macro synthetic fibres will bend around the aggregate. After pouring the concrete into moulds the wall-effect will occur. The length of the fibres will determine the width of the wall-effect zone, where the orientation of the fibres will change, which will be different for both steel and macro synthetic fibres. Steel fibres are rigid, and the fibres will rotate as they touch the mould, while synthetic fibres will bend. This difference can be seen in our numerical models. The length of the steel fibres were optimised with the help of the critical length (l_c), which is the maximum length where fibres are pulled-out from the matrix instead of rupturing. In this paper it was shown that for the macro synthetic fibre this length is different. The optimal fibre length where the fibre will have the maximum *fibre-work* (W_f) is longer than the optimal length calculated according to the critical length.

Acknowledgments

The Authors would like to express their gratitude to Elasto Plastic Concrete (BarChip) for the fibre samples and Fiberguru Ltd. for providing the Collomatic XM-2 forced action mixer. The research was done at the Laboratory of Czako Adolf, Department of Mechanics, Materials and Structures, Budapest University of Technology and Economics.

References

[1] Juhász K P 2013 Modified fracture energy method for fibre reinforced concrete in: *Fibre Concrete 2013* ed Prof. A Kohoutková et al. pp 89-90 ISBN 978-80-01-05238-9

[2] Kooiman A G 2000 *Modelling Steel Fibre Reinforced Concrete for Structural Design*. TU Delft, Delft University of Technology

[3] Martinie L, Rossi P and Roussel N 2010 Rheology of fober reinforced cementitious materials: classification and prediction *Cement and Concrete Research* **40(2)** pp 226-234

[4] Alberti M G 2015 Polyolefin fibre-reinforced concrete: from material behaviour to numerical and design considerations, Thesis doctoral, Technical University of Madrid

[5] MacDonald C N et al. 2009 Case Histories Using Synthetic Fiber Reinforced Concrete, in: *Shotcrete for Underground Support XI* ed F Amberg and K F Garshol, ECI Symposium Series

IOP Conf. Series: Materials Science and Engineering **246** (2017) 012027 doi:10.1088/1757-899X/246/1/012027

[6] Wang Y, Backer S, Li VC 1988 A special technique for determining the critical length of fibre pull-out from a cement matrix *Journal of Materials Science Letters* **7(8)** pp 842-44

[7] Vandewalle L, Nemegeer D, Balazs L, Barr B, Barros J, Bartos P et al. 2003 RILEM TC 162-TDF: Test and design methods for steel fibre reinforced concrete' - sigma-epsilon-design method - Final Recommendation. *Materials and Structures*. **36(262)** pp 560-67

[8] Romualdi J P and Mandel J A 1964 Tensile Strength of Concrete Affected by Uniformly Distributed and Closely Spaced Short Lengths of Wire Reinforcement *ACI Journal* **61(6)** pp 657-71

[9] Soroushian P and Lee C D 1990 Distribution and orientation of fibres in steel fiber reinforced concrete *ACI Materials Journal* **87(5)** pp 433-39

[10] Dupont D and Vandewalle L 2005 Distribution of steel fibres in rectangular section *Cement and Concrete Composites* **27(3)** pp 391-98

[11] Juhász K P 2015 Evaluation of fibre reinforced concrete beam test results based on the examination of the real fibre distribution in: *Fibre Concrete 2015*

[12] Krenchel H 1975 Fibre spacing and specific fibre surface in Fibre Reinforced Cement and Concrete, *Proc. RILEM Conf.*, ed A Neville pp 69-79

IOP Conf. Series: Materials Science and Engineering **246** (2017) 012028 doi:10.1088/1757-899X/246/1/012028

Creep behaviour of macro glass fibre reinforced concrete beams

P Löber[1], B Heiden[1] and K Holschemacher[1]

[1]Structural Concrete Institute, HTWK Leipzig, University of Applied Sciences, Karl-Liebknecht-Straße 132, Leipzig Germany

E-mail: philipp.loeber@htwk-leipzig.de

Abstract. This paper aims to present a creep study on structural concrete reinforced with macro glass fibres and wants to contribute to the understanding of creep behaviour of fibre reinforced concrete (FRC). Fibre reinforced concrete beams have been subjected to bending and tested in cracked state under defined stress levels. Therefore, a four-point- bending test setup was chosen and the creep period was 372d. The aim was to determine creep coefficients and to test residual strength values afterwards. Results show a dependence of residual strength and applied stress level. It turned out, that the beams failed due to tertiary creep at stress levels between 65 and 70% of residual crack load at 0.5 mm pre-crack deflection. Nevertheless, all remaining specimens showed increased loads after creep period. Finally, the evaluation is conducted in comparison to other fibre types.

1. Introduction

Regarding the long-term performance and durability of concrete members it is important to consider shrinkage and creep phenomena. Creep behaviour of plain concrete under compressive stresses is anchored in the codes [1]. Nevertheless, the durability of structural concrete members is mainly influenced under service conditions. In case of FRC this means in cracked state because macro fibres start contributing to load bearing after concrete has cracked. Whereas creep behaviour of FRC under compression is similar to plain concrete, in cracked state additional factors like creep of fibre material and bond of fibre and matrix must be considered next to compressive creep strains in concrete. Therefore fibre only reinforced members show higher structural creep behaviour than reinforced concrete structures [2].

The general creep behaviour of FRC has not been considered in most codes until yet. The only exception is the Austrian ÖBV Guideline "Fibre reinforced concrete", where a test setup for polymer fibre reinforced concrete beams is proposed in the informal part of the guideline. Since the early 2000 there have been investigations on creep behaviour of cracked steel and polymer fibre reinforced concrete [2–12]. Nevertheless, because of the lack of a regulation it is complicated to compare the results of different investigations due to different test setups, concrete compositions and stress levels. Since 2014 the RILEM technical committee CCF collects relevant information of FRC creep behaviour.

Most researchers chose a defined crack opening between 0.5 and 2.5 mm as starting point for the creep tests. Investigated stress levels are often scaled on either flexural strength value or the respective residual strength value corresponding to the chosen crack opening.

Macro glass fibres (length: 36 mm, $\varnothing_{equ}$=0.54 mm) are relatively new material used to strengthen structural concrete applied in industrial floors [13, 14]. Made of alkali-resistant (AR) glass, they have some advantages in applications where steel reinforcement is not preferred like stables for animals, where glass fibres reduce harming due to surface near fibres or laboratories where undisturbed electromagnetic fields are required. Furthermore, they cannot rust and disperse very well during mixing without tending to segregate or float. It has to be mentioned that in comparison to other fibre materials, glass itself has nearly no viscoelastic behaviour and therefore very little to no tendency to creep [15].

According to international codes like the fib Model Code 2010 [16] glass fibres can substitute mesh reinforcement in statically highly undetermined structures up to a certain extend [13, 14, 17]. In Germany, only steel and polymer fibres may be used in structural concrete. The requirements for these fibres are given in DIN EN 14889-1/2 [18, 19]. Steel fibres may be used to substitute steel bar or mesh reinforcement in serviceability and ultimate limit state according to DAfStb Guideline "Steel fibre reinforced concrete" [20].

2. Test method

The general post cracking performance of FRC is recommended to be tested on bending beams. Therefore, according to codes mainly two different setups are allowed. Firstly the three point bending test on notched specimen according to fib Model Code 2010 [16], EHE-08 [21], ACI 318 [22] and EN 14651:2007 [23]. Secondly some countries like Germany or Austria test FRC in four point bending of unnotched specimens according to DAfStb Guideline „Steel fibre reinforced concrete" [20] or ÖBV Guideline "Fibre reinforced concrete" [24]. It seems suitable to test FRC creep behaviour also according to one of these guidelines.

Both, notched and unnotched beams have advantages and disadvantages when it comes to determine residual strength values. In three-point bending tests on notched specimens crack development is forced at one specific section. Residual strengths acquired on notched beams tend to be overestimated, because the specimen cracks not necessarily at its weakest point. In unnotched beams crack position and therefore crack widths corresponding to certain deflections are not equal in all specimens. The disadvantage of this test comes with the uniformly distributed bending moment in the middle of the specimen between the two loads. Nevertheless, the differences in crack openings at small deflections are very low. Therefore, the authors of this paper decided to use the four-point bending setup to test creep behaviour of macro glass fibre reinforced concrete in cracked state, as it is described in the German Guideline for the four-point-bending test. Only deflections and no crack openings have been measured during creep tests.

With the test setup shown in **Figure 1**, two beams could be tested simultaneously. All in all, there have been six stands, which allowed to test twelve FRC specimens. The specimens have been loaded with a lever mechanism.

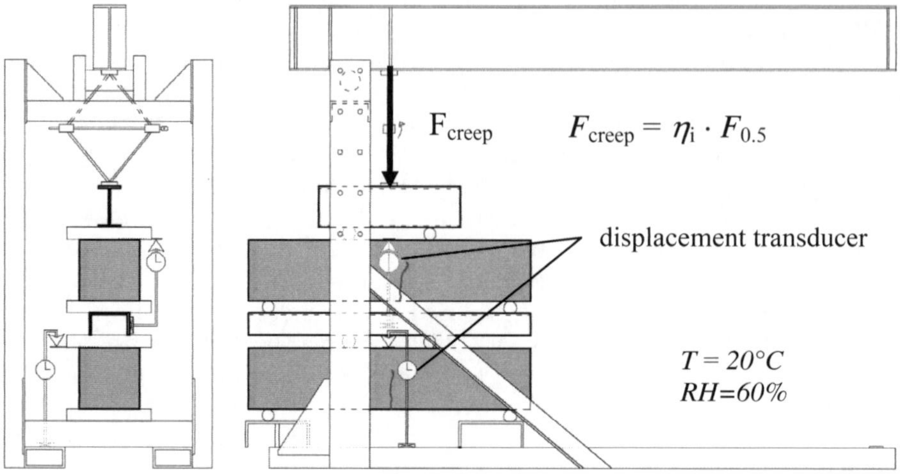

Figure 1. Test setup for creep tests

Before undergoing the creep procedure all beams have been tested in four-point-bending until the defined deflection of 0.5 mm which equals approximately 0.4 mm crack opening. After creep test, all remaining specimen have been tested further on their residual strength values. It must be mentioned that in case of unnotched specimen the crack not necessarily appears in the middle of the beam, which causes the crack width to scatter between around 0.35 and 0.45 mm. Figure 2 shows the complete test procedure. The stress level η_i of each specimen is the ratio between creep load and on the respective residual load $F_{0.5}$ at 0.5 mm deflection.

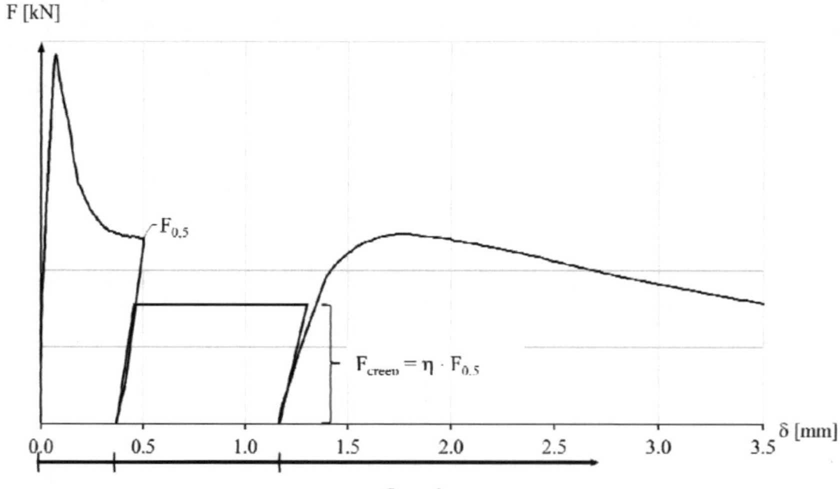

Figure 2. Scheme of the complete test procedure

3. Materials and investigated parameters

The aim of the creep tests was to determine a dependency of different stress levels on the creep deformations and post creep residual strengths. To allow different stress levels, only one glass fibre reinforced concrete mixture has been tested, which is given in Table 1.

Table 1. Material composition.

	Type	Content
Cement	CEM II/A-LL 32,5R	350 kg
Water		209 kg
Additions	Fly ash (FA)	80 kg
W/(C+0.4 FA)*		0.5
	Macro glass fibres (36 mm)	20 kg
Additives	Superplasticizer (PCE)	0.25 % of cement
Aggregates	0/2	656 kg
	2/8	328 kg
	8/16	656 kg

* equivalent water/cement ratio according to DIN EN 206 [25]

All specimens were cured in the same way. The beams were kept in moulds for two days. Afterwards they were wrapped in foil and stored in a climate room at 20°C, 60% RH until the age of 28 d. Then they were tested in four-point-bending and stored again in climate room until an age of 48 d. Afterwards creep tests with a duration of 372 d were performed. Some of the specimen failed after short time and were replaced. Table 2 gives an overview about all tested specimens in the respective test setups. The respective stress level for each beam has been calculated including self-weight.

Table 2. Arrangement of creep test setups

Stand No.	1	1*	2	2*	3	4	5	6
Age at start [d]	48	32	48	32	48	48	48	48
Stress level η_i top	0.7	0.5	0.7	0.535	0.5	0.6	0.5	0.6
Stress level η_i bottom	0.74	0.469	0.8	0.5	0.546	0.645	0.559	0.615

* after failure of the beams in the first series, new specimens have been produced

The respective stress levels were in range between 46.9% and 80% and have been chosen to exceed the maximum design value according to DAfStb Guideline "Steel fibre reinforced concrete" [20] which may be calculated utilizing different coefficients used for FRC design:

$$\eta_{\max} = \alpha_c^f / \gamma_{ct}^f \cdot \kappa_F^f \cdot \kappa_G^f \cdot 0{,}51 / \gamma_R^f = 0{,}85 / 1{,}5 \cdot 1{,}0 \cdot 1{,}7 \cdot 0{,}51 / 1{,}4 = 0{,}421 \quad (1)$$

For the design of FRC members under permanent load maximum 42.1% of residual flexural strength values will be considered.

4. Results and Evaluation

As it can already be seen in Table 2 specimens with stress levels 70% and higher failed after around two days. Although steel fibres show higher resistance during very high stress levels like this, the results are in accord with macro polymer fibre reinforced concrete [2]. In the following, not all test data is presented. Only six beams across tested stress levels are shown to provide a clearer view.

Directly after applying the creep load, the elastic deformation $\delta_e(t_0)$ can be measured. Together with the creep deformation $\delta_t(t)$ the creep coefficient $\varphi(t, t_0)$ can be calculated for a defined time point per equation (2).

$$\varphi(t,t_0)=\frac{\delta_t(t)}{\delta_e(t_0)} \tag{2}$$

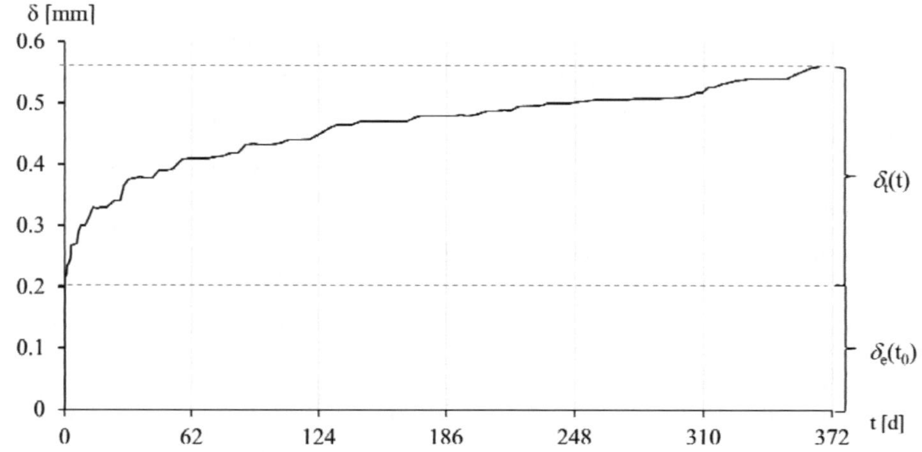

Figure 3. Example for elastic and creep deformations

In the following creep deformations of the beams are presented. It is obvious, that higher stress levels cause larger deformations over time. Although not failed within 372 days of creep stresses it is likely that specimen with stress level over 50% may fail soon due to tertiary creep which is characterized by a change from linear to exponential creep rate until failure.

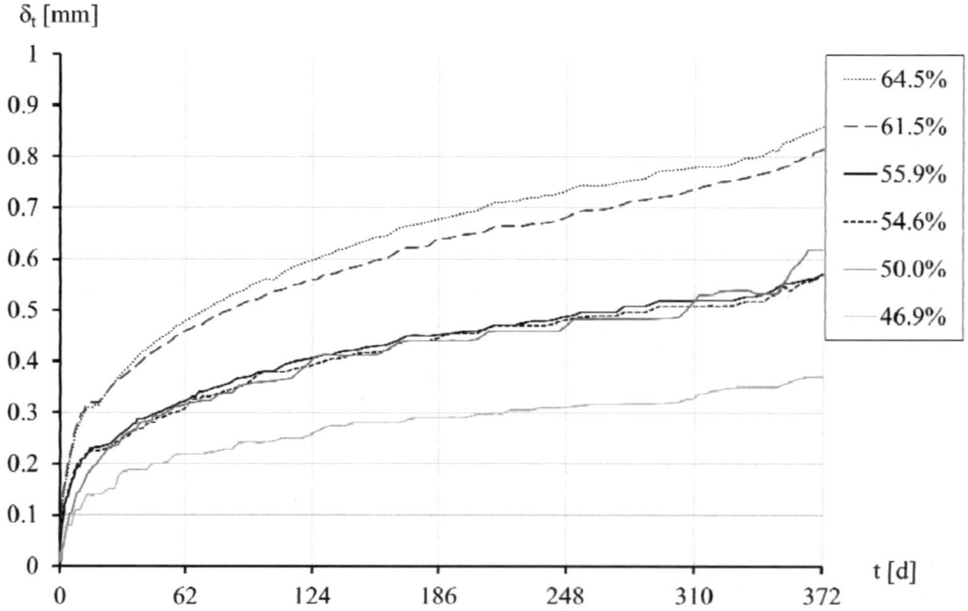

Figure 4. Creep deformations of six specimens

Eurocode 2 (EC2) [1] provides equations to calculate creep coefficients in dependence of cement type, compressive strength, age of concrete at start of creep, creep duration, moisture and temperature. Although the formulas only consider compressive creep stresses, it seems possible to estimate approximate creep coefficients. In case of stress levels above 45% of concretes characteristic compressive strength, non-linear creep must be considered. **Figure 5** shows calculated creep coefficients per equation (2) in comparison with the theoretical creep coefficients according to EC2. The theoretical creep curves are shown for stress levels below 45% and for 64.5% which is the highest stress level which did not fail during the test.

Although some experimental data exceeds the theoretical values after around 320 d, it can be seen that creep behaviour of beams with a stress level below 50% correspond good with the EC2 curve for 45% and below. It is worth to mention that the results from EC2 incorporate a coefficient of variation of 30%.

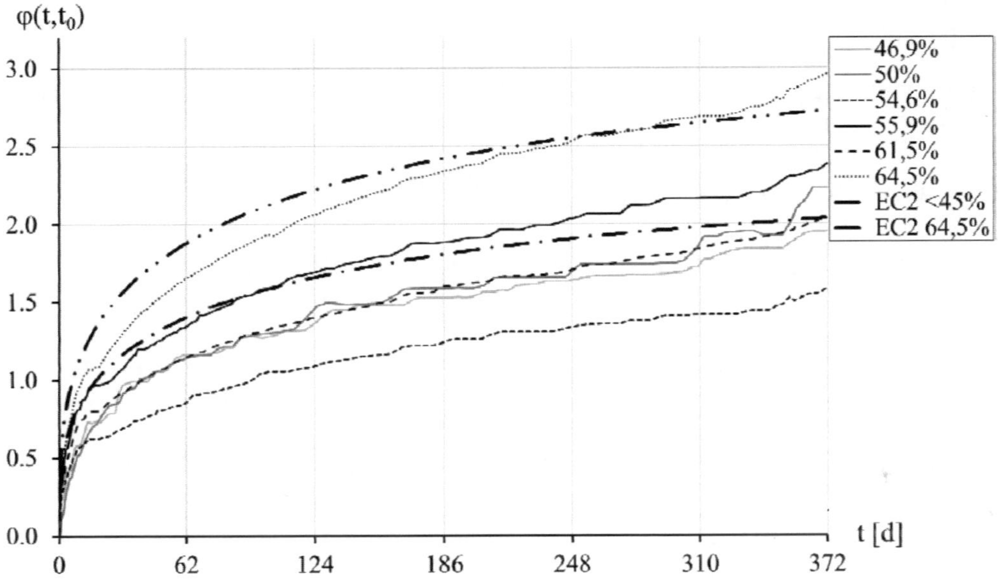

Figure 5. Creep coefficients of specimens compared to calculated creep coefficients from EC2

The results of pre- and post-creep flexural tests are provided in **Figure 6**. The measured creep deformations are also included between those test curves.

All specimen showed increased maximum load after creep duration. This may result from higher concrete compressive strength resulting in better bond between glass fibres and matrix. In contrast, the ratio between post- and pre-creep loads is smaller with increased stress level as shown in **Figure 7**. Obviously higher stress levels cause larger creep deformations as well as bigger crack widths resulting in lesser residual strength.

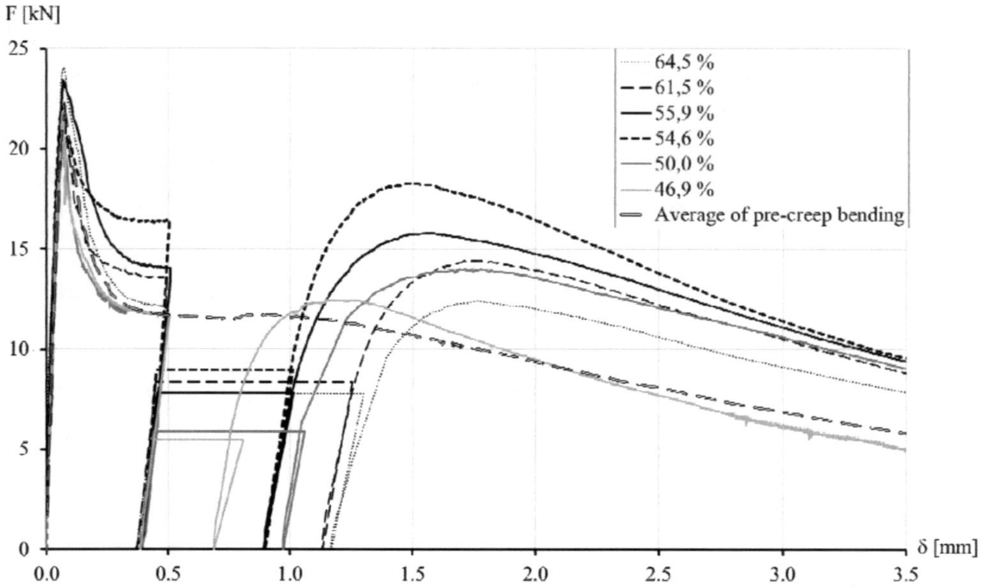

Figure 6. Load vs. deflection of pre-creep, creep and post creep flexural tests

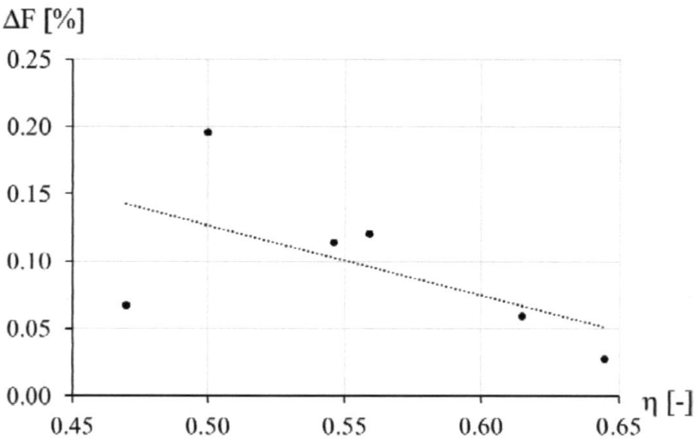

Figure 7. Relative load increase in the post creep flexural tests vs. stress level

5. Comparison

To compare creep performance with other fibre types is very difficult due to the lack of sufficient data. Results from other researchers scatter across a wide range of different concrete types, strengths, specimen types and stress levels. Therefore, it was complicated to find appropriate numerical data in the tested range of parameters (compressive strength, crack opening, stress levels). Nevertheless **Figure 8** shows some results taken from literature [2, 4] in comparison to own results and calculated values according to EC2 [1]. In the presented range, glass fibres show biggest creep coefficients. It has to be considered that these values are still significantly lower than expected by EC2 and range of possible stress levels is larger compared to polymer fibres. In case of presented results for steel fibres, concretes compressive strength was at 70 N/mm² which is much higher than the investigated strength of 40 N/mm² from structural glass fibre reinforced concrete. This leads to significantly reduced creep coefficients. Compared to other results from literature it can be told, that steel fibre reinforced concrete shows least creep deformations.

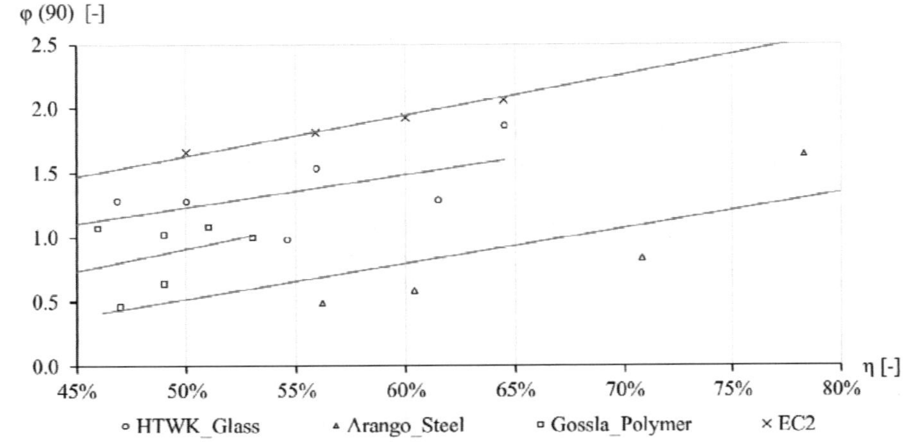

Figure 8. Comparison of creep coefficients after 90 d for different fibre types

As it is visible in **Figure 9**, polymer fibre reinforced concrete tends to show bigger creep deformations with increasing stress level compared to glass or steel fibre reinforced concrete. This can be seen in the gradient of regression graphs. Although low creep deformations at low stress levels, polymer fibre reinforced concrete tends to fail earlier due to tertiary creep especially if stress levels exceed 55%.

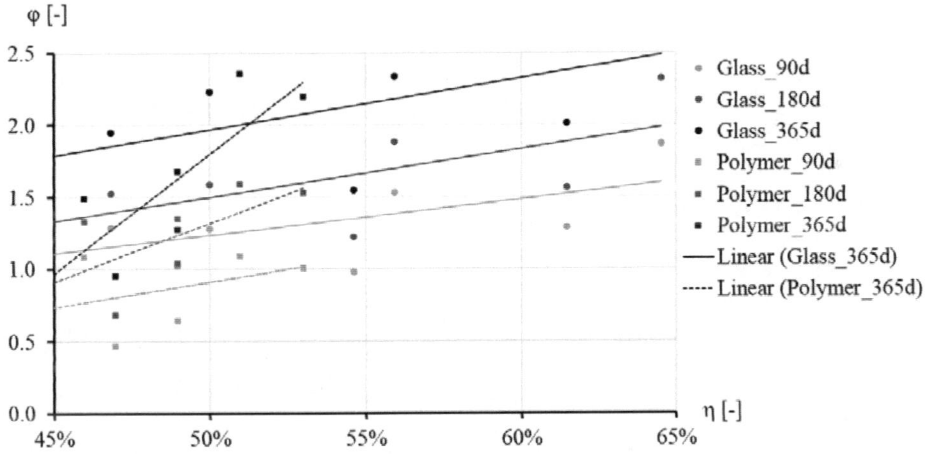

Figure 9. Creep coefficients of macro glass and polymer fibres after different periods

A huge influence is also given by the creep behaviour of the fibre material itself and the bond between fibre and matrix. Whereas glass has nearly no tendency to creep, polymer fibres may show very large creep deformations greatly dependent on their stress level [2, 7, 8]. Steel fibres show very low creep especially when bond between fibre and matrix is enhanced due to hooked ends [4, 8]. This physical bond is a further disadvantage of polymer as well as glass fibres, as it can practically not be improved mechanically. In contrast glass fibres have a good chemical bond to the surrounding matrix because of their mineral properties. However, they may lack in cracked state, because of their composition of many filaments where bond is established only in surface near filaments, where they are penetrated by cement particles up to a certain extend.

IOP Conf. Series: Materials Science and Engineering **246** (2017) 012028 doi:10.1088/1757-899X/246/1/012028

6. Summary

With the present study creep as well as post creep bending behaviour of structural glass fibre reinforced concrete beams have been investigated, following other researchers approaches for FRC. The authors want to give a contribution to the increasing interest of creep behaviour of FRC. It could be shown that structural concrete reinforced with macro glass fibres does not show increased tendency to creep. The results are comparable with high quality polymer fibres, although the pull-out procedure of fibres in concrete are different. Furthermore, it has to be mentioned, that creep deformations are greatly dependent of concretes properties such as compressive strength, age and cement type. There is a big need to investigate different matrix-fibre combinations at several stress levels. Parametric studies would be relevant to fully understand FRC creep behaviour. Furthermore, it is important to establish international rules and recommendations for a uniform test setup, specimen type, stress level and climate conditions.

Acknowledgements

We would like to thank the Federal Ministry for Economic Affairs and Energy (BMWI) for the financial support of the research project in the Central Innovation Programme (ZIM).

References

[1] DIN EN 1992-1-1:01-2011 Eurocode 2: Bemessung und Konstruktion von Stahlbeton- und Spannbetontragwerken – Teil 1-1: Allgemeine Bemessungsregeln und Regeln für den Hochbau; German version EN 1992-1-1:2004 + AC:2010 (Berlin: Beuth Verlag GmbH)
[2] Gossla U and Rieder K-A 2009 *Beton- und Stahlbetonbau* **104** 76–87
[3] Arango S E, Serna P, Martí-Vargas J R and García-Taengua E 2012 *Exp Mech* **52** 1067–78
[4] García-Taengua E, Arango S, Martí-Vargas J R and Serna P 2014 *Construction and Building Materials* **65** 321–9
[5] Blanco Álvarez A 2013 Characterization and modelling of SFRC elements *Doctoral Thesis* Universitat Politècnica de Catalunya, Departament d'Enginyeria de la Construcció
[6] Fládr J, Vodička J and Krátký J 2012 Methodology of measuring of creep of concrete specimen reinforced with polymer fibres subjected to bending *BEFIB2012 – Fibre reinforced concrete* ed J Barros (UM, Guimarães)
[7] Kurtz S and Balaguru P 2000 *Cement and Concrete Research* **30** 183–90
[8] Kusterle W 2015 Creep of fibre reinforced concrete - flexural test on beams *8th international Conference: Fibre Concrete 2015: Technology, Design, Application (September 10-11, 2015)* ed CTU Prague, Faculty of Civil Engineering
[9] Raul L. Zerbino and Bryan E. Barragan 2012 *ACI Materials Journal* **109** 215–24
[10] Serna P, Llano-Torre A and Cavalaro S H P (eds) 2017 *Creep behaviour in cracked sections of fibre reinforced concrete: Proceedings of the International RILEM Workshop FRC-CREEP 2016 (RILEM bookseries* Volume14*)* (Dordrecht: Springer)
[11] Vasanelli E, Micelli F, Aiellom M. A. and Plizzari G 2012 Long term behaviour of fiber reinforced concrete beams in bending *BEFIB2012 – Fibre reinforced concrete* ed J Barros (UM, Guimarães)
[12] Pujadas P, Blanco A, Cavalaro S, La Fuente A de and Aguado A 2017 *RILEM Bookseries* **14** 77–87
[13] Löber P, Heiden B, Holschemacher K and Barragán B 2015 Glass Fiber Reinforced Concrete For Slabs On Ground - Material Characterization And Application *8th international Conference: Fibre Concrete 2015: Technology, Design, Application (September 10-11, 2015)* ed CTU Prague, Faculty of Civil Engineering
[14] Löber P, Holschemacher K, Heiden B and Barragán B 2015 *Concrete* **49** 48–50
[15] Knippers J, Waimer F and Oppe M 2015 Faserverbundwerkstoffe im Bauwesen *Stahlbau-Kalender 2015: Eurocode 3 - Grundnorm, Leichtbau (Stahlbau-Kalender)* ed U Kuhlmann 17th edn (Berlin: Ernst & Sohn a Wiley brand) pp 464–516

[16] fib Model Code 2010: Final draft **66** *(Bulletin / International Federation for Structural Concrete Draft model code)* (Lausanne: International Federation for Structural Concrete)
[17] Löber P and Holschemacher K 2014 *WJET* **2** 48–54
[18] DIN EN 14889-1:11-2006 Fasern für Beton – Teil 1: Stahlfasern – Begriffe, Festlegungen und Konformität; Deutsche Fassung EN 14889-1:2006 (Berlin: Beuth Verlag GmbH)
[19] DIN EN 14889-2:11-2006 Fasern für Beton – Teil 2: Polymerfasern – Begriffe, Festlegungen und Konformität; Deutsche Fassung EN 14889-2:2006 (Berlin: Beuth Verlag GmbH)
[20] DAfStb 2012 Richtlinie Stahlfaserbeton (Berlin: Beuth Verlag GmbH)
[21] Ministerio de Fomento 2011 EHE-08 Instrucción de Hormigón Estructural
[22] American Concrete Institute 2014 ACI 318-14 Building Code Requirements for Structural Concrete and Commentary
[23] DIN EN 14651:12-2007 Prüfverfahren für Beton mit metallischen Fasern – Bestimmung der Biegezugfestigkeit (Proportionalitätsgrenze, residuelle Biegezugfestigkeit); Test method for metallic fibre concrete – Measuring the flexural tensile strength (limit or proportionality (LOP), residual)
[24] Öserreichische Bautechnik Vereinigung 2008 Richtlinie "Faserbeton"
[25] DIN EN 206:2017-01 Concrete – Specification, performance, production and conformity; German version EN 206:2013+A1:2016

IOP Conf. Series: Materials Science and Engineering **246** (2017) 012029 doi:10.1088/1757-899X/246/1/012029

Application of trilinear softening functions based on a cohesive crack approach to the simulation of the fracture behaviour of fibre reinforced cementitious materials.

A Enfedaque[1], M G Alberti[1] and J C Gálvez[1]

[1] Departamento de Ingeniería Civil: Construcción, E.T.S de Ingenieros de Caminos, Canales y Puertos, Universidad Politécnica de Madrid. C / Profesor Aranguren, s/n, 28040, Madrid, Spain

E-mail: jaime.galvez@upm.es

Abstract. The relevance of fibre reinforced cementitious materials (FRC) has increased due to the appearance of regulations that establish the requirements needed to take into account the contribution of the fibres in the structural design. However, in order to exploit the properties of such materials it is a key aspect being able to simulate their behaviour under fracture conditions. Considering a cohesive crack approach, several authors have studied the suitability of using several softening functions. However, none of these functions can be directly applied to FRC. The present contribution analyses the suitability of multilinear softening functions in order to obtain simulation results of fracture tests of a wide variety of FRC. The implementation of multilinear softening functions has been successfully performed by means of a material user subroutine in a commercial finite element code obtaining accurate results in a wide variety of FRC. Such softening functions were capable of simulating a ductile unloading behaviour as well as a rapid unloading followed by a reloading and afterwards a slow unloading. Moreover, the implementation performed has been proven as versatile, robust and efficient from a numerical point of view.

1. Introduction and background

Cementitious materials share a common characteristic among them which is the preponderantly proportion of cement in their composition. Therefore, most of their properties are highly dependent on those of cement being its good compressive strength and its high modulus of elasticity two of them. Nevertheless, some other characteristics of concrete are not so adequate for constructing such as its limited flexural and tensile strength. Concrete, which has been the most widespread construction material, is a clear example of what has been previously mentioned. It boasts a noticeable compressive strength but its tensile strength might be considered about the tenth of such value. Consequently, in structural elements subjected to bending moments, the tensile stresses may exceed the tensile strength of concrete and cracks would appear, producing eventually the fracture of the element and causing both an economic damage and increasing the risk of physical harm in humans. One possibility to solve that issues is adding a continuous reinforcement to the cementitious material, in the way of randomly distributed fibres. This approach was firstly considered in the nineteenth century but due to its success it is still applied.

When randomly distributed fibres are added to the concrete mix the material is commonly known as fibre-reinforced concrete (FRC). Traditionally, the most employed fibres in concrete structural elements have been steel ones and consequently the material formed has been named steel fibre reinforced concrete (SFRC). However, due to the rise of the steel price and the possible corrosion that steel might suffer when in potentially hazardous environments, the use of structural macro-polyolefin fibres as a concrete addition has become a possibility. The latter material has been termed polyolefin fibre reinforced concrete (PFRC). Another cementitious material that has been reinforced with fibres is cement mortar. It has been mainly used in elements that have a reduced thickness which tend to crack as a result of the shrinkage. In order to avoid such cracking, randomly distributed glass fibres have been added to the cement mortar forming glass fibre reinforced cement (GRC). The presence of glass fibres increase not only the flexural and tensile strength but also provide certain degree of ductility.

In order to fully exploit the mechanical properties of fibre reinforced cementitious materials several models that deal with the physical process of cracking in concrete under tensile stresses have been postulated. One remarkable one is the smeared crack approach [1] that has been commonly used when there is no localisation of the cracks and when the crack opening is reduced. Another, used when crack is localised is the so-called discrete approach which has provided more accurate results. One of the most used cracking models for plain concrete was developed by Hillerborg [2] and named the cohesive crack model. The applicability of this model has been analysed in depth by several authors and has been applied not only to plain concrete, but also to other brittle or quasi-brittle materials such as brick masonry [3-5]. In addition, in certain circumstances this model represents with accuracy both fracture under Mode I and a fracture process generated under a Mode II without the need of using a tracking algorithm [6]. A more detailed explanation of this model may be found in any of the references provided.

The cohesive crack model is able to reproduce the fracture behaviour of brittle materials by using as an input several mechanical properties that can be obtained by performing standard laboratory tests. In the case of plain concrete, these properties are the tensile strength (obtained by means of the indirect tensile strength test) [7] and the fracture energy (by following any of the several recommendations commonly adopted) [8]. Using these parameters, the cohesive crack model replicates the fracture tests of plain concrete only if another characteristic of concrete is defined, that is to say, the type of softening function [9]. Several authors have studied the suitability of using diverse softening functions, which may vary from an exponential function to a linear or bilinear one. The latter has been profusely used due both to its simplicity and the accurate results that it provides.

However, none of these functions can be directly applied to reproduce the fracture behaviour of the various types of FRC produced nowadays. Recently, the application of multilinear linear softening functions to FRC and GRC have been shown as a suitable and simple option that provides accurate results [10-14]. However the changes that should be applied to such softening functions in order to widen their applicability to those materials have not studied in depth. Therefore, the present paper seeks to fulfil that gap by analysing the suitability of such softening functions in order to obtain accurate simulation results in numerical analysis of fracture tests of a wide variety of FRC and GRC mixes. In that sense, the changes introduced by the variation of behaviour of the amount of fibres added in a PFRC will be studied. The analysis of the softening functions of three types of GRC will be also analysed. Lately, all functions will be compared and some conclusions from such comparison will be extracted.

2. Softening behaviour of fibre reinforced cementitious materials

The embedded cohesive crack model that has been extensively used for simulating the fracture of concrete is based on a central forces model and has been extensively explained in references [13 14].

Once this implementation has been performed, the most important issue to address the fracturing behaviour of the fibre reinforced cementitious material is defining the softening function. The cohesive fracture approach considers the softening function of concrete as a material property [15].

Such function is defined between the strain were the tensile strength of the material is reached, where the width of the crack is still zero and the critical crack opening where the material is no longer capable of withstanding any loads. The area below the softening function set the fracture energy and therefore the shape of the function determines the behaviour of the material. Such area can be obtained by integrating the softening function from a zero crack width and the critical crack width, w_c where cohesive stress becomes zero [15] being *f(w)* the tensile stress at *w* crack opening. This expression can be seen in (1)

$$G_F = \int_0^{w_c} f(w)dw \quad (1)$$

When the maximum strength of the material equals the tensile strength (f_{ct}) the cracking behaviour onsets and expression (2) is validated.

$$f_{ct} = f(0) \quad (2)$$

Consequently, it is the softening function the main characteristic of the material while fracturing and therefore the one that defines the behaviour while cracking. In this regard it has to be outlined that when dealing with plain concrete cracking there are several types of functions that have been used successfully. For instance, linear, bilinear or exponential curves have been used obtaining accurate results [2, 9, 16-18]. One possibility of an exponential softening function can be seen in (3).

$$\boldsymbol{\sigma = f_{ct} \cdot e^{\left(-\frac{f_t \omega}{G_F}\right)}} \quad \mathbf{(3)}$$

where G_F is the specific fracture energy and f_{ct} is the tensile strength. A way of obtaining such softening functions is by what has been termed inverse analysis, adjusting experimental response of a notched specimen by trial-and-error optimisation through use of finite element methods [19]. Based on the promising results obtained using a bilinear softening function both in accuracy to the experimental results and in numerical calculus efficiency, the material improvements provided by the presence of fibres has been introduced in several studies as modifications of multilinear functions [our references of simulation articles]. Such relations were implemented in a commercial finite element programme by means of a user subroutine for material. Hence, the numerical simulations were performed using ABAQUS code and one UMAT subroutine to model the fracture behaviour of PFRC and GRC. In such a sense, the non-linear fracture process zone emerges in the elements placed on the crack. Given that the behaviour of the fracturing elements depends on a constitutive relation that needs to be iteratively fit the scheme that can be seen in Figure 1 was followed.

Applying the multilinear approach to the softening functions that are capable of introducing the effect in the fracture behaviour of PFRC and GRC such functions can be defined as in expressions (4) and (5) respectively. In these expressions the only difference can be observed in the first unloading part. In this regard, both approaches; the one related with PFRC using an exponential function and the one related with GRC which uses a linear function provide a remarkable degree of accuracy.

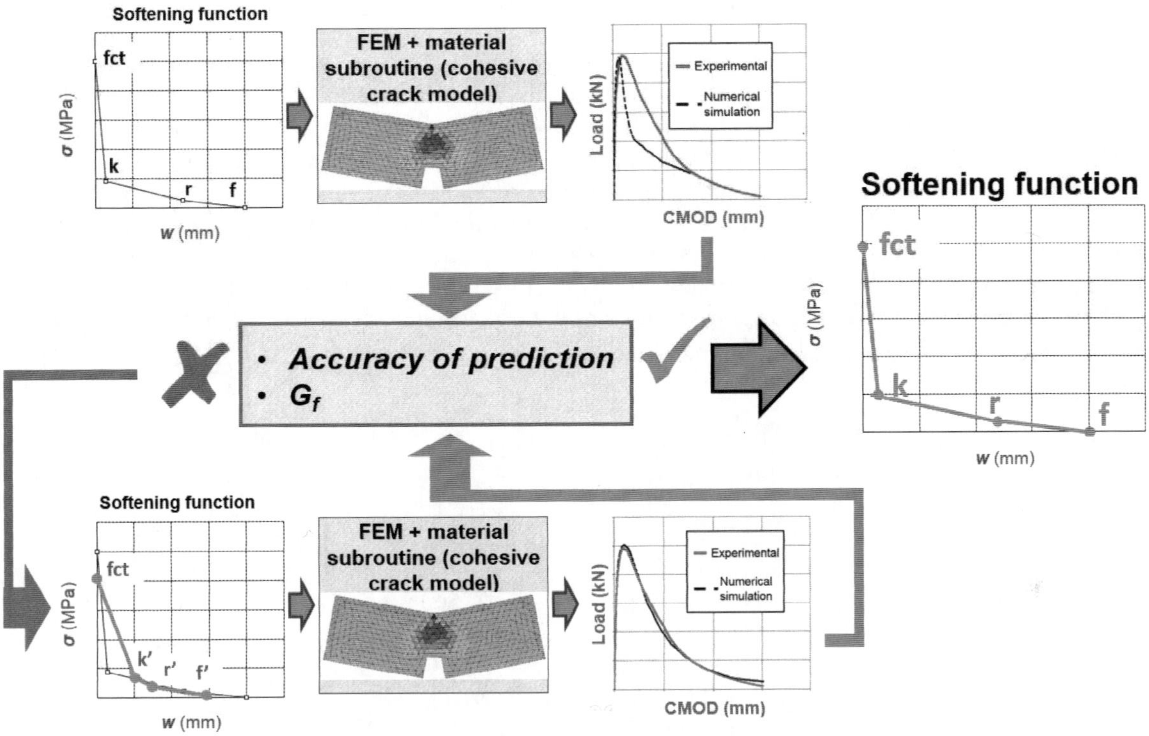

Figure 1: Inverse analysis used to reproduce the fracture test results

$$\begin{cases} \sigma = \boldsymbol{f_{ct}} \cdot \boldsymbol{e}^{\left(-\frac{f_{ct}w}{G_F}\right)} & \boldsymbol{if} \quad \boldsymbol{0} < w \leq \boldsymbol{w_k} \\ \sigma = \sigma_k + \left(\frac{\sigma_r - \sigma_k}{w_r - w_k}\right)(\boldsymbol{w} - w_k) & if \quad w_k < w \leq w_r \\ \sigma = \sigma_r + \left(\frac{-\sigma_r}{w_f - w_r}\right)(\boldsymbol{w} - w_r) & if \quad w_r < w \leq w_f \\ \sigma = 0 & if \quad w > w_f \end{cases} \tag{4}$$

$$\begin{cases} \sigma = \boldsymbol{f_{ct}} + \left(\frac{\boldsymbol{\sigma_k} - \boldsymbol{f_{ct}}}{\boldsymbol{w_k}}\right) \boldsymbol{w} & \boldsymbol{if} \quad \boldsymbol{0} < w \leq \boldsymbol{w_k} \\ \sigma = \sigma_k + \left(\frac{\sigma_r - \sigma_k}{w_r - w_k}\right)(\boldsymbol{w} - w_k) & if \quad w_k < w \leq w_r \\ \sigma = \sigma_r + \left(\frac{-\sigma_r}{w_f - w_r}\right)(\boldsymbol{w} - w_r) & if \quad w_r < w \leq w_f \\ \sigma = 0 & if \quad w > w_f \end{cases} \tag{5}$$

3. Discussion

Several experimental results of fracture tests performed in cementitious materials were reproduced using the aforementioned implementation. The test reproduced in the case of the PFRC tests had been conducted in $100\text{x}100\text{x}430\text{mm}^3$ specimens. The span of the three-point bending tests was $3D$ and the notch $1/3D$, being D the height of the square cross section (100mm). Such specimens had been manufactured adding 3, 4.5, 6 and 10kg/m^3 of 60mm-long polyolefin fibres to a self-compacting concrete. The details of the manufacturing process as well as the testing setup and procedure followed can be found in [20]. The recommendation followed was RILEM-TOC 187 [12]. The curves shown are the average of at least three successful tests. Although it could be argued that a larger amount of test would be required to provide sound conclusions it has to be highlighted that the careful handling and preparation of the test setup provided results with a limited degree of scatter. Regarding the simulations, if Figure 2 is observed it can be seen how the softening function implemented was capable of reproducing with a remarkable accuracy the behaviour of PFRC. This implementation was found both robust and numerically efficient besides of the great adaptability that it boasts. This can be clearly perceived in Figure 2. By changing the points that define each stretch of the softening functions the several implementations were able to reproduce all the features of the experimental results. The latter is of high relevance because the minimum post-peak registered changes noticeably between the formulations. Similarly, the values of the maximum post peak loads of the simulations appear in similar values in the simulated curves. Not only these values were reproduced but also the slopes of the unloading and reloading branches of the experimental curves are accurately obtained.

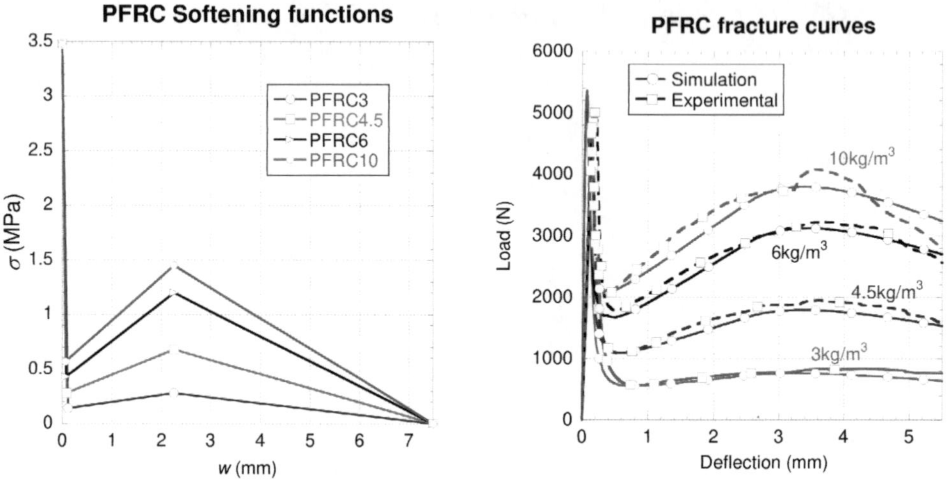

Figure 2: Softening functions (left) and comparison between simulated and experimental results (right).

In order to check the adaptability of the multilinear softening curves another fibre reinforced cementitious material of a totally different nature was simulated. In this case the material chosen was a glass fibre reinforced cement (GRC). This material is manufactured by merging cement mortar and randomly distributed short glass fibres. Similarly to the case previously explained of PFRC the multilinear softening functions were implemented in this case for reproducing the fracture test results obtained in previous research [21]. Three GRC formulations were manufactured, being among them the use of some chemical additives the only difference. Such additives seek to prevent the change of mechanical properties that undergo the traditional GRC formulations as times passes. Such formulations were named GRC, GRC-M and GRC-P which correspond to traditional GRC, GRC with a Metaver additive and a GRC with Powerpozz additive. In contrast to PFRC the fracture tests conducted with the several formulations of GRC could not be performed following any recommendation as at the time of such research there was no one available. Nevertheless, the closest adaptation of the previously mentioned RILEM-TOC 187 recommendation was followed. Similarly to PFRC the curves shown in Figure 3 are the average of at least three successful tests and a remarkably low degree of scatter was registered.

As in the previous case of PFRC, the multilinear softening functions implemented were divided into three stretches. However, in the case of GRC the slope of the stretches were in all cases negative. The changes in the values that define the stretches of the softening functions were able to reproduce with remarkable accuracy the fracture behaviour of GRC including the noticeable changes in the maximum load sustained, ductility and in the unloading rate that the material boasted. In the first case it can be seen that there is no clear connection between the maximum load obtained in the test and the tensile strength of the material. It seems that the slope of the material in the first unloading branch is also of high importance to determine the maximum load of the test. This can be seen comparing the behaviour of GRC-P and GRC. Both softening functions have the same tensile strength but on the contrary the slope of the first unloading branch of the GRC-P provides the material a higher loading capacity.

Regarding the changes in the ductility, the crack width of the softening functions is the major factor that define it. However there is also influence of the slope of the third stretch of the softening function because as can be seen in Figure 3, GRC-M and GRC have the same maximum crack width but the maximum crack mouth opening displacement (CMOD) is different in both formulations.

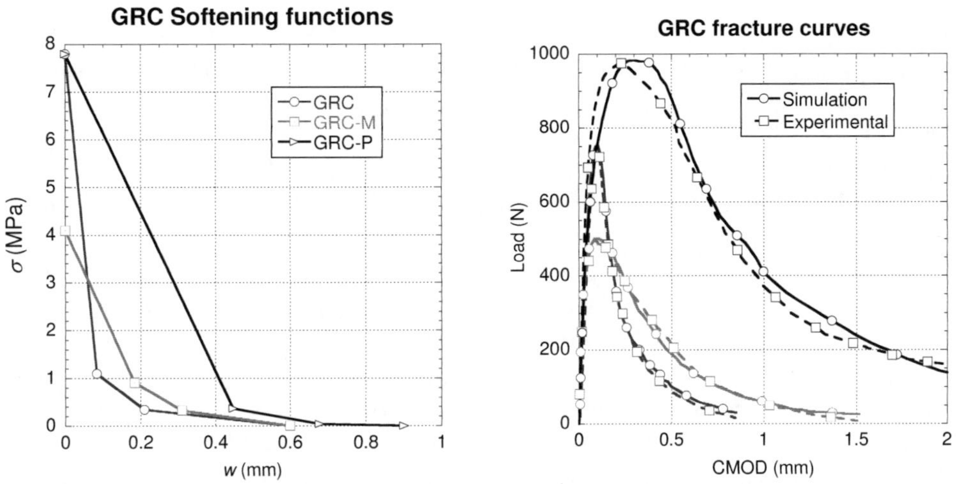

Figure 3: Softening functions (left) and comparison between simulated and experimental results (right).

Comparing the features of the softening functions implemented it can be stated that when the first unloading branch has a great negative slope the maximum load that the material can sustain in the fracture tests is mainly determined by the tensile strength of the material as can be seen in all the cases of PFRC and also in GRC formulation. Nevertheless, as in the case of GRC-P when there is a more gradual loss of stiffness the material is still able to increase its capacity of sustaining load although the closer parts of the tip of the notch are already damaged and the stiffness of the sample decreased noticeably before reaching the maximum load. This phenomenon was also perceived in the numerical models performed where several elements closer to the notch tip were already damaged when the maximum load was reached.

4. Conclusions

The implementation of multilinear softening functions has been performed successfully in a material user subroutine in a commercial finite element code. Using this implementations, fracture tests of PFRC and GRC have been reproduced with a remarkable degree of accuracy. The implementation has been proven as versatile, robust and efficient from a numerical point of view.

The changes in the tensile strength of the material and the points that define the stretches of the softening functions have been shown suitable to simulate the progressive unloading that appear in the GRC formulations fracture tests and the reload events that take place in the PFRC ones.

The variations of the length and slope of the parts of the softening functions enable to analyse the changes of the material behaviour that are introduced by several the amount of fibres in PFRC and the effects of the chemical additives in the case of GRC. It has to be underlined that this approach was able to simulate a ductile unloading behaviour, such the one that appears in GRC, as well as the rapid unloading followed by a reloading and afterwards a slow unloading as the one that appears in PFRC.

When there is a first unloading branch with a high slope the peak load registered in the tests is mainly determined by the tensile strength of the material. Nevertheless, if such slope is more gradual the combination of the slope and tensile strength of the material determine the peak load registered experimentally. In what regards the material ductility, it is mainly influenced by the maximum crack width that the material sustains.

Acknowledgements
The authors gratefully acknowledge the financial support provided by the Ministry of Economy and Competitiveness of Spain by means of the Research Fund Project BIA 2016-78742-C2-2-R. They also offer their gratitude to SIKA SAU for supporting the Enterprise University Chair: Cátedra Sika-UPM.

References

[1] M. Polanco-Loria, Numerical modelling of plain and reinforced concrete by damage mechanics, Trondheim : Department of Structural Engineering, Norwegian University of Science and Technology , 1997.

[2] A. Hillerborg, M. Modéer and P. E. Petersson, "Analysis of crack formation and crack growth in concrete by means of fracture mechanics and finite elements," *Cement and concrete research,* vol. 6(6), pp. 773-78, 1976.

[3] J. C. Gálvez, J. Planas, J. M. Sancho, E. Reyes, D. A. Cendón and M. J. Casati, "An embedded cohesive crack model for finite element analysis of quasi-brittle materials," *Engineering Fracture Mechanics,* vol. 109, pp. 369-386, 2013.

[4] E. Reyes, J. C. Gálvez, M. J. Casati, D. A. Cendón, J. M. Sancho and J. Planas, "An embedded cohesive crack model for finite element analysis of brickwork masonry fracture," *Engineering Fracture Mechanics,* vol. 76(12), pp. 1930-1944, 2009.

[5] J. M. Sancho, J. Planas, D. A. Cendón, E. Reyes and J. C. Gálvez, "An embedded crack model for finite element analysis of concrete fracture," *Engineering Fracture Mechanics,* vol. 74(1), pp. 75-86, 2007.

[6] J. C. Gálvez, Č. J., D. A. Cendon and V. Saouma, "A discrete crack approach to normal/shear cracking of concrete," Cement and concrete research, vol. 32(10), pp. 1567-1585, 2002.

[7] EN 12390-6, "Testing hardened concrete. Part 6. Tensile spliting strength of test specimens," 2009.

[8] RILEM TC-187-SOC, "Indirect test for stress-crack opening curve," 2007.

[9] V. Slowik, B. Villmann, N. Bretschneider and T. Villmann, "Computational aspects of inverse analyses for determining softening curves of concrete," Computer Methods in Applied Mechanics and Engineering, vol. 195(52), p. 7223–7236, 2006.

[10] Caggiano, A., Etse, G., & Martinelli, E. Zero-thickness interface model formulation for failure behavior of fiber-reinforced cementitious composites. Computers & Structures, 98, pp. 23-32, 2012.

[11] S. T. Kang, Y. Lee, Y. D. Park and J. K. Kim, "Tensile fracture properties of an Ultra High Performance Fiber Reinforced Concrete (UHPFRC) with steel fiber," Composite Structures, vol. 92(1), pp. 61-71, 2010.

[12] S. T. Kang and J. K. Kim, "The relation between fiber orientation and tensile behavior in an Ultra High Performance Fiber Reinforced Cementitious Composites (UHPFRCC)," Cement and Concrete Research, vol. 41(10), pp. 1001-1014, 2011.

[13] A. Enfedaque, M.G. Alberti, J.C. Gálvez, J. Domingo, Numerical simulation of the fracture behaviour of glass fibre reinforced cement, Construction and Building Materials, Volume 136, 1 April 2017, Pages 108-117, ISSN 0950-0618, http://dx.doi.org/10.1016/j.conbuildmat.2016.12.130.

[14] M.G. Alberti, A. Enfedaque, J.C. Gálvez, E. Reyes, Numerical modelling of the fracture of polyolefin fibre reinforced concrete by using a cohesive fracture approach, Composites Part B: Engineering, Volume 111, 15 February 2017, Pages 200-210, ISSN 1359-8368, http://dx.doi.org/10.1016/j.compositesb.2016.11.052.

[15] Z. P. Bazant and J. Planas, Fracture and size effect in concrete and other quasibrittle materials, CRC press, 1997.

[16] P. E. Roelfstra and F. H. Wittmann, "Numerical methods to link strain softening with failure of concrete," FractureToughness and Fracture Energy , pp. 163-175, 1986

[17] G. V. Guinea, J. Planas and M. Elices, "A general bilinear fit for the softening curve of concrete," Materials and Structures, vol. 27(2), pp. 99-105, 1994.

[18] A. G. Kooiman, C. Van Der Veen and J. C. Walraven, "Modelling the post-cracking behaviour of steel fibre reinforced concrete for structural design purposes," vol. 45(4), pp. 275-308, 2000

[19] J. L. A. De Oliveira e Sousa and R. Gettu, "Determining the tensile stress-crack opening curve of concrete by inverse analysis," *Journal of engineering mechanics,* vol. 132(2), pp. 141-148, 2006

[20] M. G. Alberti, A. Enfedaque and J. Gálvez, "On the mechanical properties and fracture behavior of polyolefin fiber-reinforced self-compacting concrete," Construction and Building Materials, vol. 55, pp. 274-288, 2014

[21] A. Enfedaque, J.C. Gálvez, F. Suárez, Analysis of fracture tests of glass fibre reinforced cement (GRC) using digital image correlation, Construction and Building Materials, Volume 75, 30 January 2015, Pages 472-487, ISSN 0950-0618, http://dx.doi.org/10.1016/j.conbuildmat.2014.11.031

IOP Conf. Series: Materials Science and Engineering **246** (2017) 012030 doi:10.1088/1757-899X/246/1/012030

How to predict the orientation factor of non-rigid macro-synthetic fibre reinforced concrete

M G Alberti[1], A Enfedaque[1] and J C Gálvez[1]

[1] Departamento de Ingeniería Civil: Construcción, E.T.S de Ingenieros de Caminos, Canales y Puertos, Universidad Politécnica de Madrid. C / Profesor Aranguren, s/n, 28040, Madrid, Spain

E-mail: jaime.galvez@upm.es

Abstract. Polyolefin fibre reinforced concrete can met the requirements set in the standards that enable to consider the residual strengths in structural design. Such residual load-bearing capacity of fibre reinforced concrete is assessed by flexural tensile tests in which the presence of fibres can bridge the crack formed and provide strengths that are directly related with the number of fibres and their positioning in the fracture surface. Therefore, the orientation and distribution of the fibres is decisive in the mechanical behaviour of fibre-reinforced concrete and this can be estimated by means of the orientation factor. Several classical models have been extensively used for the case of rigid steel fibres. The increasing interest in structural synthetic fibres that can bend demands new considerations in this matter. A probabilistic model that considers the previous research with stereographical assumptions has been performed allowing the use of fibres that can bend. This study has developed significant tools for design with the aim of predicting such number of fibres crossing a vertical surface using fibre reinforced concrete with steel and polyolefin fibres. The model has been verified with experimental data and represents with accuracy the existence of boundaries, the type of concrete and compaction methods used to cast the moulds.

1. Introduction and background

Given the increasing interest on the use of fibre reinforced concrete (FRC) and the advances on improved synthetic materials, macro polymer fibres that provide concrete with properties and behaviour analogous of those of steel fibre reinforced concrete (SFRC) have become an alternative to consider in structural design. Essentially, polyolefin-based macro-polymer fibres, can be used for structural design with lower dosages in terms of weight. Such polyolefin fibres (PF) are chemically stable and their promising properties have been the centre of significant amount of research in recent years [1, 2, 3, 4, 5]. Post-cracking behaviour of FRC has become the reference property in order to consider tensile strengths in structural design. Such behaviour is typically assessed by flexural tensile tests on notched specimens and the results have shown to be strongly influenced by the number and positioning of the fibres crossing the fracture surface [6, 7, 8]. In such a sense, with the aim of providing trustworthy tools for designers, it is necessary to find predictive models that may allow the prediction of the number and positioning of the fibres placed in the critical sections of a structural piece [9]. At the time of writing there was limited research about the orientation factor in PFRC elements. Some results [10] have shown that there are several differences that should be considered before assuming the validity of the models of SFRC. The most remarkable one is the effect of fibre-flexibility. Besides, macro-synthetic fibres are remarkably different in the physical properties with

IOP Conf. Series: Materials Science and Engineering **246** (2017) 012030 doi:10.1088/1757-899X/246/1/012030

lower density and flexibility when compared with steel fibres. Thus, models that deal with SFRC are questionable.

FRC in the modern concept entails by definition [11, 12] randomly added fibres to concrete during the mixing stage. Thereby, the position and orientation of each isolated fibre depends on the pouring process, formwork geometry and rheological properties of concrete. Some studies have assessed the problem by using stereological tools [13], statistical ones [14] or rheological measures [15]. Regarding the orientation of fibres, it should be noted that some models are based on the angles that fibres take in the concrete bulk material. These seem to be appropriate when using methods such as CT-scan [16], X-Ray [17], optical methods [18], methodologies which seek to imitate concrete by using a translucent fluid [19], and others that exploit the electrical properties of concrete in measuring real angles of the fibres [20]. However, most of the works performed for this purpose have measured the number of fibres placed in a sawn surface or in the resulting surfaces after fracture tests [21, 22]. In such cases, the orientation factor (θ) proposed by Krenchel in 1975 [23] enables the use of a factor that offers a coupled value of the orientation and the distribution of the fibres. By means of θ, the models have even assessed the improvements in SFRC combined with self-compacting concrete (SCC) [24] and the effects of several types of vibration on conventional concrete (VCC) [25].

Fibres could be idealized as a number of fibres with the same cross section crossing the whole piece completing the final volume fraction (V_f) that was introduced in the mixer. Hence, the theoretical number of fibres crossing any section would be the one obtained by means of expression (1). Then, θ is defined as the relation of the number of fibres counted (n) in a certain surface over its theoretical number. That is to say that it is possible to obtain the orientation factor once it has been counted by using the expression (2) with, A being the cross section of the sample, A_f the section of one fibre and n the number of fibres actually counted in situ.

$$th = \frac{A \cdot V_f}{A_f} \tag{1}$$

$$\theta = \frac{n}{th} = n\frac{A_f}{V_f A} \tag{2}$$

If fibres were perfectly aligned, the number of fibres counted would match the theoretical value and, therefore, the ratio θ would be the unity. Subsequently, the counting process supplies a value that deals with the orientation of the fibres and which is not a measure of fibre efficiency. The relation θ takes values in general below one and the rotation angles of the fibres are strongly related with such reduction. That is to say, given certain distance from the gravity point of the fibre to a counting surface, such a fibre would cross the surface for limited rotations until reaching a critical angle. A summary of the values taken by previously mentioned models and the model presented in this study by Alberti, Enfedaque and Gálvez and with more detail in references [26, 27], presented in this paper, is summarized in Table 1. The orientation factor θ is a common measure in research with values that ranging from 0.41 to 0.82 [28].

Table 1. Values provided by the models and its main reference.

Probabilistic boundaries	2-D	3-D
Buffon´s needle problem	$2/\pi = 0.637$	-
Soroushian and Lee [28]	0.637	0.405
Dupont and Vandewalle [29]	0.600	0.500
Alberti, Enfedaque and Gálvez [26, 27]	0.637	0.500

With the aim of addressing flexible and lower density fibres in the existing models as well as making new considerations about the pouring and compaction methods given the fast development of SCC and PFRC, this study has developed new model built for the theoretical prediction of the positioning of fibres adapted for rigid and non-rigid fibres alike. The flexible nature of macro-synthetic fibres has been explicitly considered. The model is based on understandable geometrical concepts, solved by numerical integration, and not only fits the most accepted values of the orientation factor but also improves the possibilities of applying counting methods to assess the homogeneity distribution of fibres in concrete elements.

2. Proposed model in isotropic conditions

In isotropic conditions, the stereographical concept proposed in this study is to relate the sphere described by a single fibre that rotates around its gravity point and the intersection produced with a vertical plane. That plane would be the base of counting in a real specimen. The probability of a fibre intersecting a vertical plane would vary with the distance to the vertical plane as it is coupled with its rotation angle, as depicted in Figure 1. Thus, if the fibre is included in a cube cell of size l_f, the intersection with the plane that moves along its horizontal axis becomes a circumference that reaches its maximum surface of probability in an equatorial circumference of diameter equal to l_f. The complete mathematical apparatus can be seen in detail in reference [27]. As a matter of length, this study shows a brief overview of the main concepts and the most useful results obtained from the model.

An intermediate situation would describe a cone as having at its base the aforementioned circumference and a height of the distance between the plane and the fibre gravity point. With such an assumption, the probability of counting one fibre in a vertical plane could be given as a function of the two parameters: the fibre orientation angle (θ) and distance of the projected fibre to the cracked or cut surface that would be represented by the plane. The extreme positions are an equatorial plane that contains fibres rotated 90º or at the cutting of a fibre in its extreme position by its last cross-section. The intermediate positions relate the probability of occurrence as the result of a fraction of the base of the cone of such a position and the equatorial one with the fibre length as its radius. Once the cone is defined, the graphical solution of the probability of occurrence of a single case would be the relation between the circumference described by the fibre in that case (A_S) and the base one of probability that would be the circumference (A_1) described by the fibre. These are described in Figure 2.

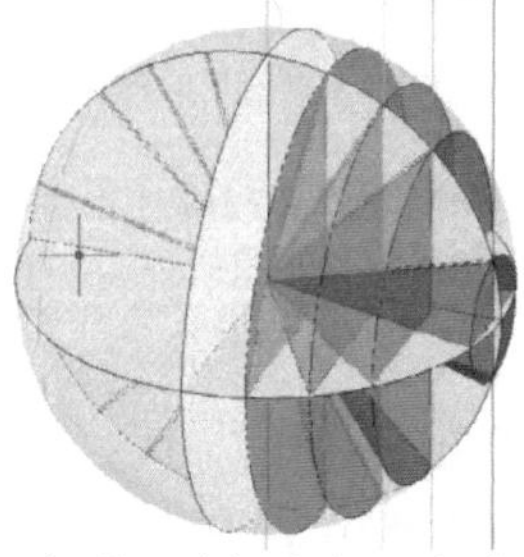

Figure 1. Graphical description of the cones formed by the probability situations

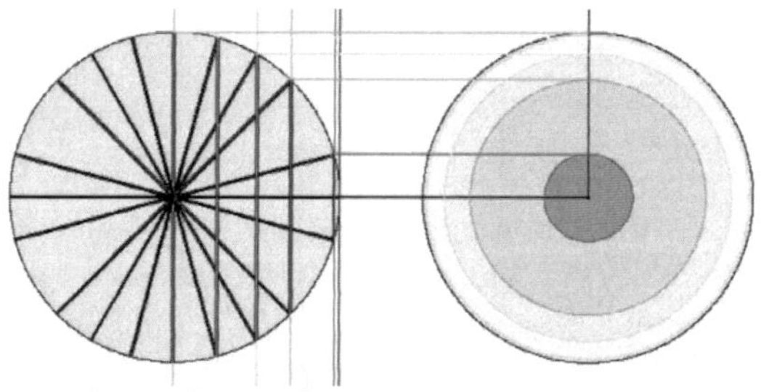

Figure 2. Surfaces of probability obtained as intersection of the vertical plane and the sphere

Therefore, the probability in a three-dimensional isotropic situation could involve the relation at a given distance of the surface drawn by the fibre that crosses the cut surface in all the angles of rotation and the total surface probability. If we consider the space probability as the average obtained in all the possible geometrical situations, expressions (3) would provide the solution. The area under the probability curve can be an average in the space probability in the value of 0.500. This value could be considered as the mean orientation factor found in the bulk area.

$$P(\theta_{isotropic}) = \frac{\int_0^{\frac{\pi}{2}} sin\,(\theta)^2\ d\theta}{\frac{\pi}{2}} = 0.500 \qquad (3)$$

3. Proposed model under real conditions: the presence of walls

Isotropic conditions would take place in the bulk zone of the specimen if the compaction is made with conventional times of vibration and no other of the previously summarized effects take place.

IOP Conf. Series: Materials Science and Engineering **246** (2017) 012030 doi:10.1088/1757-899X/246/1/012030

Concrete elements need to be poured in some mould or formwork. Therefore, the areas surrounding the walls should be considered separately due to the wall effect. The presence of a boundary condition impedes the fibre from taking certain positions, as can be noticed in Figure 3. If we consider the fibre distance to the wall from its gravity point, the symmetrical positions would not be reached. If the wall reaches the position of the centre of the sphere, in which the fibre would be located in the wall, given that the only possible rotations would be in the angle perpendicular to the paper the situation would be equal to a two-dimensional one. Therefore, the same conceptual basis were used in the presence of on wall but new geometrical problems appeared in order to compute expression (4).

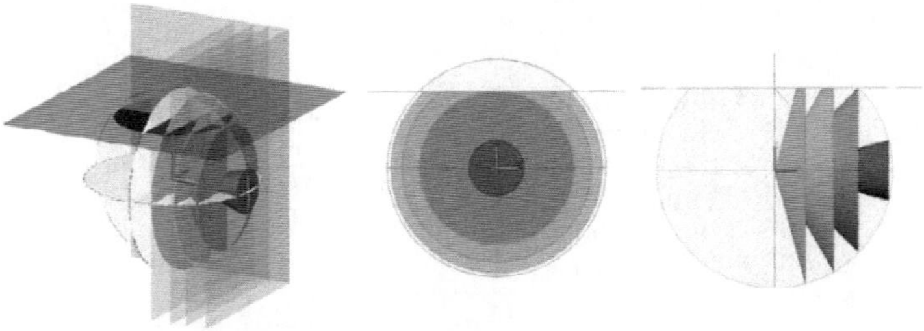

Figure 3. The presence of one wall

If Figure 4 is analysed, two significant changes in the calculation emerge in order to compute expression (4). The first is that the surface of probability one ($A_{1-1wall}$) is smaller than in the isotropic case given that in the surface of the circumference two sectors have to be subtracted (one in the side of the wall and another in the symmetrical position). The second one is that there are two formulations for the drawing of the possible cut on the counting surface ($A_{i-1wall}$). In the positions of the cone not reaching a limit angle of rotation, the base of the cone remains as a circumference of similar dimensions. Once the distance allows rotation angles that would allow positions that are impeded by the existence of a boundary, the obstructed sector and its symmetrical position have to be subtracted.

$$P(\theta_{1wall}(h)) = \frac{A_{i-1wall}}{A_{1-1wall}} \tag{4}$$

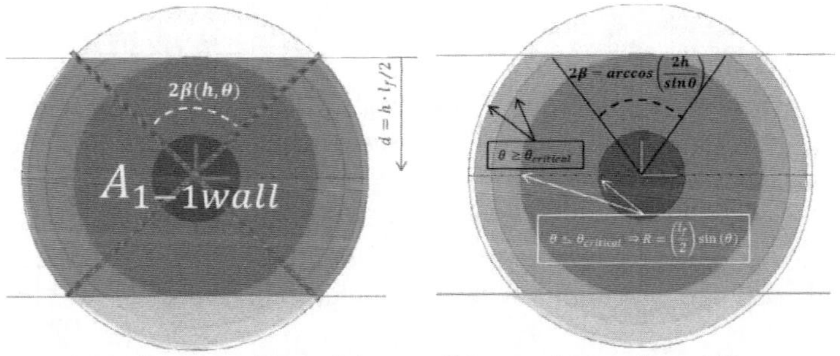

Figure 4. Surfaces resulting of the possible rotation angle and distances

The numerical solution of this integral provides a value for each distance, varying from the case of a distance equal to half the fibre length to the one in which the fibre is in the wall plane. In such a sense, the probability distribution is unique for each distance. The area of probability when $h = l_f/2$ produces a mean value of probability of $2/\pi$. It is possible to build a profile of the orientation factor by giving values to h as a function of l_f. In doing so, the maximum value at the edge of the wall is $\theta_{1wall} = 2/\pi$, with the rest resulting in a continuously increasing curve from the value of the bulk up to what could be named as θ_{1wall}. Table 2 offers the values of the orientation factor in several intermediate relative positions of the gravity point and the wall.

Regarding the presence of two walls, it would be merely an extension of the previous concept developed for one wall. Thus, analogous formulation would apply as can be seen in expression (5) although new geometrical problems must be solved. If the position of the first wall is fixed in a position at a certain distance h defined as a fraction of $l_f/2$, the position of the new wall needs to define a new parameter in order to integrate the possible positions. That is to say, the distance with the second wall, named u in the study, can be defined as a fraction of $l_f/2$ and varies u from 0.0 to 0.5. With this new parameter, it is possible to find the surfaces generated to compute the probability, since the position of the first wall remains fixed.

$$P(\theta_{2walls}(h,u)) = \frac{A_{i-2walls}}{A_{1-2walls}} \tag{5}$$

Table 2. Orientation factor with one wall

$h(l_f)$	θ
0.500	0.500
0.400	0.531
0.250	0.584
0.100	0.625
0.000	0.637

However, there is not only a new limit angle but also a key point for a general situation in which both planes intersect in a straight line idealized in a two-dimensional point. That point can be either outside or inside the circumference. Thus, the singular point (k) in which such a key point belongs to the circle, border of the probability one circumference, is the limit between the two situations. Therefore, the reference surface of probability one $(A_{1-2walls})$ changes in shape and it should be computed in parts. Then, probability is defined with analogous processes to the previous probabilities. With such assumptions, limits and solving of the numerical integration, the probability surfaces and the mean values that vary the distance to the second wall (and for each fixed value of the first) would show a three-dimensional orientation factor map. It is evident that for the lowest values of (u,h), the mean orientation factor has to tend to one and for values of (u,h)close to 0.500, the orientation factor has to be close to 0.500. All the mathematical elaboration of the model can be studied in more detail in references [26, 27].

4. Orientation profile from the model

As may be seen in previous explanations, the values that the orientation factor take for a decreasing distance to the wall are continuously increasing up to the values obtained in the edges. That is to say, the profile of orientation factor that considers the presence of only one wall varies from 0.500 in the bulk to 0.637. In the case of the values of two walls when the first wall is considered in the edge, it would vary from 0.637 to 1.000. It should be highlighted that the model presented in this paper can provide the full profile for each section by numerical integration which can be seen in Figure 5.

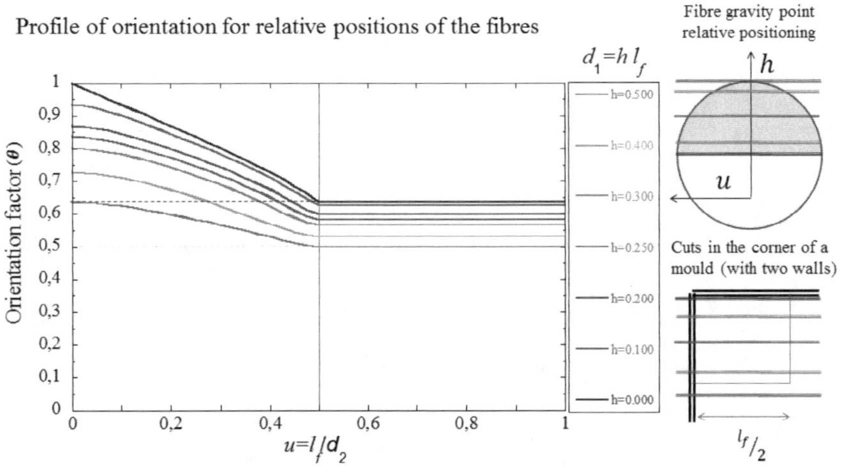

Figure 5. Profile of θ as a function of the distance to the two walls.

Research on this matter has provided a variety of functions representing these variations of the orientation factor as a function of the thickness of the piece (or the relation of fibre length with the thickness of the piece). Exponentially decreasing relations such as the presented in references [30] and [31] related the orientation factor with the specimen thickness (h). Nonetheless, Martinie [32] provided expressions and graphic descriptions that relate the orientation factor with the relation between fibre length and specimen thickness. Laranjeira [24] also proposed the use of a bilinear model to adjust the results of his theoretical model.

5. Non-rigid fibres: folding effect

The possible folding effects have been considered as a reduction of the sphere drawn by the fibre in the model. That is to say, the possible positions are the same but for a reduction of the projected fibres due to its own bending. This can be done by assuming a reduction of the diameter of the sphere considering the fibre length as the projected length when it is folded and introducing such value in the model. In such a way, the expression needed has a geometrical relation with the cord (l_{fr}) of the circumference drawn by the fibre in each plane. The expression (6) relates the fibre length with the bend radius, with the probability of the reduction of the fibre being obtained for a maximum bending radius limited with half the fibre length. Expression (7) provides the projected length of the fibre for each curvature angle (φ). The resulting reduction of probability is in the case of the average fibre having a length of 94.6% of the original length, as obtained by expression (8). These values were set as an initial hypothesis in order to propose the model as shown in Figure 6. The possible values would vary with the geometrical and material properties of each type of fibres and the procedure sought allows adjusting the model to any other type of non-rigid fibres. The procedure description for the adjustment to 60mm-long polyolefin fibres can be read in the verification of the model and in more detail in reference [27].

$$R = \frac{l_f}{2\varphi} \tag{6}$$

$$l_{fr} = \frac{l_f}{\varphi} sin\,\varphi \tag{7}$$

$$l_{fr} = \int_0^1 l_f \cdot \frac{1}{\varphi} \cdot sin\,\varphi \cdot d\varphi = 0.946 l_f \tag{8}$$

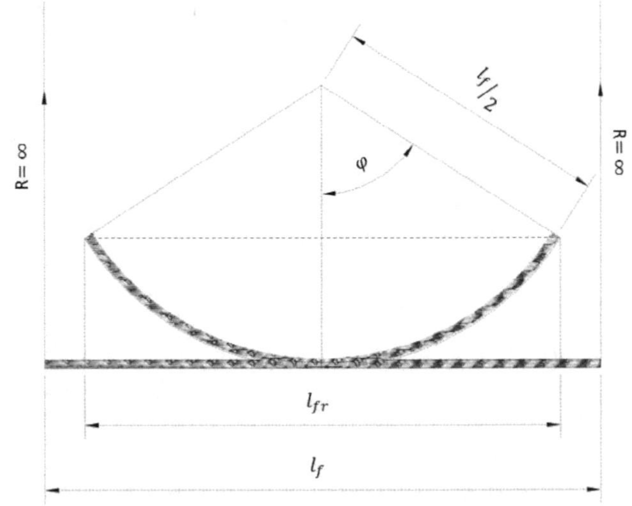

Figure 6. Fibre bending and reduction of its mean projection.

6. Discussion and verification of the model

It should be highlighted that models proposed by Soroushian [28], Stroeven [13] or Dupont and Vandewalle [29] have been successfully used for the case of steel fibres and having suitable results, even though there were several computed values that were used adequately for a varied casuistry. In any event, there was a demand of an updated model which allowed considering the use of flexible fibres. In addition, some other facts such as the use of SCC, the relative positions of the walls, the sizes of the beams and notches or the treatment suggested for the topped surfaces could be reconsidered and revised.

Regarding the verification of the model, it is worth noting that the variety of solutions, models and procedures to assess the orientation factor, makes it difficult to use reported experimental data to validate theoretical models. Even though this reveals the need of a standard procedure that would boost the possibility of sharing data, the values provided by using the presented model with regard to steel fibres can be validated with the data published in references [29, 22, 21, 28]. In addition, for smaller sizes of 100x100 mm² square cross-section, and when using SCC with steel hooked-fibres 35 mm long, the experimental data in reference [26] provided an average orientation factor of 0.590. Computation of the orientation factor with the proposed model, with the assumption of the bulk zone, as having a boundary condition due to the flux (0.578 in the central sector) gives 0.593.

Regarding synthetic fibres, validation has been made in references [1, 2, 3, 26]. In such a sense, the factor that affects the fibre length obtained by means of expression (35) was initially set on 0.946. With such a value, the ratio computed over experimental number of fibres counted takes a value 1.009. Since the value of the projected chord required verification, it was possible to pitch the limits of the integration placed in expression (8). The limit for a maximum curvature radius that fits the model was 0.895 radians. Therefore, this change in expression (8) would be that presented in expression (9).

$$l_{fr} = \int_0^{0.895} l_f \cdot \frac{1}{\varphi} \cdot \sin\varphi \cdot d\varphi = 0.856 l_f \quad (9)$$

One more validation may be of interest regarding synthetic fibres: the percentage of distribution in nine equal sectors counted. In order to provide practical tools, the two sizes of cross-sectional specimens placed on the test standards EN 14651 [33], RILEM TC 162-TDF [34] and ASTM C 1609 [35] have been taken as an example. The orientation factor in each of the nine equal sectors of a specimen with squared cross section and with a notched beam of sizes 100mm² and 150 mm² and for fibres 35, 48, 60 and 70 mm long were published in reference [27]. With the values of orientation computed, it is possible to achieve a number of fibres to be counted in each of the sector by fixing the fibre dosage and the cross-section surface of a single fibre. However, beyond such a simple and powerful possibility, a new concept appears that can be useful. Once the distribution of the orientation factor is not uniform, the number of fibres counted on similar surfaces would not be uniform either due to the orientation. In summary, a uniform distribution of the fibres across the cross-sectional surface and along the beam would lead to a non-proportional distribution of the fibres counted in equal squares, having an uneven percentage due to the distinct orientation of the sectors that are close to a wall or any other boundary condition. In addition to this, it is worth noting that the assumption of Laranjeira [24] establishes the value of the bulk in the case of SCC as being in the presence of one wall and thus idealizing the flux as taking place in a layer of fibre-length thickness. Figure 7 shows the results obtained in reference [26] for the sawn surface, which detailed counting results. The table shows the mean experimental distributions of fibres in 20 surfaces with a 150x150 mm² square cross-section produced with PFRC with 6 kg/m³ polyolefin fibres 60 mm long.

PFRC with 6 kg/m³ of 60mm long polyolefin fibres

Mean distribution of fibres in VCC			Mean distribution of fibres in SCC			Theoretical distribution of fibres		
11,20%	11,20%	9,50%	10,70%	10,20%	12,70%	10,80%	9,90%	10,80%
11,30%	10,00%	11,00%	10,20%	9,90%	10,20%	10,80%	9,90%	10,80%
13,70%	10,50%	11,60%	12,30%	11,10%	12,70%	13,00%	10,80%	13,00%

Figure 7. Experimental distributions of fibres in 150x150 mm² square cross-section produced with PFRC using 60 mm-long fibres.

7. Conclusions

The review of the previous models has allowed an initial validation of the model for rigid fibres. Moreover, the most accepted values for the orientation factor in distinct positions of an isolated fibre inside the concrete were fully complied with by the results obtained by means of a numerical integration. In comparison with other classical models that have been extensively used, the proposed model supplies the possibility of considering the bending of the non-rigid fibres. It allows the fitting of the model for any type of fibres that should not be considered as remaining straight inside the concrete piece. Although wider research on this matter would provide reliability, due to the typical scatter of fibre reinforced concrete, the proposed concept of reducing the overall fibre length by a bending factor would appear to be suitable.

Additionally, the model allowed the study to prepare curves with the profile of orientation when one and two walls affect the fibre positioning as a function of the relation of the fibre length and the distance to a wall.

The model presented in this study permits comparison of the number of fibres predicted to cross a vertical plane with the most common structural fibres and lengths. Beyond such a possibility, it allows this to be done for non-rigid fibres. Moreover, it is a powerful tool gathered with the developed distribution tables also presented to evaluate the behaviour of new types of fibres or shapes in research. This also applies to the assessment of the influence of the placing conditions, pouring methods and rheological properties of the concrete in the final positioning of the fibres. In summary, this achieves not only a more reliable use of the fibres but also provides a valid and sound measurement.

Acknowledgements
The authors gratefully acknowledge the financial support provided by the Ministry of Economy and Competitiveness of Spain by means of the Research Fund Project BIA 2016-78742-C2-2-R. They also offer their gratitude to SIKA SAU for supporting the Enterprise University Chair: Cátedra Sika-UPM.

IOP Conf. Series: Materials Science and Engineering **246** (2017) 012030 doi:10.1088/1757-899X/246/1/012030

References

[1] M. G. Alberti, A. Enfedaque and J. C. Gálvez, "On the mechanical properties and fracture behavior of polyolefin fiber-reinforced self-compacting concrete," Construction and Building Materials, vol. Volume 55, pp. 274-288, 2014.

[2] M. G. Alberti, A. Enfedaque, J. C. Gálvez, M. F. Cánovas and I. R. Osorio, "Polyolefin fiber-reinforced concrete enhanced with steel-hooked fibers in low proportions," Materials & Design, vol. 60, pp. 57-65, 2014.

[3] M. G. Alberti, A. Enfedaque and J. C. Gálvez, "Comparison between polyolefin fibre reinforced vibrated conventional concrete and self-compacting concrete," Construction & Building Materials, vol. 85(15), pp. 182-194, 2015.

[4] M. N. Soutsos, T. T. Le and A. P. Lampropoulos, "Flexural performance of fibre reinforced concrete made with steel and synthetic fibres," Construction and Building Materials, vol. 36, pp. 704-710, 2012.

[5] K. Behfarnia and A. Behravan, "Application of high performance polypropylene fibers in concrete lining of water tunnels," Materials & Design, vol. 55, pp. 274-279, 2014.

[6] S. Grünewald, F. Laranjeira, J. Walraven, A. Aguado and C. Molins, "Improved Tensile Performance with Fiber Reinforced Self-compacting Concrete," in High Performance Fiber Reinforced Cement Composites 6, Springer, 2012, pp. 51-58.

[7] M. G. Alberti, A. Enfedaque and J. C. Gálvez, "Fracture mechanics of polyolefin fibre reinforced concrete: Study of the influence of the concrete properties, casting procedures, the fibre length and specimen size," Engineering Fracture Mechanics, 2016.

[8] M. G. Alberti, A. Enfedaque, J. C. Gálvez and V. Agrawal, "Reliability of polyolefin fibre reinforced concrete beyond laboratory sizes and construction procedures," Composite Structures, vol. 140(15), pp. 506-524, 2016.

[9] R. Zerbino, J. M. Tobes, M. E. Bossio and G. Giaccio, "On the orientation of fibres in structural members fabricated with self compacting fibre reinforced concrete.," Cement and Concrete Composites, 34(2), p. 191–200, 2012.

[10] M. G. Alberti, A. Enfedaque, J. C. Gálvez and V. Agrawal, "Fibre distribution and orientation of macro-synthetic polyolefin fibre reinforced concrete elements.," Construction and Building Materials, , vol. 122, pp. 505-517, 2016.

[11] fib Model Code, Model Code, Paris: Fédération Internationale du Béton fib/International Federation for Structural Concrete, 2010.

[12] EHE-08, Spanish Structural Concrete Code, Spanish Minister of Public Works, 2008.

[13] P. Stroeven, "Stereology of concrete reinforced with short steel fibres," Heron, vol. 31(2), pp. 15-28, 1986.

[14] F. Laranjeira, S. Grünewald, J. Walraven, C. Blom, C. Molins and A. Aguado, "Characterization of the orientation profile of steel fiber reinforced concrete," Materials and structures, vol. 44(6), pp. 1093-1111, 2011.

[15] P. Stähli, R. Custer and J. G. M. Mier, "On flow properties, fibre distribution, fibre orientation and flexural behaviour of FRC," Materials and Structures, vol. 41(1), p. 189–196, 2008.

[16] T. Ponikiewski, J. Gołaszewski, M. Rudzki and M. Bugdol, "Determination of steel fibres distribution in self-compacting concrete beams using X-ray computed tomography," Archives of Civil and Mechanical Engineering, 2014.

[17] L. Vandewalle, G. Heirman and F. Van Rickstal, "Fibre orientation in self-compacting fibre reinforced concrete.," in Proc. of the 7th Int. RILEM Symp. on Fibre Reinforced Concrete: Design and Applications (BEFIB2008), 2008.

[18] A. Bernasconi, F. Cosmi and P. J. Hine, "Analysis of fibre orientation distribution in short fibre reinforced polymers: A comparison between optical and tomographic methods," Composites Science and Technology, vol. 72(16), pp. 2002-2008, 2012.

[19] B. Boulekbache, M. Hamrat, M. Chemrouk and S. Amziane, "Flowability of fibre-reinforced concrete and its effect on the mechanical properties of the material," Construction and Building Materials, vol. 24(9), pp. 1664-1671, 2010.
[20] H. Al-Mattarneh, "Electromagnetic quality control of steel fiber concrete,," Construction and Building Materials, vol. 73, pp. 350-356, 2014.
[21] B. I. G. Barr, M. K. Lee, E. J. de Place Hansen, D. Dupont, E. Erdem, S. Schaerlaekens and L. Vandewalle, "Round-robin analysis of the RILEM TC 162-TDF beam-bending test: Part 3—Fibre distribution," Materials and Structures, vol. 36(9), pp. 631-635, 2003.
[22] M. Di Prisco, M. Colombo and D. Dozio, "Fibre reinforced concrete in fib Model Code 2010: principles, models and test validation," Structural Concrete, vol. 14(4), pp. 342-361, 2013.
[23] H. Krenchel, "Fibre spacing and specific fibre surface," Fibre reinforced cement and concrete, pp. 69-79, 1975.
[24] F. Laranjeira, A. Aguado, C. Molins, S. Grünewald, J. Walraven and S. Cavalaro, "Framework to predict the orientation of fibers in FRC: a novel philosophy," Cement and Concrete Research, vol. 42(6), pp. 752-768, 2012.
[25] R. Gettu, D. R. Gardner, H. Saldivar and B. E. Barragán, "Study of the distribution and orientation of fibers in SFRC specimens," Materials and Structures, no. 38(1), pp. 31-37, 2005.
[26] M. G. Alberti, "Polyolefin fibre-reinforced concrete: from material behaviour to numerical and design considerations," 2015.
[27] M. G. Alberti, A. Enfedaque and J. C. Gálvez, "On the prediction of the orientation factor and fibre distribution of steel and macro-synthetic fibres for fibre-reinforced concrete," Cement and Concrete Composites, 2016.
[28] P. Soroushian and C. D. Lee, "Distribution and orientation of fibers in steel fiber reinforced concrete," ACI Materials Journal, vol. 87(5), 1990.
[29] D. Dupont and L. Vandewalle, "Distribution of steel fibres in rectangular sections," Cement and Concrete Composites, vol. 27(3), pp. 391-398, 2005.
[30] J. Rosenbusch, "Einfluss der Faserorientierung auf die Beanspruchbarkeit von Bauteilen aus Stahlfaserbeton," Beton und Stahlbetonbau, vol. 99(5), pp. 372-377, 2004.
[31] J. Michels, D. Waldmann, S. Maas and A. Zürbes, "Steel fibers as only reinforcement for flat slab construction–Experimental investigation and design," Construction and Building Materials, vol. 26(1), pp. 145-155, 2012.
[32] L. Martinie and N. Roussel, " Simple tools for fiber orientation prediction in industrial practice.," Cement and Concrete Research, vol. 41(10), p. 993–1000, 2011.
[33] EN 14651:2005+A1:2007, Test method for metallic fibre concrete. Measuring the flexural tensile strength (limit of proportionality (LOP), residual), 2007.
[34] RILEM- TC 162-TDF, "Test and design methods for steel fibre reinforced concrete," Materials and Structures, vol. 33, pp. 75-81, 2000 .
[35] ASTM C 1609/C 1690M-07, "Standard test method for flexural performance of fiber reinforced concrete (using beam with third-point loading)," p. 1–8, 2007.

IOP Conf. Series: Materials Science and Engineering **246** (2017) 012031 doi:10.1088/1757-899X/246/1/012031

Effect of Different Matrix Compositions and Micro Steel Fibers on Tensile Behavior of Textile Reinforced Concrete

J Esmaeili[1], I Sharifi[2], K Andalibi[3] and J Kasaei[4]

[1]Associate Professor, Department of Civil engineering, The University of Tabriz, Iran
[2] MSc student, Department of Civil engineering, The University of Tabriz, Iran
[3] PhD student, Department of Civil engineering, The University of Tabriz, Iran
[4] MSc, Department of Civil engineering, The University of Tabriz, Iran

E-mail: J-Esmaeili@tabrizu.ac.ir

Abstract. This paper presents results of a research on uniaxial tensile behavior of textile reinforced concrete (TRC) prepared with different matrix compositions containing different contents of micro steel fibers. TRC exhibits very favorable stress–strain behavior, high Load-carrying capacity and a certain ductility which results in a strain-hardening behavior. At this paper, different Glass-TRCs were prepared using different commonly used normal and also innovative matrix compositions containing different volume fractions of micro steel fibers. Three commonly used matrices, a polymer-based composite and also a UHPC mixture were prepared containing different percentages of micro steel fibers. The direct tensile tests were applied on all specimens to study the tensile properties (first crack stress and ultimate tensile strength) and strain-hardening behavior. Considering the stress-strain curves of all specimens, it has been found that the tensile properties and strain-hardening behavior of Glass-TRC can be considerably improved by using steel micro fibers in an appropriate matrix composition.

1. Introduction

Textile reinforced concrete (TRC) has emerged in recent years as an attractive cement composite which can provide numerous advantages for developing new type of thin sheet component for a variety of building systems [1-9]. It is a composite material consisting of fine-grain concrete and multi-axial textile reinforcement. By using corrosion-free fibers such as alkali-resistant glass (AR-glass) fiber, carbon fiber, aramid fiber, and basalt fiber, TRC possesses superior durability [1, 5-8]. Furthermore, the high strength fiber along the main stress direction of TRC makes this composite have high effectiveness in strengthening of plain concrete members [1, 5-8]. To ensure the performance, the mechanical behavior of TRC has been investigated. Besides the properties of fiber and matrix themselves, the mechanical behavior of TRC members is also significantly associated with the interface bond characteristics between fiber and matrix, which is influenced by fabric geometry, knitting type, mesh size and fiber type, textile surface treatment methods and matrix properties. Jesse et al. [1] summarized major factors which influence the load-bearing capability of TRC, including reinforcing ratio, fabric filament damage during production, fabric reinforcing direction, and interface bond characteristics between textile and matrix. Peled et al. [2, 3, 9, 10] investigated the effect of fabric geometry on the bonding performance and reinforcing effectiveness of TRC. It was discovered that the type of knitting (or weaving) had important influence on the interfacial bond strength.

To obtain a complete interfacial bond characterization between matrix and fiber, many researchers have performed a lot of experiments. Colombo et al. [11] presented experimental investigation of TRC on tensile tests, considering the influence of reinforcement ratio, fabric geometry, curing conditions, displacement rate, and specimen size. Cevallos and Olivito [12] studied the effects of the fiber type, fabric geometry, physical and mechanical properties of fabrics and the volume fraction of fibers on the tensile stress–strain response and crack propagation of cementitious composites reinforced with natural fabrics. It was found that the fabric geometry and the volume fraction of fibers were the parameters that have the greatest effects on the tensile behavior of these composite systems. Mumenya et al. [13] conducted uniaxial tensile testing on weathered textile concrete samples to analyze the multiple cracking patterns. Peled et al. [14] used closed-loop uniaxial tensile tests to study the tensile behavior of pultruded cement composites with various fabric types. It was found that hybrid composite made of PE and AR glass sustained strains better than 100% AR glass composite, and was stronger than single PE fabric composite. Larrinaga et al. [15] investigated the experimental and numerical modeling of basalt textile reinforced mortar to deepen the knowledge of this composite material in terms of tensile behavior. Santis and Felice [16] presented an experimental study on the tensile behavior of strengthening systems comprising two different textiles and five mortar matrices.

The stress–strain behavior of TRC under tensile loading can be subdivided into three states [17], as presented schematically in Fig. 1. The first state (I) is the free-crack state. In this state, the stiffness of both matrix and fibres determines the slope of this part of the curve, where TRC shows nearly linear-elastic behavior up to the point at which the increase in stress leads to the formation of the first crack. The second state (IIa) is defined by further crack formation, leading to pronounced quasi-ductile behavior of the composite. In this state more relatively fine cracks form due to the increase in the tensile stress. The slope, the length, and the coarseness of this part depend on the quality of the bond between textile and matrix [18] as well as on the volume proportion of the fibers in the composite activated for load transfer. Indeed, they can be related to the number of cracks and crack widths. The crack-widening state (IIb) is the final state in the stress–strain relationship. In this state either no new cracks or only a few appear, but the existing cracks grow and become wider until the ultimate stress is reached. The load in this state is carried only by the multifilament-yarns of the textile reinforcement and increases until the tensile strength and the strain capacity of these yarns are reached and the TRC fails. The sequence of these states can be observed on the specimen's surface by the naked eye during testing under tensile load.

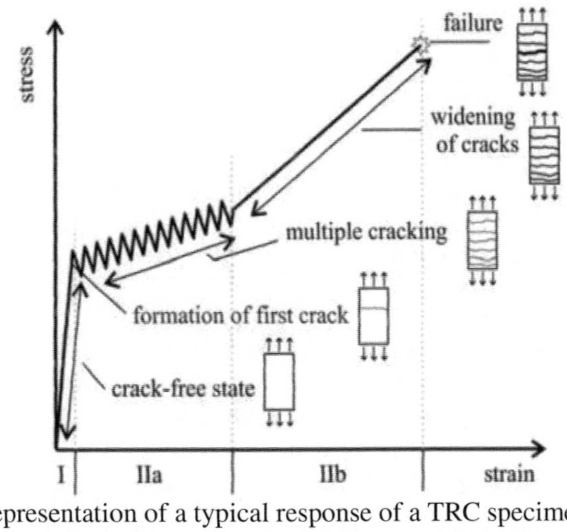

Figure 1. Schematic representation of a typical response of a TRC specimen subjected to uniaxial tensile loading with indication of cracking states.

This research present the results of a research on uniaxial tensile behavior of textile reinforced concrete (TRC) produced with different matrix compositions. At this paper, different Glass-TRCs were prepared using different commonly used normal and also innovative matrix compositions containing different percentages of steel micro fibers. Three commonly used matrixes, a polymer-based composite and also a UHPC mixture were prepared containing different percentages of micro steel fibers. The direct tensile tests were applied on all specimens to study the tensile properties (first crack stress and ultimate tensile strength) and strain-hardening behavior.

2. Experimental program

2.1. Materials

2.1.1. Matrices

Different Glass-TRC specimens are prepared using different commonly used normal and also innovative matrix compositions containing different percentages of steel micro fibers. PZ, M3 and M7 as commonly used matrices [19] and a polymer-based composite (PC) and also a UHPC mixture as innovative matrix compositions are prepared containing different percentages of steel micro fibers [20]. Table 1 gives the compositions and compressive strength of the mentioned commonly used and also innovative matrices. The workability of cement-based matrices is evaluated through flow table test and is kept in the range of 180-200 mm.

Table 1. Mixture proportions and compressive strength of TRC specimens

Mix No	Cement (kg/m^3)	Fly ash (kg/m^3)	Silica fume (kg/m^3)	Pumice (kg/m^3)	W/b	Water (kg/m^3)	Resin-epoxy suspension (kg/m^3)	Sand (kg/m^3)	Hyper plasticizer (kg/m^3)	Compressive strength (MPa)
PZ-0%	490	175	35	---	0.4	280	---	715	4	63
PZ-1%	490	175	35	---	0.4	280	---	715	4	70
PZ-2%	490	175	35	---	0.4	280	---	715	4	73
M3-0%	549	246	25	---	0.3	246	---	1092	5.5	75
M3-1%	549	246	25	---	0.3	246	---	1092	5.5	80
M3-2%	549	246	25	---	0.3	246	---	1092	5.5	83
M7-0%	839	---	---	---	0.33	277	---	1189	5	80
M7-1%	839	---	---	---	0.33	277	---	1189	5	84
M7-2%	839	---	---	---	0.33	277	---	1189	5	85
UHPC-0%	757	---	265	300	0.2	264	---	789	9.5	145
UHPC-1%	757	---	265	300	0.2	264	---	815	9.5	160
UHPC-2%	757	---	265	300	0.2	264	---	842	9.5	163
PC-0%	600	---	---	---	---	---	350	1098	---	88
PC-1%	600	---	---	---	---	---	350	1098	---	95
PC-2%	600	---	---	---	---	---	350	1098	---	102

2.1.2. Glass textile reinforcement

A particular type of polymer-coated, biaxial fabric made of alkali-resistant (AR) glass was used as textile reinforcement for the TRC specimens. The spacing between yarns was 13 mm and the average tensile strength of the yarns was approximately 1200 MPa, as reported by the manufacturer.

2.1.3. Micro steel fiber

Brass-coated micro steel fibers with an average diameter of 0.2 mm and a length of 6 mm were applied as additional reinforcement. Two fiber contents of 1% and 2% by volume of concrete were investigated. The fibers had a density of 7.8 g/cm3 and a tensile strength of 2800 MPa.

2.2. Preparation of specimens and test setup

For uniaxial tensile tests, rectangular plates with 400 mm long, 80 mm wide and 6 mm thick were prepared. First, glass textile sheets were laid and fixed at the end of the moulds and then the matrices spread in and finally smoothed. Schematic view of specimens' preparation technique and uniaxial tensile testing setup are illustrated in Figure 2.

In addition to the TRC specimens, which were reinforced with 2 layers of glass textile, a series of textile reinforced specimens were also produced containing different contents of micro steel fibers.

Specimens were demoulded after one day and then stored in water until reaching an age of 28 days. UHPC specimens were cured through accelerating method for 48 hours in boiling water.

The uniaxial tensile tests on the TRC plates were conducted with a controlled deformation rate of 0.5 mm/min. In order to better transfer the force to the specimens, steel plates with 4 mm thickness were glued to their ends by epoxy resin to avoid the failure at the connection during the uniaxial tensile tests. Deformation was measured using two linear variable differential transformers (LVDTs). The connection between the plates and the testing machine was designed as hinges to eliminate bending moment in the specimens (Figure 2).

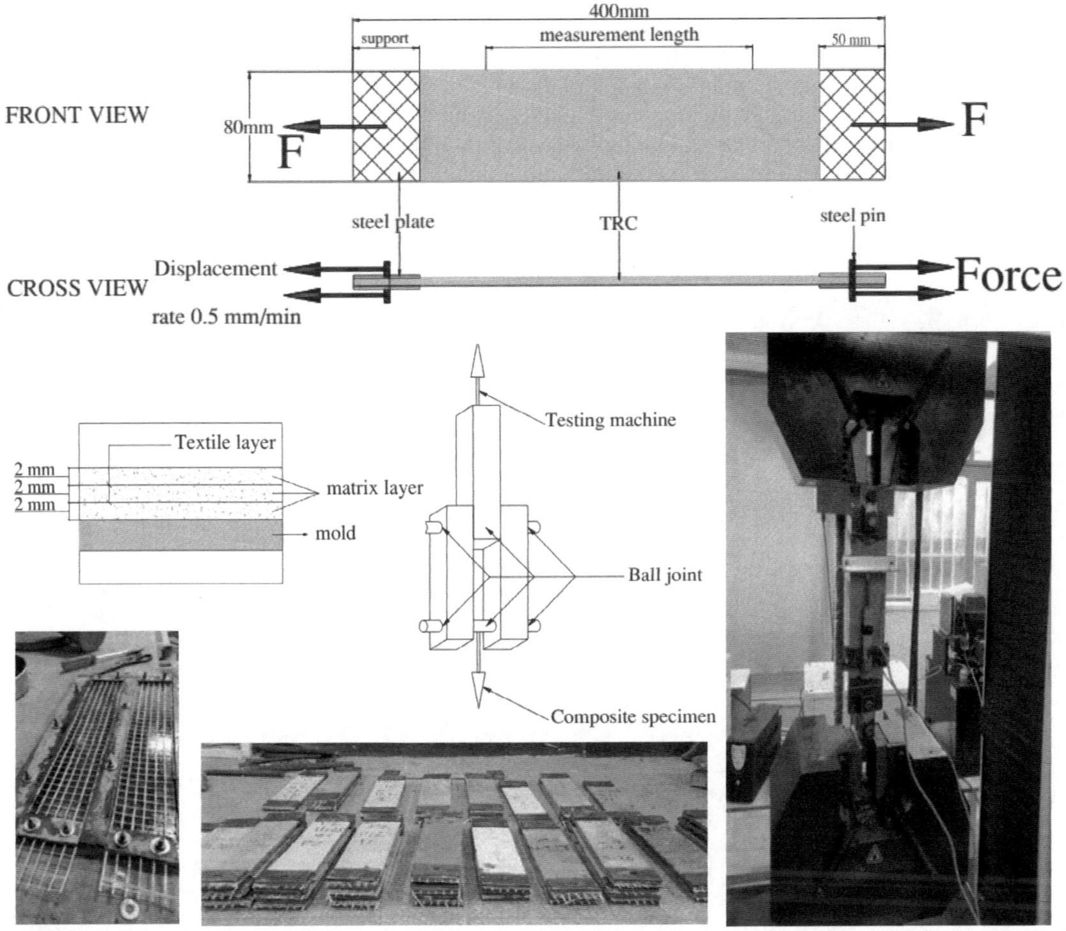

Figure 2. Schematic representation, testing device setup and specimens' preparation of TRCs subjected to uniaxial tensile test

3. Results and discussion

The stress-strain behavior of TRC specimens made with different matrices containing different volume fractions of micro steel fibers is illustrated in Figure 3. Figure 4 shows the effects of both matrix type and also the use of micro steel fibers on first-crack stress and ultimate tensile strength of TRCs.

As seen in Figure 3 and Figure 4, matrix composition has considerable influence on first-crack stress, ultimate tensile strength and tensile stress-strain behavior of TRC specimens. Tensile behavior of TRC specimens made with commonly used matrices (PZ, M3 and M7) varies considerably. At this series of specimens the maximum increase of app. 50% is observed in first-crack stress of M7-0% in comparison with that of the PZ-0%. The maximum enhancement of app. 31% is seen in ultimate tensile strength of M3-0% when compared with that of the PZ-0%. In fact, lower porosity of matrix in textile-matrix interface at commonly used matrices with lower water/binder ratios results in further bonding between matrix and textile and then improvements in tensile properties and strain-hardening behavior of TRC specimens are achieved.

Strain-hardening behavior of TRC specimens made with UHPC matrix containing different volume fractions of micro steel fibers is clearly observed at Fig 3. Despite an increase in first-crack stress of TRC specimens prepared with UHPC matrix in comparison with that of the specimens made with commonly used matrices, the ultimate tensile strength of UHPC specimens is slightly lower than that of the corresponding M3 and M7. It seems that accelerated water curing (in boiling water) of UHPC specimens might be the main reason for this reduction in ultimate tensile strength.

As seen in Fig 4, first-crack stress of TRC specimens made with PC matrix is considerably higher than that of the rest of specimens. Indeed, first-crack stress of PC specimens is approximately 4 times more than that of the PZ specimens. However, in TRC specimens made with PC matrix the first crack stress of PC matrix was even more than tensile capacity of glass textile and then a sudden failure of the specimen was occurred following the first crack and then the strain-hardening behavior was not observed. It seems that the strain-hardening behavior can be achieved at PC specimens by increasing the number of glass textile layers or by decreasing the thickness of specimens' cross-section.

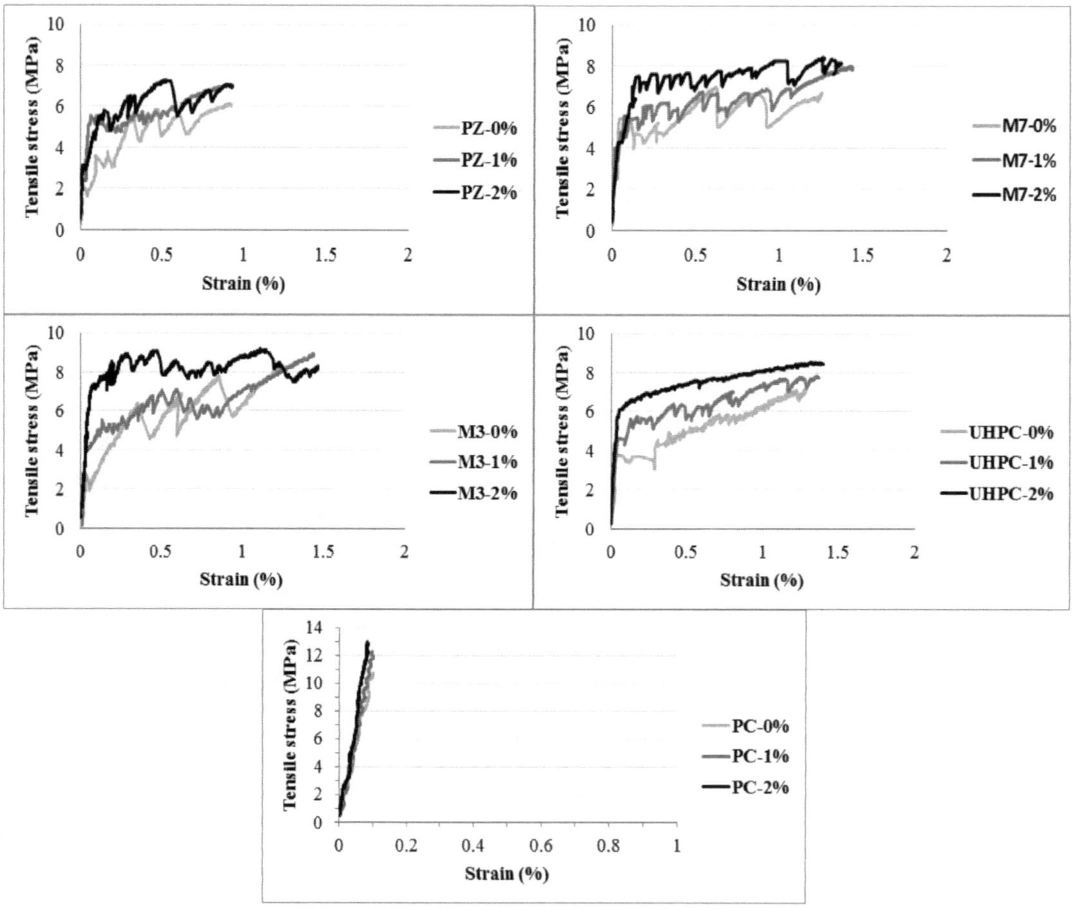

Figure 3. Tensile Stress-strain curves of TRC specimens prepared with different matrices containing different contents of micro steel fibers

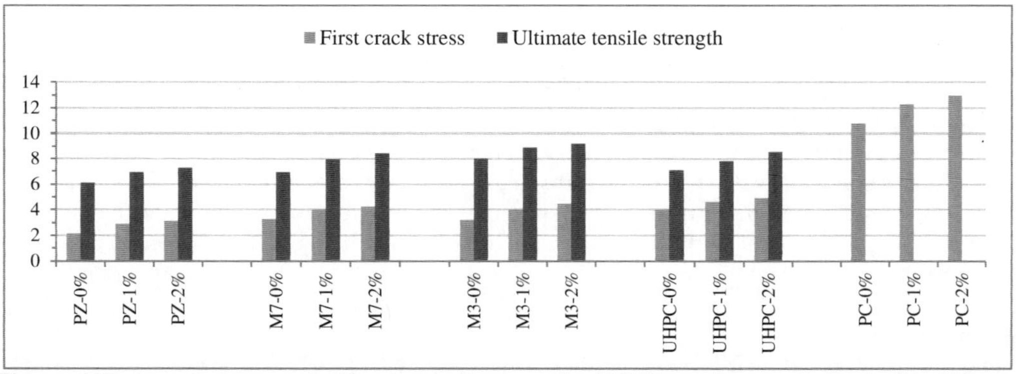

Figure 4. First crack stress and ultimate tensile strength of TRC specimens prepared with different matrices containing different contents of micro steel fibers

As observed in Figure 4, the first crack stress and ultimate tensile strength of TRC specimens made with different matrices are increased by raising the volume fractions of micro steel fibers. Adding micro steel fibers by 1% volume leads to an increase of about 14% to 35% in first crack stress of TRC specimens prepared by different matrices.

In TRC specimens prepared with commonly used matrices, adding micro steel fibers by 1% volume at PZ-1%, M7-1% and M3-1% results in increases of about 35%, 23% and 25% in first-crack stress, respectively in comparison with corresponding plain specimens. Increasing the micro steel fibers volume content up to 2% leads to an increases of about 44%, 31% and 41% in first-crack stress of PZ-2%, M7-2% and M3-2%, respectively when compared with that of corresponding plan specimens. Increases of about 14%, 15% and 11% in ultimate tensile strength are generated when PZ-1%, M7-1% and M3-1% are compared with corresponding plain specimens, respectively. Moreover, adding micro steel fibers by 2% volume at PZ-2%, M7-2% and M3-2% results in increase of about 19%, 21% and 15% in ultimate tensile strength, respectively when compared with corresponding plain specimens.

Increase of 16% in first-crack stress and 10% in ultimate tensile strength are observed when UHPC-1% is compared with UHPC-0%. An increase of 24% in first-crack stress and 20% in ultimate tensile strength is observed When UHPC-2% is compared with UHPC-0%. First crack stress of PC-1% and PC-2% are increased app. 14% and 20%, respectively in comparison with plain specimens.

As seen in Fig 3, adding micro steel fibers to TRC specimens' matrix results in a higher number of tensile cracks in all types of matrices. As the improved first crack stress and ultimate tensile strength are along with a higher number of cracks, a considerable improvement in strain-hardening behavior and then a significant increase in area under the stress-strain curves and more energy absorption are observed.

4. Conclusions

At this experimental research, effects of different matrix compositions containing different contents of micro steel fibers on tensile properties and strain-hardening behavior of glass TRC were investigated. Commonly used matrices such as PZ, M3 and M7 and also two innovative matrices of UHPC and PC were used to prepare TRC specimens. Micro steel fibers were added into the mentioned matrices in different contents of 0%, 1% and 2% by volume. The following conclusions can be drawn from the present research work.

- Both the first crack stress and the ultimate tensile strength of TRC specimens were significantly influenced by the matrix composition type. It seems that, different w/b ratio of cement-based matrices was the main factor influenced on bonding of matrix-textile and matrix-steel micro fibers that resulted in significant changes in tensile behavior of TRC specimens.
- Usage of micro steel fibers in matrices led to an increase in first-crack stress and ultimate tensile strength of TRC specimens made with different matrix compositions. The tensile properties of TRC were more improved with increasing the volume fractions of micro steel fibers. A significant increase in number of cracks was observed by adding micro steel fibers to matrices that followed with more improved strain-hardening behavior of TRC specimens.

- It can be clearly concluded that tensile properties of TRC specimens are improved by using an appropriate matrix containing micro steel fibers. As the improved first-crack stress and ultimate tensile strength are along with an increase in number of tensile cracks, a considerable improvement in strain-hardening behavior and then a significant increase in area under the stress-strain curves and more energy absorption are observed.
- Considering the acceptable tensile performance of specimens prepared by UHPC matrix, textile reinforced ultra-high performance concrete seems to be introduced as an innovative material especially for structural applications which is well worth to be more investigated in upcoming researches.

Acknowledgments
The authors gratefully acknowledge the Solidian company, specially Dr. Ali Shams for their kindly support in providing textile sheets.

References

[1] Jesse, F., et al. 2008, *Special Publication,* Load-bearing behavior of textile-reinforced concrete, 250: p. 59-68.
[2] Bentur, A., A. Peled, and D. Yankelevsky, 1997. Cement and concrete research ,Enhanced bonding of low modulus polymer fibers-cement matrix by means of crimped geometry, 27(7): p. 1099-1111.
[3] Peled, A., A. Bentur, and D. Yankelevsky,1998, Advanced Cement Based Materials, Effects of woven fabric geometry on the bonding performance of cementitious composites: mechanical performance.,7(1): p. 20-27.
[4] Banholzer, B. 2004, Bibliothek der RWTH Aachen, Bond behaviour of a multi-filament yarn embedded in a cementitious matrix,
[5] Hegger, J., et al.,2006. Materials and Structures, Load–bearing behaviour and simulation of textile reinforced concrete., 39(8): p. 765-776.
[6] Dilthey, U., et al., 2006. in 1st International RILEM Conference on Textile Reinforced Concrete. RILEM Publications SARL. Application of polymers in textile reinforced concrete from the interface to construction elements.
[7] Li, Q. and S. Xu, 2010. Journal of composite materials, Experimental research on mechanical performance of hybrid fiber reinforced cementitious composites with polyvinyl alcohol short fiber and carbon textile.
[8] Xu, S., et al., 2004, Journal of materials in civil engineering, Bond characteristics of carbon, alkali resistant glass, and aramid textiles in mortar. 16(4): p. 356-364.
[9] Peled, A., A. Bentur, and D. Yankelevsky, 1999, Journal of materials in civil engineering, Flexural performance of cementitious composites reinforced with woven fabrics. 11(4): p. 325-330.
[10] Peled, A., et al. 2008, Cement and concrete composites, Influences of textile characteristics on the tensile properties of warp knitted cement based composites. 30(3): p. 174-183.
[11] Colombo, I.G., et al., 2013, Materials and structures, Erratum to: Textile Reinforced Concrete: experimental investigation on design parameters. 46(11): p. 1953-1971.
[12] Cevallos, O. and R. Olivito, 2015, Composites Part B: Engineering, Effects of fabric parameters on the tensile behaviour of sustainable cementitious composites. 69: p. 256-266.
[13] Mumenya, S., R. Tait, and M. Alexander, 2011, Materials and structures, Evaluation of toughness of textile concrete. 44(1): p. 279-289.
[14] Peled, A., B. Mobasher, and Z. Cohen, 2009, Cement and Concrete Composites, Mechanical properties of hybrid fabrics in pultruded cement composites. 31(9): p. 647-657.
[15] Larrinaga, P., et al. 2014, Materials & Design, Experimental and numerical modeling of basalt textile reinforced mortar behavior under uniaxial tensile stress. 55: p. 66-74.
[16] De Santis, S. and G. de Felice, 2015, Composites Part B: Engineering, Tensile behaviour of mortar-based composites for externally bonded reinforcement systems. 68: p. 401-413.

IOP Conf. Series: Materials Science and Engineering **246** (2017) 012031 doi:10.1088/1757-899X/246/1/012031

[17] Jesse, F. and I.M. Curbach, 2004, TechnischeUniversität Dresden, Dresden, Tragverhalten von Filamentgarnen in zementgebundener Matrix Load Bearing Behaviour of Filament Yarns in a Cementitious Matrix.

[18] Hartig, J., U. Häußler-Combe, and K. Schicktanz, 2008, Cement and concrete composites, Influence of bond properties on the tensile behaviour of Textile Reinforced Concrete. 30(10): p. 898-906.

[19] Esmaeili, J. And J. Kasaei, 2016, in First International Interactive Symposium on UHPC. Proceedings. Effect of Different Curing Regimes on Strength and Transport Properties of UHPC Containing Recycled Steel Tire Wires as Micro Steel Fibers.

[20] Mechtcherine, V., Schneider, K. and W. Brameshuber, 2016, Woodhead Publishing, Textile Fibre Composites in Civil Engineering: Mineral-based matrices for textile-reinforced concrete.

IOP Conf. Series: Materials Science and Engineering **246** (2017) 012032 doi:10.1088/1757-899X/246/1/012032

Influence of steel fibers on the shear and flexural performance of high-strength concrete beams tested under blast loads

O Algassem[1], Y Li[1] and H Aoude[1]

[1]Department of Civil Engineering, University of Ottawa, Ottawa, ON, K1N5S5, Canada

E-mail: haoude@uottawa.ca

Abstract. This paper presents the results of a study examining the effect of steel fibres on the blast behaviour of high-strength concrete beams. As part of the study, a series of three large-scale beams built with high-strength concrete and steel fibres are tested under simulated blast loading using the shock-tube testing facility at the University of Ottawa. The specimens include two beams built with conventional high-strength concrete (HSC) and one beam built with high-strength concrete and steel fibres (HSFRC). The effect of steel fibres on the blast behaviour is examined by comparing the failure mode, mid-span displacements and, overall blast resistance of the specimens. The results show that the addition of steel fibres in high-strength concrete beams can prevent shear failure and substitute for shear reinforcement if added in sufficient quantity. Moreover, the use of steel fibres improves flexural response under blast loading by reducing displacements and increasing blast capacity. Finally, the provision of steel fibres is found to improve the fragmentation resistance of high-strength concrete under blast loads.

1. Introduction

High-strength concrete (HSC) may be defined as concrete having a compressive strength which exceeds 60 MPa. The high-strength properties of HSC are achieved by reducing the porosity in the concrete matrix through careful selection and proportioning of mix constituents (cement, sand, aggregate), use of low water-cement ratio and use of chemical and mineral admixtures (e.g. superplasticizers and silica fume). The high-strength of HSC makes it ideal for heavily-loaded structures (e.g. columns in high-rise buildings) while its high stiffness and reduced porosity make it well-adapted for use in long-span structural components, and structures where durability has increased importance (e.g. bridges). On the other hand, the high strength of HSC results in reduced ductility, with brittle failure in compression and poor tensile resistance.

The provision of steel fibres is a practical solution for the brittle response of HSC. The ability of fibres to arrest and redistribute cracking allows high-strength fibre-reinforced concrete (HSFRC) to show more ductile failure in compression, while also allowing the material to have increased tensile resistance and post-cracking capacity. The provision of fibres also results in increased toughness and fragmentation resistance. The high-performance properties of HSFRC make it ideally-suited for use in extreme loading applications, such as impact and blast.

Over the years, several researchers have investigated the effect of steel fibres on the behavior of concrete under high stress-rates. Most of the previous studies have focused on material tests, conducted on smaller-scale specimens, using various impact setups [1-5]. A relatively fewer investigations have focused on the impact and blast performance of larger-scale structural components. Most previous studies in this area have focused on slabs [6-8], with a limited number of investigations examining behavior of beams and columns [9-10]. This paper presents the results from an experimental program examining the effect of steel fibres on the blast behaviour of high-strength concrete beams. As part of the study three beams built with and without steel fibres are tested under simulated blast loads using a high-capacity shock-tube. The effect of HSFRC on blast behaviour is examined by comparing the failure mode, mid-span displacements and overall blast resistance of the beams.

2. Experimental program

2.1. Specimen designs

Three reinforced concrete beams were built and tested in this research program. Table 1 summarizes the design details of the specimens. As shown Figure 1, the beams had cross-section of 125 mm x 250, with a total length of 2440 mm. The beams were simply supported over a span of 2232 mm, with a constant moment region of 750 mm and two equal shear spans of 741 mm. Two of the beams were cast with plain high-strength concrete (HSC), with the remaining beam cast with HSC and 1% of steel fibers by volume of concrete (80 kg/m^3). Longitudinal reinforcement consisted of 2-20M Canadian size bars (d_b = 19.5 mm, A_s=300 mm^2). One of the HSC beams and the HSFRC beam were designed without stirrups. The remaining HSC beam was reinforced with U-shaped stirrups, made from 6.3 mm diameter smooth steel wire, spaced at 100 mm in the shear spans. Two 6.3 mm bars were also provided at the top of the beams in the shear spans only. The specimen nomenclature reflects the test variables which include concrete type (where "HSC-F0" refer to plain HSC and "HSC-F1" refers to HSFRC) and shear reinforcement details ("S" for presence of stirrups and "0" for lack of stirrups).

Table 1. Beam test matrix.

			Steel fibre properties		Steel reinf. properties	
Beam I.D.	Concrete Mix	Concrete Strength f'$_c$ (MPa)	L_f / D_f (mm/mm)	V_f (%)	Transverse Steel	Flexural Steel
HSC-F0-20M-0	HSC	110	-	-	-	2 - 20M
HSC-F1-20M-0	HSFRC	104	30/0.55	1.0	-	
HSC-F0-20M-S	HSC	105	-	-	6.3 mm stirrups @ s = 100 mm	

2.2. Materials

The high-strength concrete in this study had a target strength of 100 MPa. The mix constituents included Portland cement, slag, silica fume, coarse aggregate (13 mm and 20 mm), sand and liquid admixtures (super-plasticizer and set retarder). The HSFRC used the same mix, with the addition of 30 mm hooked-end steel fibres at a dosage (V_f) of 1% by volume of concrete (80 kg/m^3). The properties of the HSC and HSFRC concrete in terms of compressive strength, obtained by testing 100 mm x 200 mm cylinders, are summarized in Table 1. Sample stress-strain curves are shown in Figure 2, along with sample flexural test results obtained from testing 100 mm x 100 mm x 400 mm prisms in accordance with the ASTM C1609 Standard. The normal-strength 20M longitudinal steel reinforcement had an average yield strength of 460 MPa, while the 6.3 mm steel wire used for the transverse reinforcement had an average yield strength of 540 MPa (sample stress-curves curves are also shown in Figure 2).

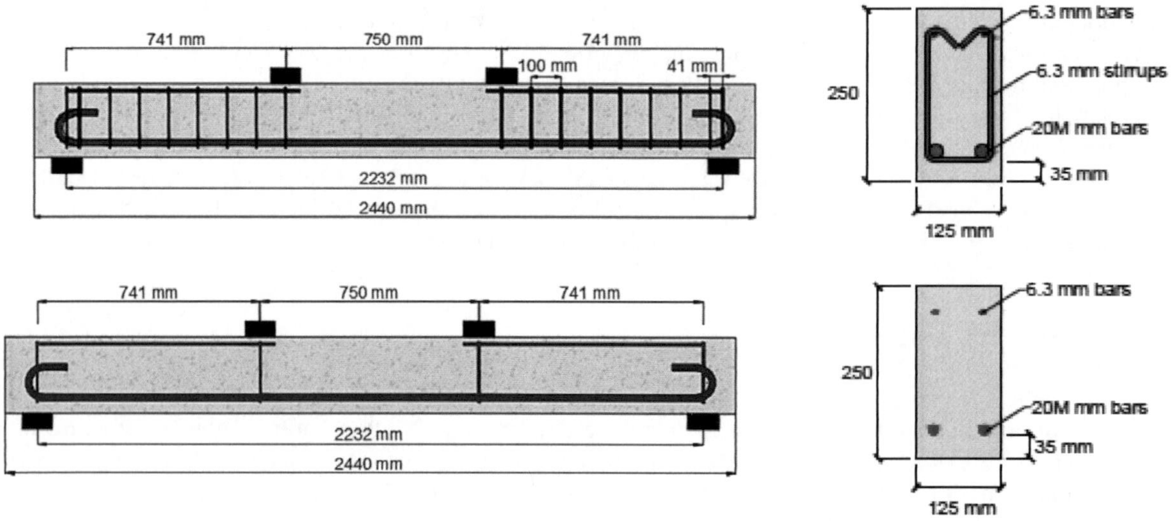

Figure 1. Dimensions and reinforcement for beams with & without shear reinforcement.

2.3. Test setup

The University of Ottawa Blast Research Laboratory is equipped with a high-capacity shock-tube that can simulate the shockwaves generated by blasts without the need for live explosives. As shown in Figure 3a, the shock-tube consists of four main components: (1) a variable length driver section, (2) a spool section, (3) an expansion section and (4) a rigid end test frame. The driver section generates the blast energy, while the firing of the shockwave is controlled using the spool section. The rigid end test frame has a 2m x 2m square opening where test specimens can be attached. For non-planar elements, such as beams, a load transfer device (LTD) is used to collect the shockwave pressure at the shock-tube opening and redirect it onto the test specimens. In the current study the LTD resulted in the application of blast loading under four-point bending. Figure 3b shows a typical beam prior to testing. Complete displacement-time histories were recorded using two linear variable differential transducers (LVDT) placed at 1/2 (mid-height) and 1/3rd span of the beam (see Figure 3b). A high-speed camera was placed at the side of the beams during testing, and recorded response at a frame rate of 500 frames per second.

Testing of the beams was conducted under gradually increasing blast pressures until failure. Blast 1 was meant to test the specimens under elastic conditions, while Blast 2 was meant to bring the steel reinforcement in the beams to yielding. Blasts 3, 4 and 5 were applied to cause damage and failure in those specimens which reached flexural capacity. Examples of the shockwaves for Blasts 1 to Blast 5 are shown in Figure 3c. Shockwave properties are also summarized in Table 2.

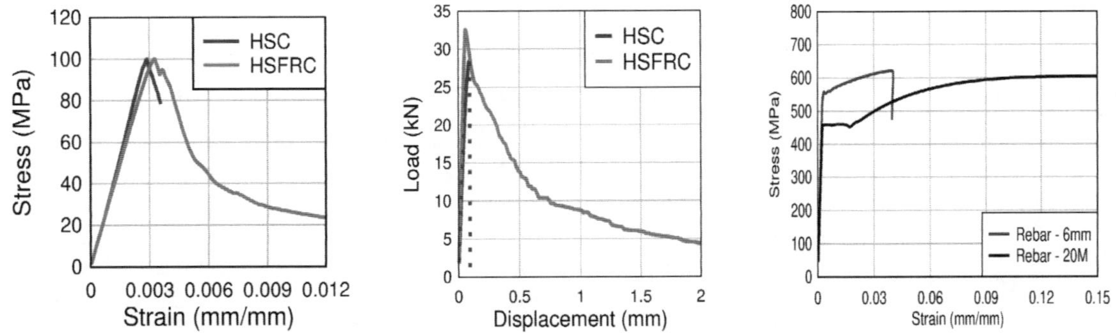

(a) (b) (c)

Figure 2. Sample stress-strain curves for: (a) HSC and HSFRC in compression, (b) HSC and HSFRC in flexure and (c) steel reinforcement in tension.

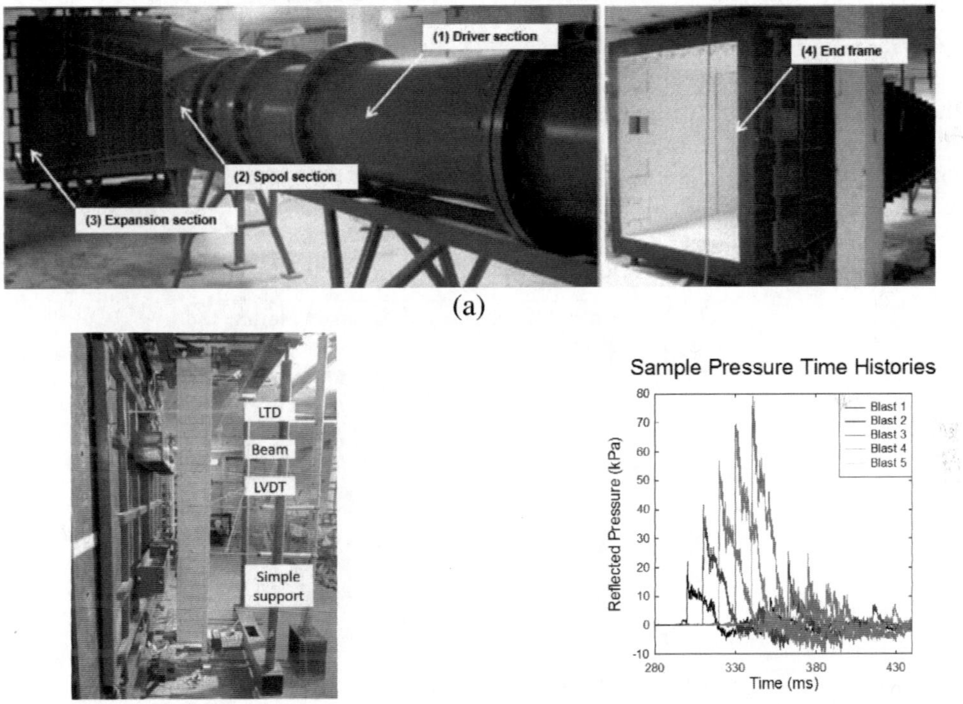

(a)

(b) (c)

Figure 3. (a) Shock-tube, (b) beam test setup and (c) sample shockwaves for Blasts 1-5.

3. Results

Table 2 summarizes the experimental results for the three beam specimens, including the shockwaves properties for each test (reflected pressure P_r, reflected impulse I_r, positive phase duration t_p) as well as beam response in terms of maximum (d_{max}) and residual (d_{res}) mid-span displacements. Table 3 to 5 show the progression of damage in the beams during testing, while Figure 4 compares the displacement response of the beams at selected blasts.

Table 2. Blast test results.

Beam I.D.	Blast	Shockwave Properties			Displacement	
		Pressure	Impulse	Duration	d_{max}	d_{res}
	#	P_r (kPa)	I_r (kPa*ms)	t_p (ms)	(mm)	(mm)
HSC-F0-20M-0	1	23.9	239.5	20.0	12.6	3.4
	2	38.7	370.8	19.2	19.3	9.2
HSC-F0-20M-S	1	23.6	244.3	20.7	10.4	2.0
	2	39.2	360.0	18.4	15.1	0.2
	3	57.4	538.2	18.8	32.9	12.4
	4	68.8	702.6	20.4	118.1	71.7
HSC-F1-20M-0	1	23.4	244.0	20.9	8.4	1.4
	2	41.8	387.0	18.5	14.7	3.4
	3	57.1	571.2	20.0	20.8	4.1
	4	69.6	734.5	21.1	42.1	19.3
	5	75.3	738.0	19.6	62.4	37.4

3.1. Beam HSC-F0-20M-0

Beam HSC-F0-20M-0 was constructed with plain high-strength concrete and 20M normal-strength steel. The specimen was built without shear reinforcement. Blast 1 was meant to test the beam within the elastic range, and resulted in cracking of concrete and small residual displacements. Under Blast 2, the beam suffered a brittle shear failure with a diagonal crack forming in the top shear span as shown in Table 3 and resulted in maximum & residual displacements of 19.3 and 9.2 mm.

Table 3. Progression of blast damage in beam HSC-F0-20M-0

Blast 1	Blast 2	Close-up of damage at failure

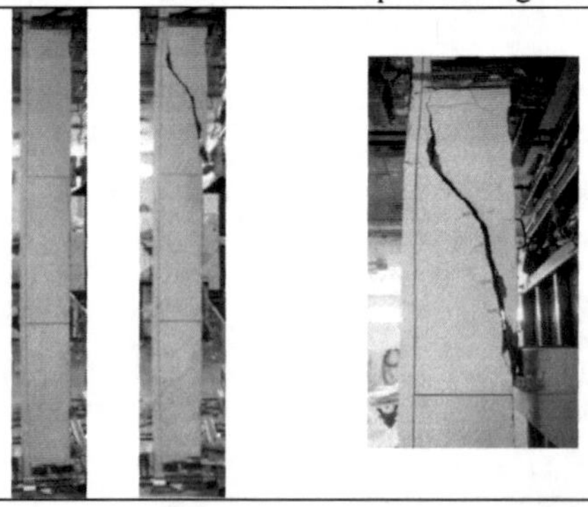

3.2. Beam HSC-F0-20M-S

Beam HSC-F0-20M-S was constructed with plain high-strength concrete and 20M normal-strength steel, however unlike the previous specimen the beam was reinforced with transverse reinforcement spaced at 100 mm in the shear spans. Blast 1 was meant to test the beam within the elastic range, and resulted in cracking of concrete and small residual displacements. Blast 2 brought the 20M steel close to yield levels and resulted in maximum displacement of 15.1 mm with minimal residual deformations. The provision of shear reinforcement in this beam prevented shear failure. The 20M reinforcement in the beam went into the plastic range after Blast 3, resulting in a maximum displacement of 32.9 mm and larger residual displacement of 12.4 mm. Cracking became more prominent after this test with a flexural crack propagating the full depth of the beam. Blast 4 (70 psi) was the last shot for this specimen and resulted in maximum and residual displacements of 118 mm and 72 mm. As shown in Table 4, failure occurred in the compression zone due to severe concrete crushing at mid-span. High-speed video shows the generation of important secondary fragments at failure.

Table 4. Progression of blast damage in beam HSC-F0-20M-S

Blast 1	Blast 2	Blast 3	Blast 4	Close-up of damage at failure
	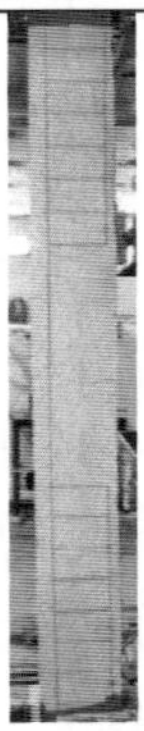			

3.3. Beam HSC-F1-20M-0

The design of beam HSC-F1(ZP)-20M-0 matched that of HSC-F0-20M-0 (20M bars, no shear reinforcement), except for the addition of steel fibres at a ratio 1.0% by volume of concrete. Blast 1 and Blast 2 resulted in limited deformations and very limited damage in the beam. Blast 3 resulted in the formation of a prominent crack at mid-span, however the beam prevented shear failure, with the 20M steel reinforcement reaching post-yield levels. The maximum and residual displacements after this blast were 20.8 mm and 4.1 mm which are reduced when compared to the previous beam. Blast 4 resulted in maximum and residual displacements of 42 mm and 19 mm, and resulted in an increase in crack widths, however, the beam survived this blast. Blast 5 resulted in major opening of the critical crack at mid-span, with the steel fibres showing signs of pull-out, and therefore the beam was deemed to have failed after this shot (see Table 6). Nonetheless, concrete crushing and fragmentation at failure was limited. The maximum and residual displacements were 62 mm and 37 mm, after this blast.

Table 5. Progression of blast damage in beam HSC-F1-20M-0

Blast 1	Blast 2	Blast 3	Blast 4	Blast 5	Close-up of damage at failure
					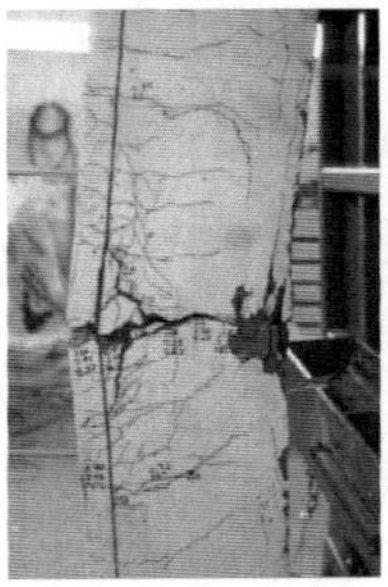

4. Discussion of results

Comparison between the behaviour of beams HSC-F0-20M-0 and HSC-F1-20M-0 allows for an investigation into the effect of steel fibres on shear resistance since both beams were built without stirrups. The plain HSC beam failed in shear at Blast 2 due to the lack of transverse shear reinforcement. In contrast, the use fibres in specimen HSC-F1-20M-0 prevented shear failure, and allowed the HSFRC beam to survive up to Blast 5 pressures, with failure occurring in flexure.

Comparison of the responses of beams HSC-F1-20M-0 and HSC-F0-20M-0, shows that the HSFRC beam was to not only able to match, but out-perform the companion HSC beam which contained stirrups. As shown in Figure 4, beam HSC-F1-20M-0 reduced maximum displacements by margins of 37% and 64% when compared to beam HSC-F0-20M-S after Blast 3 and Blast 4, respectively (d_{max} = 20.8 vs. 32.9 mm and 42.1 vs. 118.1 mm, respectively). Residual displacements were also significantly reduced for the HSFRC beam, with reductions of 67% and 73% when compared to beam HSC-F0-20M-S (d_{res} = 4.1 vs. 12.4 mm and 19.3 vs. 72 mm, respectively). More importantly, the provision of fibres allowed for an increase in blast capacity with failure delayed from Blast 4 in the plain HSC beam to Blast 5 in the HSFRC companion. Damage tolerance was also significantly affected, with failure in the HSC specimen associated with complete disintegration of concrete in the compression zone, while the HSFRC beam showed limited crushing with failure occurring due to fibre pullout in the mid-span tension region. Because of the ability of the fibres to hold the concrete together, secondary blast fragments where also significantly reduced.

In summary, the results demonstrate clear benefits associated with the use of steel fibres in high-strength concrete beams. The provision of steel fibres was able of effectively substitute for transverse shear reinforcement, prevented shear failure and promoted ductile flexural response. Moreover, the use of fibres allowed for a better control of maximum and residual displacements at equivalent blasts, and led to an increase in blast capacity (ability to sustain larger blast loads).

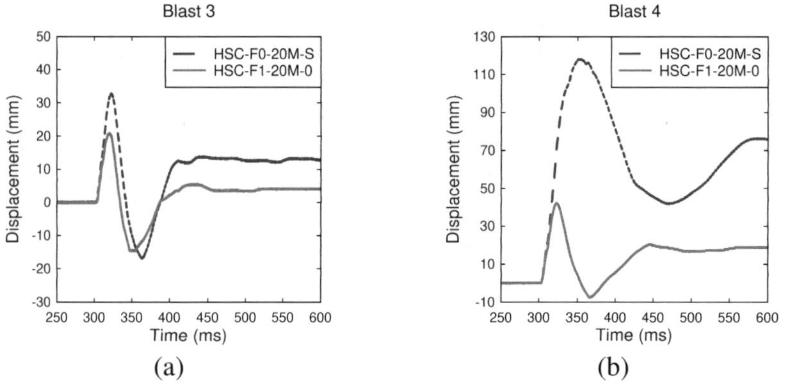

Figure 4. Mid-span displacement time-histories at: (a) Blast 3 and (b) Blast 4.

5. Conclusions

This paper presented the results from three high-strength concrete beams, built with and without steel fibres, and tested under blast loading. The following conclusions can be drawn from this study:

- The provision of steel fibers in the high-strength concrete beams increased shear resistance and prevented brittle shear failure under blast loading;
- The addition of steel fibers in the high-strength concrete beams improved flexural performance by reducing maximum and residual displacements and increasing overall blast resistance;
- The addition of steel fibers to high-strength concrete had an resulted in better control of damage and reduced secondary blast fragments.

References

[1] Naaman A E and Gopalaratnam V S 1983 Impact properties of steel fiber reinforced concrete in bending, *Int. J. Cem. Compos. Lightweight Concr.* **5(4)** 225-33

[2] Gopalaratnam V S and Shah S P 1986 Properties of steel fiber reinforced concrete subjected to impact loading *ACI J.* **83(1)** 117-26

[3] Wang Z L, Liu Y S and Shen R F 2007 Stress-strain relationship of steel fiber-reinforced concrete under dynamic compression *Construct. & Build. Mater.* **22(5)** 811-19.
[4] Banthia N, Mindess S, and Trottier J F 1996 Impact resistance of steel fiber reinforced concrete *ACI Mat. J.* **93(5)** 472-479
[5] Banthia N P 1987 *Impact resistance of concrete* (PhD dissertation) University of British Columbia, Vancouver
[6] Ong K C G, Basheerkhan M and Paramasivam P1999 Resistance of fibre concrete slabs to low velocity projectile impact *Cem. and Conc. Comp.* **21** 391-401
[7] Lok, T S and Xiao J R 1999 Steel-fibre-reinforced concrete panels exposed to air blast loading *Struct. and Build.* **134(4)** 319-31
[8] Yi N, Kim J J, Han T, Cho Y and Lee J 2012 Blast-resistant charcteristics of ultra-high strength concrete and reactive powder concrete *Construct. & Build. Mater.* **28** 694-707
[9] Magnusson J, Hallgren M and Ansell A 2010 Air-blast-loaded, high-strength concrete beams Part I: experimental investigation. *Mag. Concrete Res.* **62(2)** 127-36
[10] Burrell, R P, Aoude H and Saatcioglu M 2015 Response of SFRC columns under blast loads *ASCE J. Struct. Eng.* **141(9)** 1-15

IOP Conf. Series: Materials Science and Engineering **246** (2017) 012033 doi:10.1088/1757-899X/246/1/012033

Characterization of polymer concrete with natural fibers

M Barbuta[1], A A Serbanoiu[1], R Teodorescu[2], B Rosca[1], R Mitroi[1] and G Bejan[1]

[1] Technical University "Gh. Asachi" from Iasi, Romania
[2] University of Agronomical Sciences and Veterinary Medicine in Bucharest, Faculty of Land Improvements and Environmental Engineering, Bucharest, Romania

E-mail: serbanoiu.adrian@tuiasi.ro

Abstract. In the study are presented the experimental results obtained for polymer concrete prepared with epoxy resin, aggregates, fly ash as filler and two types of fibers: wool and hemp. The influence of type and dosage of fibers were studied. The density and mechanical characteristics were determined: compressive strength, flexural strength and split tensile strength. For both types of fibers, with increasing the fiber dosage the density decreases. The studied dosages had not an important influence on mechanical strengths. The fibers improved especially the tensile strength and the compressive strength presented generally smaller values than the control mix.

1. Introduction

In the domain of sustainable building materials a lot of wastes are used for obtaining new materials, for improving the characteristics or as addition in the mix [1,2]. The wastes are used for obtaining cement concrete, polymer concrete, cement, plaster, composites which are used in civil engineering, bridges, building services, highways, hydraulic constructions, etc. [3-7]. Generally concrete is sensitive to cracking and in tension it presents a brittle failure. The fibers help to a better behavior of concrete in tension by stopping the cracking processes and improving ductility at failure. Several types of fibers are used in building materials such as: synthetic fibers (glass, carbon, metallic, etc.), natural fibers (cellulose, sisal, rice husk, etc). [8-14]. Polymer concrete presents some advantages in comparison with cement concrete related to higher mechanical strengths, low permeability, a better resistance to aggressive agents, etc. Although the costs of polymer concrete are higher than cement concrete it is very useful for using in severe conditions from industry, public buildings, transport, etc.[15-22]. Natural materials like wool or hemp are abundantly available materials in many countries. And when they are not used they are dumped in landfills with high costs or are burned polluting the environment. Their use in building materials industry can be a possibility of consuming these natural materials [23-25].

The present study analyses the influence of fibers from wool or hemp on the mechanical characteristics of polymer concrete with fly ash as a possibility of obtaining new eco-friendly concrete.

2. Experimental program

The polymer concrete used as control sample was obtained with epoxy resin type Ropoxid, a Romanian product which in combination with hardener type Romamid was the binder of the mix. Epoxy resin was in a dosage of 12.4% from the mix, minimum quantity that could be used from the condition of workability. Two sorts of aggregates were used: sort I (sand) and sort II-river gravel (4-8 mm) both in the same dosage of 37.4% from the total volume of the mix. A dosage of 12.8% from the total volume mix was fly ash used as filler. The experimental mixes with fibers type wool, noted PCW1 to PCW5, and hemp, noted PCH1 to PCH5, were prepared with the same components as the control mix and the fibers were added in dosages of 1%, 2%, 4%, 5% and 6% from the resin weight.

2.1. Samples

According the European standards for determining the mechanical characteristics such as compressive strength (fc), flexural strength (f_{ti}) and split tensile strength (f_{td}) [26-28] of both type of polymer concrete, the following types of samples were poured: cubes of 70 mm sizes and prisms of 70x70x210 mm sizes. The experimental tests on samples were effectuated at 14 days according to standards [26-28].

3. Results and discussions

3.1. Physical characteristic

The density of polymer concrete with fiber varied between 1996 kg/m^3 for PCW5 (with 6% wool fibers) and 2218 kg/m^3 for PCH1 (with 1% hemp fibers), Figure 1.The mixes with wool fibers presented the majority of densities under the control mix. For both types of fibers, with increasing the fiber dosage the density decreases.

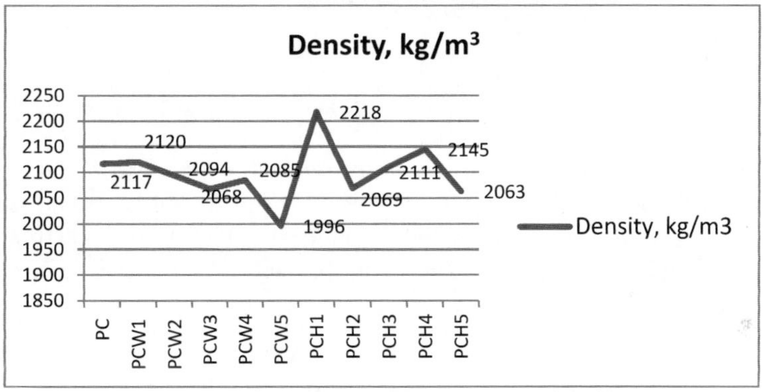

Figure 1. Density of polymer concrete with fibers

3.2. Mechanical strengths

The experimental results were determined for compressive strength (f_c), flexural strength (f_{ti}) and split tensile strength (f_{td}).

Compressive strength obtained on polymer concrete with wool fiber waste was with values between 31.7 MPa and 57.1 MPa and with hemp fibers was with values between 27.8 MPa and 53.1 MPa, Figure 2.

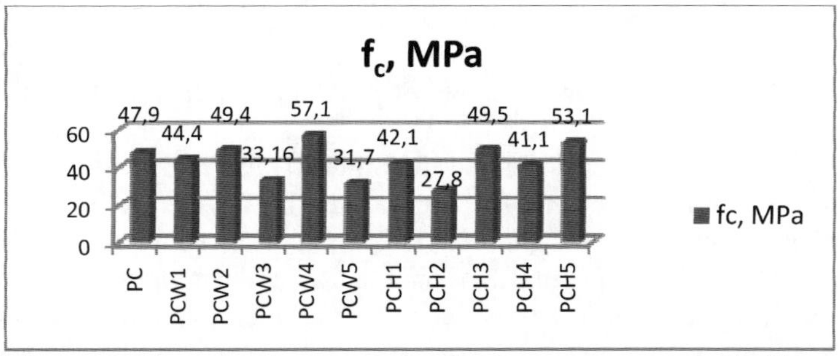

Figure 2. Compressive strength of polymer concrete with fibers

Comparing the both types of fibers with control polymer concrete it can observe that:

- The polymer concrete with 1% wool fiber presented the highest value of compressive strength, an increase with 19.2% in comparison with control mix (PC) and that with 6% wool fibers presented a decrease of compressive strength of about 33.8%.
- The concrete with 6% hemp fiber presented the highest value of compressive strength, an increase with 10.8% in comparison with the control mix (PC) and that with 2% hemp fibers presented a decrease of compressive strength of about 42%.

As conclusion, the influence of wool and hemp fiber on compressive strength of polymer concrete is that of decreasing this characteristic.

Comparing the influence of type of fiber on the compressive strength it can observe that polymer concrete with wool fiber presented higher values that polymer concrete with hemp fibers for dosages of 1%, 2% and 5%, Figure 3.

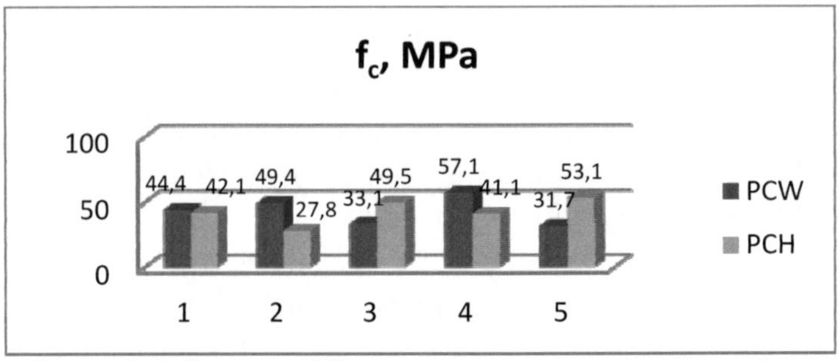

Figure 3. Influence of fiber type on compressive strength

Flexural strength obtained on polymer concrete with wool fiber waste was with values between 14.4 MPa and 19.3 MPa, and with hemp fibers was between 8.6 MPa and 17.8 MPa, Figure 4. The polymer concrete with 2% wool fiber presented the highest value of flexural strength, an increase with 19.1% in comparison with the control mix (PC) and that with 5% wood fibers presented a decrease of flexural strength of about 42%.

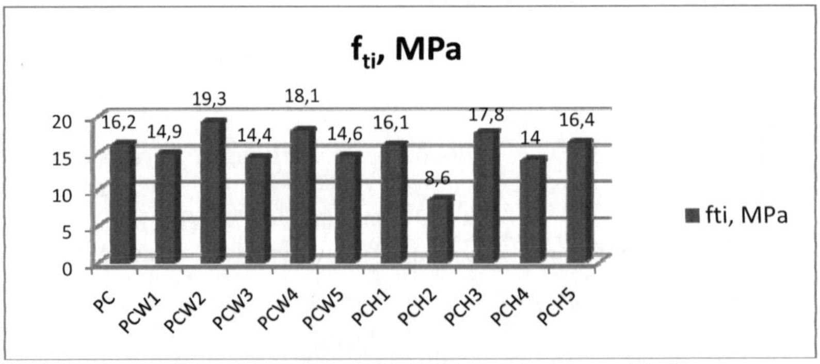

Figure 4. Flexural strength of polymer concrete with fibers

The concrete with 4% hemp fiber presented the highest value of flexural strength, an increase with 9.8% in comparison with the control mix (PC) and that with 2% hemp fibers presented a decrease of flexural strength of about 46%.

Comparing the influence of type of fiber on the flexural strength it can observe that polymer concrete with hemp fiber presented higher values that polymer concrete with wool fibers for dosages of 1%, 4% and 6%, Figure 5.

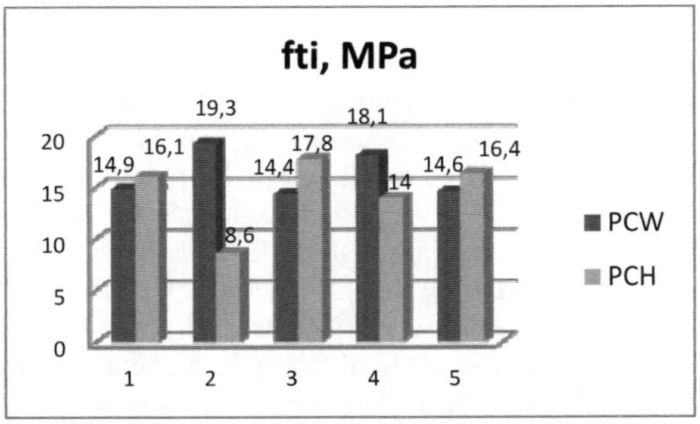

Figure 5. Influence of fiber type on flexural strength

Split tensile strength obtained on polymer concrete with wool fiber waste was with values between 4.87 MPa and 7.7 MPa, all values are bigger than that of control mix (PC). The maximum increase was for mix PCW2 that had an increase of about 60% in comparison with the control mix (PC). The polymer concrete with hemp fibers presented values of split tensile strength between 3.5 MPa and 6.9 MPa, Figure 6. The polymer concrete with 4% hemp fiber presented the highest value of split tensile strength, an increase with 43.4% in comparison with the control mix (PC) and that with 2% hemp fibers presented a decrease of split tensile strength of about 27.3%. The polymer concrete with hemp fibers presented also higher values of f_{td} than the control mix, with an exception in the case of mix PCH2.

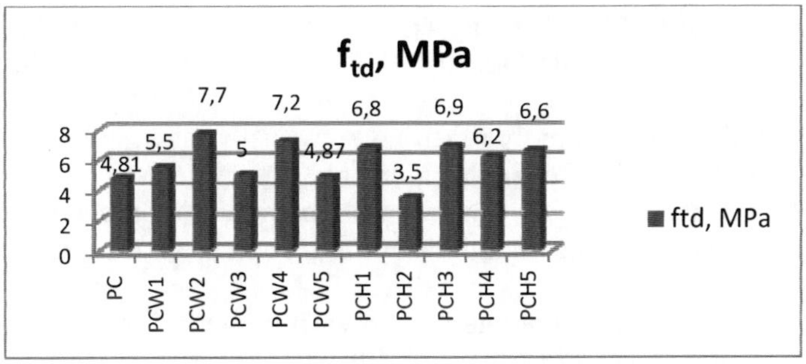

Figure 6. Split tensile strength of polymer concrete

Comparing the influence of type of fiber on the split tensile strength it can observe that polymer concrete with hemp fiber presented higher values that polymer concrete with wool fibers for dosages of 1%, 4% and 6%, Figure 7.

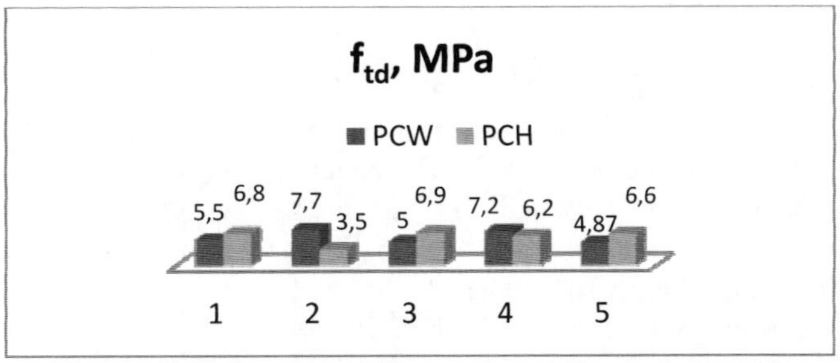

Figure 7. Influence of fiber type on flexural strength

The mix PCW4 (with 5% wool fiber) presented the highest value of f_c and the mix PCW2 (with 2% wool fibers) presented highest values of f_{ti} and f_{td} in comparison with all mixes.

4. Conclusions

In the study are presented the experimental results obtained for polymer concrete prepared with epoxy resin, aggregates, fly ash as filler and two types of fibers: wool and hemp. For both types of fibers, with increasing the fiber dosage the density decreases.

The values of mechanical strengths were influenced by the fiber type and dosage. The studied dosages between 1% and 6% had not an important influence on mechanical strengths and higher dosages must be studied. The compressive strength obtained with wool and hemp fiber was generally smaller than that of control mix. Comparing the influence of type of fiber on the compressive strength it can observe that polymer concrete with wool fiber generally presented higher values that polymer concrete with hemp fibers. In the case of flexural strength the hemp fiber presented generally higher values in comparison with wool fibers. In the case of split tensile strength only the polymer concrete with wool fiber presented all values bigger than that of control mix, but for hemp fibers only some values of split tensile strength were bigger than that of control mix.

From the experimental researches was observed that a better influence of wool and hemp fibers is beneficial on the tensile strengths and density. Function other characteristics such as thermal conductivity, acoustic properties, etc. the polymer concrete can be used in obtaining eco-friendly lightweight concrete, concrete with improved thermal properties, etc.

References

[1] Bolden, J.; Abu-Lebdeh, T.; Fini, E. Utilization of recycled and waste materials in various construction applications. Am. J. Env. Sci. 9 [1], (2013) 14-24, doi:10.3844/ajessp.2013.14.24

[2] M. Barbuta, R. D. Bucur, S. M. Cimpeanu, G. Paraschiv, D. Bucur-Agroecology, Chapter 3, Wastes in Building Materials Industry, INTECH, Croatia, ISBN 978-953-51-2130-5 (2015) 81-99.

[3] J. Bydzovsky, Utilization of rapidly renewable raw materials in building materials, Proceedings of the 12th International Scientific Conference. Brno, Czech Republic (2009), pp. 43–46

[4] L.P. Singh, S.K. Bhattacharyye, S. Ahalawat, Preparation of sizecontrolled silica nano particles and its functional role in cementitious system, Journal of Adv. Concr. Technol. 10 (2012) 345-352.

[5] A. Garbacz, J.J. Sokolovska, Concrete like polymer composite with fly ash-Comparative study, Constr. and Build. Mater. 38 (2013) 689-699.

[6] C. Bilim, C.D. Atiş, H. Tanyildizi, O. Karahan, Predicting the compressive strength of ground granulated blast furnace slag concrete using artificial neural network, Adv. Eng. Softw. 40 (2009) 334-340.

[7] M. Nadeem, A.D. Pofale, Utilization of Industrial Waste Slag as Aggregate in Concrete Applications by Adopting Taguchi's Approach for Optimization, Open Journal of Civil Engineering 2 (2012) 96-105.

[8] Soroushian, P.; Ravanbakhsh, S. (1998) Control of plastic shrinkage cracking with specialty cellulose fibres. ACI Mater. J. 95 [4], 429–435

[9] P.B. de Bruijn, K.H. Jeppsson, K. Sandin, C.H. Nilsson, Mechanical properties of lime-hemp concrete containing shives and fibres, Biosystems Engineering, 103 (2009), p. p474

[10] Awwada, E.; Mabsout, M.; Hamad, B.; Farran, M. T.; Khatib, H. (2012) Studies on fiber-reinforced concrete using industrial hemp fibers. Constr. Build. Mater. 35 [10], 710–717. doi:10.1016/j.conbuildmat.2012.04.119.

[11] Awwad, E.; Mabsout, M.; Hamad, B.; Khatib, H. (2011) Preliminary studies on the use of natural fibers in sustainable concrete. Leban Sci. J. 12 [1], 109–17.

[12] Bilba, K.; Arsene, M.A.; Ouensanga, A. (2007) Study of banana and coconut fibers botanical composition, thermal degradation and textural observations. Bioresour. Technol. 98 [1], 58–68. doi:10.1016/j.biortech.2005.11.030.

[13] Pelisser, F.; Neto, A.B.D.S.S.; La Rovere, H.L.; de Andrade Pinto, R.C. (2010). Effect of the addition of synthetic fibers to concrete thin slabs on plastic shrinkage cracking. Constr. Build. Mater. 24 [11], 2171–2176. doi:10.1016/j.conbuildmat.2010.04.041.

[14] Bayasi, Z.; Mc Intyre, M. (2002) Application of fibrillated polypropylene fibres for restraint of plastic shrinkage cracking in silica fume concrete. ACI Mater J. 99 [4], 337–344.

[15] M. Barbuta, M. Rujanu, A. Nicuta, Characterrization of polymer concrete with different wastes additions, Procedia Technology 22 (2016), 407-412

[16] Lin, C.; Hong, Yu-J.; Huc, A. Using a composite material containing waste tire powder and polypropylene fiber cut end to recover spilled oil. Waste Manage 2010, 30, 263–267.

[17] J. Chris Carroll, N. Helminger, Fresh and hardened properties of fiber-reinforced rubber concrete, Journal of Materials in Civil Engineering, 28(7), 2016

[18] M. Barbuta-Influence of fly ash dosage on the properties of polymer concrete", 9th International Scientific Conference , VSU'2009, Sofia-Bulgaria,Tom 2, Vol. 2 (2009) 55-59.

[19] M. Golestaneh, G. Amini, D. Najafpour, M.A. Beygi, Evaluation of mechanical strength of epoxy polymer concrete with silica powder as filler, World Appl Sci J. 9 (2010) 216-220.

[20] S.-C. Kou, C.-S. Poon, A novel polymer concrete made with recycled glass aggregates, fly ash and metakaolin, Constr. and Build. Mater.,2013, 41 146-151

[21] G. Martinez-Barrera, C. Menchaca-Campos, O. Gencel, Polyester polymer concrete: Effect of the marble particle size and High gamma radiation doses, Constr. and Build. Mater. 41 (2013) 204-208.

[22] M. E. Tawfik, S.B. Eskander, Polymer concrete from marble waste and recycled polyethylene terephtalate. J. Elastomers Plast 38 (2006) 65-79.

[23] V. Placet, Characterization of the thermo-mechanical behaviour of hemp fibres intended for the manufacturing of high performance composites, Compos. Part A, 40 (2009), p. p1111

[24] P.B. de Bruijn, K.H. Jeppsson, K. Sandin, C.H. Nilsson, Mechanical properties of lime-hemp concrete containing shives and fibres, Biosystems Engineering, 103 (2009), p. p474

[25] A. Korjenic, S. Klarić, A. Hadžić, S. Korjenic, Sheep Wool as a Construction Material for Energy Efficiency Improvement, Energies 2015, 8, 5765-5781; doi:10.3390/en8065765

[26] Romanian Standard Association, SR EN 12390-3:2009/ AC:2011, Testing hardened concrete. Part 3: Compressive strength of test specimens (2011).

[27] Romanian Standard Association, SR EN 12390-5:2009 – Testing hardened concrete. Part 5: Flexural strength of test specimens (2009).

[28] Romanian Standard Association, SR EN 12390-6:2010, Testing hardened concrete. Part 6: Split tensile strength of test specimens (2010).

IOP Conf. Series: Materials Science and Engineering **246** (2017) 012034 doi:10.1088/1757-899X/246/1/012034

Feasibility of using fibrous waste in cement-based material

M Beddar[1], A Meddah[1] and L Belagraa[1]

[1)] Laboratory of materials and structures mechanics (LMMS), Civil Engineering Department, Faculty of Technology, M'sila University, B.P.n° 166 Ichbilia, , M'sila 28000, Algeria

E-mail: beddarm@yahoo.fr

Abstract. The use of fibers into cement–based materials has been studied extensively and as a result of this research and development, a wide range of fibers is now available to be used in cementitious materials. However, little work has been done on the addition of fibrous waste to the cement, mortar and concretes to investigate this by-product fiber as reinforcing material due the absence of industrialized fibers in many countries.
The main aim of this experimental study is to investigate the suitability of using fibrous waste as a reinforcing material for cement-based materials. Presentation of such fibrous waste is presented in this paper. The influence of type and content of fibrous waste is evaluated. The results obtained, from series of trails, shown that some physical and mechanical properties of the fibrous waste's composite material are improved

1. Introduction

The application of composite materials in engineering field is growing rapidly. Most of the technical activity or objects produced for everyday life where composites are based on composites materials. Fibre reinforced cement matrices, which contains concretes and mortars reinforced with short fibres, are perhaps the most important group of modern materials applied since 1940's in various fields [1] from building construction up to concrete layers on runways and highways.
The mechanism of reinforcement of concrete by fibres consists of distributing short fibres regularly in the concrete matrix. This network of fibres opposes, as much as it is denser, to the widening of the crack and acts as crack arrestors by producing a pinching force which tend to make its propagation slower and causes transfer of stress across cracked sections allowing the affected parts of the composite to retain some post-crack strength and to withstand deformations much greater than what can be sustained by the matrix alone.

Many studies [2, 3,4] reported that the influence of fibres on the composite behaviour of elements subjected to loading is complex. In the limit of elastic deformation the fibres are not active and their role may derive from the law of mixtures [5]. When the micro-cracks are open, the fibres act as crack-arrestors and control their propagation. The total strength is increased due to the fibre contribution. The load corresponding to the first crack is slightly increased, but the main effect is a considerable increase in deformability and in the amount of energy of the external load.

Although, the use of different types at different amount of fibres in cement, mortar and concrete, are reported in many reviews and reports [6], limited study has been conducted on fibre reinforced cementitious materials using industrial waste fibres. Therefore, little information is available about the use of this kind of fibres as reinforcing materials.

In developing countries, the available of different categories of fibres in market is considerable expensive or sometimes inexistent. To eliminate this lack of availability of fibres at reasonable prices, fibrous waste is used to reinforce cement based-materials. Furthermore, with increase in population and industrial activities, the quantity of fibres waste generated from various industries will increase manifold in the coming years. So that, these industrial waste can be effectively be used for making good-strength and low cost fibre reinforced materials after exploring their feasibility. Otherwise this fibrous waste can be representing a danger to the environment and mankind by leaving it the industrial areas, especially in the absence of clear policy for the successful management of industrial waste.

This investigation is an attempt to study the feasibility of using two kinds of locally available industrial fibrous waste for making fibre reinforced cementitious materials (mortar and concrete). The types of fibrous waste considered in this study include: steel fibrous waste and polypropylene fibrous waste are obtained from machining waste of steel part "chips" and polypropylene factories respectively

2. Proposed approach

Due to the absence of a fibres industry in Algeria, and consequently the possibility of the integration of such fibrous waste as material of reinforcement for the cement-based matrices, our reflection was oriented to the valorisation of local steel and polypropylene fibrous waste (figures 1and 2) which are the most plentiful are susceptible to be integrated as material for strengthening the cement-based composites. This approach contributes effectively on one hand to the cleanup of the environment and on the other hand, to the reduction of cost of composite, while respecting the mechanical performances needed by the fibres addition.

In this context, this experimental study was carried out in order to get a rational exploitation of this industrial waste generally available in steel workshops, local lathes and steel transformation units. For this purposes of this study, two types of fibrous waste (steel and polypropylene) were used while a manufactured and commercialised fibres were used for comparison purposes.

Figure 1. Steel fibrous waste before mechanical treatment.

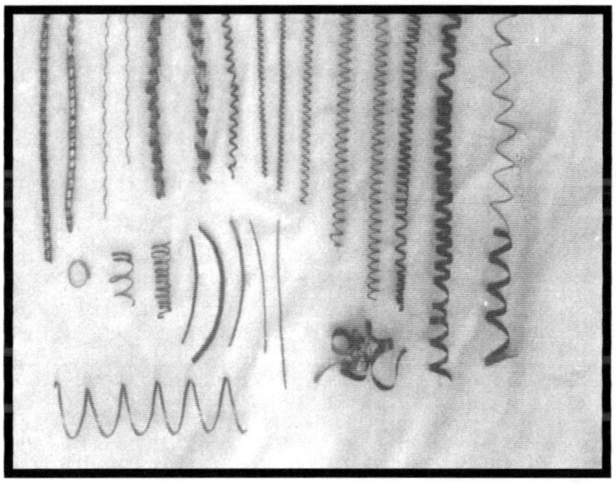

Figure 2. Different shapes of steel fibrous

3. Geometrical characteristics of steel fibrous waste

To study the concrete properties in the fresh and in the hardened state, steel fibrous waste was used in cementitious materials. A preliminary study was made to have a general idea on the proportions of the existing forms in a sample of fibrous waste, taken randomly. This statistical study aims to show the proportion of every existing shape. The table below gives the results and show that 38.45 % of fibrous waste is curved in shape.

IOP Conf. Series: Materials Science and Engineering **246** (2017) 012034 doi:10.1088/1757-899X/246/1/012034

4. Materials and experimental procedure

4.1. Materials

4.1.1. Cement. The cement used was type CPJ 45 (Cement Portland with Additive). This cement was chosen because of its wide availability and largely used in the concrete construction sector in Algeria. Its density is 3.1, and its Blaine's specific surface area is 3600 cm2/g. Their chemical and mineralogical compositions are given in table 2 and 2 respectively.

4.1.2. Dune Sand (Fine Aggregate). The sand's equivalent measured by the NF P18 standard [7] shows that the dune sand used in this experimental study is clean, siliceous and contains very few fine dust or clay elements. The specific gravity and fineness modulus calculated was respectively 2.64 and 1.73.

4.1.3. Crushed Gravel (Coarse Aggregates). Crushed gravel is obtained by crushing limestone rock from the quarry; two fractions 3/8 mm and 8/15 mm of coarse aggregates have been used in this experimental study. The specific gravity was 2.54.

4.1.4. Fibrous waste (Fibres). Different types of waste fibers from different industries, namely, lathe industry , wire winding , wire drawing industries , steel transformation industrial unit were collected and used in this investigation after mechanical treatment (pressing , cutting to small fibrous pieces). They have approximately 1.5 mm in diameter, 20÷50 mm in length and with an average specific gravity of 7.5, while the polypropylene fibrous waste were collected from textile industries, with cut lengths varying from 2cm to 5 cm and a specific gravity of 0.90-0.91.

Table 1. Average proportion of every existing shape

Shape	Straight	Plate-shaped 	Drawn 	Corrugated shape 	Curved-shape 	Arbitrary shape
Proportion %	5.97	16.40	2.69	27.75	38.45	8.74

Table 2. Chemical composition (%, by weight) of CPJ

Constituent, %	SiO2	Al2O3	Fe2O3	Ca O	Mg O	Ca O free	SO3	P.F	Insoluble
CPJ 45	19.48	5.06	3.72	61.95	0.85	1.63	1.1	3.44	1.26

Table 3. Mineralogical composition (%) for the cement used (according to BOGUE potential)

Compounds,	C_4AF	CSH_2	C_3A	C_3S	BC_2S	C (Free)
CPJ 45	11.31	2.37	7.12	55.03	14.64	1.63

4.1.5. Admixture. The admixture product used in our research (less than 1%) is a super-plasticizer manufactured by the Algerian Granitex Company. This super-plasticizer, named "Medafluid SFA", has a density of 1.2 ; a content of dry matter of 36.16 % and a PH of the order of 7.

IOP Conf. Series: Materials Science and Engineering **246** (2017) 012034 doi:10.1088/1757-899X/246/1/012034

5. SPECIMEN PREPARATION AND CONDITIONNING

5.1. Mixing, Casting and Curing
Three concrete series of trials were carried out to investigate the suitability of fibrous waste as a reinforcing material for normal concrete. The first one concerns the concrete control in which the various parameters were illustrated in the table 4.

Twenty four test specimens of control concrete were made to choose a composition which gives the best results in order to estimate the benefits brought by the fibrous waste on the compressive strength.

Table 4. Different parameters of control concrete

W/C	S/G	Slump(cm)	Air content (%)
0.5	0.51	6	1.75
0.5	0.49	6	1.6
0.57	0.70	6	1.78
0.5	0.76	6	1.65

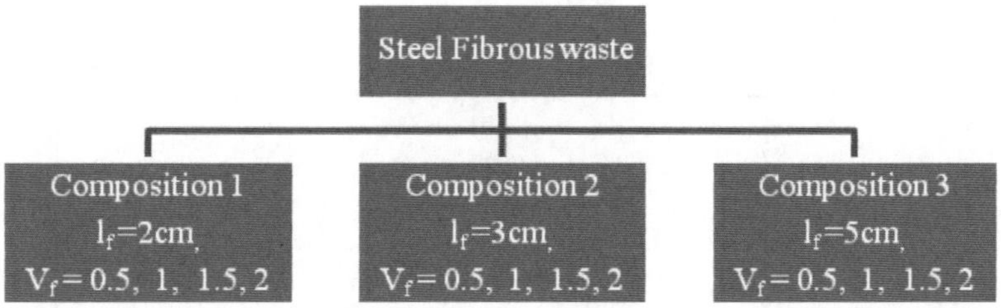

Figure 3. Flowcharts of the

The second series consists in estimating the addition of steel fibrous waste in the concrete made from local materials according to the experimental programme shown in Fig.3, while the third series concerns the addition of commercialized fibres marketed in this concrete. Twenty four test specimens were also made for every composition

A total of twenty four plains for control and seventy two steel fibrous waste reinforced mixes were prepared. Materials were mixed in a linear-cum-flow mixer type 'O' which had a capacity of 0.043 cubic metres and a power driven rotating pan and paddle. This mixer was sturdy enough for fibrous mixes, and a good uniformity of fibrous waste distribution could be achieved. The mixer was first loaded with the coarse aggregate and a portion of the mixing water; after starting the mixer; sand, cement and the rest of water were added and mixed for 5 min. The fibres, in the case of fibrous mixes, were added following the addition of all mix ingredients.

Six 16×32 mm cylinders were cast from each mix. They were cast in cylindrical steel moulds and compacted on a vibrating table. Specimens were placed under plastic sheet membrane for 24 hours in an ambient temperature. Following removal from the moulds, they were individually sealed in plastic bags and stored at room maintained at 24°C and 65% relative humidity until testing at an age of 28 days for compressive.

5.2. Compressive tests
Compressive strength test was carried out according to NF.P.18-406 [8], in a testing machine of 3000KN at a loading rate of 0.5 KN/S. The different values of the compressive strength of concrete at 28 days are depicted in Figures 4a, 4b and 4c. The results show that addition of steel fibrous volume waste in the ratios of 0.5, 1, 1.5, 2 by volume in the concrete will increase the compressive strength at

different percentages. It is seen that the use of steel fibrous waste significantly increased the compressive strength of the normal concrete made of local materials by using an admixture.

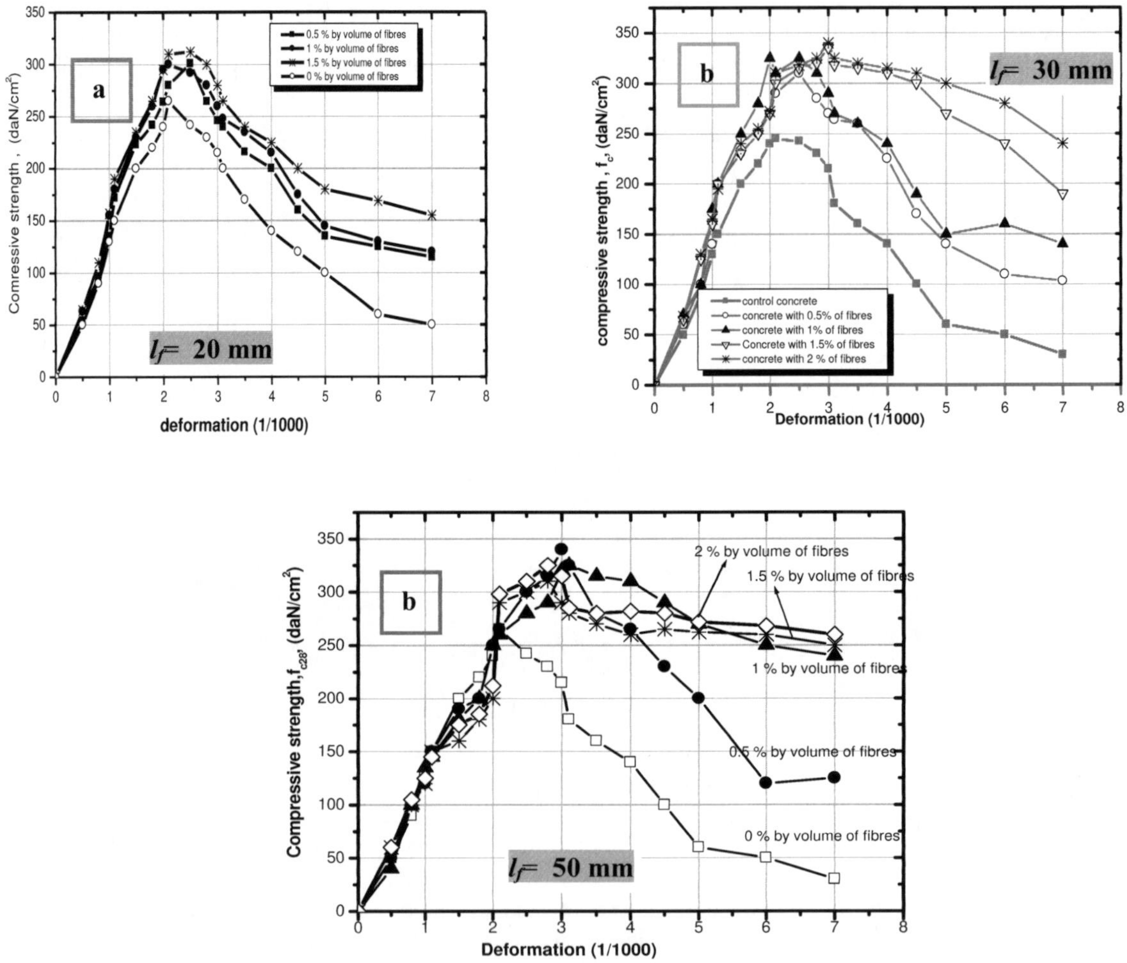

Figure 4. Relationship between compressive strength and deformation of steel fibrous waste reinforced concrete

5.2.1. Effect of steel fibrous waste content on compressive strength. The curves obtained for long steel fibrous waste (l_f = 50 mm) and for short steel fibrous waste (l_f = 20 mm, l_f = 30 mm) are shown Figures 4a,4b and 4c, which three distinct phases are appeared 5.

5.2.2. The first phase corresponds to an elastic behaviour of fibre concrete. In this part, the addition of the fibres does not bring any significant modification either in terms of the shape of the curve or in terms of modulus of elasticity. In other words, the modulus of elasticity is not modified and this irrespective of the quantity of fibres introduced into the concrete.

The explanation given for this is that the fibres are not added to improve the concrete resistance; but to affect the behaviour of the material once the matrix is cracked and thus control the cracking of steel fibrous waste concrete. This is possible due the sewing effect created by the fibres through the

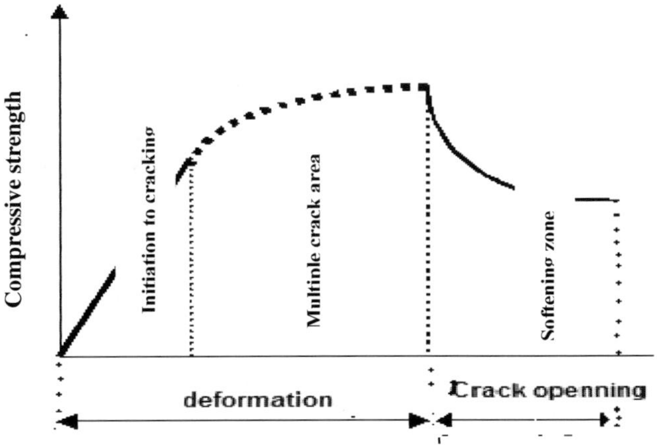

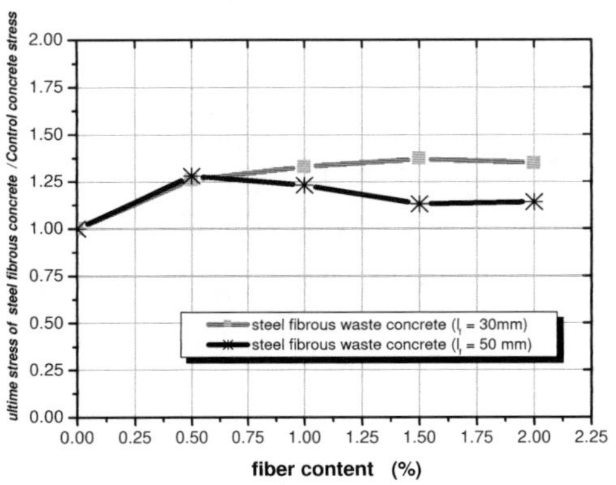

Figure 6. Rate of increase in% of the ultimate stress for l_f=50mm, l_f = 30mm

cracks. This effect improves the energy absorption capacity of the material. During this relatively short phase, there is a rapid progression of the deformation due to propagation of the pre-existing defects in the matrix.

In the second phase which occurs immediately after the cracking of the matrix, the fibres contribute not only to the improvement of the ductility of the concrete material but also to a significant increase in the strength of the composite. This increase in stress is associated with the appearance of multiple cracking through the specimen, which requires a greater energy input to open the micro- cracks. Once the rupture is reached, the third phase begins with a rapid propagation of the cracks, the stress decreases until the total break of the specimen.

It is noted that, in the second phase, a marked improvement in the ductility of the concrete is observed for the different quantities of fibres introduced. Figure 6 shows that with 0.5% fibre volume, a 25% increase in the ultimate fibre concrete stress is reached for both lengths (l_f = 30 mm, l_f = 50 mm), 0.5% volume, the continuous increase for both fibres in a different way.

For fibres of length, l_f = 30 mm, the increase is 32.6% with the fibre addition of 1% by volume, 36.7% with the addition of 1.5% volume of waste fibres. This increase remains within about 36% with the addition of 2% volume of waste fibre. With long fibres (l_f = 50mm), this increase is 28% with the addition of 0.5% fibre in volume, 22.6% with the addition of 1% fiber volume. These percentages of the increase are decreased towards 13.2% with the addition of 1.5%.

The addition of a percentage greater than 2.5% of fibre volume creates a problem of maneuverability of waste fibre concrete and implicitly a problem of the implementation. For this reason, every precaution must be taken when adding quantities exceeding 2.5% fibre volume with $l_f \geq$ 50mm. It will be noted that the rates of increase of the ultimate compressive strength of the waste fibre concrete are shown in Figs 7a, 7b and 7c for l_f = 2cm, 3cm and 5cm respectively.

It can be said that the presence of fibres derived from the fibrous waste brings a gain to the rupture strength of concrete reinforced with waste fibres. This gain, which depends on the concentration and the geometry of the fibres, reaches a maximum value of 25% for a length of 20 mm with a concentration of 2% by volume, 37% for fibres of 30 mm length with a concentration of 2% volume and a value of 28.8% for fibres of length 50 mm with a concentration of 0.5% of volume. This means that the longest fibres have the lowest efficiency. In other words, with a length of 50 mm and a concentration of 2% of volume, the gain recorded does not exceed 13.25%.

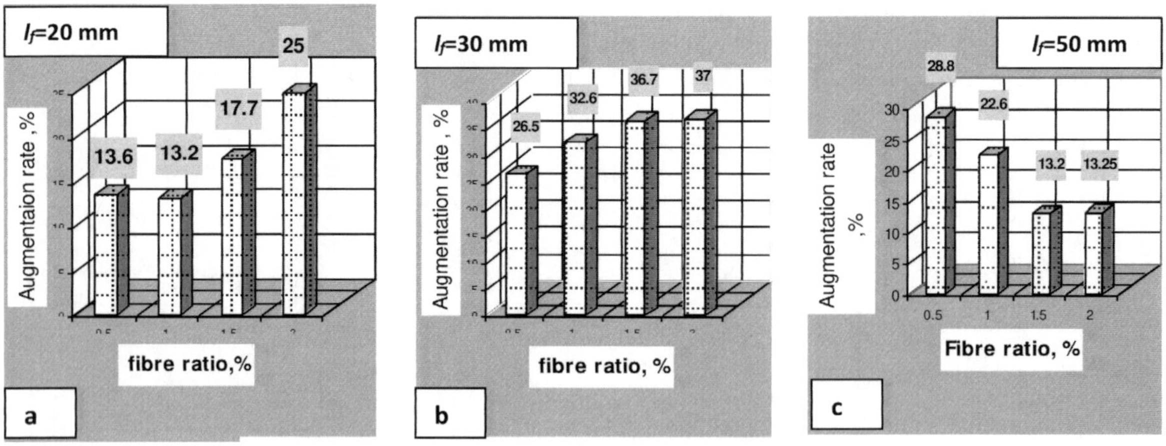

Figure 7. Augmentation rate in compressive strength

5.2.3. Influence of fibrous length on the compressive strength Influence of fibre length. By comparing the results obtained with the fibre length of 20 mm, 30 mm and 50 mm, it can be seen that the parameter (fibre length) is of great influence on the behaviour of the concrete subjected to compression.

Curve 8 shows the tests prepared with waste fibres of length 20, 30 and 50 mm with a volume concentration of 0.5, 1 and 1.5%. It shows that the length of fibre affects the behaviour of concrete in compression. The breaking strength increases with the length of the fibres.

Table 5 presents the results obtained with the three lengths used in our work. It will be noted here that the deformation corresponding to the maximum resistance increases by 19% for l_f = 30mm, more than 42% for fibres of length l_f = 50mm whereas for l_f = 20mm no increase was observed.

To explain this, it must be said that the addition of fibres, in general, and waste fibres in particular modifies the behaviour of concrete in the fresh state (decreased of workability, increase in air content, etc.) and a modification of the microstructure of the concrete in the cured state. Thus, the addition of a high percentage of fibre will increase the stiffness of the solid phase. This will lead to contacts and clutter reactions which increase the workability time, which translates into a slowing down of the flow of the fresh concrete. In this case, there are two cases: either sufficient energy is used to be able to distribute all the fibres in the matrix with an excess of vibration or in this case an increase in the ultimate compressive strength will be obtained. Either the same operating conditions as the control concrete are used, or in this case a slight decrease in the ultimate compressive strength will be observed, as several researchers have reported [9,10].

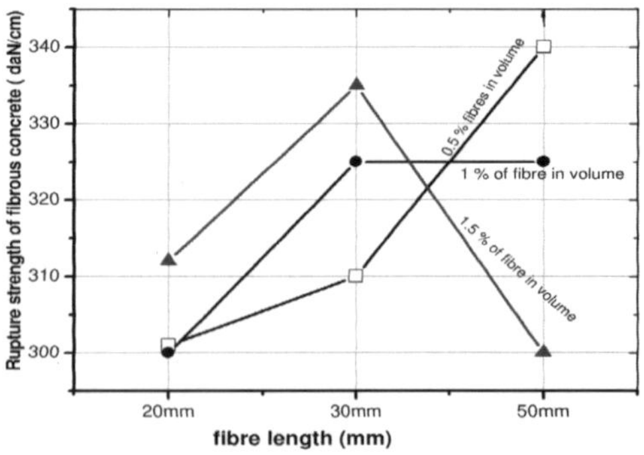

Figure 8. Effet de la longueur de fibres sur la résistance de rupture de béton de fibres

Table 5. Some results of steel fibrous waste concrete

	Fiber length								
	l_f = 20 mm			l_f = 30 mm			l_f =50 m		
Fibre content %	0.5	1	1.5	0.5	1	1.5	0.5	1	1.5
Slump (cm)	7	6	4	6	4	3	6	3	1.5
air content (%)	5.4	5.1	5	5.1	4.8	4.6	4.9	4.8	3.5
Rupture strength (daN/cm^2)	301	300	312	310	325	335	340	325	300
Corresponding deformation $\times 10^{-3}$	2.5	2.1	2.5	2.5	2.5	3	3	3.1	2.5

6. Shrinkage of mortar by using fibrous waste

6.1. Materials and procedure

The main aim of this part of the study is to evaluate the influence of the fibrous waste used (metallic and polypropylene) on the phenomenon of cracking caused by shrinkage in a mortar prepared from sand of dune (important shrinkage). The method described in this part to evaluate the restrained shrinkage consists in using a rigid ring made of steel that is coated of mortar. After hardening the sample is submitted to conditions wanted to evaluate the effect of restrained shrinkage .In a given moment, the resistance to the cracking, the number of cracks and the total maximal widths, are measured and compared to what obtained with a plain mortar.

The free shrinkage tests were carried out on prismatic test specimens of dimension (4×4×16 cm) whose Plexiglas plates were fixed. To improve the contact mortar-plates, the latter have been perforated (in the middle) and studs were fixed. On the first stud (the upper end) will come to butt a comparator, the second (lower end) will be arranged on other plate glued to a motionless frame and ensuring a free rotation of the test specimen submitted for testing (Figure 9). The precision of our measurement comparator is 0.01 mm.

The restrained shrinkage has been evaluated by the test with the ring which consists in using a steel ring (Figure 10) of thickness 8mm, external diameter143mms, internal diameter 127 mm and height of 44 mm, while cementing material ring has an external diameter of 127mm, internal diameter of 83 mm; and 44 mm height and 22 mm thickness (Figure 10).

The casting in two layers in this circular mould is carried using a steel shaft. Three rings were made for each test. The specimens prepared were cured at 20°C during 24 ±1hour in a wet condition. The apparatus allows a release from the mould and an easy cure of the test specimens.

Figure 9. Free shrinkage measuring apparatus

Figure 10. Mortar ring under (black) and after test (white)

IOP Conf. Series: Materials Science and Engineering **246** (2017) 012034 doi:10.1088/1757-899X/246/1/012034

Those specimens submitted for testing with the ring were cured during 24 hours, and then they were placed in an environment similar to that of free shrinkage until the formation of cracks. For better viewing of the cracks and the microscopic cracks propagation, the test specimen submitted for testing with the ring were covered by a white painting.

Cementitious mortar was employed for these tests using superplasticizer admixture to maintain a constant workability, without varying the quantity of mixing water among the various mixes. Materials used in the tests to evaluate the free and restrained shrinkage are : Sand of dune grade 0/5 ; Cement CPJ45; Mortar of fraction 1/3 (C:S = 1:3)E : C = 0.58 (to have a constant and normal workability), Fibres : (fibrous waste of metal and polypropylene), Superplasticizer : 1 %

6.2. Results and discussions

6.2.1. Free Shrinkage.

The average curves of the shrinkage according to the time (days) are presented on figures 11a and 11b. It is noticed, according to the figures 11a and 11b, that the steel fibrous waste decreases the shrinkage. This reduction is about 16 % compared to a mortar without fibres at 28 days, for a proportion of 0.5 % of volume of fibres. But, with 1% of volume of fibres, this reduction is ranged from 11 to 13 %. Regarding the experimental tests results, we have noted the maximum percentage of fibres to reduce the effect of shrinkage is 1.5 %, beyond this percentage of volume of fibres; the presence of fibres in a mortar is not significant.

These results confirm those already obtained from former studies [11] by using fibres marked under various types, to know that deformations of shrinkage are generally smaller for the fiber reinforced mortars compared to those of the mortar without fibres. This reduction is affected by several parameters like the duration of drying, the shape and the quantity of fibres. The loss of weight of the test specimens was checked. The results are illustrated in figure 11c. It is noted that the loss in weight is more significant for the fibre mortars than for the mortar without fibres. The presence of dispersed fibres increases the size of pores in material as well as volume and decreasing, thus, intensity of the capillary pressures by facilitating the evaporation of water. On the other hand, according to the experimental tests, the use of fibrous polypropylene waste, with percentage higher than 1% in volume of fibres, do not give any improvement to the free shrinkage of mortar. This result confirms those obtained by other researchers [12, 13, 14]. It is noted, according to figure8, that there is an increase in the shrinkage of approximately 20% for a content of 1 % in volume of fibres. But this increase reaches the 30 % for a content of 1.5% in volume of fibres. The increase in the free shrinkage of mortar reinforced with polypropylene fibres may be explained by the fact that these fibres are thin, very flexible and generally coated by layers of fines which causes the increase in free shrinkage by increasing the percentage of fibres.

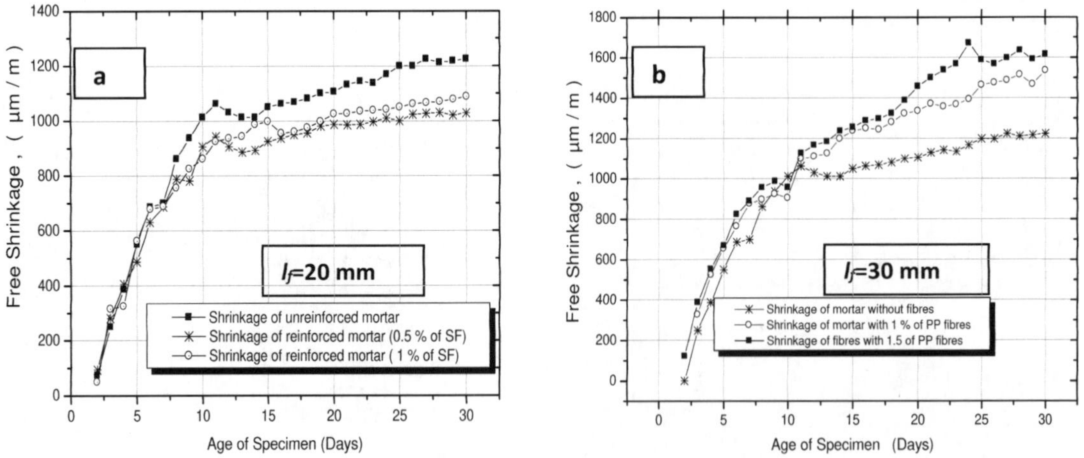

Figure 11a, b. Free shrinkage of mortar without and with fibrous waste

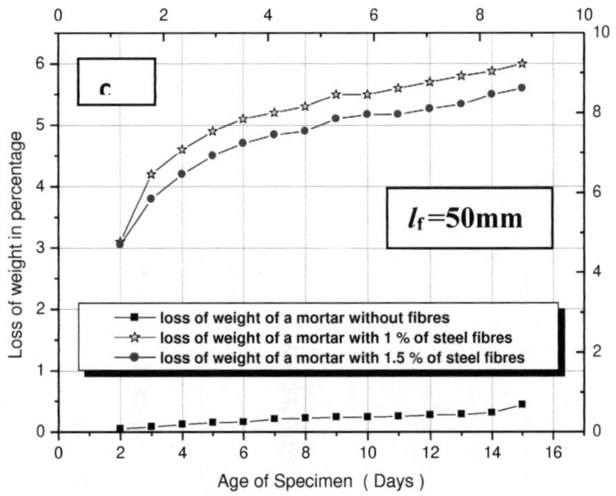

Figure 11c. % loss weight of mortar without and with fibrous

6.2.2. Restrained shrinkage

The second part of the experiment programme was oriented towards the study of the effect caused by the introduction of fibre content on the cracking tendency of the mortar. The mixes of mortar were prepared with constant workability by using a superplasticizer in order to have always the same quantity of mixing water, and therefore the same water/cement ratio.

Cracking resistance of mortar was evaluated by the ring test. It was evaluated by the number of days and the thickness of the crack. Figures 12 and 13 give an idea on the effect of the fibre addition in a cementing matrix like that of a mortar. In all the cases we observed a formation of macro-cracks with thickness varies between 0.6 to 0.8 mm. Also, two deep macro-cracks on the fourth day have been remarked as seen in Figure 10. But with 0.5 % of steel fibres, microscopic cracks have been observed (seen with a magnifying glass) after 9 days. Whereas with a content of 1 % of steel fibres the appearance of the microscopic cracks was not observed at early age less more than two weeks. By adding 1.5 % of steel fibres many microscopic cracks have been appeared after more than 4 weeks. The appeared cracks are very thick (less than 0.1mm) and less deeper.

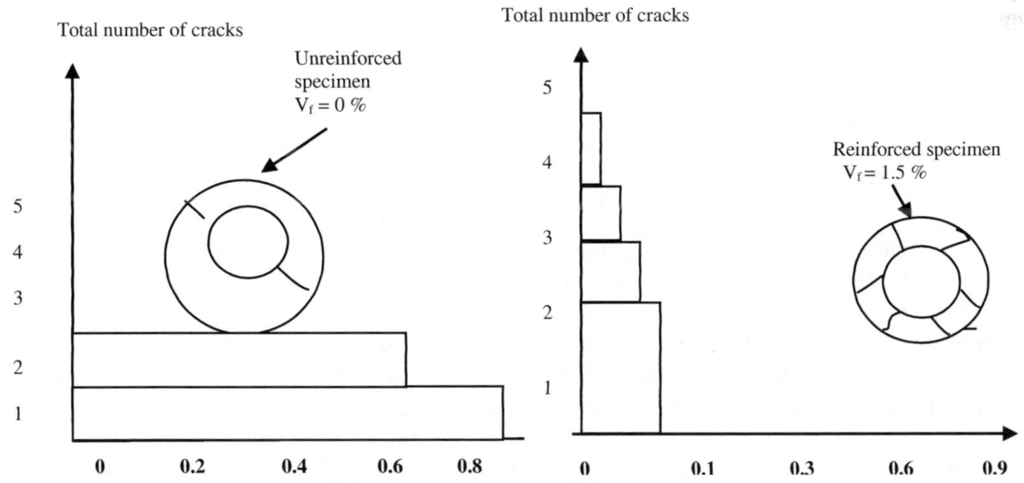

Figure 12. Results of the tests showing the number and the thickness of the cracks of a mortar without (left) and with (right) fibrous waste.

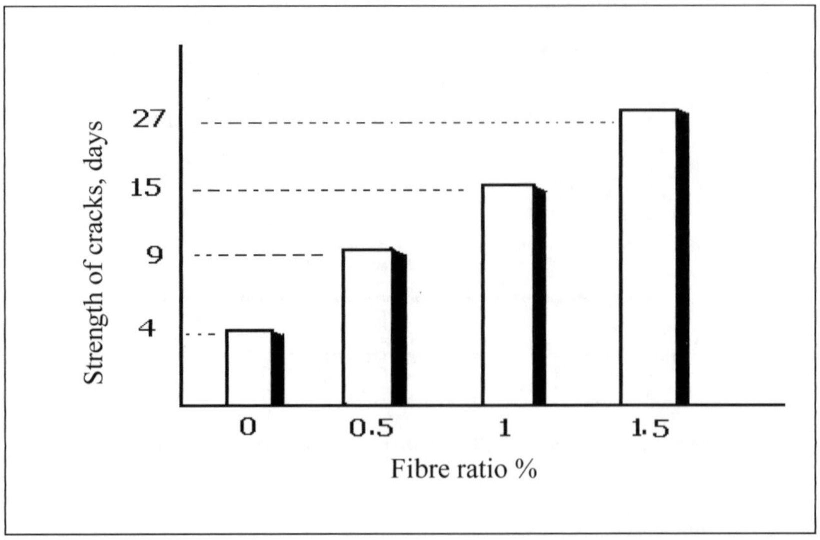

Figure 13. Typical example from the tests showing the strength of cracks,

With the same mixture of mortar, specimens were prepared by adding polypropylene fibres. It is announced here that the tests of restrained shrinkage (ring tests) were carried out by using percentages of fibres of 0.5 and 1 %. The observation from the test series confirm that the microscopic cracks in this kind of shrinkage did not appear before 4 weeks. This fact confirm also the results of a recent research[11] which shows that the introduction of a fibre content equal or higher than 1 % make in evidence that cracking processes is completely illuminated. The results obtained are illustrated in the table 6.

Table 6. Effect of polypropylene fibrous waste on restrained shrinkage with constant workability

N° of mix	Cement	sand	Water	Admixture %	Fibres	Cracking strength (days)
M01EP	1	2	0.5	1	0 %	Micro-cracks on 4 days
M02EP	1	2	0.5	1.25	0.5%	No cracks after 35 days
M03EP	1	2	0.5	1.75	1 %	No cracks after 56 days

7. Conclusion

Steel fibres are commonly used in cement-based materials to improve their mechanical properties. But; little information is available on the use of fibrous waste to reinforced cementitious materials.

The main conclusions drawn from this experimental work are:

- It is possible to use steel and polypropylene fibrous waste as a reinforcing element in the concrete to improvie some mechanical properties on the one hand , and of developping fibrous waste on the other hand which is, in the majority of the cases , throuwn in the urban zones leading to an environmental and human damage.
- Compressive and shrinkage tests were performed on the concrete and mortar specimens with and without those fibrous waste inclusions and the testing results were reported and compared. From this study, it was found that the compressive strength and both free and restrained shrinkage were improved with fibrous waste addition. This means that the benefit brought by this kind of fibrous waste was not only to remove this kind of waste but also to have great potentials to become a cost-effective resource and provide performance enhancement for concrete material in building industry where the use of these fibrous wastes is possible and can brought enough strength to the mechanical properties.
- In addition to the improvement in compressive strength and shrinkage phenomena ; graeater deformation ability was obsreved for consrete and mortar reinforecd by these fibrous waste.

- Unlike many researches in the field of fiber reinforced concrete, which resulted in a compressive strength decrease, this experimental research carried out by using steel and polypropylene fibrous waste leads to an improvement in both compressive strength and shrinkage phenomena . However, this conclusion is based on a limited number of test specimens; further research is being done with more testing specimens and dosage rates to confirm these benefits.

References

[1] Beddar, M, 2002, Steel fibre reinforced concrete: Past, Present and Future . *Technical report,* National Project, pp 9

[2] Fattuhi, N, I, April 1974 Properties of steel fibre reinforced cemen- based matrices Thesis submitted to the university of shellield for degree of *M.Eng* pp 357.

[3] Alghamdhi , M, 1984 , Effect of matrix composition on SFRC properties, *PhD thesis* submitted to the university of Michigan, U.S.A, pp 276.

[4] Hughes, B.P et al, sept 76, The workability of steel reinforced concrete *Magazine of concrete* Vol **28**, N° 96 , pp 157-61 .

[5] Brandt, A.M, 1995, Cement –based Composites: materials, Mechanical properties and performance. E & FN Spon, London, pp 470.

[6] Beddar, M. April 2004 Fibre reinforced Concrete : Past, Present and Futur *Concrete magazine,* Vol **38**, N° 4, U.K

[7] AFNOR N.F.P. 18-301, dec 1982, Granulats naturels pour le béton hydraulique,

[8] AFNOR N.F.P. 18-406, dec 1981, Essai de compression,

[9] Dardare, J. Contribution à l'étude du comportement mécanique des bétons renforcés avec des fibres de polypropylène *RILEM Neville (A),* 120, communication 5.2, pp.227-235.

[10] Gopalaratnan**,** V.S.; Shah, S.P**,** May-June 1985 , Softening Response of plain concrete in Direct Tension. *ACI Journal*, Vol. **82**, N° 3, pp 310-323

[11] Mesbah, H.A; Buyle, B. and Siwak, J.M, 1996 , Influence de l'ajout des fibres sur le retrait des mortiers a base de granulats recycles. 2e *Colloque International Francophone sur les bétonsbrenforcés de fibres métalliques*, Toulouse, pp163-172.

[12] Edington, J.; Hannant, D. J; Williams.R.I.T. 1974. Steel Fibre Reinforced Concrete; G.B.Building . *Res.EST-Current paper*, c.p.69/74

[13] Elhachem , M. , 1990 . Etude de l'influence de la fibre de fonte sur le retrait et le fluage du béton.*Thèse de Doctorat*, Ecole Centrale, Paris.

[14] Houari , H. , 1993 Contribution à l'étude du comportement du béton renforcé de fibres soumis à l'action des charges maintenues et cycliques. *Thèse de Doctorat*, INSA Lyon,

IOP Conf. Series: Materials Science and Engineering **246** (2017) 012035 doi:10.1088/1757-899X/246/1/012035

Probabilistic Analysis of Structural Member from Recycled Aggregate Concrete

I Broukalová[1] and K Šeps[1]

[1] Department of Concrete and Masonry Structures, Faculty of Civil Engineering, Czech Technical University in Prague

E-mail: iva.broukalova@fsv.cvut.cz

Abstract. The paper aims at the topic of sustainable building concerning recycling of waste rubble concrete from demolition. Considering demands of maximising recycled aggregate use and minimising of cement consumption, composite from recycled concrete aggregate was proposed. The objective of the presented investigations was to verify feasibility of the recycled aggregate cement based fibre reinforced composite in a structural member. Reliability of wall from recycled aggregate fibre reinforced composite was assessed in a probabilistic analysis of a load-bearing capacity of the wall. The applicability of recycled aggregate fibre reinforced concrete in structural applications was demonstrated. The outcomes refer to issue of high scatter of material parameters of recycled aggregate concretes.

1. Introduction

Growing demand for cutting carbon emissions, reduction of pollution emitted from building industry, saving natural resources and raw materials should target to reduction of construction waste disposal in landfills, re-use of construction materials, utilization of non-traditional materials or non-traditional use of common materials.

Recycled aggregate accounts for ca 8% of aggregate use in the European production (European Aggregates Association, 2013). Our efforts aim to increase this ratio and consume a significant quantity of crushed concrete from construction and demolition waste (CDW) in production of recycled aggregate concrete (RAC) and minimizing of natural aggregate depletion.

Use of RCA (recycled concrete aggregate) in concrete is restricted by code prescriptions on aggregate testing, quality of aggregate and limits of maximal amount of recycled aggregate which can be used as a substitution of natural aggregate. The restrictions are caused by the reason of worse quality and high dispersion of recycled aggregate characteristics.

The purpose and target of our aspirations was to change the understanding in attempts to apply RCA in concrete. In our opinion, it should not be regarded that RCA is used to make concrete with downgraded properties compared to common concrete. Under a new perspective RCA is used to make a new composite, which is relevant in certain applications.

The keynotes of the new composite design are maximizing of recycled aggregate consumption, minimizing of cement consumption (low cement dose), simplification of the composite manufacturing by minimizing the demanded pre-treatment of recycled aggregate and minimizing required accompanying tests. These provisions are supposed to provide competitive advantage of the composite made from recycled concrete.

IOP Conf. Series: Materials Science and Engineering **246** (2017) 012035 doi:10.1088/1757-899X/246/1/012035

As a result of this approach in the concrete mix designing, a low-strength porous composite is manufactured. Efficiency of the composite for application in a load-bearing member was studied in the testing program and probability analysis.

2. Mixture composition

The governing intention of the mixture proposal is maximising of the recycled aggregate consumption. Basis of the mixture design was the grain size distribution curve of the delivered recycled aggregate. Natural aggregate was added to replenish missing fractions of recycled aggregate so that the ideal grain size distribution was approximated.

2.1. Aggregate. The dose of 1 310 kg of recycled concrete aggregate was combined with 200 kg of greywacke and 150 kg of sand per cubic meter.

2.2. Cement binder. Dosage of cement relates to economic and ecology aspects of the research project and it was determined with respect to the European Standard EC 206-1 requirements for minimal cement content.

Type I hydraulic Portland cement was used, with 95 % of Portland cement clinker (chemical composition is in the table 3). Specific surface area of cement binder was 381 m^2/kg.

Table 1. Main chemicals in the cement binder

Component	**Content** [%]
CaO	64.2
SiO_2	19.5
Fe_2O_3	3.2
MgO	1.3
SO_3	3.2

2.3. Fibres in the mixture strengthen the structure of the composite and provide tensile strength and toughness. Activation of fibres after macrocracking increases the ductility of the composite.

In the choice of materials for fibres, high porosity and permeability of hardened concrete should be considered. To prevent deterioration of fibres due to penetrating moisture, synthetic fibres should be selected; here 4.55 kg of polypropylene macrofibres with strength ca 600 MPa were used.

2.4. Water. The water absorption capacity of aggregate influences the properties of the fresh concrete such as its workability.

The dosage of mixing water depends on water absorption of recycled aggregate and aggregate moisture state. The mixing water was dosed to reach the mixture consistency convenient for compaction by ramming or tamping (vibration cannot be used because of the high porosity of composite).

Table 2. Fibre reinforced composite from recycled concrete aggregate mixture design

Component	**Dosage** [kg/m^3]
Portland cement 42,5 R	310
water	150
recycled concrete aggregate	1 310
natural aggregate 8/16	200
natural aggregate 0/4	150
polypropylene macro-fibres	4.55

IOP Conf. Series: Materials Science and Engineering **246** (2017) 012035 doi:10.1088/1757-899X/246/1/012035

3. Testing program

Material parameters were determined in common tests used for concretes. Cubes 150/150/150 mm were used for determination of compressive strength (table 3). Tensile strength was specified in a three-point bending test of prisms 100/100/400 mm (table 4).

Table 3. Compressive strength of the composite

Specimen	Weight	Density	Compression	
			Load	Strength
	g	kg.m^{-3}	kN	**MPa**
1	6 985.0	2 052	843.33	**37.2**
2	6 420.0	1 876	427.85	**19.2**
3	6 730.0	2 050	784.13	**35.7**
4	6 465.0	1 890	369.62	**16.6**
5	6 905.0	2 017	815.80	**35.9**
6	6 780.0	2 003	611.53	**27.2**
Average value		1 981		**28.6**
Standard deviation		79		**9.1**

Table 4. Tensile strength of the composite

Specimen	Weight	Density	Tensile	
			Load	Strength
	g	kg.m^{-3}	kN	**MPa**
1	8 055.0	1 986	9.09	**2.69**
2	7 865.0	1 978	9.06	**2.74**
3	7 475.0	1 934	7.20	**2.30**
4	7 780.0	1 921	8.16	**2.40**
Average value				**2.53**
Standard deviation				**0.22**

4. Analysis of the structural member

One of the targets of the investigations was a feasibility study of the proposed composite in a structural element.

For application of any material in load-bearing structural elements a reliable design is an essential condition. Reliability is defined as ability of the structure to withstand loads with corresponding reliability level and fulfil given requirements during the whole lifetime of the structure. The reliability level is assessed on a statistical basis, by defining the probability of collapse that is reputed acceptable as a function of the consequences of the collapse itself and the nominal lifetime of the construction.

Present designing and verification of reliability are based on limit state methodology that includes randomness of actions, material parameters, geometry of the structure and uncertainties in structural models. The stochastic nature of material parameters and actions on structures in common design approach is covered in partial safety factors. Entirely probabilistic approach to the design is advanced method of structure assessment.

4.1. Probabilistic analysis

The proposed fibre reinforced composite is a novel material; hence it is essential to prove its reliability for structural applications in thorough probability analysis. The probabilistic verification was performed for a wall made from recycled concrete composite of 30 cm thickness and 2.8 m height. The choice of the wall dimensions and a manner of loading and its eccentricity relates to referential masonry wall from solid bricks.

The probabilistic analysis was performed with software system SARA – FREET – ATENA. FREET (Feasible Reliability Engineering Tool) is multipurpose probabilistic software tool for statistical, sensitivity and reliability analyses. ATENA is a program developed especially for non-linear analyses of concrete and reinforced concrete structures. SARA (Structural Analysis and Reliability Assessment) is computing environment providing communication between FREET and ATENA. Randomness and uncertainties are described as random variables by probability functions. Random input parameters are generated by LHS method (Latin hypercube sampling). The stochastic analysis refers to deterministic simulation of the composite wall behaviour assembled in ATENA program, which modelled behaviour of a wall loaded vertically with eccentricity 15 mm until failure. The simulations in ATENA stemmed from material model "Nonlinear Cementitious 2" implemented in the software tool and material parameters measured in laboratory tests.

In the probabilistic analysis the material characteristics of the composite were chosen as random variables. The mean values, variance, standard deviation and variation coefficient were derived from scatter of laboratory tests results. Table 5 shows material parameters assumed as random variables in the probabilistic analysis.

Table 5. Random material parameters

Material parameter	Distribution	Mean value	Standard deviation
Elastic modulus	normal (Gauss)	25 GPa	8.25
tensile strength	normal (Gauss)	2.5 MPa	0.22
compressive strength	normal (Gauss)	28.5 MPa	9.24
fracture energy	normal (Gauss)	3.10^{-5} MN/m	1.10^{-5}

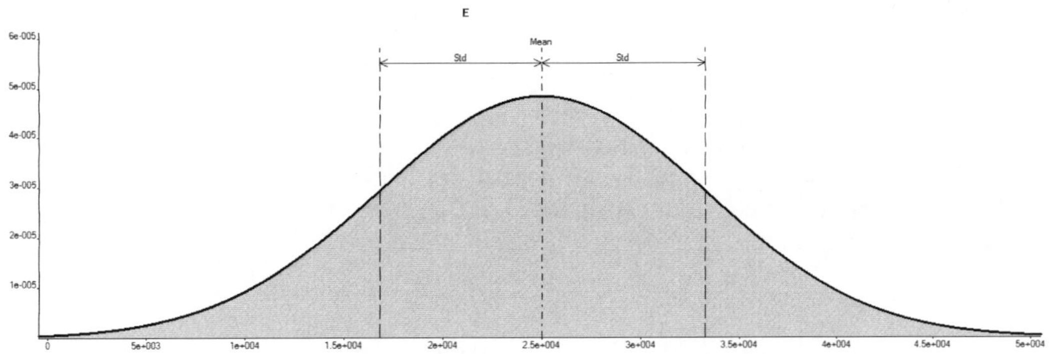

Figure 1: Distribution of random variable (elastic modulus)

Twenty simulation tasks were analysed with random material parameters. The scatter of resistance calculated in particular tasks is rather high (figure 2).

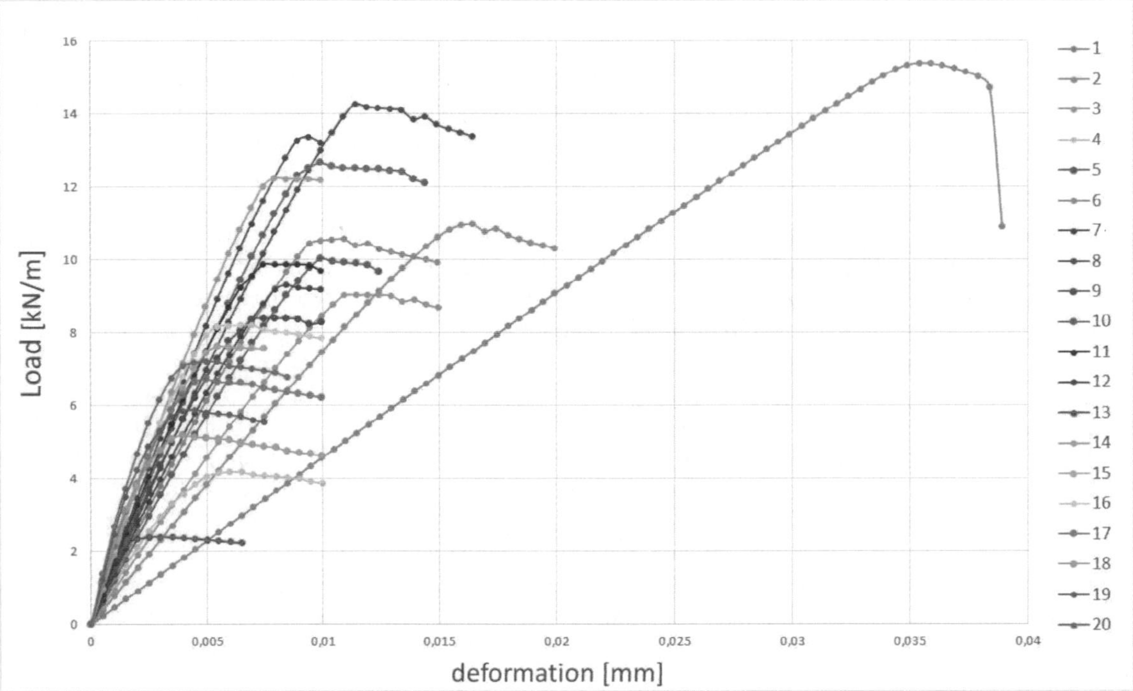

Figure 2. Bundle of load – deformation diagrams from probabilistic analysis

Summary of random resistance of the wall loaded with vertical load applied with eccentricity equal to 15 mm assessed in probabilistic calculation is shown in table 6.

Table 6 Statistical resistance of recycled aggregate composite wall

Mean value of the maximum load	9 105 kN/m
Variation coefficient	0.23
standard deviation	2 110
distribution function	normal

Based on the results of analysed simulation tasks statistical evaluation of resistance was performed. The action applied on wall was assumed with normal distribution and variance 0.1. The results in terms of reliability index are plotted in the figure 3. The diagram shows decreasing reliability index related to increasing value of load in kN/m plotted on horizontal axis.

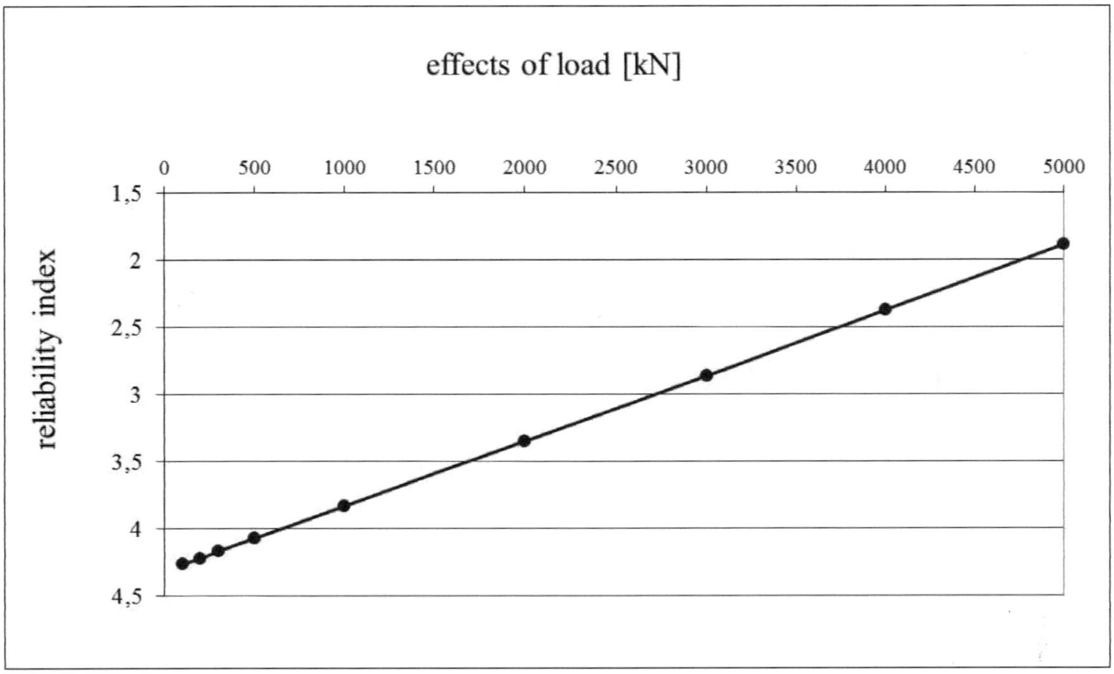

Figure 3: Reliability index of the recycled aggregate composite wall determined in probabilistic analysis

As the wall is described with structural resistance function with its mean value and standard deviation and the load effect function can be described in a similar manner, the structure can be designed to have a prescribed reliability index. A target value of the reliability index is specified in Eurocode 1990, annex B according to reliability classes table 7.

Table 7 Recommended minimum values for reliability index β at ultimate limit state

Reliability Class	**Minimum values for β**	
	1 year reference period	50 years reference period
RC3	5.2	4.3
RC2	4.7	3.8
RC1	4.2	3.3

The reliability class level relates to dividing of structures to consequence classes according to type, function of the building and damage, injury impacts or loss of lives caused by failure of the structure. The structural member investigated in the feasibility study is intended for use in simple buildings as family houses and small office or agricultural buildings. Thus it was classified RC2 according to the table 8 given in EC 1990.

Table 8 Consequences classes in Eurocode, EN 1990

Class	Description	Examples
CC1	Low consequence for loss of human life, and economic, social or environmental consequences are small or negligible	Agricultural buildings where people not normally enter (e.g. storage buildings), greenhouses
CC2	Medium consequence for loss of human life, economic, social or environmental consequences are considerable	Residential and office buildings, public buildings where failure consequences are medium (e.g. office building)
CC3	High consequence for loss of human life, or economic, social or environmental consequences are very great	Grandstands, public buildings where the consequences of failure are high (e.g. concert hall)

The target reliability index is reached for load 1 000 kN/m, what signifies quite satisfactory resistance for a loadbearing wall. The value of resistance is bigger compared to resistance of common masonry walls. (Design resistance of 30 cm thick masonry wall from solid bricks determined by Eurocode 1996 calculation procedure is ca 360 kN/m.)

Undesirable factor is a big divergence of the maximum load value and targeted resistance. For the investigated composite wall the ratio of the mean resistance and targeted satisfactory resistance is 9.

The high variance of resistance determined in the probabilistic analysis is caused by high scatter of material properties measured in laboratory tests. It is evident that the scatter of compressive strength relates to scatter of density; specimens with the lowest compressive strength have at the same time the lowest value of density (table 3). The small strength is caused by bad compaction of the composite; consequently technology discipline should be the principal focus in manufacturing of the fibre reinforced composite with recycled concrete aggregate for structural use.

5. Conclusions

The resistance, safety and the economic aspects are the most important factors in the decision-making process when the construction material for the planned structure is selected. However, new aspects of the designing process such as environmental issues and depletion of natural resources are becoming more and more important. Increased utilization of recycled concrete would contribute to saving of raw materials, elimination of the need for disposal of construction waste and low carbon production.

The analysis proved feasibility of recycled aggregate concrete in structural applications. On the other hand confirmed commonly known issue of the high scatter of material parameters of recycled aggregate concrete. It is a challenge for the future investigations to overcome the drawbacks, benefit from features of the fibre reinforced composite with recycled concrete aggregate and consider the material properties in the choice of appropriate application. Only approaches that will embrace challenges and overcome obstacles, can tribute to wider application of wastes and reduce the amount of C&D waste being diverted to landfill.

Acknowledgement
This outcome has been achieved with support of the Technology Agency of Czech Republic, project No. TH02030649 Environmentally Efficient Construction and Demolition Waste for Structures (EESDOK).

References

[1] Sagoe-Crentsil, K. K., Brown, T., & Taylor, A. H. Performance of concrete made with commercially produced coarse recycled concrete aggregate. Cement and Concrete Research 2001, 31, 701–712

[2] Ajdukiewicz, A., Kliszczewicz, A.: Influence of recycled aggregates on mechanical properties of HS/HPC, Cement and Concrete Composites, Volume 24, Issue 2, April 2002, Pages 269-279, ISSN 0958-9465

[3] ISO 2394, General principles on reliability for structures

[4] Broukalová, I. - Šeps, K. - Vodička, J. - Novák, J.: A Feasibility Study on the Structural Use of Fibre Reinforced Concrete with Recycled Aggregate. In Proceedings of the Fifteenth International Conference on Civil, Structural and Environmental Engineering Computing. Stirling: Civil-Comp Press Ltd, 2015, ISSN 1759-3433. ISBN 978-1-905088-63-8.

[5] Bílý, P.; Fládr, J.; Ryjáček, P.; Vodička, J.: Development of special fibre reinforced concrete for exposed concrete pavements on bridges, In: Fibre Concrete 2015 - Technology, Design, Application. Praha: České vysoké učení technické v Praze, Fakulta stavební, 2015. ISBN 978-80-01-05683-7.

[6] Novák, J.; Brejcha, V.; Kohoutková, A.; Fládr, J.; Broukalová, I.: Precast Fibre Reinforced Concrete Element of Retaining Wall, In: Fibre Concrete 2015 - Technology, Design, Application. Praha: České vysoké učení technické v Praze, Fakulta stavební, 2015. ISBN 978-80-01-05683-7.

[7] Bílý, P.; Fládr, J.: Verification of compressive strength on different sized hsc specimens, In: Fibre Concrete .2013 - Technology, Design, Application. Praha: České vysoké učení technické v Praze, Fakulta stavební, 2013, pp. 47-52. ISSN 2336-338X. ISBN 978-80-01-05240-2.

IOP Conf. Series: Materials Science and Engineering **246** (2017) 012036 doi:10.1088/1757-899X/246/1/012036

Studies for understanding effects of additions on the strength of cement concrete

R D Bucur[1], M Barbuta[2], P Konvalina[3], A A Serbanoiu[2] and J Bernas[3]

[1] University of Agricultural Sciences and Veterinary Medicine in Iasi, 3, Mihail Sadoveanu, 700490, Iasi, Romania
[2] "Gheorghe Asachi" Technical University of Iasi, Faculty of Civil Engineering and Services, 45 Mangeron Blvd., 700050, Iasi Romania
[3] University of South Bohemia in České Budějovice

E-mail: rbucur@uaiasi.ro

Abstract. The paper analyzes the effects of different types of additions introduced in concrete mix on the compressive strength. The current studies show that additions contribute to improve some characteristics of concrete and to reduce the cement dosage, so it can obtain concretes which are cheaper and friendlier with environment. In the experimental mixes were introduced: crushed natural aggregates, slag aggregates, silica fume, fly ash, chopped tire, polystyrene granule, glass fibers and metallic fibers. The experimental values of compressive strengths were compared for two concrete grades (C20/25 and C25/30). The study shown that near the well-known possibilities of improving mechanical strengths of cement concrete by increasing cement dosage and strength, by using crushed aggregates and by reducing the water/cement ratio, there are other methods in which less cement is used by replacing it with different wastes or by adding fiber.

1. Introduction

In construction industry one of the most used materials is concrete, which, in the last decades has improved its mechanical and durability properties due to its versatility of combining with different materials. The properties of cement concrete can be improved in different ways, which can contribute to preserving the natural resources and to the consumption of wastes that are polluting the environment. The researchers have replaced parts or totally the cement or even the aggregates with other materials (by-products, polymers, wastes, etc.) [1, 2, 3, 4, 5]. Also in the composition were added powders, fibers, nano-materials, etc., and so new eco-friendly concretes were developed. Some additions such as silica fume, fly ash, fibers, rice husk, etc. had contributed to obtaining high strength and high performance concretes [6, 7, 8, 9, 10, 11].

Concrete properties depend on a lot of factors, such as: components, admixtures, pouring technology, treatment after pouring, etc. [12, 13, 14, 15]. The influence of different components on the properties of cement concrete is studied for new concretes because each material has its own effect in the mix and each depends on a lot of factors. For the application domain of concrete with additions preliminary studies must be done for characterizing and recommending the specific use.

Generally, the constructions realized of concrete do not require all performing characteristics. A concrete with high strength can be realized by using active additions. The influence of additions wastes on concrete properties was analyzed in the paper for two grades of concrete with cement. It must mention that each result represents the best value obtained from an experimental study made separately for each type of concrete sample [16, 17, 18].

Experimental program

The samples from the experimental study were realized of three types of aggregates (river, crushed and slag), different types of cement and with/ without additions (fly ash, silica fume, chopped tire, polystyrene granule, glass fiber or metallic fibers). All additions are wastes which pollute the environment. An admixture type superplasticizer was used for all mixes.

1.1. Series 1- Concrete grade C20/25.

The samples for the first grade of concrete C20/25 were prepared with different cements type: Pa32.5, (concrete sample A1 and A2, with different dosages), P40, (concrete samples A3 and A4), and CEM-II-32.5 [19] (concrete samples A5, A6, A7, A8), Table 1. The aggregates were type: river (all samples except A5 which was with slag aggregate) and slag, in three sorts: 0-4 mm, 4-8 mm and 8-16 mm, Table 1. The additions types: silica fume (SUF), chopped tires (CT) and polystyrene granule (PG) were added in the compositions (A6 with SUF, A7 with CT and A8 with PG).

Table 1. Components dosage for series 1.

Sample	Dosages								W/C	Addition			f_c N mm^{-2}
	Cement kg m^{-3}	River aggregate kg m^{-3}			Steel slag aggregate kg m^{-3}			Water l m^{-3}		SUF %	CT %	PG %	
		0-4	4-8	8-16	0-4	4-8	8-16						
A1:C20/25 Pa35-L1	343	715	458	659	-	-	-	182	0.53	-	-	-	35.5
A2:C20/25 Pa35-L4	457	517	408	643	-	-	-	242	0.53	-	-	-	39
A3:C20/25 P40-L1	319	783	464	667	-	-	-	182	0.57	-	-	-	33
A4:C20/25 P40-L4	425	683	399	575	-	-	-	242	0.57	-	-	-	33.5
A5:C20/25 CEM II.32.5	350	-	-	-	495	330	1010	185	0.53	-	-	-	36.1
A6:CEM-II-32.5	360	803	384	559	-	-	-	172	0.48	10	-	-	41.2
A7:C20/25 CEM II-32.5	350	495	330	1180	-	-	-	166	0.47	-	25	-	27.67
A8:C20/25 CEM II-32.5	385	671	430	621	-	-	-	130	0.34	-	-	15	27.48

1.2. Series 2- Concrete grade C25/30

The samples for the series 2 grade C25/30 were prepared with two types of aggregates: river (samples B1, and B5 to B12) and with crushed aggregates (samples B2, B3 and B4); three types of cement: Pa35 (samples B5 and B6), P40 (samples B7, B8, but with different dosages of cement) and for all other samples the cement was CEM I-42.5 [19]; different additions were added: B9 with fly ash (FA), B10 with silica fume (SUF), B11 with fly ash and disperse glass fiber (FS), B12 with fly ash and disperse metallic fibers (FM) also with fly ash, Table 2. The glass fibers were of 30 mm length and the metallic fibers were of 50 mm length with deformed ends.

The cubic samples of 150 mm sizes were poured and kept in standard conditions. The tests were effectuated at 28 days according to Eurocode SR EN 12390 - 3:2009 [20].

Table 2. Components dosage for series 2.

Sample	Dosages									Addition				fc N mm^{-2}
	Cement kg m^{-3}	River aggregate kg m^{-3}			Crushed aggregate kg m^{-3}			Water l m^{-3}	W/C	FA %	SUF %	FS %	FM %	
		0-4	4-8	8-16	0-4	4-8	8-16							
1	360	803	384	559	-	-	-	172	0.48	-	-	-	-	36.62
2	360	803	-	559	-	384	-	172	0.8	-	-	-	-	44.73
3	360	803	384	-	-	-	559	172	0.48	-	-	-	-	43.12
4	360	803	-	-	-	384	559	172	0.48	-	-	-	-	47.06
5:C25/30Pa35-L1	444	575	436	232	-	-	-	182	0.41	-	-	-	-	45
6:C25/30Pa35-L4	617	460	348	586	-	-	-	253	0.41	-	-	-	-	48
7:C25/30-P40	387	592	466	735	-	-	-	182	0.47	-	-	-	-	43
8:C25/30-P40	538	488	381	601	-	-	-	253	0.47	-	-	-	-	46
9:10% FA	324	803	384	559	-	-	-	180	0.55	36	-	-	-	44.98
10:10%SUF	324	803	384	559	-	-	-	180	0.55		36	-	-	42.4
11:10%FA+FS	324	803	384	559	-	-	-	180	0.55	36	-	0.25	-	45.95
12:10%FA+FM	324	803	384	559	-	-	-	180	0.55	36	-	-	0.25	45.82

The compressive strength on cubes of 150 mm size was determined for experimental mixes according to European standards at 28 days. The test results experimentally obtained are the medium value of three tests [20].

Results and discussion

1.3. Series 1.

The experimental results which are selected from the studies on each type of concrete in previous researches are presented in Figure 1 [16, 17, 18].

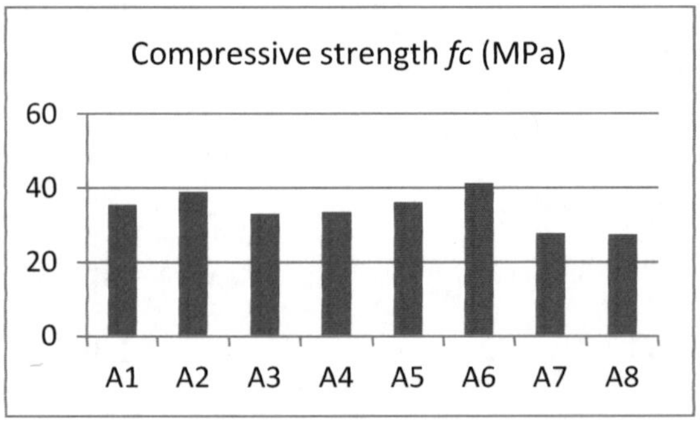

Figure 1. Variation of compressive strength of cement concrete function of components (series 1).

Analyzing the results, it can observe that an improve of compressive strength can be obtained by: increasing the cement dosage, which is not economically, the use of SUF addition in concrete mix (an increase in f_c of about 16% was obtained in comparison with witness A1), and also by the use of slag aggregates (A5), in which case an increase of about 1.7% was obtained in comparison with witness A1. Other additions used in the experiments decreased the compressive strength, but created a more elasticity behavior under load of concrete. The mixes A7 and A8 with chopped tires and respectively with polystyrene granules had a decrease of f_c of about 22.1% in the case of A7 and of about 22.6% respectively, in comparison with the witness A1.

As conclusion for these types of additions it resulted that for increasing the compressive strength except the classical methods of increasing the cement dosage, using high strength aggregate and reducing W/C ratio, fine addition (SUF) can be used without affecting the compressive strength. Other types of additions such as chopped tires and polystyrene granules can be used in situations in which compressive strength is less important than other characteristics such as elasticity, ductile failure, thermal and acoustical properties, etc.

1.4. Series 2

For the series 2 the experimental results are given in Figure 2.

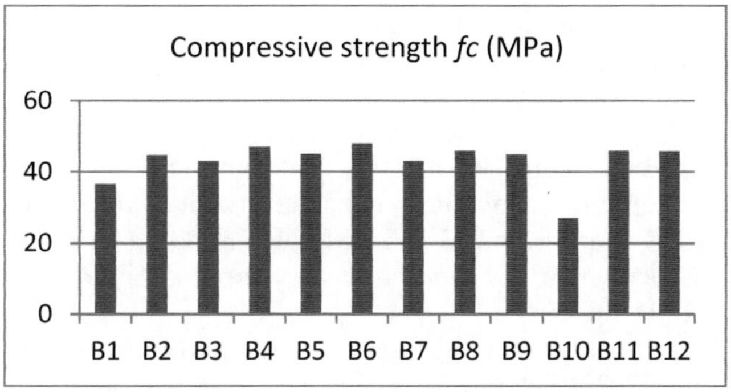

Figure 2. Variation of compressive strength of cement concrete function of components (series 2).

From the graph - Figure 2 - it can observe that in the case of a higher grade of concrete, an improvement of compressive strength can be obtained by using all the additions which were studied. Evidently, the clearest influence had cement dosage and crushed aggregates that are not economical solutions. For improving the compressive strength of concrete, wastes type FA, SUF, fibers can be used as addition or by replacing parts of cement. The crushed aggregates (mix B4) highly influenced the compressive strength, in which case the maximum value of f_c was obtained, that means an increase in compressive strength of about 28.5% in comparison with that of the mix with river aggregate (mix B1), which was considered as witness. The type and dosage of cement also influences the compressive strength, the highest value of f_c being obtained for a dosage of 617 kg m^{-3} for mix B6 (the increase in strength was of about 31.1%). The substitution of 10% cement with fly ash, mix B9, resulted in an increase of compressive strength with 16% in comparison with the compressive strength of witness concrete. An addition of 10% of fly ash as replacement of cement and 1% glass fibers, mix B11, resulted in an increase of compressive strength with 25% of compressive strength in comparison with that of witness concrete.

The conclusion of compared experimental results it is evident that near the well-known possibilities of improving mechanical strengths of cement concrete by increasing dosage and strength of cement, by using crushed aggregates or by reducing the W/C ratio, there are other methods in which less cement is used by replacing it with powder type SUF or FA or by adding disperse fibers type glass or metallic.

Conclusions

The addition of different wastes in cement concrete is a possibility of obtaining new building materials with specific requirements and to consume the wastes. In experimental tests there were studied some types of additions such as silica fume and fly ash, chopped tires, polystyrene granules and disperse fibers (glass and metallic), the results were compared for a better understanding of properties of new concretes. From the study it can observe that some additions had improved the mechanical strengths and other had changed the physical properties of cement concrete. It resulted that the additions such as: silica fume, fly ash and disperse fibers type glass and metallic improved the compressive strength. Other additions, such as: chopped tire and polystyrene granules had improved the elasticity of concrete and a more ductile failure was observed.

In conclusion, different types of additions can be used for improving properties of concrete and to obtain building materials friendlier with environment. Function the predominant characteristic of the new concrete there are necessary experimental tests for a better understanding of effects of additions and clearer definition of the specific properties of each concrete for recommending the use domain.

Acknowledgements
This work was supported by a grant of the Romanian National Authority for Scientific Research and Innovation, CCCDI - UEFISCDI, project number 109 BM / 2017.

References

[1] Juenger C G M and Siddique R 2015 Recent advances in understanding the role of supplementary cementitious materials in concrete *Cement and Concrete Research* **78** 71-80

[2] Hamza A R, Haggar S E and Khedr S 2011 Marble and Granite Waste: Characterization and Utilization in Concrete Bricks *International Journal of Bioscience, Biochemistry and Bioinformatics*, **1** (4) 286-291

[3] Kayali O. 2008 Fly ash lightweight aggregates in high performance concrete *Construction* and *Building* Materials **22**, 2393-2399

[4] Barbuta M. Harja M. Ciobanu G. 2014 Mechanical properties of polymer concrete containing tire waste powder *Journal of Food, Agriculture & Environment* **12** (2) 1186-1190

[5] Boháč M, Palou M, Novotný R, Másilko J, Všianský D and Staněk T 2014 Investigation on early hydration of ternary Portland cement-blast-furnace slag-metakaolin blends Construction and Building Materials **64** 333-341 http://dx.doi.org/10.1016/j.conbuildmat.2014.04.018

[6] Hunchate S R Chandupalle S Ghorpode V G and Reddy T C V 2014 Mix Design of High Performance Concrete Using Silica Fume and Superplasticizer *International Journal of Innovative Research in Science, Engineering and Technology* **3**(3) 10735 - 10742

[7] Yu R, Spiesz P and Brouwers H J H 2015 Development of an eco-friendly Ultra-High Performance Concrete (UHPC) with efficient cement and mineral admixtures uses *Cement & Concrete Composites* **55** 383-394

[8] Wang W, Liu J, Agostini F, Davy C A, Skoczylas F and Corvez D. 2014 Durability of an Ultra High Performance Fiber Reinforced Concrete (UHPFRC) under progressive aging. *Cement and Concrete Researches* **55** 1-13

[9] Rossi P. 2013 Influence of fibre geometry and matrix maturity on the mechanical performance of ultra-highperformance cement-based composites *Cem Concr Comp* **37** 246-248

[10] Tuan N V, Ye G, Breugel K and Copuroglu O. 2011 Hydration and microstructure of ultra-high performance concrete incorporating rice husk ash *Cem Concr Res* **41** 1104-1111

[11] Barbuta M, Bucur R D, Cimpeanu S M, Paraschiv G and Bucur D. 2015 *Wastes in building materials industry*, In: *Agroecology*, Pilipavicius V. ed., 81 - 100, Intech, ISBN 978-953-51-4193-8

[12] Yazici H, Deniz E and Baradan B. 2013 The effect of autoclave pressure, temperature and duration time on mechanical properties of reactive powder concrete *Construction and Building Materials* **42** 53-63

[13] Ghafari E, Costa H, Júlio E, Portugal A and Durães L 2014 The effect of nanosilica addition on flowability, strength and transport properties of ultra high performance concrete *Materials & Design* **59** 1-9

[14] Fabbri B, Gualtieri S and Leonardi C 2013 Modifications induced by the thermal treatment of kaolin and determination of reactivity of metakaolin *Applied Clay Science* **73** 2 - 10
[15] Marin E, Barbuta M, Ciobanu L and Cioara I 2014 Practical issues related to processing of fiberreinforced concrete made with recycled materials, 14th International multidisciplinary scientific geoconference SGEM 2014, Energy and Clean Technology, Nuclear Technology, recycling, air pollution and climate change, vol. II. 191-198, 17-26 June, Albena, Bulgaria, ISBN 978-619-7105-16-2, ISSN 1314-2704, DOI:10.5593/sgem2014B42
[16] Barbuta M 2006 *Aggregate influence on concrete strengths* Academic Society Publishing House Mihai Teiu Botez Iasi Romania (in Romanian)
[17] Barbuta M 2010 *Concrete with additions* Academic Society Publishing House Mihai Teiu Botez Iasi Romania (in Romanian)
[18] Barbuta M 2015 *Experimental Studies in Characterizing and Obtaining New Modern Concretes* Habilitation Thesis, PIM Publishing House Iasi Romania
[19] Romanian Standard Association, SR EN 197-1:2011. Cement, Part 1: Composition,specifications and conformity criteria for common cements
[20] SR EN 12390-3:2009 Testing hardened concrete. Part 3: *Compressive strength of test specimens*. British Standard

IOP Conf. Series: Materials Science and Engineering **246** (2017) 012037 doi:10.1088/1757-899X/246/1/012037

The influence of concrete mixture`s rheological properties on the quality of formed concrete surfaces

M Daukšys[1], A Klovas[1], L Venčkauskas[1]

[1]Department of Construction Technologies, Kaunas University of Technology, Studentu str. 48, LT-51367 Kaunas, Lithuania

E-mail: mindaugas.dauksys@ktu.lt

Abstract. This study mainly lays emphasis on examining the influence of concrete mixture rheological properties on the quality of formed concrete surfaces. Mixture`s fine aggregate change was taken into the consideration. Over the course of concrete mixture preparation the inner ratio of fine aggregate (sand: fraction of 0/1 and 0/4) was changed. The idea was to increase the quantity of fine particles in the total aggregate`s volume therefore quantity of sand (fraction 0/1) was increased. Six different concrete mixture`s compositions were designed as well as three specimens (concrete piles of 1m^2 surface area) were casted. Rheological properties of concrete mixtures were analytically obtained and the quality of formed concrete surfaces was evaluated using image analysis method "BetonGUY 2.0". As can be obtained from the dependence between concrete mixture rheological properties and its formed surface quality, the increase of concrete mixture`s yield stress and plastic viscosity reduces the quantity of air pores on formed concrete surfaces.

1. Introduction

Problems regarding concrete surface`s quality become very important because the demand of high level of aesthetics has spread across a wide variety of construction works [1]. Clients are no longer willing to accept poor quality structures which incur enormous repair and maintenance costs during the life-cycle of the building. It is obvious that producers of construction materials are offering many types of coatings for formed surfaces in order to achieve better appearance as well as durability characteristics [2-5]. The flatness of concrete surface is less important factor as long as it does not extremely exceed the certain limits. Tint and air bubbles are often a subject of conflict between owners, architects and building contractors as there are no strict national standards covering the problem [6]. A framework for dimensional and surface quality assessment of precast concrete elements using BIM and 3D laser scanning is one of the most recent designed to assess the quality of concrete surface [7]. This system, despite having a potential in subjective and reliable quality assessment of precast concrete elements can be hardly utilized. The method may be used only with precast elements in rectangular form also the thickness must be uniform. These factors result in a big limitation since concrete surfaces have complex geometries. Authors [8] offered 3 input parameters to assess the quality of concrete surface: bughole area; maximum bughole diameter; and bughole size distribution curve. This system proved to be effective in classifying the concrete surfaces depending on the degree of treatment required.

It must be noted that there are a lot of ways to improve concrete`s surface quality: suitable concrete mixture compositions, utilization of controlled-permeable formworks, correct usage of form release agents, careful workmanship and decent vibration technology [9-10]. In order to obtain suitable concrete mixture compositions one should investigate its rheological properties. Scientists [11] presented the case that high quality concrete surfaces may be obtained with low values of plastic viscosity and yield stress for SCC concrete. Rheological properties are directly influenced by the constituents of concrete mixture [12-13]. Scientists [14] conducted a wide study on the influence of various mineral admixtures on rheological properties of self-compacting concrete. They put forward the fact that the ground granulated blast furnace slag can reduce the yield stress without resulting in the large decrease in plastic viscosity only up to 70%. Another study was conducted in order to improve concrete surface quality of precast concrete plant in Brazil. The subject of interest was the amount of surface honeycombs. Results showed that the increase of cement to 75% resulted in a significant decrease of their size and number. However the different type of form release agent that was used did not show a significant influence on surface quality. There are some recommendations on concrete surface`s evaluation but most of them leave plenty of room for subjective appreciation. Concrete Innovation Centre (COIN) is currently working on the tool which would provide an objective evaluation of formed surface`s quality. The tool covers grey scale and distribution of pores. These two main parameters of concrete surface`s quality must be evaluated separately.

This study was carried out due to the important questions raised whether the rheological properties of concrete mixture influence the formed concrete surface quality. The authors of this research assed the significant of fine aggregate usage as one of the main methods to change the rheological properties of concrete mixture. The dependence between concrete mixture`s rheological properties and its formed surface quality was also taken into the consideration. Norwegian evaluation methodology of concrete surfaces was used as the main frame [15].

2. Materials and methodology

JSC "Akmenes cementas" (Lithuania) Portland cement CEM II/A-LL 42.5 R was used for the test. Physical and mechanical properties of Portland cement are presented in Table 1.

Table 1. Physical and mechanical properties of Portland cement

Property	Value
Specific surface area, m^2/kg	410
Particle density, kg/m^3	3.05
Normal consistency of cement paste, %	26.5
Volume stability, mm	0.8
Initial setting time, min.	195
Compressive strength after 2 days / 28 days, MPa	27.1 / 54.0
Loss on ignition, %	5.05
Insoluble materials, %	-
SO_3, %	2.48
Cl^-, %	0.015
Alkalis, calculated by Na_2O equivalent, %	<0.8

Sand from "Kvesu" quarry fraction 0/1 and 0/4, bulk densities 1520 kg/m^3 and 1710 kg/m^3 and fineness modules 1.78 and 2.62 was used as fine aggregates. Gravel fraction 4/16 and bulk density 1327 kg/m^3 was used as the coarse aggregate. Granulometric composition of aggregates is conducted according to LST EN 12620 and presented in Table 2.

Superplasticizer Glenium SKY 628 based on polycarboxylate (BASF „Construction Chemicals Italia Spa") was used as plasticizing admixture. The density of solution 1.06 kg/l, yellow liquid. The total dosage of admixture was 1.2% of cement.

Table 2. Granulometric composition of aggregates

Diameter of the sieve's mesh, mm	The amount of poured out material, %		
	Sand fraction 0/1	Sand fraction 0/4	Gravel fraction 4/16
16.0	100.00	100.00	98.80
8.0	100.00	100.00	42.10
4.0	100.00	95.10	4.30
2.0	99.80	81.80	1.00
1.0	99.10	54.60	0.52
0.500	77.40	12.40	0.44
0.250	2.20	0.70	0.36
0.125	0.50	0.30	0.32
0.000	0.00	0.00	0.00

Concrete mixtures were mixed for 3 minutes in the laboratory in forced type concrete mixer. Only dry aggregates were used for concrete mixtures. Cement and aggregates were dosed by weight while on the other hand, water and chemical admixtures were dosed by volume. 90% of water was instantly poured to the mix. Plasticizing admixture was mixed with 10% of remaining water and poured into the mixer. Remaining admixtures were dozed directly to the mix.

The slump of concrete mixture was determined according to standard LST EN 12350-2, the density according - to standard LST EN 12350-6 and the air content - according to standard LST EN 12350-7.

Concrete mixture`s rheological property yield stress was evaluated according to equation (1). Yield stress of non-vibrated concrete mixture can be calculated according to its fresh technological parameters: density and slump [16]. This equation is adjusted according to standard cone which is used to evaluate the slump of concrete mixture.

$$\tau_0 = \frac{0.00815 \cdot \rho_m}{\left(\sqrt{\frac{0.498}{30-SL}} - 0.001724 - 0.024\right)^2}; \tag{1}$$

here: τ_0 – yield stress of concrete mixture, Pa; ρ_m – density of concrete mixture, kg/m^3; SL – slump of concrete mixture, cm.

Second mixture`s rheological property – plastic viscosity was evaluated according to equation (2). Plastic viscosity is established by concentration of aggregates which are used in concrete mixture.

$$\eta_b = \eta_v \cdot \exp\left(\frac{a_c \cdot \rho_v}{\rho_v + \frac{V}{C} \cdot \rho_c - b_c \cdot \rho_v} + \frac{a_{sm}\left(1-\varphi_{st}-\varphi_o-\frac{V}{\rho_v}-\frac{C}{\rho_c}\right)}{1-\varphi_{st}-b_{sm}\left(1-\varphi_{st}-\varphi_o-\frac{V}{\rho_v}-\frac{C}{\rho_c}\right)} + \frac{a_{st} \cdot \varphi_{st}}{1-b_{st} \cdot \varphi_{st}} \right); \tag{2}$$

here: φ_o – air content of concrete mixture, %; V, C – water and cement quantities in concrete mixture (1 m^3); ρ_v, ρ_c – densities of water and cement, kg/m^3; φ_{sm}, φ_{st} – volume concentration of fine aggregate in mortar and coarse aggregate in concrete mixture; a_c, a_{sm}, a_{st} – factors describing the form of cement, fine and coarse aggregate`s particles ($a_c = 2.6$; $a_{sm} = 2.5$; $a_{st} = 2.6$); b_c, b_{sm}, b_{st} – factors describing the density distribution of cement, fine and coarse aggregate respectively in mortar and concrete mixture.

It is utmost important to assess the significance of concrete casting process. PERI LM0303S formwork system (Figure 1a) was used to hold the concrete mixture. Formwork`s surface (plywood) was carefully cleaned after the application of form release agent Rheofinish 218 (BASF „Construction Chemicals Italia Spa"). The consolidation of concrete mixture was not taken into the consideration therefore the method and time of vibration was not changed (internal vibrator, 6 seconds). In order to evaluate the quality of formed concrete rectangular columns (1.2 × 0.2 × 0.2 m) were casted (Figure 1b). Each side of the column was divided into two separate regions (0.6 × 0.2 × 0.2 m). The average results of 4 different surfaces of each concrete specimens were evaluated in this study.

Figure 1. Formwork system (a) and formed concrete rectangular columns (b)

High-end photography equipment was used for this study. All pictures were taken by Nikon D3200 with 24 megapixels camera. The lens Nikon DX AF-S NIKKOR 18-55 mm was used to improve quality of the images. Separate mounted flashlight "Quantuum" was also used. Sensor's parameters: max resolution – 6016 × 4000; image ratio w:h – 3:2; effective pixels – 24 megapixels; sensor photo detectors – 25 megapixels; sensor size – APS-C (23.2 × 15.4 mm).

The parameters of image taking equipment used in this study: aperture, 8 F; shutter speed, 1/160 s; flash intensity, 15 GN; ISO number 100. Figure 2 indicates the positioning of image taking equipment.

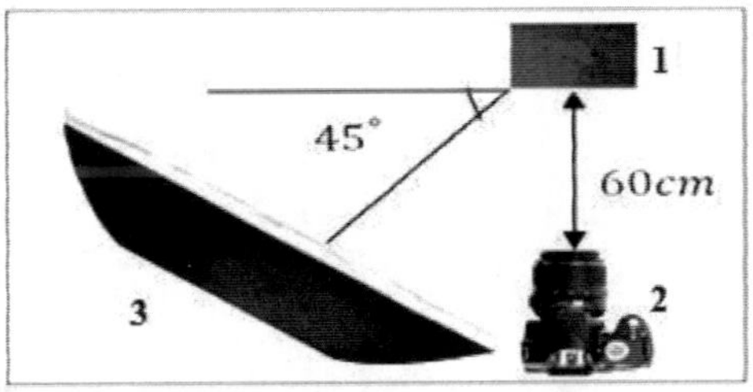

Figure 2. Positioning of image taking equipment
(1- flash light; 2- camera; 3- concrete surface)

Flash equipment had to be moved in both sides (left and right) in respect to camera's position in order to get the best possible illumination (shadows) of surface bubbles. Firstly, it had to be moved to the left and secondly to the right side in 450 towards the testing surface. It must be noted that a distance between the surface and the camera should not exceed 70 cm in order to get height quality images. After that the quality of formed concrete surfaces was evaluated using image analysis method "BetonGUY 2.0". The main parameters for evaluation of surface quality were the number of air pores per $1 m^2$ and their size. Air pores were classified not only by their quantity but by biggest dimensions as well: 1-5 mm, 5-10 mm, 10-15 mm and bigger than 15 mm. The biggest attention was paid to bigger air pores: 10-15 mm and >15 mm. Bigger pores are more noticeable from the distance and therefore more important in terms of assessing the quality of formed concrete surface.

3. Results and discussions

In order to evaluate the influence of fine aggregate – sand (fraction 0/1) quantity to the technological and rheological properties of concrete mixtures, the quantity of sand (fraction 0/1) was increased while the quantity of sand (fraction 0/4) was decreased. The quantity of sand (fraction 0/1) was changed from 11 to 36% according to mass in respect to total fine aggregates quantity. The compositions of concrete mixtures (BA1-0 – BA1-5) used in this research are presented in Table 3.

Table 3. Compositions of concrete mixture

Materials	Unit	Concrete mixture compositions. Amount of materials for 1m^3 of concrete mixture					
		BA1-0	BA1-1	BA1-2	BA1-3	BA1-4	BA1-5
Portland cement	kg	330	330	330	330	330	330
Water	l	178	178	178	178	178	178
Gravel fraction 4/16	kg	986	986	986	986	986	986
Sand fraction 0/4	kg	701	607	512	417	322	227
	(%)	37	32	27	22	17	12
Sand fraction 0/1	kg	208	303	398	493	588	682
	(%)	11	16	21	26	31	36
Superplasticizer	l	3.74	3.74	3.74	3.74	3.74	3.74
Water and cement ratio	–	0.54	0.54	0.54	0.54	0.54	0.54

As there is shown at Table 4, with the increase of sand (fraction 0/1) quantity from 11 to 36%, the slump of concrete mixture decreases from S4 to S3 class according to LST EN 12350-2 standard. To sum up, the slump of concrete mixture reduces about 1.03. By increasing the quantity of fine aggregate, surface area increases. In order to keep the mixture‘s consistency at the same level, the liquid phase must be increased. The highest values of concrete mixture slump (~270 mm) was obtained with the usage of 11 % of 0/1 fraction sand.

Table 4. Technological properties of concrete mixtures

Technological property	Unit	Value of technological property of concrete mixture					
		BA1-0	BA1-1	BA1-2	BA1-3	BA1-4	BA1-5
Density	kg/m^3	2380	2310	2330	2310	2310	2320
Slump	mm	270	260	260	250	250	200
Air content	%	4.5	4.3	4.1	4.1	4.5	4.9

The density of concrete mixture decreases while increasing the quantity of fine aggregate (sand, fraction 0/1) from 11 to 36%. On the other hand, the air content slightly increases. The density of concrete mixture varies from 2380 to 2320 kg/m^3, while air content – from 4.5 to 4.9%. The lowest air content rate (~4.1 %) was obtained with the usage of 21-26% of 0/1 fraction sand.

The dependence between concrete mixture`s composition and its rheological properties is presented in Figure 3.

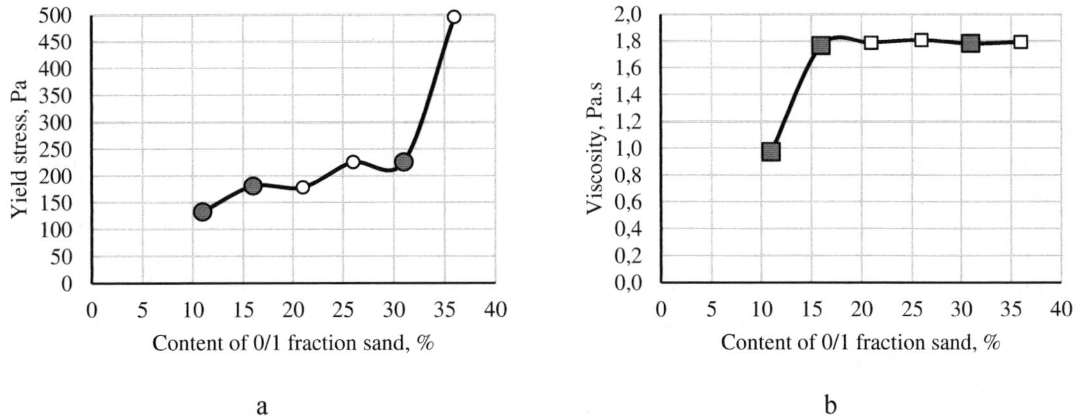

Figure 3. The dependence between mixture`s composition and rheological properties: yield stress (a) and viscosity (b)

It is obvious that plastic viscosity increased by about 1.8 when the quantity of fine aggregate was increased from 11 to 16%. Further increase of fine aggregate`s quantity did not have a significant influence on plastic viscosity. On the other hand, yield stress constantly increased while increasing the quantity of fine aggregate from 11 to 31%. Yield stress significant increased while increasing the quantity of fine aggregate from 31 to 36%.

The influence of sand (fraction 0/1) quantity on formed concrete surface quality was established by 3 specimens formed of different concrete mixtures. Three different quantities of fine aggregate were chosen according to previously established rheological properties: 11% – reference point with the lowest quantity; 16% – an intermediate point with medium quantity; 31% – limit point with the biggest quantity. Following graphs show the ratio between area of surface air pores and total specimen`s area, which named Ap/As. Figure 4 presents the dependence between mixture`s rheological properties and area of surface air pores and total specimen`s area while changing the quantity of fraction 0/1 sand. As can be obtained from Figure 4, the quantity of surface air pores per 1 m^2 significantly declined by increasing the yield stress (Figure 4a) and plastic viscosity (Figure 4b). Despite the further decrease of surface air pores, the optimal ratio between 0/1 mm and 0/4 mm fractioned sand was 0.5. The bigger ratio resulted in a stiff mixture which was not technological to work with. It must be noted that vibration was not taken into the consideration while preparing concrete mixtures. This is another very important factor and should be covered in more detail in future researches.

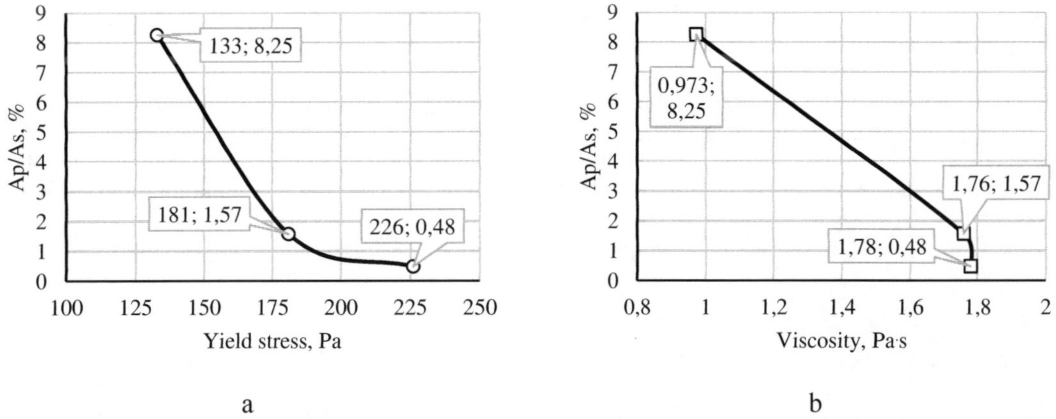

Figure 4. The dependence between mixture`s yield stress (a), plastic viscosity (b) and area of surface air pores and total specimen`s area (Ap/As)

Provided mixture`s parameters give the perfect insight on the dependence between rheological properties and number of surface air pores. The optimal concrete mixture composition was found to be BA1-1 with the values of yield stress – 180.54 Pa and plastic viscosity – 1.764 Pa·s. Yield stress and plastic viscosity values were found to be increased while increasing the amount of fine aggregate. With increased amount of fine aggregate, water demand became higher and therefore concrete mixtures became stiffer. To sum up, the stiffer concrete mixture is, more power is needed in order to initiate the material`s flow (yield stress) and to keep that flow going (plastic viscosity).

Figure 5 shows how formed concrete`s surface quality had changed. It indicates the distribution of differently sized concrete surface air pores. The increase of fine aggregate from 208 kg to 588 kg per 1 m^3 of concrete mixture reduced the number of surface air pores: 1÷5 mm reduced by 2.7 times; 5÷10 mm by 5.8 times; 10÷15 mm by 8.3 times and > 15 mm reduced by 51 times. The conclusions can be drawn that more fine aggregate (more fine particles) are added the higher quality of concrete surface can be expected.

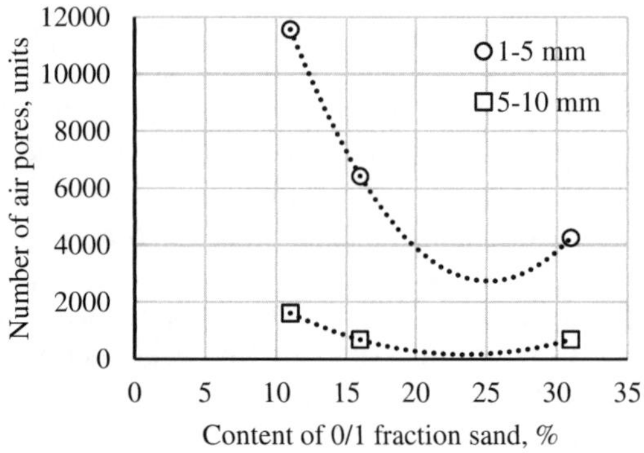

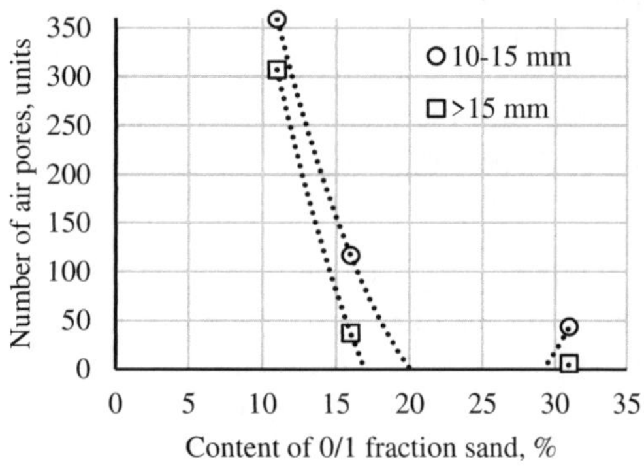

Figure 5. The influence of quantity of fine aggregate on the formed concrete surface air pores number: 1-5 mm and 5-10 mm (a); 10-15 mm and >15 mm (b)

The best dependence of dispersion of values of concrete mixtures composition and formed concrete surface air pores number is presented in Figure 5. Equations of dispersion of the results, empirical and correlation coefficients values are presented in Table 5. Correlation coefficient (Pirson), which is evaluating the strength of linear relationship, was calculated according to coefficient of empirical equation. The closer correlation coefficient is to 1, the better representation of the dispersion of values in the curve. According to the obtained correlation coefficient, it was determined which equation describes the best distribution of statistical data.

Table 5. The dependence of dispersion of data

Size of surface air pores, mm	Equation	Coefficients value of equation			R-squared value	Correlation coefficient r
		a	b	c		
1-5	$y = ax^2\text{-}b{\cdot}x+c$	44.080	2217.5	30626	1.0	1.0
5-10	$y = ax^2\text{-}b{\cdot}x+c$	9.2600	435.25	5279.3	1.0	1.0
10-15	$y = ax^2\text{-}b{\cdot}x+c$	2.1754	107.14	1274.0	1.0	1.0
>15	$y = ax^2\text{-}b{\cdot}x+c$	2.5967	124.10	1357.8	1.0	1.0

The dispersion of values of dependence of concrete mixture composition on the quantity of concrete surface air pores could be described by a polynomial dependency. The calculated values of correlation coefficients are 1.0. It shows, that variables are statistically dependent and based on values of correlation coefficients – the relationship between variables is very strong.

The results of this study revealed that more fluid concrete mixtures with more fine particles (within normal workability limits) provides higher surface quality. The air content of concrete mixture is widely questionable in terms of concrete surface`s quality. Air content of concrete mixture falls into two main categories: entrapped air and entrained air. Entrapped air is a type of air bubbles which appears in the concrete mixture during the casting process or vibration. The second type of air appears due to the usage of air entraining admixtures. It is obvious that more entrapped air in the concrete mixture is, the lower concrete surface quality, due to air pores, is obtained. The usage of air entraining agents raises important questions whether it can reduce the amount of entrapped air in concrete mixture and thus to improve concrete surface`s quality. Scientists put forward the argument that the utilization of air entraining admixture results in reordering of the entrapped air bubbles (bigger ones) due to the micro air pores which are entrained in the concrete mixture by using certain. In this way, bigger entrapped air pores reach the surface of the formwork and escape easier [17-19]. On the other hand, scientists briefly summarize that the utilization of air entraining admixtures results in lower values of concrete mixture`s yield stress and this factor has the tendency to worsen the quality of formed concrete surface.

As can be obtained from the dependence between concrete mixture rheological properties and its formed surface quality, the increase of concrete mixture`s yield stress and plastic viscosity reduces the number of surface air pores. It is obvious that bigger amount of fine aggregate results in better concrete surface quality.

4. Conclusions

- The digital software "BetonGUY 2.0" enables to evaluate the influence of different concrete mixture`s components on the quantity and size distribution of formed surface air pores.
- The increase of concrete mixture`s yield stress and plastic viscosity reduces the quantity of air pores on formed concrete surfaces. The research was carried out with sand as fine aggregate.
- The increase of fine aggregate from 208 kg to 588 kg per 1 m^3 of concrete mixture reduced the number of surface air pores: 1÷5 mm reduced by 2.7 times; 5÷10 mm by 5.8 times; 10÷15 mm by 8.3 times and > 15 mm reduced by 51 times.
- Based on this study it can be concluded that the most significant surface`s air pores fall into two main categories: 5÷10 mm and > 15 mm. This is due to the fact that bigger pores are clearly visible and recognized from the reasonable distance and therefore they have the biggest influence on surface quality.

Acknowledgments
Albertas Klovas acknowledges support by project "Promotion of Student Scientific Activities" (VP1-3.1-ŠMM-01-V-02-003) from the Research Council of Lithuania. This project is funded by the Republic of Lithuania and European Social Fund under the 2007–2013 Human Resources Development Operational Programme's priority 3.
Albertas Klovas acknowledges support by Norwegian university of science and technology and SINTEF for the possibility to use the program "BetonGUY 2.0" for concrete surface quality evaluation for the purposes of dissertation thesis.

References
[1] Courard L, Goffinet C, Migeotte N, Martin M, Pierard J and Polet V 2012 Influence of the reuse of OSB and marine plywood formworks on concrete surface aesthetics *Mater. Struct.* **45** 1331–1343.
[2] Rwelamila P D, Wiseman G T 1995 Concrete quality management: a research study of the general contractor in South Africa *Constr. Build. Mater.* **9** 173–183.
[3] Almusallam A A, Khan F M, Dulaijan S U, Al-Amoudi O S B 2003 Effectiveness of surface

coatings in improving concrete durability *Cem. Concr. Compos.* **25** 473–481.
[4] Wilson D J 2008 *Tailor Made Concrete Structures: New Solutions for Our Society,* ed J C Walraven and D Stoelhorst (London:Taylor and Francis Group) 116.
[5] Goldammer K R 2012 DBV-guide to good practice for concrete cosmetics on exposed concrete surfaces *Beton- Stahlbetonbau* **107** 490–494.
[6] Lemaire G, Escadeillas G, Ringot E 2005 Evaluating concrete surfaces using an image analysis process *Constr. Build. Mater.* **19** 604–615.
[7] Kim M-K, Cheng J C P, Sohn H, Chang C-C 2015 A framework for dimensional and surface quality assessment of precast concrete elements using BIM and 3D laser scanning *Automat. Constr.* **49** 225–238.
[8] Da Silva W R L, Štemberk P 2013 Expert system applied for classifying self-compacting concrete surface finish *Advanc. Engineer. Softw.* **64** 47–61.
[9] Arslan M 2001 The effects of permeable formworks with sucker liners on the physical properties of concrete surfaces *Constr. Build. Mater.* **15** 149–156.
[10] Coutinho J S 2001 Effect of controlled permeability formwork (CPF) on white concrete *Aci Mater. Jour.* **98** 148–158.
[11] Report of RILEM technical committee TC 145-WSM 2002 *Workability and rheology of fresh concrete: Compendium of tests*, ed P J M Bartos, M Sonebi and A K Tamimi (RILEM Publications Sarl) 127 p.
[12] Banfill P F G 2011 Additivity effects in the rheology of fresh concrete containing water-reducing admixtures *Constr. Build. Mater.* **25** 2955–2960.
[13] Libessart L, Djelal C, de Caro P 2014 Influence of the type of release oil on steel formwork corrosion and facing aesthetics *Constr. Build. Mater.* **68** 391–401.
[14] Bauchkar S D, Chore H S 2014 Rheological properties of self-consolidating concrete with various mineral admixtures *Structur. Engineer. Mech.* **51** 1–13.
[15] Eide M B, Kaspersen K, Vikan H, Schulerud H 2011 *Nordic Concrete Research. Research Projects 2011. Proceedings of XXI Nordic Concrete Research Symposium. Hämeenlinna, Finland* (Nordic Concrete Federation: Nordic Concrete Research / 43) 521–524.
[16] Skripkiūnas G 2007 *The properties of building conglomerates* (Kaunas: Vitae Litera) 336.
[17] Libessart L, De Caro P, Djelal Ch, Dubois I 2015 Correlation between adhesion energy of release agents on the formwork and demoulding performances *Constr. Build. Mater.* **76** 130–139.
[18] Da Silva W R L, Lucena D S, Štemberk P, Prudêncio Jr L R 2014 Evaluation of the effect of concrete compositional changes and the use of ethyl-alcohol and biodegradable-oil-based release agents on the final surface appearance of self-compacting concrete precast elements. *Constr. Build. Mater.* **52** 202–208.
[19] Bouharoun S, De Caro P, Dubois I, Djelal Ch, Vanhove Y 2013 Effect of a superplasticizer on the properties of the concrete/oil/formwork interface *Constr. Build. Mater.* **47** 1137–1144.

IOP Conf. Series: Materials Science and Engineering **246** (2017) 012038 doi:10.1088/1757-899X/246/1/012038

Sustainability and durability analysis of reinforced concrete structures

A Horáková[1], I Broukalová[1], A Kohoutková[1] and J Vašková[1]

[1] Department of Concrete and Masonry Structures, Faculty of Civil Engineering, Czech Technical University in Prague

E-mail: anna.horakova@fsv.cvut.cz

Abstract. The article describes an assessment of reinforced concrete structures in terms of durability and sustainable development. There is a short summary of findings from the literature on evaluation methods for environmental impacts and also about corrosive influences acting on the reinforced concrete structure, about factors influencing the durability of these structures and mathematical models describing the corrosion impacts. Variant design of reinforced concrete structure and assessment of these variants in terms of durability and sustainability was performed. The analysed structure was a concrete ceiling structure of a parking house for cars. The variants differ in strength class of concrete and thickness of concrete slab. It was found that in terms of durability and sustainable development it is significantly preferable to use higher class of concrete. There are significant differences in results of concrete structures durability for different mathematical models of corrosive influences.

1. Introduction

In past years, the issue of sustainable development and the impact of construction activities on the environment were often discussed. Compared with other sectors, these impacts are rather significant, especially in terms of consumption of material resources and in terms of greenhouse gas production. It is desirable to minimize these environmental impacts by suitable design and material selection. The problematics of the environmental impact is closely related to the service life of buildings. Buildings with high durability required fewer repairs during the course of their existence. Furthermore, it takes a longer time before they reach the state when their demolition and construction of a new building is needed. Environmental impacts associated with construction are then compensated by longer building operation without the necessary repairs. In this work variants of the design in terms of sustainable development and durability were compared.

2. Methods

2.1. A description of the structure

The analysis was performed for a simple construction –– a concrete ceiling structure of a garage parking house for cars. The variants differ in strength class of concrete and thickness of concrete slab. One of the variants was designed from normal strength concrete (C35/45), the other variants were designed from high performance concrete (C90/105). In one case the mechanical properties of high performance concrete were used to reduce the thickness of the slab. In the other case these benefits were used to reduce the volume amount of steel. The parameters of the variants and the concrete recipes are shown in the following tables:

Table 1. The parameters of the variants.

The parameters of the variants			
Variant	**1**	**2A**	**2B**
Strength class of concrete	C35/45	C90/105	C90/105
Spacing of rebars [mm]	120	100	130
Rebar diameter [mm]	8	8	8
$A_{s,req}$ [mm^2]	408,0263	486,7253	384,4082
$A_{s,prov}$ [mm^2]	418,879	502,6548	386,6576
Cover of reinforcement [mm]	50	45	45
Distance of the centre of reinforcement and surface of the slab [mm]	54	49	49
Thickness of slab [mm]	180	140	180

Table 2. Recipe of concrete C35/45-XC4, XD3-Cl 0,2-D_{max}16-S4.

Cement 42,5 R	335 kg/m^3
Water	135 kg/m^3
Coarse aggregate (8-16 mm)	780 kg/m^3
Medium aggregate (4-8 mm)	290 kg/m^3
Fine aggregate (0-4 mm)	750 kg/m^3
Limestone	110 kg/m^3
Plasticizing admixture	3,7 kg/m^3
Silica fume	14 kg/m^3

Table 3. Recipe of concrete C90/105-XC4,XD3-Cl 0,2-D_{max}16-S4.

Cement 52,5 R	500 kg/m^3
Water	165 kg/m^3
Coarse aggregate (8-16 mm)	700 kg/m^3
Medium aggregate (4-8 mm)	220 kg/m^3
Fine aggregate (0-4 mm)	860 kg/m^3
Plasticizing admixture	4,5 kg/m^3

Reinforcement is from steel B500B. The design variants were assessed in terms of ultimate limit state and in terms of service limit state, too.

2.2. Sustainability assessment

Evaluation in terms of sustainable development was based on Life-cycle assessment (LCA) according to relevant standards. LCA is a method of assessing the environmental impact of a product, which is usually based on the whole life-cycle or at least on its significant part. So, the evaluation includes obtaining of raw materials, their transport to the place of processing, processing of raw materials, manufacturing of a final product, storage if necessary, use of the product and also maintenance or repairs if necessary, and finally removing the product, including the recycling of its parts. This assessment is known as "from cradle to grave" analysis. In some cases, the prediction of the course of the phase of use is not possible and the evaluation includes only the phases from obtaining of raw material until the time when the product leaves the factory ("from cradle to gate") or until transporting the product to the place of use ("from cradle to site"). Within the assessment the most significant environmental impacts are considered, such as consumption of raw materials, global warming and climate change, acidification and eutrophication of the environment, depletion of stratospheric ozone and photooxidation. These environmental impacts are within the LCA method called impact categories. Assessment of the designed variants is based on the part of life cycle: from obtaining of raw materials to manufacturing of a final product. So, the evaluation includes impacts associated with the manufacturing of concrete and steel, transport of these materials to the site of the building and transfer of these materials during the realization of the building. The following table shows the volume of concrete and the weight of steel for the designed variants:

Table 4. The volume of concrete and the weight of steel for the designed variations.

	Volume of concrete [m^3]	Weight of steel [kg]
Variant1	86,438	4361,946
Variant 2A	67,411	4759,971
Variant 2B	86,438	4111,496

In the assessment emissions of following substances are considered:

- carbon dioxide CO2
- sulfur dioxide SO2
- nitrogen oxides NOx
- carbon monoxide CO
- methane CH4
- non-methane volatile organic compound NMVOC
- nitrous oxide N2O
- hydrochloric acid HCl
- hydrofluoric acid HF,
- hydrogen sulphide H2S
- ammonia NH3

The effect of specific substance on each impact category was determined with using so-called characterization models. Characterization model for a specific impact category is a set of values that reflect the ability of various substances to damage the environment within this impact category. All of issued substances are converted to the equivalent amount of a reference substance by using these values (characterization factors). For this assessment, the characterization model recommended in Product category rules (PCR) for concrete products was used. [1]

Values of impact categories for 1 kg of each compound (cement, aggregate, plasticizer…) was then converted to the content in 1 m3 of concrete. Finally, the values of impact categories for a real amount of concrete and steel were calculated for all variants. Impacts associated with the manufacturing of concrete were included in the evaluation too. Furthermore, the values of environmental impacts associated with transportation of materials to the construction site was calculated. In this assessment, for the transport of concrete the nearest concrete mixing plant was considered – it was the concrete mixing plant in Benešov (the distance from the construction site is 10,6 km). Also for the steel reinforcement the nearest reinforcement manufactory was considered – the reinforcement manufactory in Prague (the distance from the construction site is 36,7 km). Finally, the values of environmental impacts associated with the transfer of these materials during the realization of the building (lifting and transport by a crane) was calculated.

2.3. Durability assessment

Building durability assessment method is not normatively determined yet, therefore the assessment was based on selected mathematical models, which describe the progress of the degradation phenomena in time. There are many different mathematical models for particular degradation phenomena, but for other degradation phenomena precise mathematical relationships were not formulated yet. The assessment included an effect of carbonation and an effect of chloride attack.

There are plenty of mathematical models for the time dependence of carbonation depth, because this degradation phenomenon is considered the most important within the framework of reinforced concrete structures. For durability of the construction the time when the steel reinforcement is depassivated due to carbonation (the time when carbonated layer reaches the level of reinforcement bars) is important. This time is in the literature referred to as the initiation time. The physical principle of carbonation mathematical models is diffusion, so they are based on II. Fick's law, which is described by the differential equation:

$$\frac{\partial c}{\partial t} = D.\frac{\partial c^2}{\partial x^2} \tag{1}$$

where c is concentration, t is time and x is the distance, in this case, the distance from the surface.
The thickness of carbonated layer can be determined from the relation:

$x_c = A.\sqrt{t}$

where x_c is the thickness of carbonated layer (mm), t is the elapsed time (years) and the A is a coefficient, which is calculated according to the used model. The coefficient A is different for different mathematical models. All models for the calculation of coefficient A include a parameter dependent on a type of used concrete. Most often, it is the value of water-cement ratio and compressive strength. Some models dependent on the weight of the components of concrete mixture (cement, aggregate, water) and their bulk densities. Sometimes also the type of used cement is taken into account. The coefficient A also sometimes depends on environmental effects, such as humidity or carbon dioxide content in air.

Model, which is currently considered the most comprehensive, includes detailed information about the composition of the concrete mixture and depends also on the concentration of carbon dioxide in air. The model dates back to 1992 and its author is V. G. Papadakis et al. [6]

$$A = \sqrt{\frac{2[CO_2].D_{e,co_2}}{[CH]+3[CSH]}} \tag{2}$$

where D_{CO2} is the effective diffusion coefficient of CO_2 in concrete (m^2/s), $[CO_2]$ is the concentration of CO_2 in the environment (mol/m^3), [CH] is the molar concentration of $Ca(OH)_2$ and [CSH] is the molar concentration of Calcium-Silicate-Hydrate.

The model is valid for the concrete, where a substantial part of hydration process is completed and whose diffusion coefficient has a constant value. So, the structure should be in a relatively constant environment. However, the model for fluctuating environmental conditions has not been created yet. If the environmental conditions are unstable, the results must evaluated with caution.

The calculation of the diffusion coefficient is quite difficult and requires detailed information about the material. However, it is possible to use a simplified version of this model, which is based on the contents of components in the concrete mixture and the humidity of the environment:

$$A = 350.\frac{\rho_C}{\rho_v}\frac{(w-0{,}3)}{(1+w.\frac{\rho_C}{\rho_v})}.f_{RH}.\sqrt{(1+\frac{\rho_C}{\rho_v}.w+\frac{\rho_C}{\rho_a}.\frac{m_a}{m_c}).c_{CO_2}} \tag{3}$$

where ρ_c, ρ_a and ρ_v are bulk densities of cement, aggregates and water (kg/m^3), m_a and m_c are weights of aggregate, and water (kg), w is the water-cement ratio, c_{CO_2} is carbon dioxide concentration in air (mol/m^3) and f_{RH} is a parameter dependent on the relative humidity of environment according to the following table:

Table 5. The values of the f_{RH} parameter.

Relative humidity RH [%]	0	7	50	93	100
Value of parameter f_{RH} [-]	0	0	0,425	0,5	0

It is possible to use an effective value of water-cement ratio w_{eff} for models dependent on water-cement ratio. It is a modified value of water-cement ratio reflecting the influence of supplementary cementing materials (SCM) on the behavior of the material. [5]

$$w_{eff} = m_v / (m_c + k.\ SCM) \tag{4}$$

where m_v and m_c are weights of cement and water, k is a parameter depending on the type of supplementary cementing material (fly ash, silica fume, slag) and SCM is the total amount of the SCM. The models are very often based on the value of water-cement ratio. The water-cement ratio has a direct effect on the porosity of the material and therefore also on its durability. However, there are not always taken into account other factors affecting the rate of carbonation. The advantage of these models is simplicity of the calculation and low requirements for input data. An example is De Sitter`s model from 1985 [7]:

$$A = \sqrt{\frac{46w - 17{,}6}{2{,}7}.R.k} \tag{5}$$

where w is the water-cement ratio, R is the coefficient depending on the type of cement and k is the coefficient reflecting the ambient humidity. Similar is Kishitani`s model from 2005 [3]:

$$\text{For } w < 0{,}6\text{: } A = R_1.\sqrt{0{,}639\,w - 0{,}244} \tag{6}$$

$$\text{For } w \geq 0{,}6\text{: } A = R_2.\sqrt{\frac{(w-0{,}25)^2}{0{,}345+w}} \tag{7}$$

where R_1 and R_2 are coefficients depending on a type of cement.
A lot of models are dependent on a compressive strength. An example is Bob`s model from 1990 [6]:

$$A = \frac{150\,.\,C\,.\,k\,.\,d}{f_c} \tag{8}$$

where f_c is the compressive strength (MPa), C is a coefficient depending on the type of cement, the k is coefficient reflecting the humidity conditions of the environment, and d is a coefficient depending on the content of CO_2 in the environment.
Duval`s model from 1992 [4] is also dependent on the compressive strength:

$$A = \sqrt{365}.\ (\frac{1}{2{,}1.\ \sqrt{f_c}} - 0{,}06) \qquad (9)$$

where f_c is the compressive strength (MPa).
Similar is Parrot`s model from 1987 [4]:

$$A = \sqrt{251}.\ exp(-0{,}05.\ f_c) \qquad (10)$$

where f_c is the compressive strength (MPa).

For the final calculation of durability a carbonation model formulated by Papadakis in year 1992 was chosen. For comparison, the service life was also calculated using the Bob`s model from year 1990. Mathematical model which was used for the chloride attack depends on concrete chloride diffusion coefficient and on surface chloride concentration in concrete. It dates back to 1972 [7].

$$c(x,t) = c_{s,0}\,[1 - erf\,(\frac{x}{2\sqrt{t\ .\ D_{e,Cl}}})] \qquad (11)$$

where $c(x,t)$ is the chloride ion concentration, $D_{e,Cl}$ is the effective diffusion coefficient for chlorides (mm^2/year), $c_{s,0}$ is the concentration of chlorides on the surface of structure, x is the distance from the surface of structure (mm) and t is the exposure time (years). *erf* is the error function.

At first, the initiation time was calculated. It is time, when the carbonated zone reached the level of the reinforcing steel or when the chloride concentration on level of the reinforcing steel reached the threshold value. After reaching this point, the reinforcing steel begins to corrode. The residual service life was calculated as the time during which the reinforcement area decreases so that it is no longer able to resist load effects.

3. Results

3.1. Sustainability assessment

The results of the sustainability assessment depend on the specific environmental impact. For most impacts the best results were reached for the variant which is designed from high performance concrete (HPC) and the mechanical properties are used to reduce the thickness of the slab. The variant in which the outstanding HPC properties are used to reduce the amount of steel is not so advantageous, according to this evaluation. In the assessment of effect on the global warming the most favourable results were calculated for the variant designed from normal-strength concrete. The same result was obtained also for the assessment of an effect on the stratospheric ozone depletion. The reason is evidently the usage of a higher volume of cement and plasticizer. The following figure shows the comparison of the designed variants for considered impact categories.

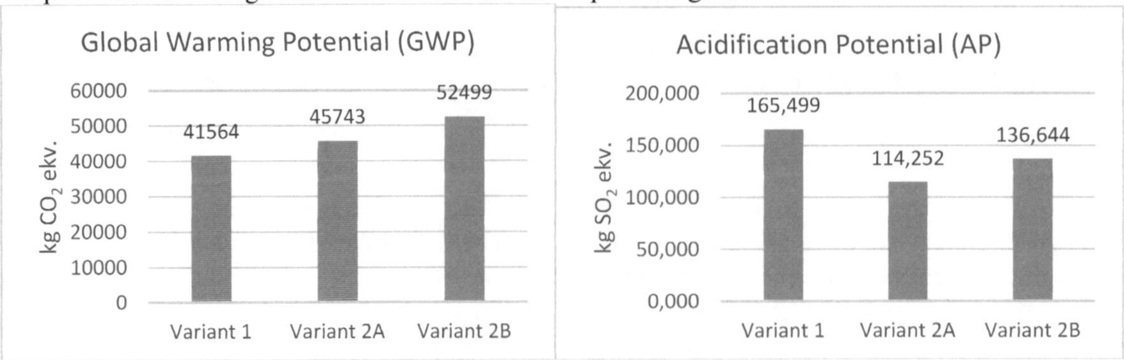

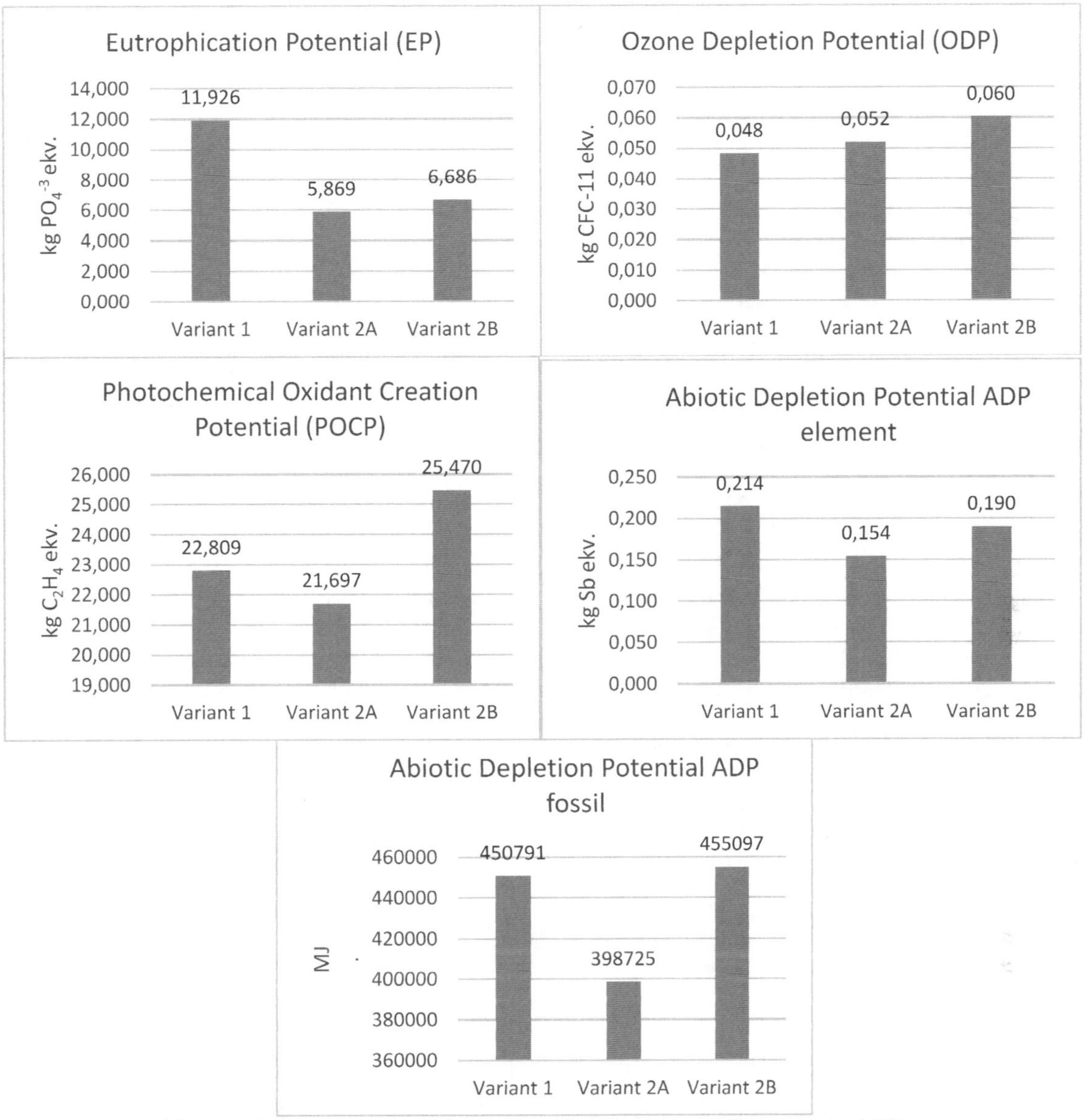

Figure 1. Comparison of the designed variants regarding sustainability.

3.2. Durability assessment

At first, several different mathematical carbonation models were exploited for durability assessment of the variant 1. The results differed significantly, depending on the model. The most favourable results were obtained for models which depend on a smaller number of parameters, for example, only on a compressive strength or water-cement ratio. The results obtained using multicriterial models (for example the Papadakis's model) were not so favourable, although they correspond with velocities of carbonation referred in literature. The following figure shows the time dependence of the carbonation depth for different mathematical models.

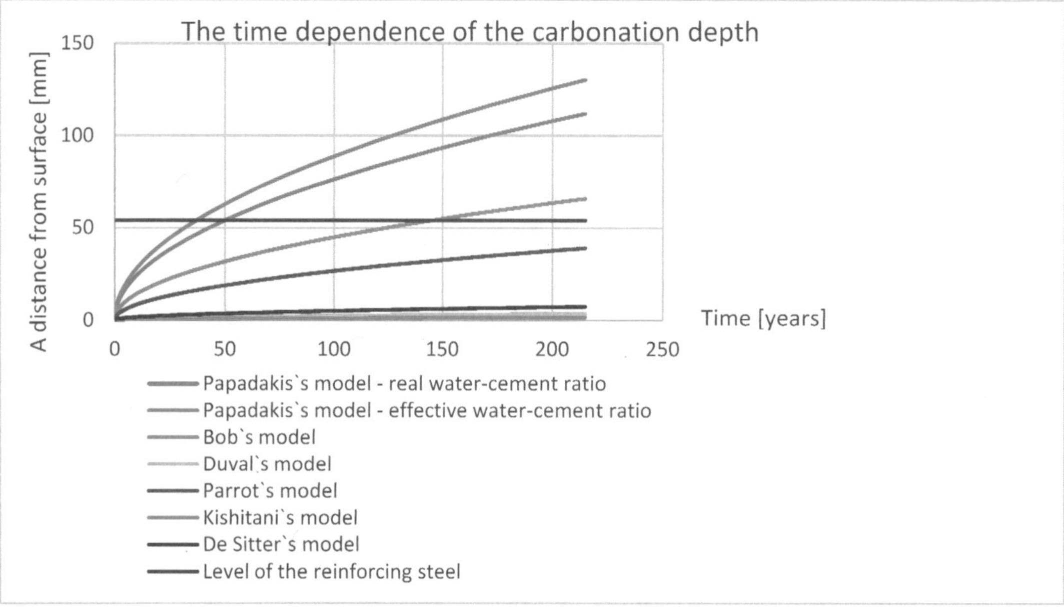

Figure 2. The time dependence of the carbonation depth according to different mathematical models.

Regardless of used mathematical models, the service life of variants from high performance concrete is significantly higher compared to the variant from normal concrete. When Bob`s model of carbonation was used, the results are quite unrealistic, especially for variants 2A and 2B (variants from HPC). So, the service life calculated with using Papadakis`s model was considered decisive. The following figure shows the service life of designed variants.

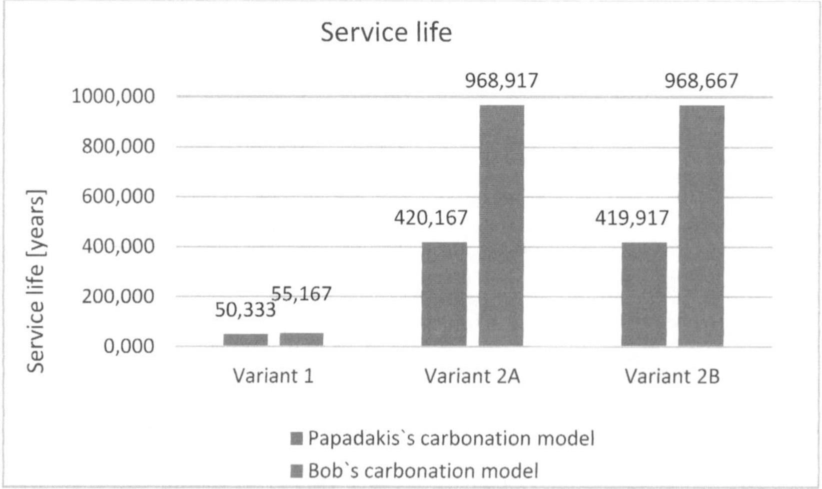

Figure 3. Service life of designed variants.

3.3. A modification of the design

Based on the results of analysis, the originally designed variants were adjusted to reach the end of service life approximately 50 years, not lower. For high performance concrete variants, the thickness of the reinforcement cover was reduced. For the variant 1, the durability was increased in two alternatives: in the first case the reinforcement cover was increased, in the second case the amount of reinforcement was increased. After that, the sustainability assessment was carried out again. For most impacts the results are better for variants from high performance concrete.

The best results were obtained for variant designed from high performance concrete with reduced thickness of the slab (variant 2A). The variant with reduced amount of steel (variant 2B) is not so advantageous, according to this analysis. For the normal-strength concrete variants, the variant, in which amount of reinforcement was increased according to durability demands, is more advantageous regarding environmental impacts.

4. Discussion

Advantages of using high performance concrete are clearly obvious from this assessment. Although environmental impacts of the unit amount of high performance concrete are due to cement and plasticizer content higher than unit amount of normal-strength concrete, the total environmental impacts of the structure from high performance concrete are usually smaller thanks to significant material savings. It was proven that the outstanding mechanical properties of high-performance concrete should be preferably utilised to reduce the thickness of the slab.

Also regarding durability the variants designed from high performance concrete are significantly advantageous. The evaluation concluded that in case of using these concretes it is possible to reduce the thickness of a cover layer much lower than values given by the standard. Especially for structures in aggressive environment very significant material savings can be achieved.

5. Conclusion

To meet sustainable development principles, it is more advantageous to design the structures from the high performance concrete. However, it is always necessary to consider all of the aspects, for example the economic efficiency and requirements for the building (e.g. the requirements for the surfaces quality). In terms of durability, the variants designed from high performance concrete are also significantly profitable. Possible reduction the thickness of a reinforcement cover brings about material savings. It is significant especially in aggressive environment. If the thickness of a reinforcement cover is designed according to standard, the service life of a HPC structure is much longer than the required service life. However, it should be considered whether it is even advisable to use such a long life of buildings, particularly in relation to its moral life service and its purpose.

An accurate durability assessment of structures is very problematic. It is often impossible to determine exactly the input values for the analysis. It is typical primarily for parameters of environment, which, moreover, may change over time. There is a large number of mathematical models for calculation of durability but the service life significantly differs depending on used mathematical model or calculation method. The results should be compared and with common rates of degradation processes known from experience and the most reliable method for analysis should be chosen.

Acknowledgements
This outcome has been achieved with the financial support of the Ministry of Education, Youth and Sports, project: The durability of concrete with recycled composites materials, SGS17/123/OHK1/2T/11.

References

[1] PRODUCT CATEGORY RULES ACCORDING TO ISO 14025:2006. *Product group classification: UN CPC 375*. The international EPD systém, 2015
[2] BADEA, Cătălin, BOB, Corneliu, BALCU, Ionel, DAN, Sorin, BUCHMAN, Iosif, GRUIN, Aurelian. The Carbonation Process Influence on Concrete Durability. *1st WSEAS International Conference on MATERIALS SCIENCE (MATERIALS'08)*, 2008. ISBN: 978-960-474-024-6.
[3] CZARNECKI, L., WOYCIECHOWSKI, P. Modelling of concrete carbonation; is it a process unlimited in time and restricted in space?. Bulletin of the Polish Academy of Sciences Technical Sciences. 2015, Vol. 63, No. 1. ISSN 2300-1917,

[4] DHIR, Ravindra K. a Michael J. MCCARTHY. *Concrete durability and repair technology: proceedings of the international conference held at the University of Dundee, Scotland, UK on 8-10 September 1999*. London: T. Telford, 1999. ISBN 0727728261.

[5] CHROMÁ, M.; ROVNANÍKOVÁ, P.; TEPLÝ, B.; STRAUSS, A. Effective water- cement ratio of concrete and its effect on the durability of concrete structures. In Proceedings pro089: Performance- based Specification and Control of Concrete Durability. Zagreb: 2014. s. 57-65.
ISBN: 978-2-35158-135- 3.

[6] PAPADAKIS, V. G, FARDIS, M. N., VAYENAS, C. G. Effect of composition, environmental factors and cement-lime mortar coating on concrete carbonation. *Materials and structures*. 1992, Vol. 25, issue 4, 293-304. ISSN: 1359-5997.

[7] ŠMERDA, Zdeněk. *Durability of reinforced concrete structures*. Prague: Czech Union of Civil Engineers, 1999. Technical library of authorized engineer and technician. ISBN 80-902697-8-8.

[8] KOHOUTKOVÁ, A.; BROUKALOVÁ, I. *Optimization of Fibre Reinforced Concrete Structural Members*. In: Concrete and Concrete Structures 2013. Žilina: University of Žilina, 2013, pp. 100-106. ISSN 1877-7058.

IOP Conf. Series: Materials Science and Engineering **246** (2017) 012039 doi:10.1088/1757-899X/246/1/012039

Evaluation of Blast Resistance of Fiber Reinforced Composite Specimens under Contact Blast Load

O Janota[1] and M Foglar[1]

[1] Department of Concrete and Masonry Structures, Czech Technical University in Prague, Faculty of Civil Engineering, Thákurova 7/2077, Prague, Czech Republic

E-mail: ondrej.janota@fsv.cvut.cz

Abstract. This paper presents results of experimental programme which took place in 2014, 2015 and 2016. Experiments were focused on the resistance of full scale concrete panels subjected to contact blast loading. Specimens were loaded by contact blast by plastic explosive. All specimens were reinforced concrete slabs made of fiber concrete. Basalt mesh and textile sheets were added to some of the experiments for creating more heterogeneous material to achieve better resistance of the specimens. Evaluation of experiments was mainly focused on the damaged area on the contact side and soffit of the specimens. Dependency of the final damage of concrete panels on the weight of explosive and concrete strength was assessed.

1. Experimental program

Experimental measurements of behaviour of concrete panels loaded by contact blast supplements the measurement of the resistance of specimens against soffit explosion. The experimental programme took place in 2014, 2015 and 2016 in the Boletice military area with cooperation of the Czech Army and University of Defence, University of Pardubice.

Specimens with dimensions 6.0 x 1.5 x 0.3 m are made of reinforced fibre concrete. The specimens were concrete slabs 6 m in length, 1.5 m in width and 0.3 m in thickness. The steel reinforcement was 11 pcs ∅ 16 mm reinforcing bars every 140 mm on both surfaces, ∅ 10 mm every 150 mm as an outer transverse reinforcement, and shear reinforcement was provided by ∅ 8 mm links (9 pcs/m^2) [3]. Cover of the stirrups was 50 mm. Explosive was made of pentaerythriol tetranitrate also known as PET or CAS 78-11-5 with TNT equivalent 1.3 [7] [4].

Each specimen can be theoretically divided into three parts according to the blast loading. The middle part was subjected to soffit blast and then inspected with ultrasound device for the internal cracks. This procedure provided information whether edge parts of specimens are affected by soffit blast or not. Edge parts of the specimen were reserved for contact blast experiment. Scheme of the experiment is presented in Figure 1. Eleven specimens were loaded in total, ten of them with two explosives on both edges of the panel. In total twentyone measurements were performed. Damage of the specimen on each side was assessed.

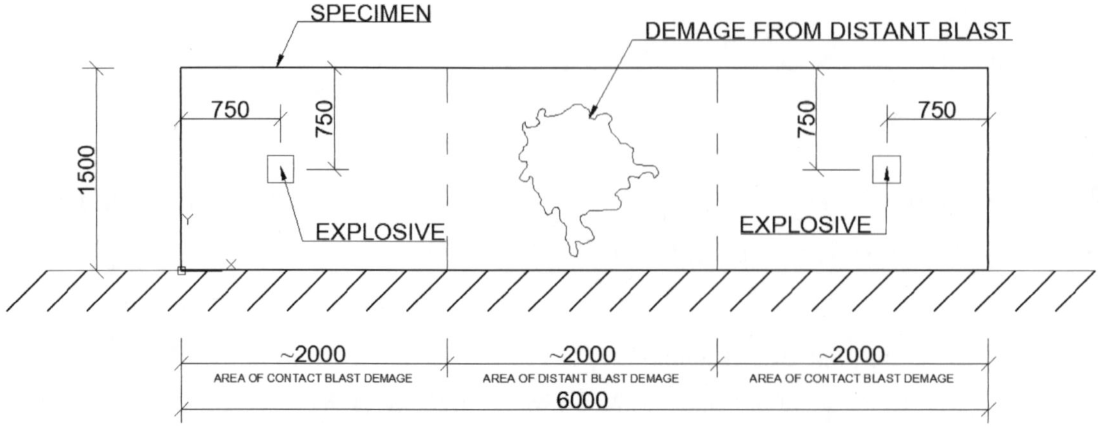

Figure 1. Scheme of the experiment

2. Blast resistance of the specimens

In this chapter, experimental results are provided; the tested specimens are assessed in terms of damaged area on both sides of the specimen, shape of the damaged area and type of the damage. More results and information are presented in [1].

The experiment was performed using eleven specimens. Specimens were numbered chronologically according the order of the experiment. Specimens No. 12 to 19 and No. 21 to 23 were tested for contact blast resistance. Ten of the specimens were loaded by two explosives, one on the left and one the right side of the specimen. Both explosives were initiated simultaneously. The weight of the explosive charge differs from 0.5 kg to 4.2 kg. Different concrete mixes were used. Cubic strength of the concrete vary from 68.5 MPa to 129.8 MPa [2]. Fibers with tensile strength 2200 MPa and length 13 and 35 mm were used in the concrete. Amount of the fibers and exact compressive strength of the each specimen is presented in Table 1.

Heterogeneity of specimens No.17 and No.18 was raised by adding basalt meshes into the structure of the concrete panel. Specimen No.17 contained only one basalt mesh in the middle of the concrete cover at the soffit of the specimen. Specimen No.18 contained five basalt meshes. Two of them were placed at the middle of the concrete cover and the rest of the meshes were placed between the concrete reinforcement layers with spacing of 50-70 mm [5] [6]. Recycled textile sheets with thickness of 50 mm were inserted into specimen No.19 to increase its heterogeneity.

Specimen No. 12 was loaded by explosives of 1.0 kg on the right side and 1.4 kg on the left side. Damage of the specimen on the both sides can be seen in Figure 2. Final damages on the left and right side of the specimen were very similar by its character. Also damaged areas on the contact and soffit of the panel were very similar. The blasts did not achieve puncture of the specimen. On the contact side, the blast damaged area is not deeper than the cover layer. Reinforcement was not damaged by the blast. On the soffit of the specimen, the damage is deeper than the cover layer, but the rebars were not damaged or bent. The damaged area on the soffit was approximately 2.5 times bigger than damage on the contact side. The damaged areas had circular shape on both sides of the panel. Concrete in the perimeter of the damaged area was affected by cracks. Some of the cracks on the left contact side reached edge of the specimen.

IOP Conf. Series: Materials Science and Engineering **246** (2017) 012039 doi:10.1088/1757-899X/246/1/012039

Figure 2. Specimen No. 12 after contact blast loading, 1.4 kg charge left left contact side, left right soffit, 1.0kg charge right left contact side, right right soffit

Specimen No. 13 was loaded by explosives with weight 1.0 kg on the right side and 0.5 kg on the left side. Damage of the specimen on the both sides can be seen in the Figure 3. As well as the specimen No.12, specimen was not punctured through. The damage on the contact side did not reach depth of cover layer and reinforcement was not deformed by the blast. Shape of the damaged area was a circle. Soffit showed only few major crack on the right side and the surface was bulged out of the specimen. On the left soffit there were not marks of the damage at all.

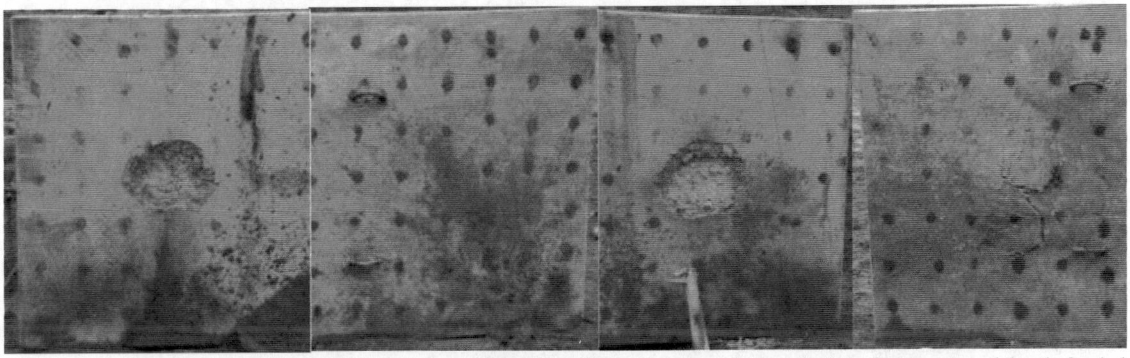

Figure 3. Specimen No. 13 after contact blast loading, 1.4 kg charge left left contact side, left right soffit, 2.8 kg charge right left contact side, right right soffit

Specimen No. 14 was loaded by explosives with weight 2.8 kg on the right side and 1.4 kg on the left side. Damage of the specimen on the both sides can be seen in the Figure 4. Specimen was punctured through on the right side. Shape of the puncture was circle and was the same as in case of the damaged area of surfaces on both sides of panel. Area of damaged surface on the contact side was approximately 43% of the soffit on the right side and 30% on the left side. There were no cracks on the contact side. The soffit on the right side was strongly disturbed by cracks, left side was not disturbed by cracks. Some of the wides cracks reach edges of the specimen. Crack were mostly perpendicular to the perimeter of the damaged area.

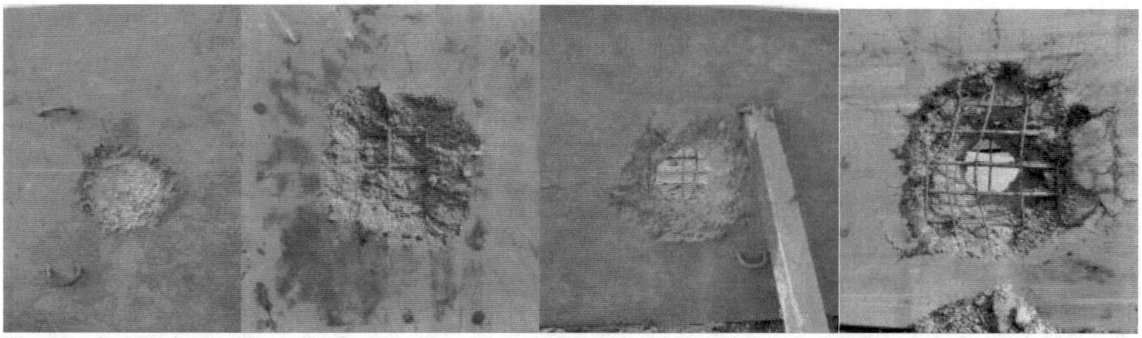

Figure 4. Specimen No. 14 after contact blast loading, 1.4 kg charge left left contact side, left right soffit, 2.8 kg charge right left contact side, right right soffit

Specimen No. 15 was loaded by explosives with weight 1.4 kg only on the left side. Damaged of the specimen can be seen in the Figure 5. The specimen was not punctured. Damaged area depth was mostly lower than the concrete layer. Damage on the contact side was approximately 25 % of the damage on the soffit. Reinforcement was not damaged on both sides of the panel. Type of damage was more close to pulling out failure than punching shear failure. On the contact side, there were cracks that reached edges of the specimen. Cracks on the soffit were perpendicular to the perimeter of the damaged.

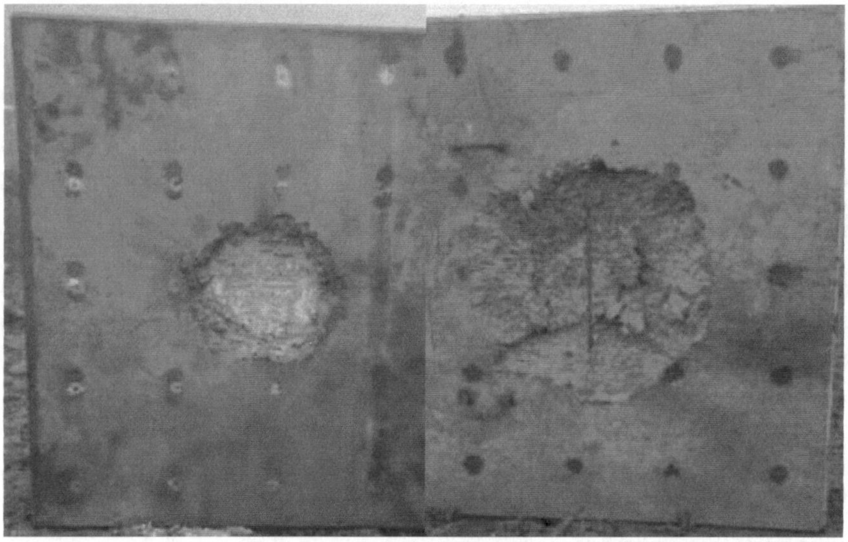

Figure 5. Specimen No. 15 after contact blast loading, 1.4 kg charge left left contact side, left right soffit

Specimen No. 16 was loaded by explosives with weight 2.8 kg on the right side and 1.4 kg on the left side. Damage of the specimen on the both sides can be seen in Figure 6. Specimen on the both sides was not punctured through. Shape of the damaged area on the contact side was circular, on the soffit it was more similar to the rectangle. Reinforcement on the contact blast was not deformed but the depth of damaged area reach the reinforcement. Steel bars were deformed and bent out of the panel on the right side of soffit, on the left side reinforcement was not damaged. Type of damage was similar to specimen No.15. There was a very dense mesh of microcracks on the contact blast. Cracks on the right side of soffit were wider and some of them reached edges of the specimen.

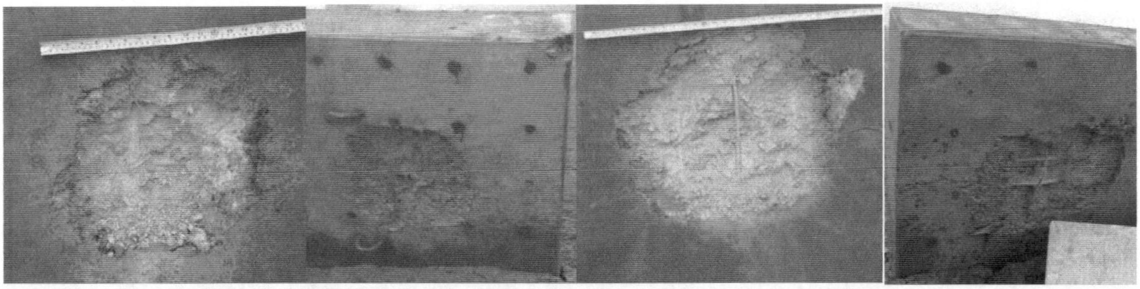

Figure 6. Specimen No. 16 after contact blast loading, 2.8 kg charge left left contact side, left right soffit, 2.0 kg charge right left contact side, right right soffit

Specimen No. 17 was loaded by explosives with weight 2.0 kg on the right side and 2.8 kg on the left side. One ballast mesh was added into the cover layer. Charge on the left side was placed on the side with basalt mesh in cover layer. Charge on the right side was placed on the opposite side of the specimen. Damage of the specimen on both sides can be seen in Figure 7. Specimen was not punctured on both sides. Damage on the right contact side was similar to the failure due to punching shear but the depth was equal to cover layer. The reinforcement was not damaged. Shape of the damaged area was circular. Area of the damage on the contact side was approximately 65% of the damaged area on the soffit. The soffit was damaged more and the shape of the damaged area was irregular. The type of the failure corresponded with the delamination of the concrete layers. Most of the damaged area was a delaminated layer to the depth of the basalt mesh. Second delaminated layer consists of concrete the depth basalt mesh to the depth of reinforcement bars. Reinforcement on the soffit was practically undamaged. Damage on the left contact side of the specimen was deeper than the cover layer. The reinforcement was unbroken but little bent. Type of the damage was similar to the punching shear failure. Damage area on the soffit was rectangle shape with the depth slightly more than the cover. Reinforcement was only little bent. Soffit side on the left side was not broken into the layers as the right soffit side was.

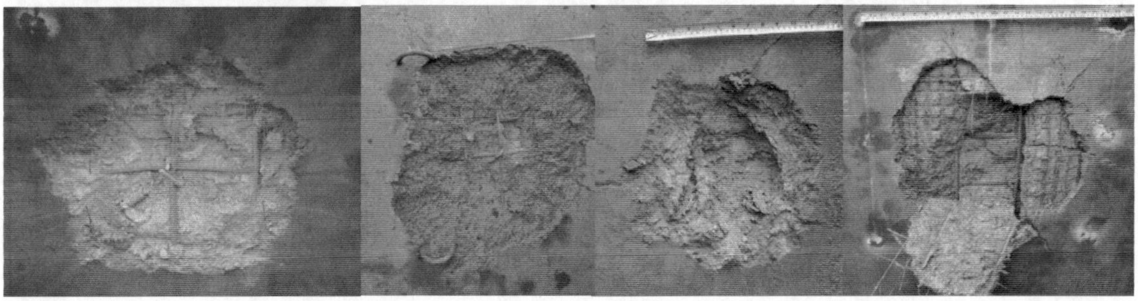

Figure 7. Specimen No.17 after contact blast loading, 2.8 kg charge left left contact side, left right soffit, 4.2 kg charge right left contact side, right right soffit

Specimen No. 18 was loaded by explosives with weight 4.2 kg on the right side and 2.8 kg on the left side. Damage of the specimen on both sides can be seen in the Figure 8. Ballast meshes were added into the cover layer in the both sides and between reinforcement. Specimen was not punctured. The damage on the left and right side was similar in its shape and type. The shape of the damaged area on contact blast was circular and was more similar to punching shear failure. The damage reached deeper than the cover layer. Reinforcement was little bent but not broken. Damaged area on the soffit was more complex than damaged area on the contact side. Shape of the damaged area was ellipse and similar to pulling out failure. Reinforcement was heavily bent but not broken. Under the reinforcement mesh there were pieces of the panel that were bigger than the spacing of the reinforcement or ballast mesh. Perimeter of the damaged area was broken by many cracks in the radial direction. Crack which crossed radial cracks and copying perimeter of the damaged area was detected approximately in the distance 220 mm from the damaged area.

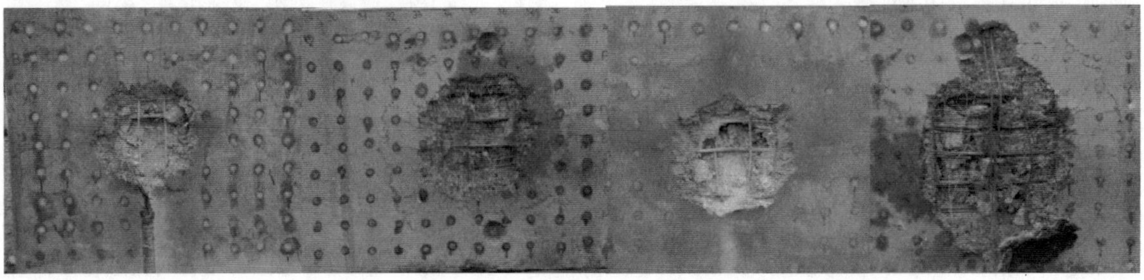

Figure 8. Specimen No. 18 after contact blast loading, 2.8 kg charge left left contact side, left right soffit, 4.2 kg charge right left contact side, right right soffit

Specimen No. 19 was loaded by explosives with weight 4.2 kg on the right side and 2.8 kg on the left side. Damage of the specimen on the both sides can be seen in the Figure 9. The recycled textile sheets were added to the specimen between reinforcement layers. The specimen was not punctured. Type of the failure on the both sides was similar to the specimen No. 18. On the contact side, it was caused by the punching shear failure, on the soffit by the tension failure. On both sides the depth of the damage was bigger than the cover layer. On the contact side the reinforcement was broken. On the soffit the reinforcement was bent but not broken. Cracks could be seen on the soffit. Cracks are radial to the damaged area. Some of them were reaching edge of the specimen. As well as in the case of No. 17 there was a crack perpendicular to the radial cracks but not continuous.

Figure 9. Specimen No. 19 after contact blast loading, 2.8 kg charge left left contact side, left right soffit, 4.2 kg charge right left contact side, right right soffit

Specimen No. 21 was loaded by explosives with weight 4.2 kg on the right side and 2.8 kg on the left side. Damage of the specimen on the both sides can be seen in the Figure 10. The specimen was punctured on both sides. Type of the failure on the both sides was similar in its type. Shape of the damage area on both sides and surfaces was circular. On the contact side and soffit, it was caused by punching shear failure. On the both sides the reinforcement was broken and on the soffit reinforcement was broken and bent out of the specimen. Cracks could be seen on the on the soffit. Cracks are radial to the damaged area. Damaged area on the right contact side is approximately 2,5 times smaller than on the soffit. On the right side damage on the contact side is approximately 3 times smaller than on the soffit.

Figure 10 Specimen No. 21 after contact blast loading, 2.8 kg charge left left contact side, left right soffit, 4.2 kg charge right left contact side, right right soffit

Specimen No. 22 was loaded by explosives with weight 4.2 kg on the right side and 2.8 kg on the left side. Damage of the specimen on both sides can be seen in the Figure 11. Ballast meshes were added into the cover layer in the both sides and between reinforcement. Specimen was punctured through on the right side. Damage on the left and right side was similar in its type. Damage on the contact side and soffit was caused by punching shear failure. Damaged area on the contact side was approximately 55% of the damaged area on the right soffit and 70% on the left side. Reinforcement on the contact side was not deformed. Reinforcement on the soffit is broken and bent out of the panel. On the soffit there was evident delamination caused by adding basalt mesh. As well as in the case of specimen No. 17 the first delaminated layer contains concrete from the surface to the depth of the first basalt mesh. Second delaminated layer was from basalt mesh to the reinforcement. Perimeter of the damaged area was broken by cracks on the soffit.

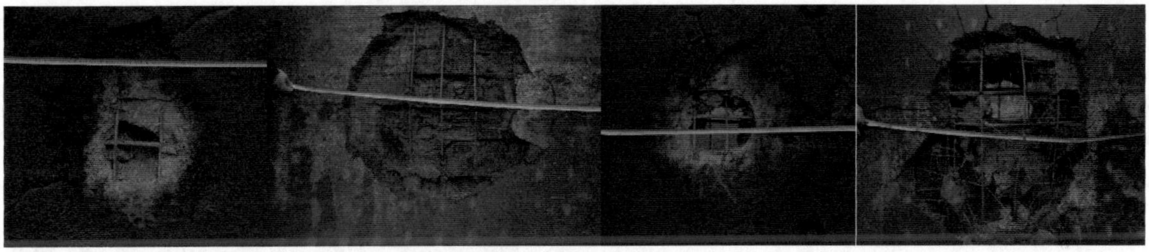

Figure 11. Specimen No. 22 after contact blast loading, 2.8 kg charge left left contact side, left right soffit, 4.2 kg charge right left contact side, right right soffit

Specimen No. 23 was loaded by explosives with weight 4.2 kg on the right side and 2.8 kg on the left side. Damage of the specimen on the both sides can be seen in the Figure 12. Textile sheets were added to the specimen between reinforcement layers. The specimen was not punctured. On the contact side, it was caused by punching shear failure, on the soffit by the tension failure. On the left contact side damage did not reach the reinforcement. On the rightside damage was deeper than the cover layer. Reinforcement was not damaged. Reinforcement on the soffit was broken on the right and left side. Damaged area on the contact side was approximately 86% of the damaged area on the right soffit and 68% on the left side. Cracks could be seen on the soffit. Cracks are radial to the damaged area. Some of them were reaching edge of the specimen.

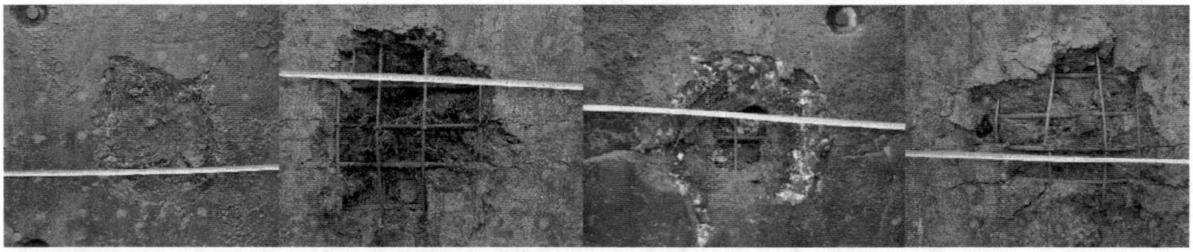

Figure 12. Specimen No. 23 after contact blast loading, 2.8 kg charge left left contact side, left right soffit, 4.2 kg charge right left contact side, right right soffit

Table 1 presents summary of the experiment in the terms of the damaged area on the different specimens. It is apparent that there is a dependency of the final damage on the value of the concrete strength and also amount of fibres in the specimen. Specimen No. 16 with strength 129.5 MPa, but much shorter fibres, was damaged little more than specimens with lower concrete strength on the soffit but the damage on the contact side was lower. Figure 13 and 14 presents dependency of damaged area to the weight of the explosive. Adding elements for increasing the heterogeneity in specimens No. 18 and No. 19 rapidly decreased the final damage of the specimens. This phenomenon can be caused by the change of the mass density within the specimen on the boundary between concrete and the added elements. On this boundary, the overpressure wave from the explosive is partially rebounded back and partially goes through. Using this system, the wave can be distributed to a wider area and the overpressure decreases. Specimens No.18, No. 19, No. 22 and No. N23 has also showed different type of the failure on the contact side than the other specimens.

Table 1. Summary of the area damaged by the explosion on the different specimens

Specimen		No. 12	No. 13	No. 14	No. 15	No. 16	No. 17	No. 18	No. 19	No. 21	No 22	No 23
Cubic strengh		68.5 MPa	66.9 MPa	73.2 MPa	76.1 MPa	129.5 MPa	125.8 MPa	77.9 MPa	78.3 MPa	127.1 MPa	121.6 MPa	91.6 MPa
Fibres FE dl.13mm f_{ct} = 2200MPa		-	-	-	-	120 kg/m^3	80 kg/m^3	120 kg/m^3	80 kg/m^3	80 kg/m^3	120 kg/m^3	80 kg/m^3
Fibres FE dl.25mm f_{ct} = 400MPa		-	-	-	-	-	-	-	-	-	-	-
Fibres FE dl.35mm f_{ct} = 2200MPa		40 kg/m^3	80 kg/m^3	40 kg/m^3	80 kg/m^3	-	80 kg/m^3	-	80 kg/m^3	80 kg/m^3	-	-
Fibres FE dl. 55mm f_{ct} = 2200MPa		-	-	-	-	-	-	-	-	-	-	80 kg/m^3
Puncture	Left side	0	0	0	0	0	0	0	0	0,15 m^2	0	0
	Right side	0	0	0.046 m^2	0	0	0	0	0	0,225 m^2	0	0
Damaged area not deeper than cover layer – soffit	Left side	0.465 m^2	0	0.361 m^2	0.402 m^2	0.433 m^2	0.122 m^2	0.16 m^2	0.12 m^2	0.48 m^2	0.567 m^2	0.313 m^2
	Right side	0.415 m^2	0.071 m^2	0.609 m^2	0	0.454 m^2	0.55 m^2	0.18 m^2	0.07 m^2	0.496 m^2	0.557 m^2	0.306 m^2

Damaged area deeper than cover layer – soffit	Left side	0.142 m^2	0	0.111 m^2	0.083 m^2	0.081 m^2	0.037 m^2	0.12 m^2	0.12 m^2	0.37 m^2	0.240 m^2	0.232 m^2
	Right side	0.077 m^2	0	0.354 m^2	0	0.115 m^2	0.102 m^2	0.14 m^2	0.11 m^2	0.38 m^2	0.400 m^2	0.216 m^2
Damaged area not deeper than cover layer – contact side	Left side	0.216 m^2	0.111 m^2	0.121 m^2	0.128 m^2	0.106 m^2	0.087 m^2	0.16 m^2	0.13 m^2	0.15 m^2	0.309 m^2	0.214 m^2
	Right side	0.165 m^2	0.122 m^2	0.266 m^2	0	0.268 m^2	0.134 m^2	0.21 m^2	0.10 m^2	0.225 m^2	0.399 m^2	0.263 m^2
Damaged area deeper than cover layer – contact side	Left side	0.037 m^2	0	0	0.023 m^2	0.005 m^2	0.010 m^2	0.11 m^2	0.08 m^2	0.075 m^2	0.056 m^2	0
	Right side	0.026 m^2	0	0.066 m^2	0	0.101 m^2	0.067 m^2	0.13 m^2	0.10 m^2	0.111 m^2	0.075 m^2	0.037 m^2

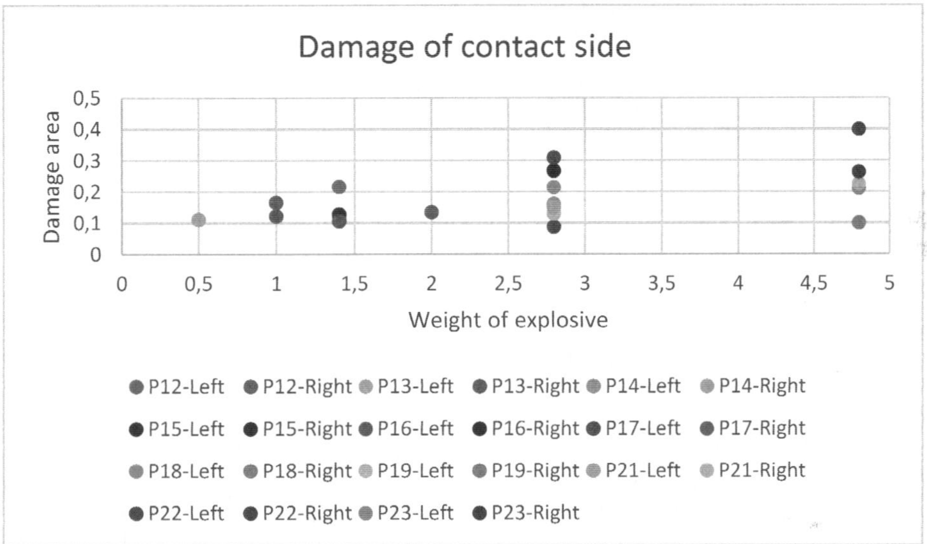

Figure 13. Dependency of the damaged area and the weight of the explosion – contact side

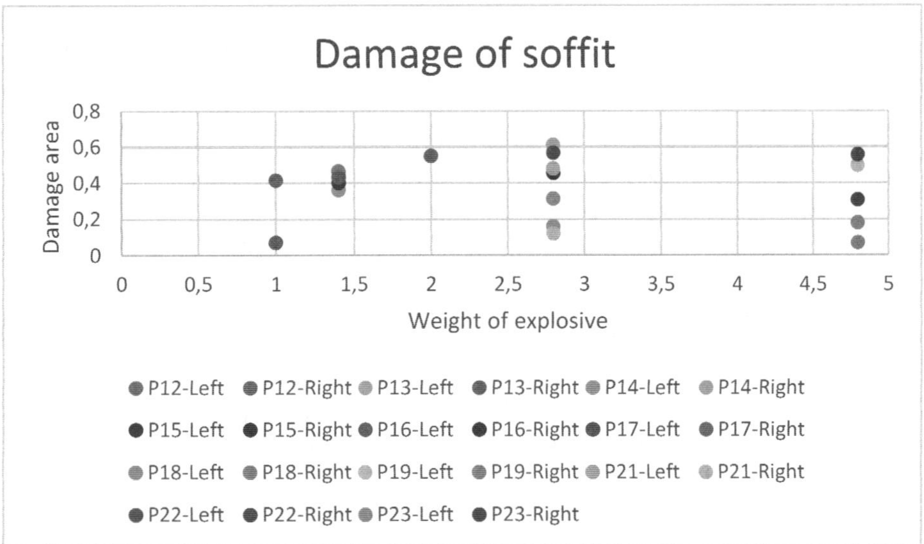

Figure 14. Dependency of the damaged area and the weight of the explosion – soffit

IOP Conf. Series: Materials Science and Engineering **246** (2017) 012039 doi:10.1088/1757-899X/246/1/012039

3. Conclusion

Experiments focused on the resistance of the concrete specimens against contact blast were presented in this paper. Different types of concrete specimens were tested. Each specimen differed by the concrete strength and amount of the fibres. Elements for increasing the heterogeneity were added to several specimens. Results of the experiments indicate that it is possible to mitigate the damage caused by the blast even if the blast load increases and the strength of concrete does not increase reasonably; the key lies in increasing the heterogeneity of the specimen.

Implementation of fibres to the concrete panel and increasing the concrete strength seems to be two of the key factors influencing the blast resistance. Results of the experiments also indicate that increasing heterogeneity of the panel is very significant for reducing its damage. Specimen with basalt meshes and recycled textile sheets experience the most significant difference in the results. Specimens with these elements were less damaged even if the concrete strength is lower than of other specimens. The type of concrete failure also varies depending on the side of the panel. Concrete surface on the contact side was often punctured. Concrete surface on the soffit was mostly spalled.

Acknowledgements

This work was financially supported by Grant Agency of the Czech Republic, project no. 17-23067S, which is gratefully acknowledged.

References

[1] Foglar, M, Hájek, R, Štefan, R and Stöhr, J. Vyšetřování vlastností cementových kompozitů vystavených působení extrémních zatížení Czech Technical University in Prague (2016) ISBN 978-80-01-05892-3

[2] Foglar, M., Kovar, M. Conclusions from experimental testing of blast resistance of FRC and RC bridge decks (2013), International Journal of Impact Engineering, 59, pp. 18-28. DOI: 10.1016/j.ijimpeng.2013.03.008

[3] Foglar, M., Hajek, R., Kovar, M., Štoller, J. Blast performance of RC panels with waste steel fibers (2015) *Construction and Building Materials*, 94, art. no. 6897, pp. 536-546. DOI: 10.1016/j.conbuildmat.2015.07.082

[4] Foglar, M., Hajek, R., Fladr, J., Pachman, J., Stoller, J. Full-scale experimental testing of the blast resistance of HPFRC and UHPFRC bridge decks (2017). Construction and Building Materials, in review process.

[5] Kovar, M., Foglar, M., Hajek, R., Blast performance of real-scale reinforced concrete specimens with varying fiber type and content. In: *WIT Transactions on The Built Environment*, 2014, 141, p. 159-169.

[6] Foglar, M., Pachman, J., Pelikan, V., Kunzel, M., Kovar, M., The structural response of a reinforced concrete specimen subjected to adjacent blast loading, In: *WIT Transactions on the Built Environment*. 2014, 141, p. 171-179.

[7] Hajek, R., Kovar, M., Foglar, M., Pachman, J., Stoller, J., Field testing of concrete members subjected to contact and adjacent blast. In: *Advanced Materials Research, 2015*, 1106, p. 164-167.

IOP Conf. Series: Materials Science and Engineering **246** (2017) 012040 doi:10.1088/1757-899X/246/1/012040

Effect of the loading rate on fibre reinforced concrete beams

K P Juhász[1, 3], P Schaul[2, 3] and L Nagy[3]

[1] Department of Mechanics, Materials & Structures, Budapest University of Technology, Műegyetem rkp. 3, 1111 Budapest, Hungary
[2] Department of Construction Materials and Technologies, Budapest University of Technology, Műegyetem rkp. 3, 1111 Budapest, Hungary
[3] JKP Static Ltd, Reitter Ferenc utca 73, 1135 Budapest, Hungary

E-mail: office@jkp.hu

Abstract. Fibre reinforced concrete has become a widely used material since the end of the 20th century. The uniformly distributed steel or macro synthetic fibres in the concrete structures can give the concrete a residual flexural strength after the first cracks. The different behaviour of materials subjected to different loading rates is a well-known phenomenon, both with steel, synthetic and concrete materials. Standards usually present a recommendation for the loading speed in for different tests. Concrete elements show higher performance due to the high speed of loading or impact loads, their fracture energy and therefore their overall capacity appears greater than the specimens loaded at standard speeds. Fibre reinforced concrete structures are widely used in tramlines and railways, where the speed of the loading is high and therefore of impact by nature. It is important to know what the effect of this high speed loading is for fibre reinforced concrete structures: do these structures have additional capacity, or has the designer overestimated their performance? This article will present an investigation into the effect of the loading rate on case of using fibres with different materials in concrete beams.

1. Introduction

The behaviour of concrete structures is well-known to be highly influenced by their loading rate [1]. Numerical and experimental studies show that by increasing the loading rate there is also an increase in resistance, failure mode, crack pattern and speed of crack propagation. The influence of the loading (strain) rate on the tensile strength and fracture energy of concrete is not yet fully understood. Therefore, it has been one of the major subjects of intensive investigation in the scientific community over many years. Also, the knowledge with regards to the response of fibre reinforced concrete subjected to dynamic loading is still incomplete.

The two types of common structural fibres used for concrete reinforcement today are the steel and the macro synthetic fibres. These fibres increase the residual flexural strength of the concrete elements by adding an additional fracture energy to the concrete material [2]. The fibres during the load start to work after the first cracks appear in the concrete during loading. The bonding mechanisms of the fibres within the concrete are different for both steel and synthetic materials are also different, the steel fibres bearing the load with the anchorage of the hooked end, and the failure mechanism is pulling out from the concrete matrix (ie. the hook straightening) and in contrast to this the synthetic fibres have an anchorage capacity over their full length of the fibre with its embossed surface so that the failure mode is both pulling out and also the rupture of the fibres. There are many specifications available to test the fibre reinforced concrete elements [3] [4], and usually these guidelines also provides a recommendation for the loading speed of the tests. Usually, the typical test for fibre reinforced concrete specimens is the three or four point bending beam tests with, or without a notch in the middle. The tests are always displacement controlled, and the recommended loading rate is 0.1 - 0.2 mm/min. According to this the determined fibre capacity is valid just for this loading rates, which corresponds to a dead loads, static live loads, soil pressures and meteorological loads.

Fibre reinforced concrete is widely used in structures where the loads are mostly impact by nature, such as railways and tracks slabs. In these cases the speed of loading and unloading the structure is very high, and thus it is very important to know what the effect of the fibres in these situations is.

Because the steel and synthetic materials have different behaviour characteristics at different speeds, it is important to analyse the effect of a particular type of fibre reinforcement on the fibre reinforced concrete acting as a composite material. This paper outlines the undertaken on steel and synthetic fibre reinforced concrete beams under different loading rates.

2. Effect of the loading rate in fibre materials

Before examining the effect of the loading rate on fibre reinforced concrete beams it is important to understand the behaviour of the materials of the fibres on a wide range of loadings.

The influence of the loading rate on the tensile strength characteristics of steel members is a well-researched field in the material science since the middle of the 20th century and such as a wide range of tests and formulas can be found in the literature which define and describe the behaviour of steel materials at high speed loading [5].

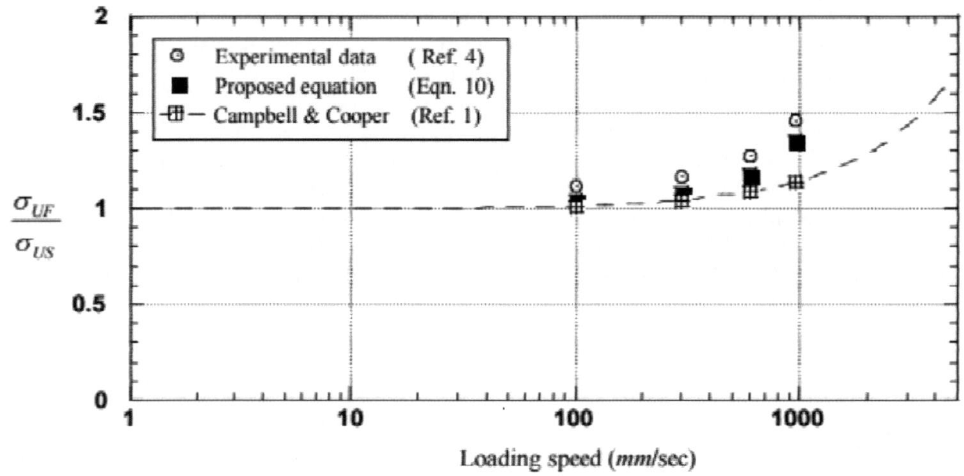

Figure 1. Effect of loading speed in steel elements [5].

Both the recommended formulas and the tests shows an increasing capacity of the steel element due to an increasing the load (figure 1). There is no significant influence of the speed up to 100 mm/sec, which means the steel structures under static loads show a constant behaviour. Over this limit the material's tensile strength starts to raise significantly. It means the structure's capacity will be higher if the type of load is impact in nature. It also means that the calculations are conservative if the designer has calculated with the low speed loading values.

The base material of the synthetic fibres is usually polypropylene (PP), which is a thermoplastic polymer and there are number of previous studies [6] available that document the test results of polypropylene at different loading rates.

Typical stress–strain curves of the tested PP at different speeds is shown in figure 2. The test showed that samples did not break even at 60% elongation and at crosshead speeds of up to 20 mm/min, but for the higher test speeds, the samples were ruptured at a lower elongation. Also, in these cases the ultimate strength of the samples was 45% higher than in the case of low speed loading.

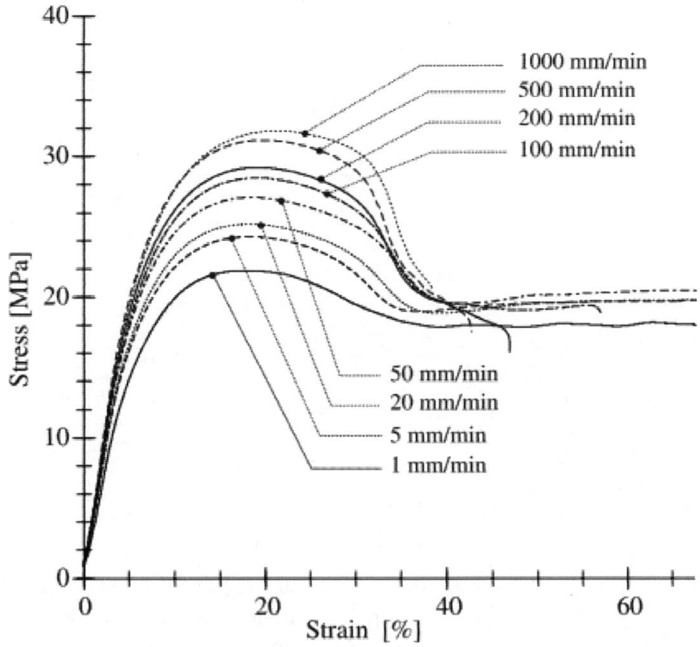

Figure 2. Effect of loading speed in polypropylene elements [6].

It can be seen, that both off the materials capacities rise if the loading speed is higher than a critical value and this critical value for steel is much higher for steel than for polypropylene. In the case of the steel there is no significant effect due to loading rate up to about 6000 mm/min (100 mm/sec).

3. Experimental program and testing method

To examine the behaviour of fibre reinforced concrete under different loading rates a three point bending beam test series was performed with steel and macro synthetic fibre reinforcement. The test matrix can be seen in table 3. The type of the reinforcement and the used dosage can be seen in table 2.

Prior to the beam tests single fibre tension test were carried out with the used fibres, to be able to verify the results of the fibre's capacity under different loading rates.

3.1. Specimen preparation and testing

Three points notched beam tests were made according to RILEM TC162 [3]. The load, the crosshead's vertical displacement and the beam's Crack Mouth Opening Displacement (CMOD) were measured for all of the beams. To be able to measure the CMOD a 25 mm deep notch was cut into the middle of the beams and two steel knifes were glued on the bottom surface of the beam. The measuring clip was positioned between the two points. The test was performed by a universal testing machine ZWICK Z150 in the Laboratory of Department of Mechanics, Materials and Structures, Budapest University of Technology and Economics. The speed of the test was 0.2 mm/min and 900 mm/min up to 4 mm central deflection. The maximum speed of the testing machine was 900 mm/min.

For the single fibre tension tests one piece of fibre was clamped to the same universal testing machine. The snapping force and elongation was measured at loading rates of 200 mm/min, 50 mm/min, 20 mm/min, 10 mm/min and 2.0 mm/min loading rate.

3.2. Concrete mix

The concrete mix was designed to model a typical concrete industrial floor. Cement type was CEM III/A-42.5-N, water-cement ratio was 0.40. The aggregate was river gravel, with size 4-16 mm. The mixture design can be seen in table 1.

Table 1. Concrete mix

Concrete name	Cement type	w/c ratio	Aggregates (kg/m^3)			Admixtures
			0-4	4-8	8-16	
A	CEM-III-A-42.5	0.400	735	294	808	Dynamon NRG 1012

Table 2. Fibre reinforcing

Fibre sign/name	Fibre type	Fibre length mm	Dosage kg/m^3	Number of fibres Number/m^3
BC48 – 4.0 Barchip48	Synthetic fibre Surface embossed	48	4.0	240 964
AF – 25.0 Armfib®	Steel fibre Hooked-end	50	25	78 150

3.3. Fibre reinforcement
Two different types of fibre reinforcement were used during the experiment: steel and macro synthetic. The dosages of the fibres were 25 kg/m^3 for steel fibres and 4.0 kg/m^3 for synthetic fibres. The type of fibre and the research matrix can be seen in table 2 and 3, respectively. Seven beams were made with each type of the fibre reinforcement.

3.4. Beam casting
The aggregate, sand and cement were first mixed while the water was added continuously. Mixing was carried out with Collomatic XM3 forced action mixer. For the fibre reinforced concrete the fibres were dispersed by hand and mixed for approximately for one minute to achieve a perfectly uniform distribution. The concrete was casted into steel formworks and was demoulded after one day. The beams were kept at a temperature of 25 °C, and at a relative humidity of 50-60% for 28 days.

Table 3. Test matrix

Name of the specimen	Concrete	Fibre – dosage [kg/m^3]	Testing speed
AF25-A-L-number	A	AF – 25.0	0.2 mm/min
AF25-A-H-number	A	AF – 25.0	900 mm/min
BC48-A-L-number	A	BC48 – 4.0	0.2 mm/min
BC48-A-H-number	A	BC48 – 4.0	900 mm/min

4. Results

4.1. Single fibre tension test
To understand the fibre reinforced concrete's behaviour under different loading rates it was important to check the capacity of the discrete fibres under different speeds of loading. Testing of the macro synthetic fibres had to be done with some circumspection, because a local failure can occur close to the clamping head of the machine which can give misleading results. For case of every loading speed 5 pieces of fibres were tested to failure to be able to calculate the average value of the fibres' capacity.

The results shows that in case of synthetic fibres there is a high effect of the loading speed both at high and low speeds. By raising the speed of the test, the capacity and also the Elastic modulus of the fibres will be higher. This increase in the capacity is significant at more than 25%. It showed that the elastic modulus changed from 7.20 GPa to 11.77 GPa. The critical value for the loading speed which

IOP Conf. Series: Materials Science and Engineering **246** (2017) 012040 doi:10.1088/1757-899X/246/1/012040

was where the strength of the fibre started to increase significantly was approximately between 20 and 50 mm/min.

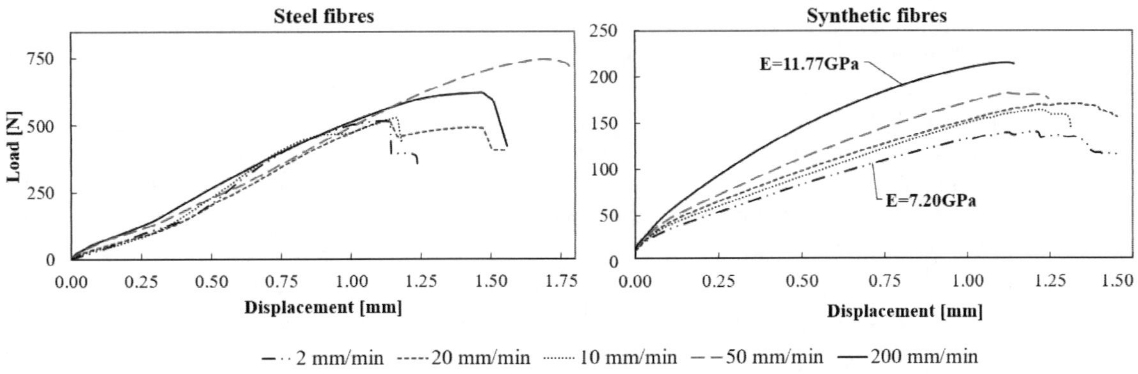

Figure 3. Results of the single fibre tests in different loading rates

The results in case of steel fibres are different. The test did not show any difference in elastic modulus at the various loading speeds. The elastic modulus and the tensile strength of the fibre is almost the same in every load rate. This is because the critical speed for steel material is much higher than what was used.

4.2. Three point bending beam test

The mixing of the concrete for the beams was good for all fibre dosages. The beams were tested at 28 days, and the necessary steel pointers for CMOD measuring were glued after 27 days. The loading schema and the geometry of the beams can be seen in figure 4. The data recovery was displacement controlled, which ensured that enough data would be available for the high speed loading test as well.

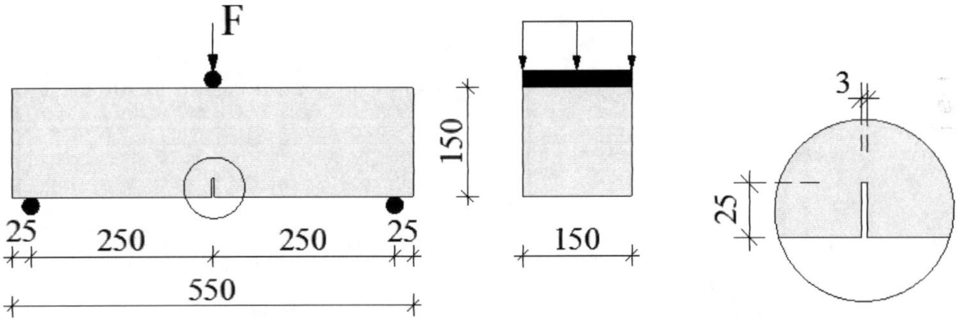

Figure 4. Three point bending beam test according to RILEM TC 162

The beams show different behaviour at different loading rates. The figure 5 shows the average values of the Load-CMOD diagram of the 7-7 beams in every case. With the standardized loading speed (0.2 mm/min) the steel (SFRC) and the synthetic (SYFRC) fibre reinforced concrete also have a drop of after the peak point. The average peak load and the elastic modulus was almost the same for both reinforcements. The residual flexural strength is higher in case of SFRC, and also the drop off is smaller. This difference can be caused by the lower elastic modulus of the synthetic fibres as well. It also can be seen, that the residual flexural strength in case of SFRC is monotone decreasing in contrast to this, the SYFRC's residual flexural strength has first a valley but then it increase as the crack opens and the fibres take load, and at 3.5 mm CMOD it is approaching the steel's curve.

By increase the speed of the load to 900 mm/min both of the curves have changed. The similarity between the steel fibre reinforced and the synthetic fibre reinforced concrete was, that the residual flexural strength increased in every case. With this the fibre-work of the FRC, which is the area under the force-CMOD diagram increased as well. This increase compared to the original curves is higher in the case of synthetic fibres. The peak load will be higher in case of steel fibres, but lower in the case of synthetic fibres. However as the figure 6 shows the elastic modulus of SYFRC increased in contrast to SFRC where it decreased. It is important to notice that this occurred in the linear phase where the fibre should not have a significant effect on the concrete's behaviour. There is also a significant changes in the curvature of the flexural strength of synthetic fibre reinforced concrete. The drop off at 0.5 mm CMOD value disappears and the curve is similar to that of the SFRC. This is very important because the normal crack width in structures is close to this value, and the synthetic fibres worked better under high speed loading. However, the increasing curve is missing after 2.5 mm CMOD and the curve starts to decrease after this CMOD.

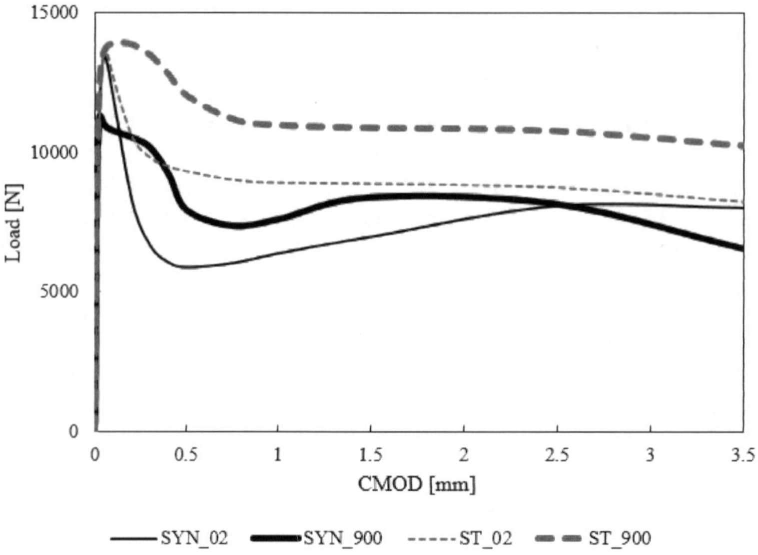

Figure 5. Result of three point bending beam test in 0.2 mm/min loading rate

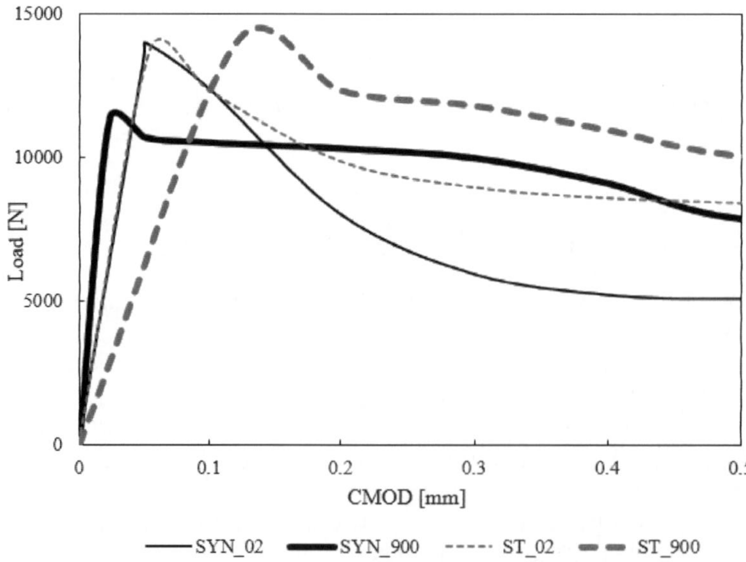

Figure 6. Result of three point bending beam test in 900 mm/min loading rate

IOP Conf. Series: Materials Science and Engineering **246** (2017) 012040 doi:10.1088/1757-899X/246/1/012040

5. Conclusion

Steel and synthetic fibre reinforcement are a well-used materials for concrete structures, where the loading rate can be significant higher (e.g. railways, tramlines etc.), than the static loading used in standard laboratory tests. It is important to know the fibre reinforced concretes' behaviour at this high loading rate.

Individual steel and synthetic fibres and fibre reinforced concrete beams were tested with different loading rates. The single fibre tension tests show a significant rise both in capacity and in case of the elastic modulus of synthetic fibres. This capacity raised by 25% and the elastic modulus changed from 7.20 GPa to 11.77 GPa. There was no significant effect of the speed on the tensile capacity or the elastic modulus in the case of steel fibres.

A three point bending beam test series was done according to the RILEM recommendation with using steel and synthetic fibre reinforcement in concrete beams. The beams were loaded with the standardized 0.2 mm/min speed and with a maximum loading capacity of the machine, i.g. 900 mm/min. Using synthetic fibre reinforcement, the elastic modulus and the residual flexural strength of the SYFRC beams increased, but the value of the peak load was lower than with the low loading rate. Also, the capacity at 0.5 mm CMOD almost reached the performance of the SFRC at low speed. The residual flexural strength of the SFRC also increased, but not as significantly as in the case of SYFRC. With steel fibres the elastic modulus of the FRC decreased significantly.

The research showed that the effect of the loading rate is significant for steel and synthetic fibre reinforced concrete. The fibre reinforced concrete's capacity increased in both cases and it can be seen that the structures under high speed loading have a higher capacity than in static loading.

References

[1] Bede N 2015 *Numerical and experimental study of concrete fracture under dynamic loading* Doctoral thesis, University of Rijeka, Faculty of Civil Engineering 142 p

[2] Juhász K P 2013 *Modified fracture energy method for fibre reinforced concrete* in: *Fibre Concrete 2013* ed Prof. A Kohoutková et al. pp 89-90 ISBN 978-80-01-05238-9

[3] Vandewalle L, Nemegeer D, Balazs L, Barr B, Barros J, Bartos P et al. 2003 RILEM TC 162-TDF: Test and design methods for steel fibre reinforced concrete' - sigma-epsilon-design method - Final Recommendation. *Materials and Structures*. **36(262)** pp 560-67

[4] Japan Society of Civil Engineers 1985 *Method of test for flexural strength and flexural toughness of SFRC, Standard JSCE SF-4*

[5] Yang S M, Kang H Y, Kim H S, Song J H, Park J M 1999 *Influence of loading speed on tensile strength characteristics of high tensile steel* Department of Mechanical Engineering 6 p.

[6] Sahin S, Yayla P 2005 *Effects of testing parameters on the mechanical properties of polypropylene random copolymer* in: *Polymer testing* **24** pp 613-619

IOP Conf. Series: Materials Science and Engineering **246** (2017) 012041 doi:10.1088/1757-899X/246/1/012041

Rigid-body-spring model numerical analysis of joint performance of engineered cementitious composites and concrete

Y Khmurovska[1], **P Štemberk**[1] **and V Křístek**[1]

[1] Department of Concrete and Masonry Structures, Czech Technical University, Thákurova 7, 166 29 Prague 6, Czech Republic

E-mail: yuliia.khmurovska@fsv.cvut.cz

Abstract. This paper presents a numerical investigation of effectiveness of using engineered cementitious composites with polyvinyl alcohol fibers for concrete cover layer repair. A numerical model of a monolithic concaved L-shaped concrete structural detail which is strengthened with an engineered cementitious composite layer with polyvinyl alcohol fibers is created and loaded with bending moment. The numerical analysis employs nonlinear 3-D Rigid-Body-Spring Model. The proposed material model shows reliable results and can be used in further studies. The engineered cementitious composite shows extremely good performance in tension due to the strain-hardening effect. Since durability of the bond can be decreased significantly by its degradation due to the thermal loading, this effect should be also taken into account in the future work, as well as the experimental investigation, which should be performed for validation of the proposed numerical model.

1. Introduction

Cover layer delamination and spalling due to concrete degradation or corrosion of reinforcement is one of the common causes of concrete structures repairing (see figure 1). An aggressive environment affect concrete significantly because of water and chemically aggressive particles (alkalis, chlorides etc.) penetration. Engineered cementitious composites (ECC) with polyvinyl alcohol (PVA) fibers can be used as repair material to reduce the negative environment effect due to its tensile ductility in the form of pseudo strain hardening, that is achieved by cement matrix multiple cracking (see figure 2) and differ from observed in metal "real" strain hardening [1, 2]. The research of tensile behavior of PVA-ECC has shown the formation of a large number of tiny cracks with a width of less than 60 μm. The presence of microcracks only, provides high durability of this material because the transport of environmental products into the concrete takes place through the cracks with a width above 0.1 mm [3].

Adhesion properties of the repair material play the crucial role in concrete cover repairing process. New material layer should provide reasonably high bond strength with old concrete layer and bars of reinforcement to ensure the joint operation of the composite structure. According to available research [4], PVA-ECC allows for much stronger bond compared to ordinary mortar, this makes PVA-ECC suitable for the described purposes. Thus, this study deals with concrete to PVA-ECC bond only.

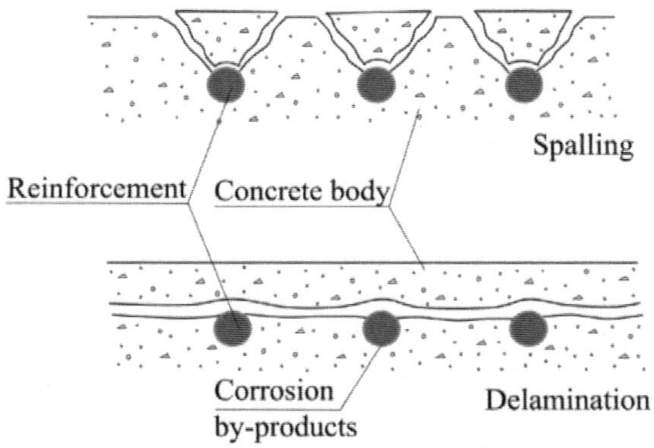

Figure 1. Type of concrete cover damage

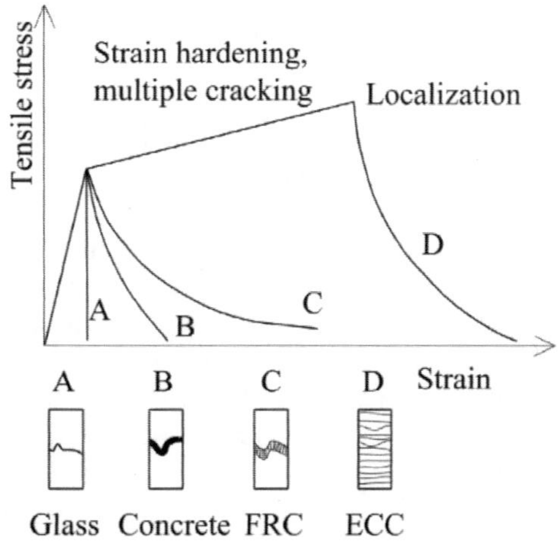

Figure 2. Tensile stress-stain behaviour of (A) brittle, (B) quasi-brittle, (C) strain softening, (D) strain hardening materials [6, 7]

A zone of special interest is represented by monolithic concrete joint of beam and column due to the concave corners presence. They may produce serious problems because of non-uniform stiffness distribution and restriction of deformation [5]. Plane concrete L-shape structure detail, strengthened by PVA-ECC layer is considered in this study to investigate PVA-ECC-concrete composite behavior.

To simplify numerical analysis and to control mechanical properties of materials and bond between them in a simple way, 3D Rigid-Body-Spring Model (RBSM) is used to simulate possible delamination and crack propagation in composite PVA-ECC-concrete L-shape structure detail.

2. Material models

RBSM is type of discrete method model, that is invented by Kawai [8]. The main benefit of RBSM is the ability to model difficult three-dimensional behavior of material, by using simple constitute model, that can be vitrificated easily by providing classical benchmark tests. The using of random geometry of particles [9] in order to simulate crack pattern and its propagation, to compute deformations and stresses of concrete and reinforced concrete structures shows reliable results.

Material is modeled as undeformable rigid-bodies interconnected by sets of springs on their surfaces. Voronoi cell shape rigid particles are generated to reduce mesh bias on cracks initiation and propagation by obtaining of random geometry.

In this model, each rigid-body has six degrees of freedom (three translational and three rotational) defined at the nuclei of Voronoi diagram. Center of gravity and vertices of the boundary surface of two neighboring particles create triangles (see figure 3). One normal and two shear springs are set at the center of these triangles to take into consideration bending and torsion without rotational springs setting [10].

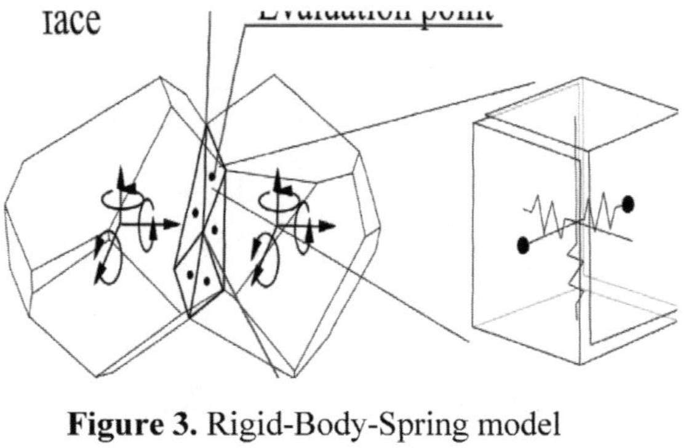

Figure 3. Rigid-Body-Spring model

2.1. Concrete material model

3D RBSM concrete material model in tension compression and shear are shown in figure 4. The tensile stress-strain relation for normal springs is linearly elastic with the slope of elastic modulus E up to tensile strength f_t with bilinear softening behavior after cracking. Tensile fracture energy is also taken into account [10-12].

The compression model for normal springs is modeled as combination of two quadratic functions with the resultant S-shape curve. Compressive failure is not considered in this model [10-12].

The shear stress-strain relation is represented as a combination of stresses and strains of two shear springs. The shear stress is increasing elastically with the slope of shear modulus $G(E,\upsilon)$ up to the shear strength $\tau_f(f_t)$, then the softening behavior is assumed. Slope of softening $K(\beta,G)$ is depended on normal spring stresses through the softening ratio $\beta(\sigma)$, that is increasing up to the maximum value with increasing of tension stresses [10-12].

The Mohr-Coulomb type criterion is assumed as the shear spring failure criteria [10-12].

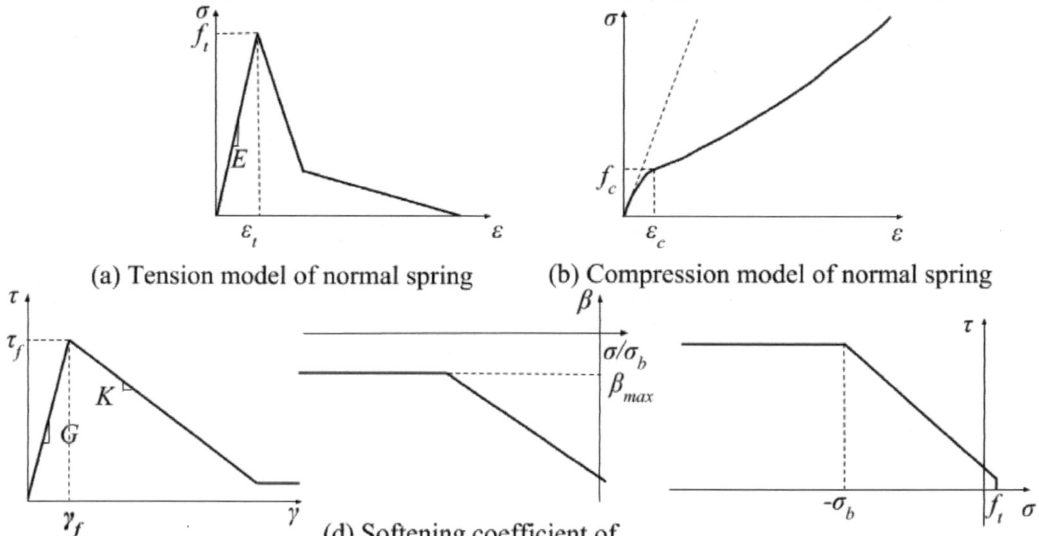

(a) Tension model of normal spring (b) Compression model of normal spring

(c) Shear spring model (d) Softening coefficient of shear spring (e) Mohr-Coulumb type criterion

Figure 4. Concrete material model

All necessary parameters and constants can be found in [10-12].

2.2. PVA-ECC material model

3D RBSM PVA-ECC tension stress-strain relation was taken from the experimental investigation of PVA-ECC with 2% fibers content [13], and simplify reasonably as shown in figure 5. The normal spring in tension is modeled as linear elastic up to the tensile strength f_t with a linear hardening branch after cracking up to the spring failure.

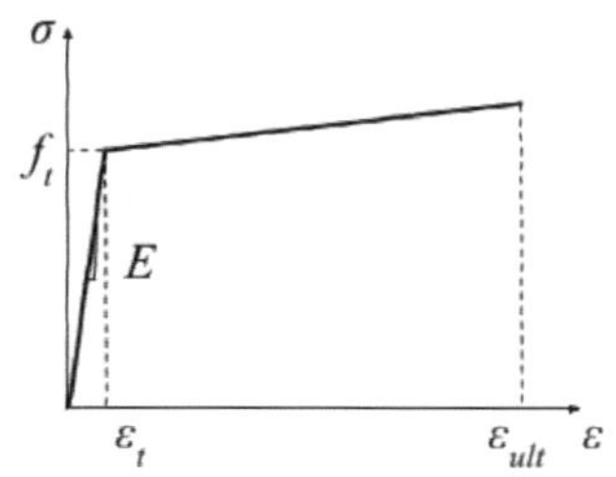

Figure 5. Tension model of normal spring of PVA-ECC

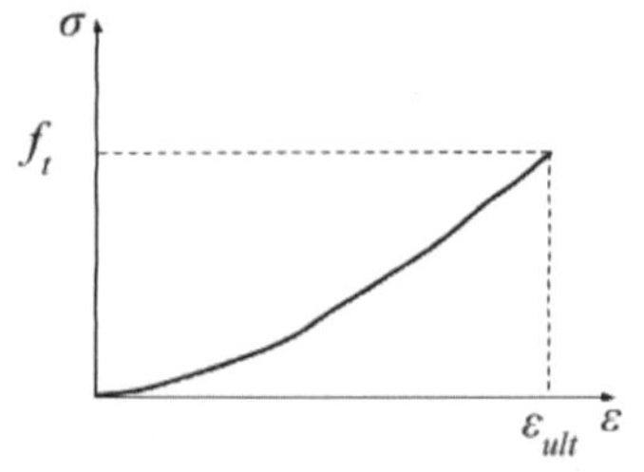

Figure 6. Tension model of normal spring of PVA-ECC to concrete bond

Since, the numerical analysis presented in this study has no compression in PVA-ECC layer, the compression model is assumed to be the same as in concrete material model.

The shear material model is constructed in the way similar to concrete shear material model from mechanical material properties (elastic modulus E, Poisson's ratio υ, etc) and material properties in tension (tensile strength f_t).

2.3. Bond material model

The tensile bond behavior is presented in previous numerical and experimental study [4] and shown in figure 6. The stress-strain tensile relation takes form of cubic parabola up to the failure without any softening brunch (see equation (1)). The trend of stress-strain curve of bond is similar to stress-strain curve of PVA fibers in tension [14], that means, that bond is provided by PVA fibers mostly.

$$\sigma = \sigma_{ult} (a \cdot \varepsilon_{norm} + b \cdot \varepsilon_{norm}^2 + c \cdot \varepsilon_{norm}^3), \qquad (1)$$

where: σ_{ult} - ultimate stresses occurring in element; $\varepsilon_{norm}=\Delta u/\Delta u_{ult}$ - normalized strain, expressed by the ratio of the relative displacement of the nodes to the ultimate relative displacement; a, b, c – shape coefficients, $a+b+c=1$.

The bond is not affected by any compression in this study example, thus the using of the concrete compression model is justified.

The bond material model in shear is created by using of tensile stress-strain curve parameters (elastic modulus E, tensile strength f_t) with the same procedure, as in concrete shear model. The corresponding mechanical parameters of bonded layers (Poisson's ratio υ) are averaged to find the unknown parameters of the bond.

3. Numerical analysis

To investigate bond properties of PVA-ECC strengthening layer on concaved concrete structure detail the L-shaped concrete sample with the layer of PVA-ECC on outer surface (see figure 7) is modeled in Matlab using non-linear 3D RBSM. Concrete with the modulus of elasticity, E_{con}, of 35 GPa, Poison's ratio, υ_{con}, of 0.18, tensile strength, $f_{t,con}$, of 3.5 MPa, compression strength $f_{c,con}$, of 40 MPa, PVA-ECC with the modulus of elasticity, E_{ecc}, of 20 GPa, Poison's ratio, υ_{ecc}, of 0.22, tensile strength, $f_{t,ecc}$, of 5 MPa, compression strength $f_{c,ecc}$, of 50 MPa, and the bond between materials with initial modulus of elasticity, E_{bond}, of 0.1 GPa, Poison's ratio, υ_{bond}, of 0 (for shear modulus calculation the value of 0.2 is used), tensile strength, $f_{t,bond}$, of 1.5 MPa, compression strength $f_{c,bond}$, of 15 MPa are considered in numerical model.

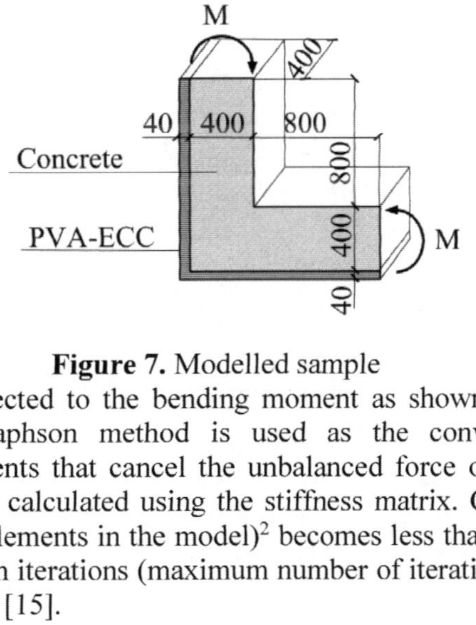

Figure 7. Modelled sample

The modeled sample is subjected to the bending moment as shown in figure 7. Displacement-controlled modified Newton-Raphson method is used as the convergence algorithm. In the convergence process, displacements that cancel the unbalanced force of elements are added to the elements. The displacements are calculated using the stiffness matrix. Convergence of the model is judged when the Σ(Reaction of elements in the model)^2 becomes less than 10^{-6}. When the model does not converge at a given maximum iterations (maximum number of iterations is 400 in this model), the analysis proceeds to the next step [15].

4. Results and discussion

The crack pattern of conrete part (see figure 8) and the crack pattern of PVA-ECC layer (see figure 9) are represented as the results of numerical analysis. Analysis of results shows that the sample destruction starts in concrete part, PVA-ECC layer is not delaminated and create the cohesion zone instead. This phenomenon is associated with compliant ductile properties of PVA fibers, that play a major role in creation of bond and providing its load-bearing capacity, while the adhesion chemical bond is destroyed in much lower deformations [4]. PVA-ECC layer is not broken due its high performance and strain hardening in tension, the microcracks are spread of all the volume of PVA-ECC layer, which corresponds to the available experimental studies [13]. Thus, the proposed numerical model shows reliable results and can be used in further studies

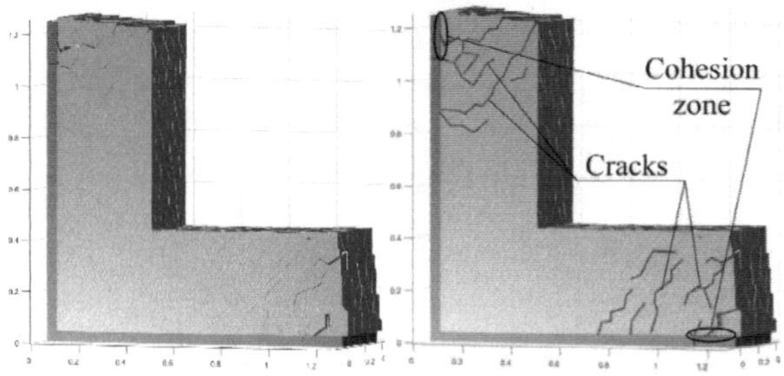

Figure 8. Crack pattern of concrete part of modelled sample (magnification=300)

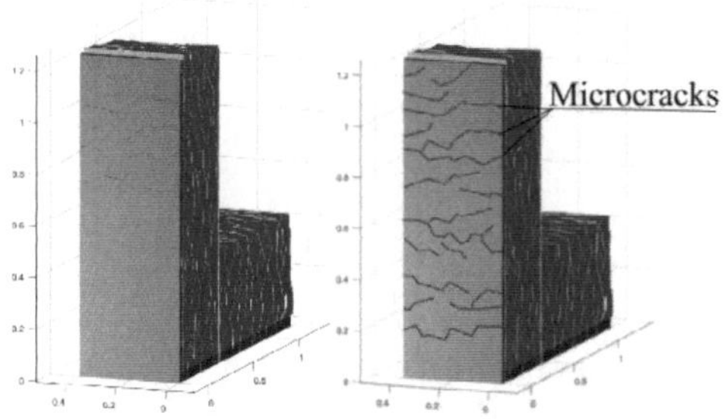

Figure 9. Crack pattern of PVA-ECC layer (magnification=300)

According to the results of numerical analysis the microcracks in PVA-ECC are too thin to provide intensive chemical particles transfer, which is approved by experimental investigation [16]. PVA-ECC provides strong bond with concrete. That properties make PVA-ECC suitable material for concrete cover repair.

The durability of PVA-ECC to concrete bond causes concern due to the degradation of mechanical properties of PVA fibers under influence of high temperature [14]. The behavior with similar trend of degradation shows PVA-ECC to concrete bond [4] under the cyclic thermal load. This type of load is one of the critical for concrete cover layer. Thus, this effect should be investigated more carefully.

5. Conclusions and future work

Based on the obtained results the following conclusion can be done:

The proposed numerical material model shows similar to experimental results and can be used in future numerical modeling.

PVA-ECC shows extremely good performance and has necessary for repair material properties, such as high mechanical properties, good hardening and setting properties, high durability and high bond strength, low permeability even after cracking. Thus, it can be used for concrete cover repairing.

Degradation of PVA fibers and PVA-ECC to concrete bond under exposure of high temperatures or thermal cyclic load can cause premature concrete cover delamination. Thus, this issue should be investigated in future work.

The proposed material model should be validated with experimental investigation.

The numerical model should be expanded and allow for PVA-ECC to reinforcement bond modeling. The experimental analysis should be provided for these purposes.

The effect of bonded surface roughness should be considered in future work.

Acknowledgment

This work was financially supported by the Czech Science Foundation, project 15-11753S, and the Czech Technical University in Prague (CTU), project SGS17/048/OHK1/1T/11, which are gratefully acknowledged.

References

[1] Fischer G and Li V C 2007 Effect of fiber reinforcement on the response of structural members. *Engineering Fracture Mechanics.* **74** pp 258-272

[2] Fishcher G 2004 Characterization of fiber-reinforced cement composites by their tensile stressstrain behavior and quantification of crack formation. *6th RILEM Symposium on Fiber-Reinforced Concretes (FRC) - BEFIB 2004.* pp 331-338

[3] Zych T 2014 New generation cementitious composites with fibres – properties and application *Technical Transactions. Architecture.* **15** pp 85-102

[4] Khmurovska Y and Štemberk P 2017 Experimental analysis of effect of cyclic thermal load on *Mechanika 2017 proc. of the 22st int. scientific conf.* May 19, 2017 Birštonas pp 183-186

[5] Yang Y, Wang Y, Fu F, Liu J 2015 Static behavior of T-shaped concrete-filled steel tubular columns subjected to concentric and eccentric compressive loads *Thin-Walled Structures.* **95** pp 374-388

[6] Fischer G and Li V C 2002 Influence of matrix ductility on tension stiffening behavior of steel reinforced engineered cementitious composites (ECC) *ACI Structural Journal.* **99** pp 104-111

[7] Matsumoto T and Mihashi H 2003 DFRCC Terminology and application concepts – JCI-DFRCC committee report *Journal of Advanced Concrete Technology.* **1** pp 335-340

[8] Kawai T 1978 New discrete models and their application to seismic response analysis of structures *Nuclear Engineering and Design.* **48** pp 207-229

[9] Bolander J E and Saito S 1998 Fracture analysis using spring network with random geometry *Engineering Fracture Mechanics.* **61** pp 569–591

[10] Yamamoto Y, Nakamura H, Kuroda I, Furuya N 2014 Simulation of crack propagation in RC shear wall using 3D Rigid-Body-Spring Model with randon geometry European *Journal of Enviromental and Civil Enginiring.* **18** pp 780-792

[11] Gedik Y H, Nakamura H, Yamamoto Y, Kunieda M 2011 Evaluation of three-dimensional effects in short deep beams using a rigid-body-spring-model *Cement & Concrete Composites.* **33** pp 978-991

[12] Nagai K, Sato Y, Ueda T 2005 Mesoscopic simulation of failure of mortar and concrete by 3D RBSM *Journal of Advanced Concrete Technology.* **3** pp 385-402

[13] Paegle I, Fischer G 2013 Evaluation of test methods used to characterize fiber reinforced composites *Proc. of the Int. Conf. "Innovative Materials, Structures and Technologies"* November 8, 2013, Riga pp 122-128

[14] Magalhaes M S, Toledo Filho R D, Rego Fairbairn E M 2013 Durability under thermal loads of polyvinyl alcohol fibers *Matéria (Rio de Janeiro)* **18** pp 1587-1595

[15] Ishikawa Y, Nakamura H and Tanabe T 2003 Development of crack propagation analysis by RBSM considering mass transfer, *Proc. of JCI meeting, Japan Concrete Institute.* **25** pp 467-472 (in Japanese)

[16] Sahmaran M, Li V C 2008 Durability of mechanically loaded engineered cementitious composites under highly alkaline environments *Cement & Concrete Composites.* **30** pp 72-81

IOP Conf. Series: Materials Science and Engineering **246** (2017) 012042 doi:10.1088/1757-899X/246/1/012042

Design of footbridge with double curvature made of UHPC

P Kněž[1], P Tej[1], D Čítek[1] and J Kolísko[1]

[1] Klokner Institute CTU in Prague, Czech Republic

E-mail: petr.knez@cvut.cz

Abstract. This paper presents design of footbridge with double curvature made of UHPC. The structure is designed as a single-span bridge. The span of the bridge is 10.00 m, and the width of the deck is 1.50 m. The thickness of shell structure is 0.03 m for walls and 0.045 m for deck. The main structure of the bridge is one arch shell structure with sidewalls made of UHPC with dispersed steel fibers with conventional reinforcement only at anchoring areas. The structure was designed on the basis of the numerical model. Model was subsequently clarified on the basis of the first test elements. Paper presents detailed course on design of the bridge and presentation will contain also installation in landscape and results of static and dynamic loading tests.

1. Introduction

The bridge should serve as a pedestrian bridge and will have a span of 10 m. The bridge is designed with double curvature. In the vertical and transverse direction it is a circular arc with a camber of 0.4 m. The cross section of the bridge has a width of 1.5 m and it is U-shaped. The bridge deck consists of a bottom plate with only 45 mm thickness and 30 mm thick side desks which serve as railings. Railing height is 1.1 m in the middle bridge span. In place of the anchor zones is the bridge reinforced with rebar networks B500B. The rest of the bridge has no reinforcement and it is only reinforced by steel fiber reinforcement.

The bridge is be made of ultra-high performance fibre reinforced concrete (UHPFRC). The main outstanding features of this material include the high compressive strength and the high tensile bending strength. High levels of strength are provided by using steel fibers. They absorb energy and control cracks growth until failure. High water impermeability is provided by a dense cement matrix and a very low level of porosity with unconnected pores. It is caused by a very low water-binder coefficient, close packing of fine grains of solid particles and by reaction of a very fine reactive material admixtures (microsilica, slag and fly ash). From high water impermeability are derived high resistance to frost and high durability.

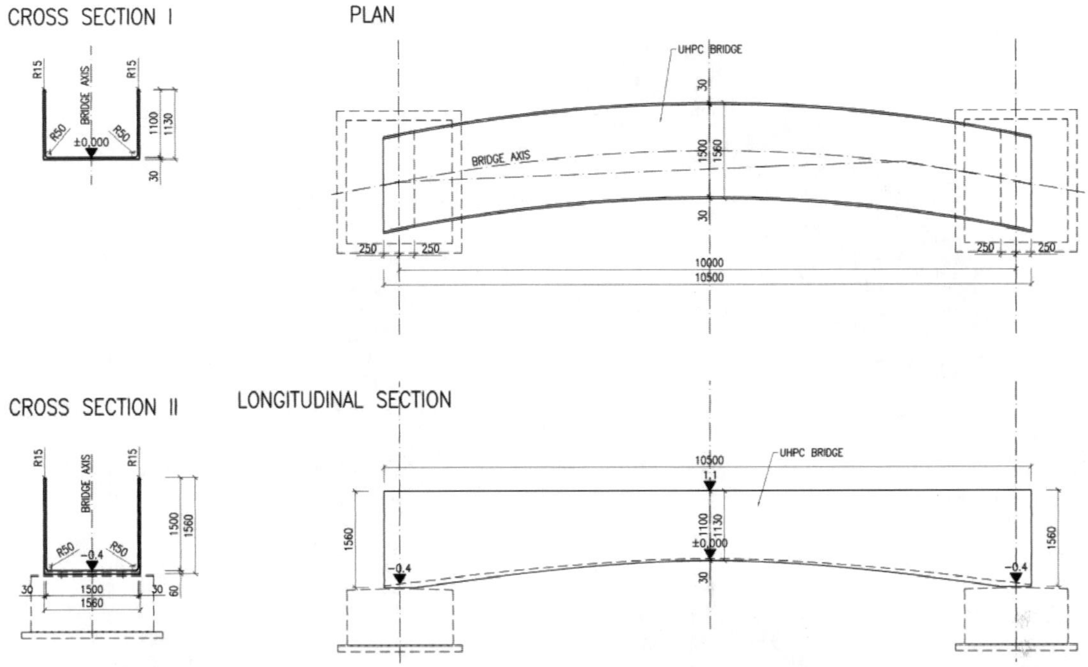

Figure 1. Plan, longitudinal and cross section of the bridge

2. Computer analysis

Bridge model was created and counted in a computer program ATENA 3D. The model is separed into 0.5 m wide parts where deck and railongs are Isoshell macro elements (bottom plate and two railings) bonded with small standard macro element, which are structured to simulate the curvature of the bridge. Anchoring deck areas are also made as standard macro elements. These macro elements have specified material item 3D Nonlinear Cementitious2 with parameters of UHPC concrete class C110/130 blended with scattered reinforcement. The tensile strength of concrete is considered as a value of 9.5 MPa. Other macro elements form surfaces for distributing the load and support elements. They consist of material 3D elastic. In anchor areas is also inserted a rod reinforcement. Material item for steel reinforcement was chosen as reinforcement with parameters for reinforcing steel B500B. Calculation is performed as a basis for the experiment, material properties are not yet reduced by material coefficients. The resulting structure is considered using one or two class better concrete.

The bridge is supported by the distributing macroelements, whose lower surface is divided by the line on which are applied such conditions that simulate fixed joint at one end and movable at the other end. Loading the bridge is except the load of the railings on the lateral force via the surface load applied to the appropriate areas of individual macro elements forming the bridge structure. The bottom plate of the bridge was loaded with whole surface load of 4.2 kN /m2, and subsequently up to structural failure, further part with the greatest eccentricity to cause torque to structural failure. Railing was loaded with a force corresponding to 1 kN /m at the top edge and vertical force until failure of the whole structure. The whole bridge was also loaded temperature change of + 40 ° C and -35 ° C.

The model was meshed by quadratic brick elements with an edge length of 0.25 meters for Isoshell macro elements. Standard macro elements are meshed by tetra elements with edgre length reduced to 40% for anchoring and supporting elements and to 20% for elements that bond deck and railings together.

As seen in Figure 6, the bearing capacity of the construction achieves satisfactory values (12.6 kN /m2), which are well above the required load capacity (4.2 kN /m2), despite its subtlety.

IOP Conf. Series: Materials Science and Engineering **246** (2017) 012042 doi:10.1088/1757-899X/246/1/012042

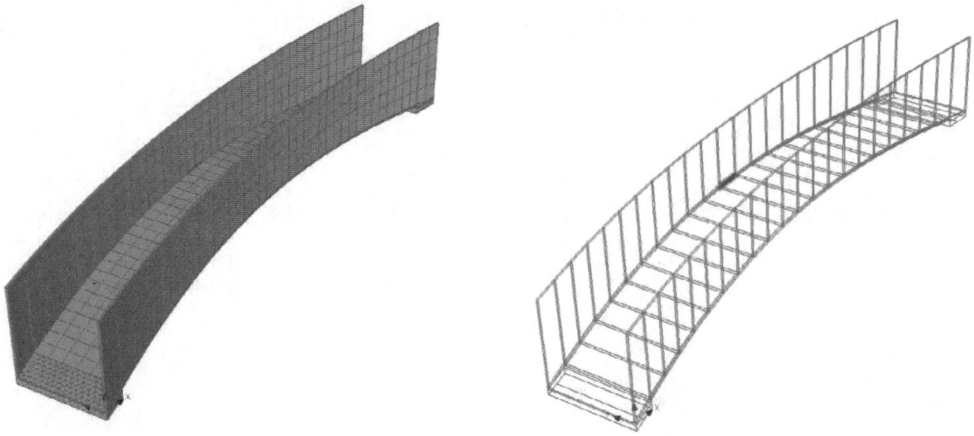

Figure 2. Finite element mesh and segmentation of macroelements to surfaces and reinforcement of the anchoring area

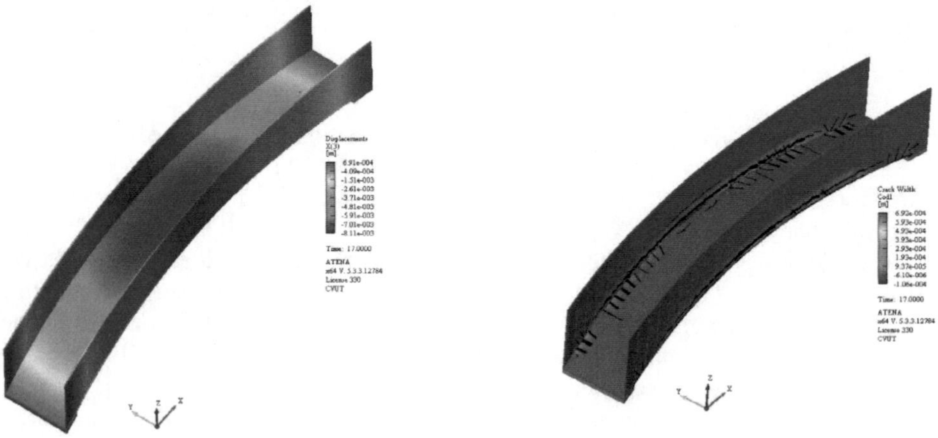

Figure 3. Vertical deflection (left) and cracks (right) of the structure before failure, the applied force of 12.6 kN /m2, the maximum deflection value -8.11 mm and the maximum crack width 0.692 mm

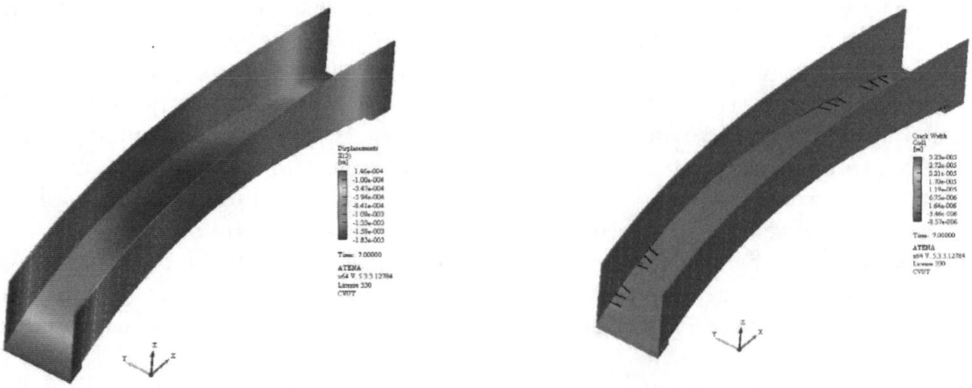

Figure 4. Vertical deflection (left) and cracks (right) of the structure, the applied force of 4.2 kN /m^2, the maximum deflection value -1.83 mm and the maximum crack width 0.032 mm

Figure 5. Wooden formwork

Figure 6. Moving with final structure out of the casting place

Figure 7. Final structure before static load test

Figure 8. Final structure before static load test

3. Conclusion

The paper presents the design of the bridge with U-shaped cross-section of ultra-high performance concrete. The bridge deck consists of a bottom plate with 45 mm thickness and 30 mm thick side desks which serve as railings.

The shape of the footbridge was based on the requirement to pass at least the standard load on this type of construction (4.2 kN/m^2) while maintaining the maximum possible subtlety of the structure without the need to insert the conventional reinforcement made possible by using UHPFRC concrete. This led to a curvature in the lateral direction, which transferred part of the load as a compressive force. The plan curvature was limited in such a way that there was no massive twisting of the structure when the external part of the footbridge was loaded.

By calculation it was shown that the load bearing capacity of the structure has a value of 12.6 kN/m^2 of continuous planar load which are well above the required load capacity (4.2 kN/m^2), despite the subtlety of the construction. The continuation of theoretical design was manufactured and experimental validation of the model of the bridge in 1:1 scale. Testing of the bridge will be presented at the conference.

Acknowledgements
This project is supported by the grant GAČR 15-22670S.

References

[1] Shi, C., Wu, Z., Xiao, J., Wang, D., Huang, Z., Fang, Z. (2015) "Review on ultra high performance concrete: Part I. Raw materials and mixture design," Concstruction and Building Materials 101, p. 741-751.

[2] Wang, D., Shi, C., Wu, Z., Xiao, J., Huang, Z., Fang, Z. (2015). "Review on ultra high performance concrete: Part II. Hydratation, microstructure and properties," Concstruction and Building Mat. 96, p. 368-377.

[3] Proceesings of the 2rd International symposium on UHPC, Kassel, March 5-7, 2008.

[4] Proceesings from HIPERMAT 3rd International symposium on UHPC and Nanotechnology for High Performance Material, Kassel, March 7-9, 2012.

[5] JSCE-USC; "Recommendations for Design and Construction of Ultra-High Strength Fiber-Reinforced Concrete Structures – Draft."

[6] AFGC/SETRA: Ultra High Performance Fibre-Reinforced Comcretes, Recommandations (06/2013). Paris CEDEX.

[7] Kolísko J., Rydval M., and Huňka P. (2013). "UHPC – Assessing the Distribution of the Steel Fibre and Homogeneity of the Matrix". Tel Aviv, Israel: fib Symposium Tel Aviv.

[8] Červenka V., Červenka J., and Pukl R. (2002). "ATENA – A tool for engineering analysis of fracture in concrete", Sadhana 27/4.

[9] Červenka J., and Papanikolaou V.K. (2008). "Three Dimensional Combined Fracture-Plastic Material Model for Concrete". Int. Journal of Plasticity 24/12.

IOP Conf. Series: Materials Science and Engineering **246** (2017) 012043 doi:10.1088/1757-899X/246/1/012043

UHPC panels utilized as permanent formwork of in-situ cast reinforced concrete deck bridges

J Kolísko[1], J Čech[1], P Tej[1] and P Kněž[1]

[1] Klokner Institute CTU in Prague, Czech Republic

E-mail: jiri.kolisko@cvut.cz

Abstract. UHPC has a number of unique properties, which stands over conventional concrete. The demand for these high performance properties predetermines some elements as typical representatives for the use of UHPC. For this reason, the range of use of UHPC elements is very wide. This article deals with the numerical design and assessment of the three variants of UHPC slabs forming permanent formwork of reinforced concrete bridge deck of the footbridge. Utilisation of outstanding properties of UHPC in this special case would enables unique and easier way of construction of this bridge.

1. Introduction

The footbridge in Týnec nad Sázavou allows multilevel transfer of pedestrian routes, respectively cyclists across the river Sazava and Jawa factory yard. The walkway leading from the railway station Týnec nad Sázavou along the concrete mix plant Eurobeton CZ, Ltd. is directly connected to this footbridge at one side and on the other side, the bridge is ended with a staircase, which begins at the walkway leading along the road III / 1068 from Týnec nad Sázavou to Kostelec. The barrier is the Sázava river of approximately 58 m width and a Jawa factory yard. The footbridge has a 3.0 m total free width.

The footbridge has four spans. In the outer spans, the supporting structure is a truss system without vertical struts with an upper deck with height of 1.75 m and a composite slab at the level of the top flange on which will be mounted the railing on the outer sides. In central fields (above the river) forms the supporting structure of the bridge also a truss system without vertical struts. This two-storey construction has a height of 3.2 m. Such solution enables the movement of pedestrians on upper level and then at the bottom are conducted all utilities. The height of the main beam is designed so that the upper edge of the upper flange also served as a handrail (Fig. 1).

The lightweight construction is made of rolled tubes and profiles. These longitudinal bearing trusses are after mounting at place connected by steel rolled I profiles, on which are placed the permanent formwork panels of UHPC for concreting of the coupling deck slabs. This article discusses the numerical design and assessment of the three variants of UHPC slabs forming permanent formwork of reinforced concrete bridge deck, which could be used for other bridge structures of similar type and arrangement.

The distance of flange of the steel beams was 2000 mm. The thickness of concrete slab was designed 150mm. For these reasons it was not possible to use conventional permanent formwork of Cetris boards, because even at the largest 40 mm thickness is the bearing capacity not sufficient. Moreover, there would be a significant weakening of reinforced concrete slabs or the Cetris board would have to be hanged beneath the flanges of the beams. Another option was to use concrete filigree. Elements should however have large thickness - min. 80 mm in the longitudinal direction and could not be counted into the thickness of the structure. The remaining option was to use complicated conventional formwork.

For these reasons it was examined the possibility to use a thin-walled ribbed slab of permanent formwork made of UHPC; while the slab thickness is only 20 mm. Its shape was designed based on the requirement of flat lower surface of slab with regard to aesthetics of the design, maintaining cover of the reinforcement of coupling slab and the thickness of coupling bridge deck slab. By the proper design of the shape is possible to include the ribs of the slab of permanent formwork in the stiffness of coupling slab and the supporting structure as a whole in both directions, so the slab height above the surface of the flanges of steel beams is calculated as a whole.

Permanent formwork made of UHPC uses a large tensile flexural strength, which is about 15 MPa, the tension zone is the whole width of the board. The stressed areas represents upper fibers of ribs where it is utilized the high strength of the material in a pressure which exceeds 150 MPa. UHPC provides a high degree of protection of the reinforcement, which is placed on the ribs of the slab, and secures 60 mm cover. Because of the span of the slab is the deflection from the load of several mm insignificant.

Slabs of UHPC will better withstand condensed humidity and deicing agents from passage of vehicles than the conventional concrete monolithic slab. The elements are not too complicated in terms of production. The board is due to its weight ~ 140 kg easily transportable and mountable. Placing on the flange of the beam was ensured by flexible band.

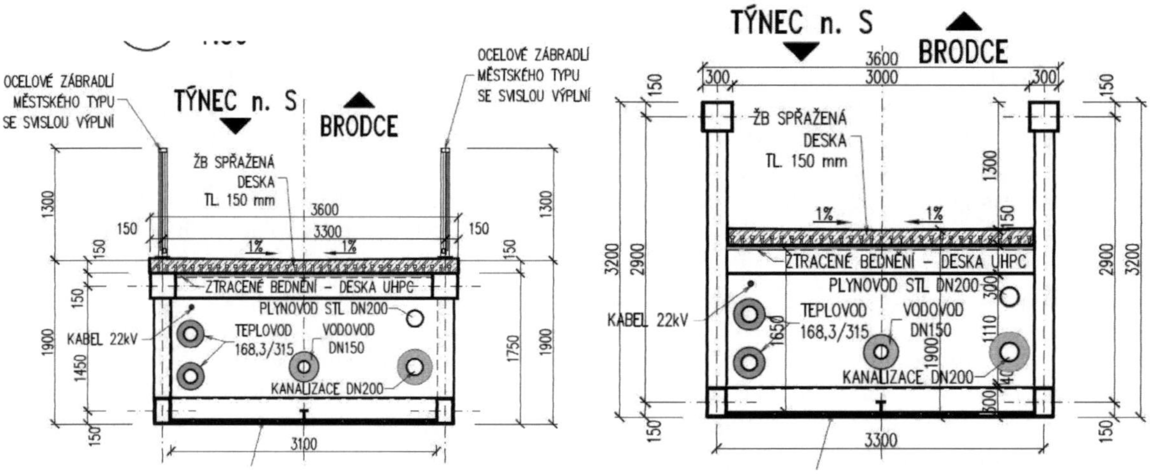

Figure 1. Cross sections of outer (left) and middle (right) span of the footbridge

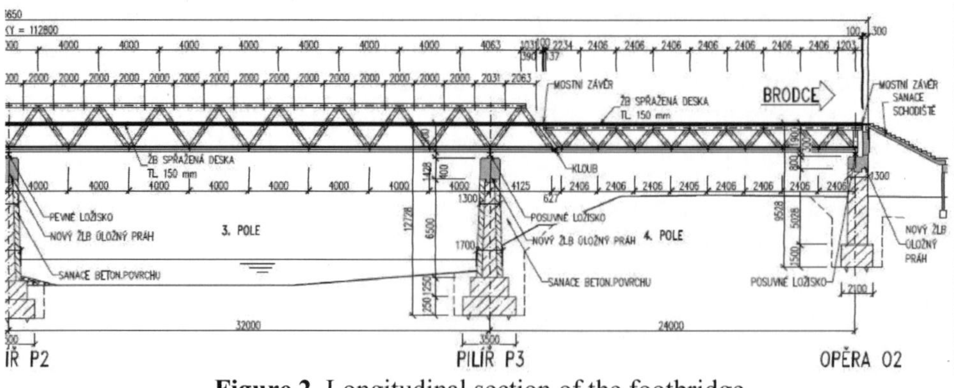

Figure 2. Longitudinal section of the footbridge

2. Numerical analysis

Numerical analysis of UHPC panels was conducted in a non-linear 3D Softwear Atena Engineering. Desk slab was modeled by slab macroelement, reinforcing ribs and the linear support were modeled using standard macroelements. Plan dimensions of slab macroelement were according to the panel variants 1850/775 mm and 1850/1550 mm with 20 mm thickness. The ribs along the perimeter of the slab have 40 mm thickness and according to variations of the panel 96, 54 and 48 mm wide. The inner lateral side of peripheral front ribs of the slab was beveled in addition, so that the width of the ribs were 75 to 115 mm. The ribs in the surface of the panel, perpendicular to the span of the panel were 20 mm high and wide according to variations of the panel 50 and 88 mm.

For all macroelements was generated by ATENA program the material model based on UHPC cube strength of 130 MPa. The model was loaded by its self weight and surface load of 1 kN / m2 at each step. Support of the model was implemented via standard macro elements of flexible material 3D Elastic from the catalog of Atena program. The model was through these macroelements supported as a simple beam - that is simply supported with the removal of all three shift options on one side and two shift options on the other side. Slab was meshed by quadratic model bricks of size 0.2 m and ribs of the model were meshed by linear finite elements with the side of 0.04 m. The model was loaded by its self weight in five steps with a coefficient of 0.2 and then in the subsequent steps of 1 kN / m2 to failure.

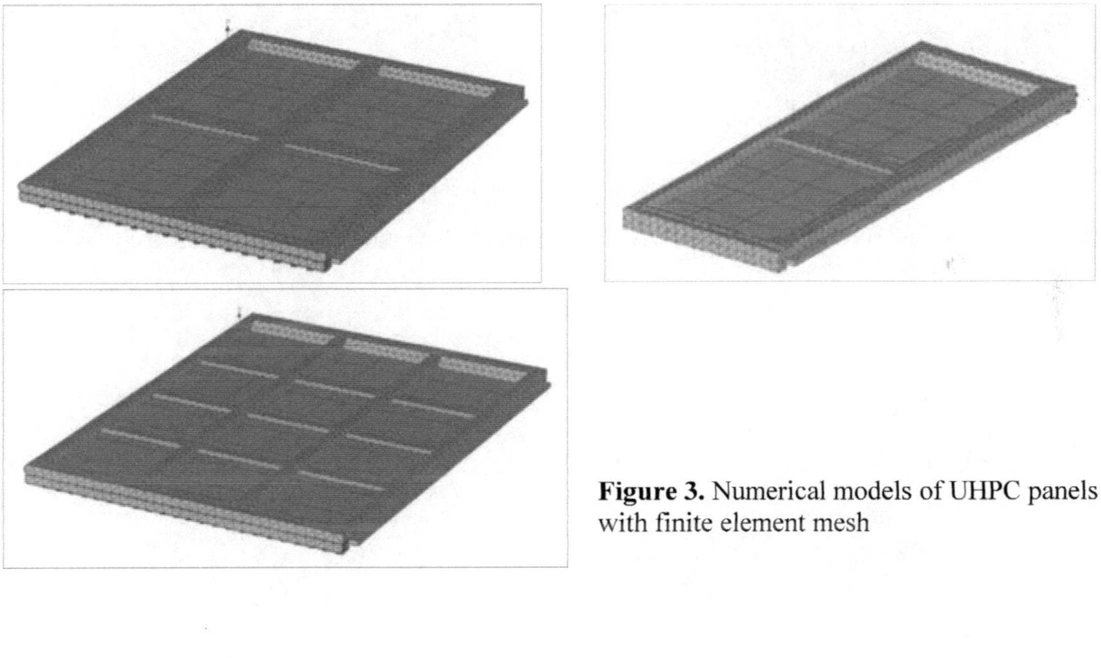

Figure 3. Numerical models of UHPC panels with finite element mesh

IOP Conf. Series: Materials Science and Engineering **246** (2017) 012043 doi:10.1088/1757-899X/246/1/012043

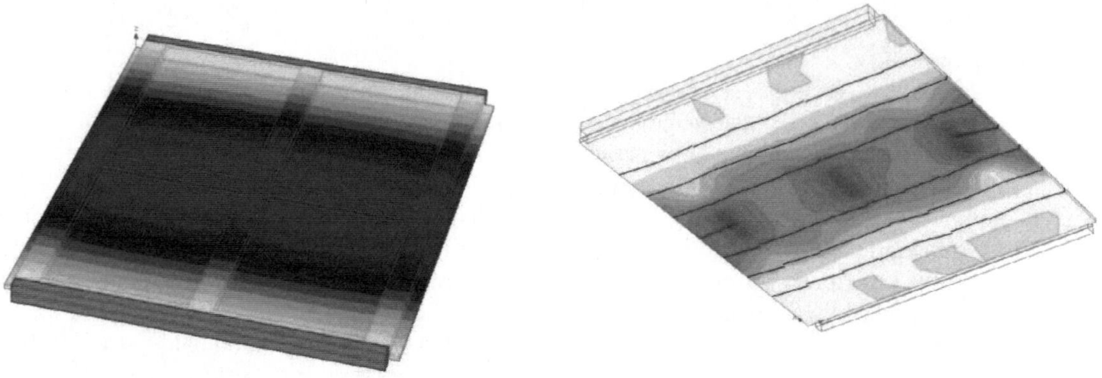

Figure 4. Deflection and cracks of UHPC panel – I. variant – max. value of deflection 4,2 mm, max. crack width 0,1 mm

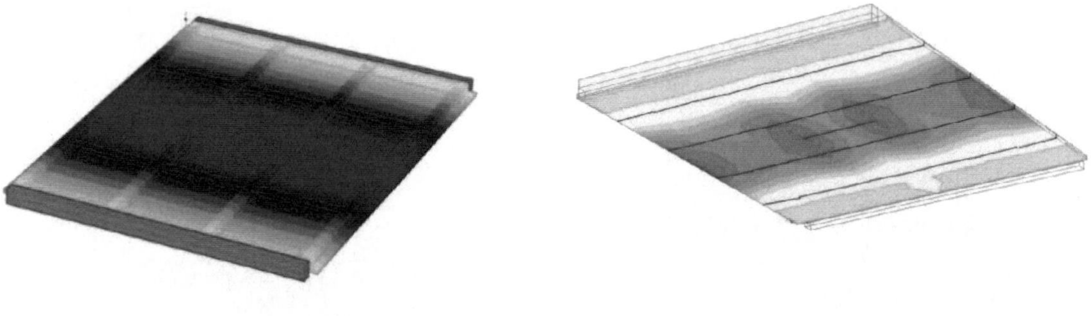

Figure 5. Deflection and cracks of UHPC panel – II. variant – max. value of deflection 3,68 mm, max. crack width 0,084 mm

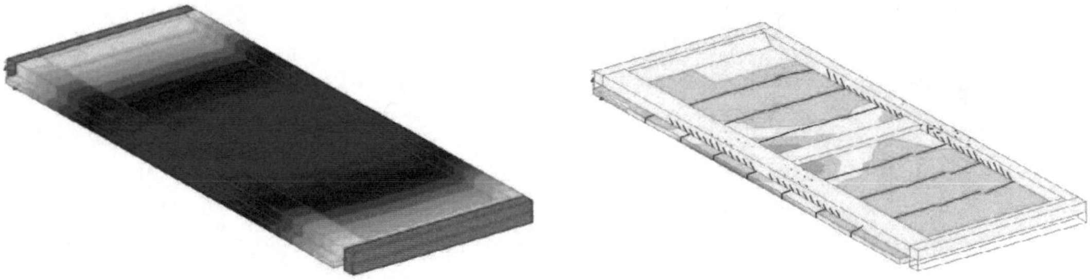

Figure 6. Deflection and cracks of UHPC panel – III. variant – max. value of deflection 4,4 mm, max. crack width 0,125 mm

IOP Conf. Series: Materials Science and Engineering **246** (2017) 012043 doi:10.1088/1757-899X/246/1/012043

3. Conclusion

For the bridge in Týnec nad Sázavou, lost formwork from UHPC boards was proposed in several variants.Variants of formwork boards with perimeter frames, which are shaped to touch the steel structure of the bridge, have been considered.

Panels of I and II. variants fail at typical load 10 respectively 8 kN / m2 (+ dead weight). The deflections and width of cracks of these variants of panels just prior to failure are shown in Figures 4 and 5. Further, an assessment was made on the ultimate (resistance) and the serviceability limit state (deflection and crack width) of panel III. variant. This panel is best suited to be used because of its smaller self weight. The aim of the numerical analysis was to determine the deformation and width of cracks that occur during loading of UHPC panel by layer of fresh concrete of 150 mm thickness. The panel will securely transmit its self weight of + 9 kN / m2 characteristic load. When analyzing the first limit state, the panel failed to design loads of self load + 7kn / m2 design load. Since the load of fresh concrete layer causes a characteristic load of 0.15 * 26 = 3,9 kN / m2 and a design load 0.15 * 26 * 1.35 = 5.265 kN / m2 it can be stated that the panel when analyzing both limit states complied.

Designing a lost formwork for the bridge in Týnec nad Sázavou is advantageous in terms of saving the construction height of the bridge board. Lost formwork will be statically coupled to the monolithic bonded board due to the peripheral and central ribs of the panel structure. The resulting panel shape is also manually manipulable by optimizing its weight.

Acknowledgements

This project is supported by the grant OP Praha DP02.

References

[1] Proceesings from HIPERMAT 3rd International symposium on UHPC and Nanotechnology for High Performance Material, Kassel, March 7-9, 2012.

[2] JSCE-USC; "Recommendations for Design and Construction of Ultra-High Strength Fiber-Reinforced Concrete Structures – Draft."

[3] AFGC/SETRA: Ultra High Performance Fibre-Reinforced Comcretes, Recommandations (06/2013). Paris CEDEX.

[4] Kolísko J., Rydval M., and Huňka P. (2013). "UHPC – Assessing the Distribution of the Steel Fibre and Homogeneity of the Matrix. Tel Aviv, Israel: fib Symposium Tel Aviv."

[5] Červenka V., Červenka J., and Pukl R. (2002). "ATENA – A tool for engineering analysis of fracture in concrete", Sadhana 27/4.

[6] Červenka J., and Papanikolaou V.K. (2008). "Three Dimensional Combined Fracture-Plastic Material Model for Concrete". Int. Journal of Plasticity 24/12.

IOP Conf. Series: Materials Science and Engineering **246** (2017) 012044 doi:10.1088/1757-899X/246/1/012044

On the effect of addition of carbon nanotubes on the electric conductivity of alkali-activated slag mortars

I Kusak[1] and M Lunak[1]

[1]Institute of Physics, Faculty of Civil Engineering, Brno University of Technology, Veveri 95, 602 00 Brno, Czech Republic

E-mail: kusak.i@fce.vutbr.cz

Abstract This paper presents basic electric properties of laboratory prepared alkali-activated composite materials on the basis of finely ground granular high furnace slag to which various quantities of carbon nanotubes (CNT) have been added. Impedance spectroscopy in the frequency range from 40 Hz to 1 MHz was used to measure the specimens. Electric resistivity ρ versus frequency and electric resistivity ρ versus CNT content relationships were examined on our specimens R&S ZNC vector analyser with DAK-12 coaxial probe (made by Speag) was used to carry out the measurements at higher frequencies (from 100 MHz to 3 GHz). Electric conductivity σ as a function of the frequency and as a function of the specimen CNT content was studied in this frequency range. Up-to-date instruments and a unique approach have evidently been employed to carry out non-destructive measurement of mortar materials.

1. Introduction

In the structure of the mortars under investigation there are air gaps, which influence the resulting properties both directly and indirectly. In a porous environment, the carbon nanotubes can behave as micro-binders, which will create a thicker and less porous structure [1]. The air gaps are reduced in number or become totally extinct in the composite material. In this way, the mix becomes more homogeneous. A cardinal effect on the physical and mechanical properties (their improvement) is evident. A change in electric parameters of the nano-composite (geopolymer), for example, the electric capacitance without and with carbon nanotubes is clearly observable. These results may contribute a lot to the further research and development of alkali-activated systems with a focus on their practical application in the building industry.

The measurements, which have been carried out to date, show that the application of carbon nanotubes in materials results in improving their strength, elasticity and a general durability [2]. The nanotubes show a low density (1.3 to 1.4 $g \cdot cm^{-3}$ depending the nanotube type), a high thermal conductivity (1 750 to 5 800 $W \cdot m^{-1} K^{-1}$) and – thanks to delocalized bonds along the entire carbon layer – an excellent electrical conductivity. Carbon nanotubes are considered to be a potential replacement of the composite reinforcement, their mechanical, electric and thermal properties exceeding those of the traditional fibres [2].

Alkali-activated materials, or, geopolymers, make a specific group of inorganic cement-free materials. They arise in a reaction of latently hydraulic substances or pozzolans (granulated slag, fly ash, meta-kaolins) with a suitable activator. Solutions of soluble compounds of alkali metals, above all, sodium and potassium, namely, carbonates, hydroxides and silicates, can be used as activators. This activation gives rise to hydraulic binders which - after water is added to them - can be used to form composite compounds featuring excellent applicability [3]. Alkali-activated materials can serve as an alternative of classical building materials, in particular, the cement concrete, but – thanks to their other features - they can also be employed in a number of other branches (such as the renovation of historic objects, etc). However, their potentiality is being exploited on a small scale in the building industry, even though the research of alkali-activated materials has been going on since the sixties of the last century [4,5].

An indisputable advantage of the alkali-activated materials consists in their economical and environmental features. These materials can substantially increase the value of secondary raw materials which show latent hydraulic or pozzolanic activity, first of all, the large-volume by-products of metallurgy and power engineering. The preparation of alkali-activated binders and composites takes place at common or slightly elevated temperatures. It requires neither previous firing of the semi-products to a high temperature (which is the case, e.g. of the portland clinker), nor high-temperature sintering, thus reducing the generation of CO_2 by combustion processes and limestone decomposition. Another benefit of the alkali-activated materials consists in the possibility of apply a wide choice of non-standard binders, which are not commonly used in the concrete technology.

This paper presents basic electric properties of laboratory prepared alkali-activated composite materials on the basis of finely ground granular high furnace slag to which various quantities of carbon nanotubes have been added.

2. Experiment setup

Several brands of the input raw material have been used to prepare the specimens. Alkali-activated finely ground granular blast-furnace slag is used to act as a binding agent. The activation was carried out using a water glass solution the silicate module of which was adjusted to Ms = 2.0 by means of 50 % sodium hydroxide. Both standard testing sand, which is used to prepare test cement specimens, and carbon nanotubes in various amounts were used as a binding agent. The composition of partial mixes is summarized in Table 2.

Three test bodies of dimensions 40 x 40 x 160 mm from each of the mixes were manufactured. The results obtained from each test body are compared with those of the reference specimen.

Table 1. Chemical composition of granular blast-furnace slag (%)

SiO_2	Al_2O_3	Fe_2O_3	CaO	MgO	K_2O	Na_2O	MnO	SO_3
39.75	6.61	0.46	39.03	10.45	0.63	0.38	0.37	0.71

Table 2. Mixture composition

Mixture	**Slag**	**Water glass**	**Sand**	**CNT**	**Water**
	[g]	[g]	[g]	[g]	[ml]
REF				0	185
0.1 % CNT				4.5	185
0.2 % CNT				9	185
0.3 % CNT				13.5	185
0.5 % CNT	450	90	1350	18	185
1.0 % CNT				22.5	185
1.5 % CNT				27	185
2.0 % CNT				31.5	185
2.5 % CNT				36	185

IOP Conf. Series: Materials Science and Engineering **246** (2017) 012044 doi:10.1088/1757-899X/246/1/012044

Thus prepared test bodies are shown in Fig. 1, their composition is in Table 2. They were characterized by means of the impedance spectroscopy method. Agilent 33220A sine waveform generator and Agilent 54645A double-channel storage oscilloscope were used. The specimens were measured in the frequency range from 40 Hz to 1 MHz. Electric resistivity ρ versus frequency and electric resistivity ρ versus CNT content relationships were studied on our specimens. To carry out the impedance analysis, the specimens were placed between brass electrodes of an area of 30 × 100 mm. To measure at higher frequencies (from 100 MHz to 3 GHz), R&S ZNC vector analyser with DAK-12 coaxial probe (made by Speag) was used. Electric conductivity σ as a function of the frequency and as a function of the specimen CNT content was studied in this frequency range.

Residual mechanical properties of the specimens were studied, too. To determine the bending forces, a standardized 3-point test was employed. The compressive strength was measured at remote edges of each of the bending test remaining specimens according to of EN 196-1 standard, "Cement testing methods".

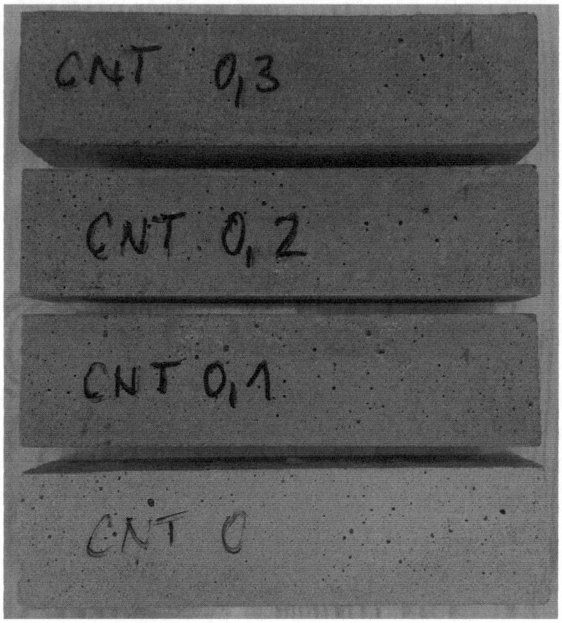

Figure 1. Specimens with respective CNT additions of 0 %, 0.1 %, 0.2 % and 0.3 %. The specimen dimensions: 40 x 40 x 160 mm.

3. Results

The electric resistivity ρ of AAS composites with CNT ranging from 0 to 2.5 % was measured at the frequencies from 40 Hz to 1 MHz (Fig. 2). The maximum values of the electric resistivity were found to be 3.3 x 10^8 Ω·cm at a frequency of 40 Hz. These values of the electric resistivity pertain to specimens to which 0, 0.1 and 0.2 % of CNT were added and are in very close proximity of each other. The electric resistivity values are decreasing in the same way for all specimens and are converging to tens of kΩ·cm at the frequencies above 10 kHz. At higher frequencies, the electric resistivity values show a sloping down trend. They almost coincide at frequencies above 10 kHz for all CNT admixture percentage values (0 through 2.5 %).

The carbon nanotubes create conducting paths in the specimens. Consequently, the electric resistivity of higher CNT concentration specimens is reduced, as is illustrated in the diagram of Fig. 2. The electric resistivity curves of specimens with CNT 0 to 1 % are mingling throughout the spectre. A more significant decrease is taking place for the specimens with CNT 1.5 %. At a frequency of 100

Hz, the electric resistivity difference between the CNT 1 and CNT 1.5 % specimens is 1.37×10^8 Ω·cm approximately.

The electric resistivity decrease with the CNT concentration will most probably stop at substantially higher CNT concentrations. In this case, however, we encounter problems when mixing fine CNT's into the geopolymer fresh mixture, as the nanotubes are clustering and, consequently, are not distributed uniformly at the specimen faces, to which conducting plate electrodes are attached. The nanotubes might get broken into fragments in consequence of blending, which also aggravates the higher CNT concentration problems.

Generally speaking, the electric resistivity differences are very low at CNT concentrations of 0.1 through 1 %. More pronounced electric parameter variations take place at a CNT concentration of 1.5 % and more. Different raw material mixing procedures as well as different calendar dates of mixing contribute also to the differences in the readings. Our objective therefore consists in comparing the trends in the different batches rather than the batches with graphite admixtures against those with CNT.

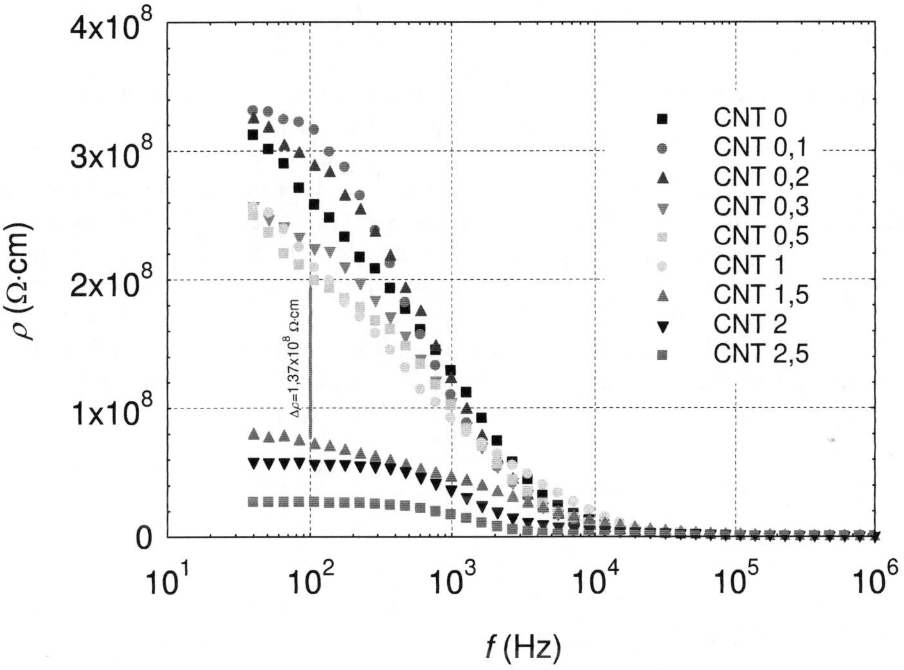

Figure 2. Electric resistivity of AAS composites with 0 through 2.5 % of CNT in the frequency range from 40 Hz to 1 MHz (the frequency is plotted in log scale).

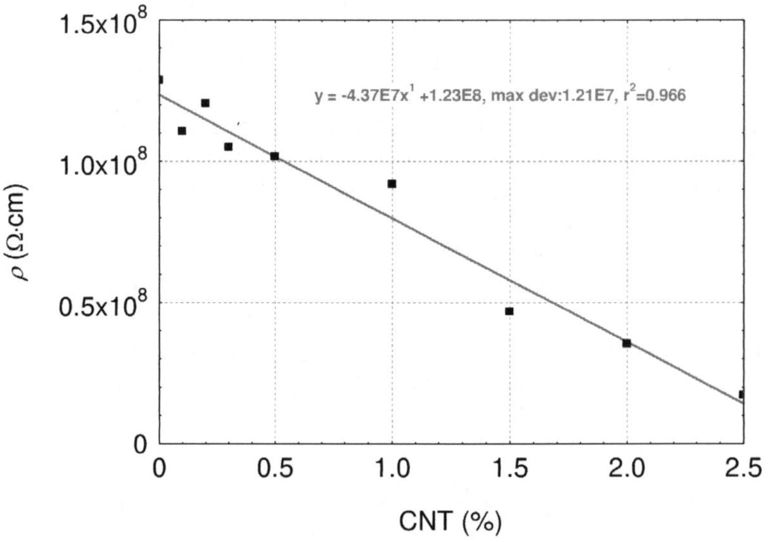

Figure 3. Variation of the electric resistivity of the composite AAS as a function of the CNT content at a reference frequency of 1 kHz.

Electric properties of AAS composites with added CNT were investigated at higher frequencies, too. Specifically, in the frequency range from 10 to 3000 MHz, where R&S ZNC vector analyser with DAK-12 coaxial probe made by Speag was used. An electric conductivity of AAS composites versus frequency plot is shown in Fig. 4. The electric conductivity initially increases with the alternating electric field frequency between 10 – 2 000 MHz. A maximum is observed to occur at a frequency of 2 GHz for all mixtures with different CNT contents as well as the reference mixture (with no CNT added). Therefore, location of the peak is constant at the above mentioned frequency, showing no shift with the frequency. Subsequently, the electric conductivity decreases.

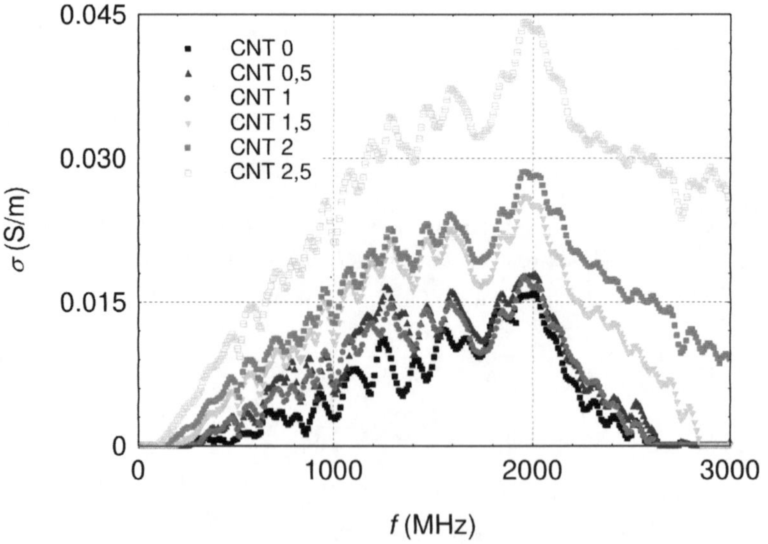

Figure 4. Electric resistivity of AAS composites with 0 through 2.5 % of CNT in the frequency range from 40 Hz to 3000 MHz.

The shapes of all curves with CNT 0 through 1 % are very similar, the differences in values being very low. A considerable change in the electric conductivity takes place at CNT 1.5 %. It is therefore in accordance with the electric resistivity. The largest differences in the electric conductivity occur at a frequency of 2 GHz, at which the above mentioned maximum is observed. For example, at a the frequency of 2 GHz, the specimen with CNT 0 % reaches the electric conductivity of 0.0156 S/m; for CNT 0.5 %, a value of 0.0175 S/m; for CNT 1 %, a value of 0.0170 S/m; for CNT 1.5 %, a value of 0.0252 S/m; for CNT 2 %, a value of 0.0281 S/m; for CNT 2.5 %, a value of 0.0433 S/m are obtained. For example, an addition of CNT 2.5 % resulted in an increase of the electric conductivity at 2.5 GHz by 0.0277 S/m against the reference specimen (CNT 0 %).

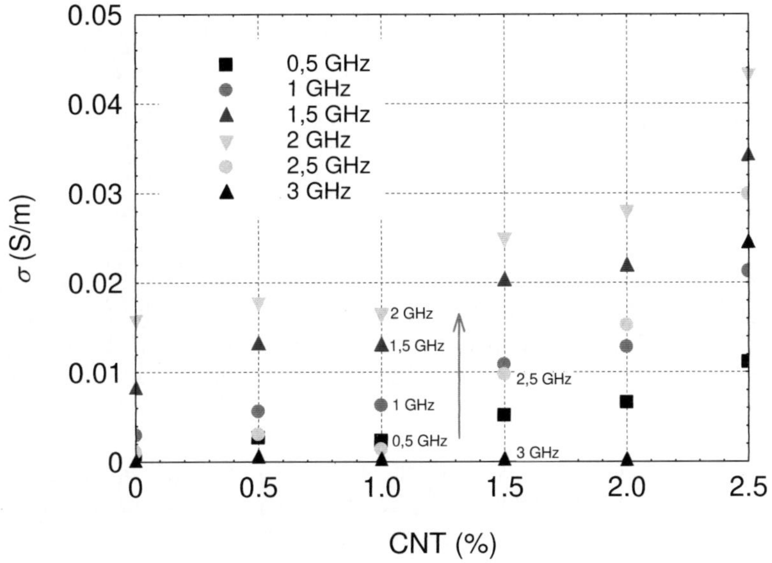

Figure 5. Electric conductivity of an AAS composite versus the addition of CNT ranging from 0 to 2.5 % (at selected frequencies of the electric field).

The diagram of Fig. 5 shows a plot of the electric conductivity versus CNT content as the material addition. A growth of the electric conductivity is observed when the CNT content is increased. At the same time, a growth of the electric conductivity can be observed when the alternating electric field frequency is increased. This corresponds to the selected frequencies of 0.5 GHz, 1 GHz, 1.5 GHz and 2 GHz. Therefore, when the frequency is increased up to 2 GHz, the electric conductivity grows up for all amounts of the CNT addition. However, at higher frequencies, i.e., 2 GHz and 2.5 GHz, the electric conductivity is observed to decrease. This can be documented by the local maximum at a frequency of 2 GHz, see Fig. 4. The growth of the electric conductivity with the frequency is due to the polarization current, which arises from the orientation oscillation movement of the dipoles. The total electric conductivity is therefore equal to the sum of the DC conductivity and the dipole polarization conductivity [6,7]. The reason for which the electric conductivity dropped at 2 GHz, will be the object of a further research.

4. Mechanical properties

Table 3 lists the compressive strength values for composites with various quantities of CNT filler. The compressive strength of an AAS composite without any CNT added equalled 58.6 MPa. Addition of CNT caused this strength to decrease dramatically. This can be accounted for the growing water to slag ratio and, also, for complications which arise on mixing the suspension with increasing CNT content. A strength increase occurred for and addition of 0.5 % CNT, namely to 51.3 MPa. A noticeable decrease took place afterwards. The compressive strength of the specimens with CNT 2.5 % is reaching about one half of the value of the specimens with CNT 0 % and 0.5 %. Our results show that only 45 % of the reference force are sufficient for the CNT 2.5 % specimens.

Table 3. Compressive strengths of AAS composites with 0 to 2.5 % of CNT added, together with the root-mean-square deviations.

Mixture	f_c [MPa]	**Root-mean-square deviation**
REF (0 % CNT)	58.6	3.2
0.1 % CNT	46.8	3.3
0.2 % CNT	48.4	5.6
0.3 % CNT	44.1	1.6
0.5 % CNT	51.3	2.0
1.0 % CNT	47.4	1.3
1.5 % CNT	45.0	1.1
2.0 % CNT	27.0	1.2
2.5 % CNT	26.2	1.8

The bending strength (Table 4) shows a similar trend. The highest bending strength value, 7.01 MPa, is reached by a CNT 0.3 % specimen. A radical reduction of the bending strength results from increasing the CNT content. Only 52 % of the reference force are sufficient for the CNT 2.5 % specimens.

Table 4. Compressive strengths of AAS composites with 0 to 2.5 % of CNT added, together with the root-mean-square deviations.

Mixture	f [MPa]	**Root-mean-square deviation**
REF (0 % CNT)	6.92	0.55
0.1 % CNT	6.19	0.91
0.2 % CNT	6.96	0.15
0.3 % CNT	**7.01**	0.25
0.5 % CNT	4.84	0.25
1.0 % CNT	4.69	0.20
1.5 % CNT	5.06	0.13
2.0 % CNT	3.74	0.18
2.5 % CNT	3.63	0.03

IOP Conf. Series: Materials Science and Engineering **246** (2017) 012044 doi:10.1088/1757-899X/246/1/012044

5. Conclusion

The present paper deals with the change of electric parameters of alkali-activated slag mortars to which carbon nanotubes have been added. This CNT admixture improves the electric conductivity of the materials in question, making them easier to measure by electromagnetic-principle-based methods. A suitable CNT admixture was described from the viewpoint of both its amount and the most effective increase of the electrical conductivity. Another quantity to study was the electric resistivity and its frequency dependence. The measurements were carried out in two frequency ranges, namely 40 Hz to 1 MHz and 10 MHz to 3 GHz, by means of two measuring systems. The specimens have subsequently been subjected to mechanical testing (compressive strength and bending strength).

Acknowledgments

This paper has been worked out under the project GAČR No. 16-02261S.

References

[1] Monica J H and Harris A T 2013 *Progress in Materials Science* **58** 1056 –1102
[2] Saafi M, Andrew K, Mcghon D, Taylor S, Rahman M, Yang S, Zhou X and Tang P L 2013 *Construction and Building Materials* **49** 46-55
[3] Lhoták P 2014 http://www.uochb.cas.cz/Zpravy/PostGrad2004/7_Lhotak.pdf
[4] Matsutani S, Shimosako Y and Wang Y 2012 *Physica A* **391** 5802
[5] Fernandez-Jimenez A 1999 *Cement Concrete Res.* **29** 1313
[6] Kusak I; Lunak M and Schauer P 2013 *Appl. Mech. Mater.* **248** 370
[7] Tupy M, Sotiriadis K, Kusak I, Lunak M, Stefkova D and Petranek V 2015 *J. of Mater in Civil. Eng.* **28** 04015171

IOP Conf. Series: Materials Science and Engineering **246** (2017) 012045 doi:10.1088/1757-899X/246/1/012045

Fibre reinforced concrete exposed to elevated temperature

J Novák[1] and A Kohoutková[1]

[1] Department of Concrete and Masonry Structures, Faculty of Civil Engineering, CTU in Prague, Thakurova 7, 166 29 Prague - 6 Dejvice, Czech Republic

E-mail: josef.novak.1@fsv.cvut.cz

Abstract. Although concrete when subject to fire performs very well, its behaviour and properties change dramatically under high temperature due to damaged microstructure and mesostructure. As fibre reinforced concrete (FRC) represents a complex material composed of various components with different response to high temperature, to determine its behaviour and mechanical properties in fire is a demanding task. The presented paper provides a summary of findings on the fire response of fibre FRC. Namely, the information on steel fibre reinforced concrete (SFRC), synthetic fibre reinforced concrete and hybrid (steel + synthetic) fibre reinforced concrete have been gathered from various contributions published up to date. The mechanical properties including the melting point and ignition point of fibres affect significantly the properties of concrete composites with addition of fibres. The combination of steel and synthetic fibres represents a promising alternative how to ensure good toughness of a concrete composite before heating and improve its residual mechanical behaviour and spalling resistance as well as the ductility after heating. While synthetic fibres increase concrete spalling resistance, steel fibres in a concrete mix leads to an improvement in both mechanical properties and resistance to heating effects.

1. Introduction

Fiber reinforced concrete (FRC) is a building material whose utilization in the concrete industry has been rapidly increasing. This development is motivated by its physical and mechanical properties which contribute to traditional concrete elements and structures various economical benefits such as structure subtlety, part or full elimination of conventional reinforcement, enhanced impact resistance, resistance to mechanical loads and environmental loads. Recently, many comprehensive studies have been undertaken with the aim to observe the mechanical behaviour of FRC exposed to elevated temperature.

Although concrete is well-known for a high degree of fire resistance, high temperature seriously damages microstructure and mesostructure which results in generalised mechanical decay of a concrete composite [1]. As a consequence, the extensive knowledge of mechanical properties of FRC exposed to elevated temperature seems to be decisive for a wider utilization of the material. There are two fundamental types of a methodological procedure used for observing the mechanical properties of specimens at elevated temperature. Most of experimental investigations [2-28] were conducted on test specimens at ambient temperature after high temperature exposure and only a few was performed on heated test specimens [29-31]. Such approach to experimental testing is used mainly due to a simple way of testing as tests are easier to conduct on test specimens at ambient temperature. However, if the results obtained from the tests on specimens after high temperature exposure correspond enough to the mechanical properties of a tested material at a certain temperature level has not been still fully understood. Bamonte and Gambarova belong to authors which very intensively deals with such issue and states in their publications [32,33] that the hot and residual (after high temperature exposure) behaviour in compression are very close, the only difference was observed in case of the peak strain in compression which is larger on heated specimens in comparison with specimens after temperature exposure.

The fire response of concrete composites is closely associated with concrete composition, particularly with a type and content of concrete components used. Generally speaking, concrete made of siliceous aggregates, such as granite, shows unfavourable mechanical properties at high temperature compared to concrete composed of calcareous aggregates such as dolomite and limestone [34]. Recently, a lot of interest is also being paid on the possible use of metakaolin, fly ash and silica fume as partial cement replacement in concrete subjected to high temperature [2,7,8]. Owing to silica fume and fly ash fineness, concrete composites with such additions have denser microstructure and as a consequence their explosive spalling tendency increases [8].

2. Fire resistance of FRC

A number of experimental investigations have been conducted up to date with the aim to observe the fire response of concrete composites. Particularly, the studies are focused on the effect of a type, shape and content of fibres on the mechanical properties of concrete composites, mostly compressive and tensile strength including elastic modulus. Namely, it concerns steel fibers [2,5,6,8-11,14,16,19-22,24 ,27-30], synthetic fibers [2,4,6-9,11,12,15,17,19,22,27,29,30] and a mix of steel and polypropylene fibres [2,6,11,13,19,22,27,29,30] which are widely used in the concrete industry. There also a few investigations which deals with carbon fibres [3,12] and glass fibers [12].

2.1. SFRC

As the melting point of steel is relatively high in comparison with other materials, the use of steel fibers seems to be beneficial for concrete composites exposed to high temperature. Incorporating steel fibers into concrete composites remains advantageous even when the concrete composites are exposed to high temperature up to 1200 °C, particularly 1% content has no deleterious effect on heated concrete. In fact, the inclusion of steel fibers in a concrete mix leads to an improvement in both mechanical properties and resistance to heating effects in comparison with unreinforced concrete [5, 16, 22, 23, 27]. Some experimental investigations [10, 14, 21] even demonstrate the compressive strength of steel fiber reinforced reactive powder concrete and geopolymer concrete gradually increases when the material is heated up to 200–300 °C (figure 1), but starts to decrease as temperatures further increase. The compressive strength of reactive powder concrete with 1% steel content is higher between 200°C and 400°C than at room temperature and subsides when temperature exceed 500°C. Furthermore, 2% and 3% steel fiber content significantly increase compressive strength from 200°C to 300°C which then gradually decreases as the temperatures reach 400 ◦C and beyond. However, a higher content of steel fibers cannot improve the compressive strength of concrete composites at elevated temperatures [24]. On the contrary, it has been also demonstrated that steel fibers have negligible effect on high temperature compressive strength and only improve tensile strength when temperature up to 400°C is considered [29]. The tensile behavior of SFRC subjected to elevated temperature is more sensitive to the volume fraction and the aspect ratio of the fiber than to its type [20].

SFRC also has the higher toughness after the high-temperature exposures when compared to the initial values of unheated concrete [2]. As temperature increases, SFRC weakens and shows reduced stiffness with the degradation depending on the type, aspect ratio, and volume fraction of the fiber [20] and the reduction in modulus of elasticity is more pronounced than the reduction in compressive strength for the same heat treatment [24]. Steel fiber reinforced recycled aggregate concrete exhibits the identical behavior and loses stiffness much faster than strength after exposure to elevated temperature [16]. As a consequence, peak strains gradually increases together with temperature. The increase in peak strains along with steel fiber content does not significantly differ between ambient temperature and 200 °C. When temperatures exceed 200 °C, a higher steel fiber content generally is associated with a higher peak strain [10]. Moreover, the addition of steel fibers does not eliminate the spalling tendency of concrete mixtures [8].

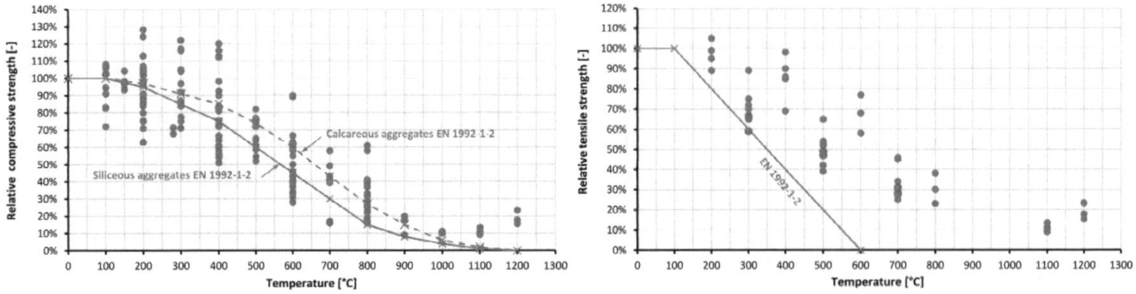

Figure 1. Residual compressive strength (left) and tensile strength (right) of SFRC

2.2. *Synthetic fibre reinforced concrete*

In the concrete industry, synthetic fibers, regardless of the type, shape and length, are mostly utilized with the aim to increase concrete spalling resistance when concrete structures exposed to elevated temperature [15,30]. As the melting point of synthetic fibers is relatively low, the presence of fibers in a concrete composite subjected to elevated temperature affects the mechanical properties of the concrete composite, particularly residual compressive strength, modulus of elasticity and splitting tensile strength [4]. While the presence of polypropylene fibers slightly increase ductility and the specific toughness, defined as the ratio of the area under the stress–strain curve, and the compressive strength of unheated concrete, after heating all the enhanced characteristics are lost [2,22]. Many contributions even demonstrate that polypropylene fibers have negative effect on the residual mechanical properties of polypropylene fiber reinforced concrete (PPFRC) after high-temperature exposure as they significantly decrease the residual compressive strength (figure 2) , elastic modulus and tensile strength as well as they increase peak strain [8,15,29]. On the other hand, some experimental investigations show that polypropylene fibers can improve the relative residual compressive strength of a concrete composite after the exposure to fire [27]. While the presence of polypropylene fibers at different dosages does not affect the residual compressive strength at 200°C and 400°C, it considerably increases the residual compressive strength of concretes after exposure to 600 °C [17]. Moreover, some even shows that compressive strength of concrete with polypropylene fiber additions is greater than unreinforced concrete without additions, when subjected to temperature up to 400 °C [23]. In comparison with SFRC, polyvinyl alcohol fibers reduce the uniaxial compressive strength but cause no appreciable change in elastic modulus in reactive powder concrete [30].

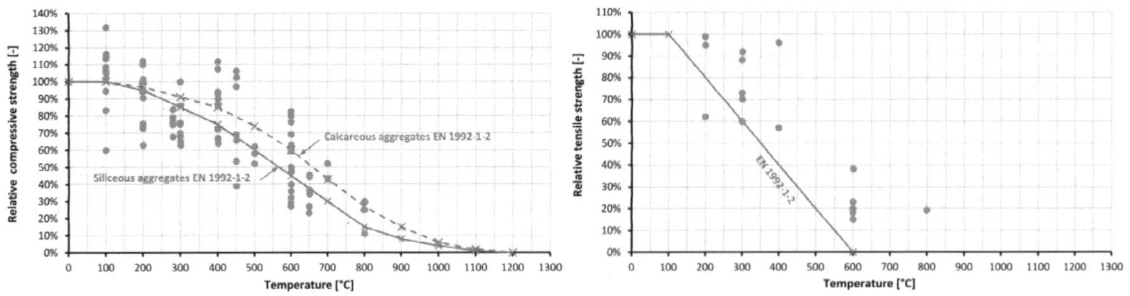

Figure 2. Residual compressive strength (left) and tensile strength (right) of synthetic fiber reinforced concrete

2.3. Hybrid (steel + synthetic) fibre reinforced concrete

The combination of steel and synthetic fibers represents a promising alternative how to ensure good toughness of a concrete composite before heating and improve its residual mechanical behavior and spalling resistance as well as the ductility after heating [11]. Although a few contributions declare that a fiber cocktail does not have much effect on high temperature compressive strength [29], the most of presented work affirm that the incorporation of steel fibers can effectively improve the compressive properties of a concrete composite when exposed to elevated temperatures while polypropylene fibers enhance concrete spalling resistance [13,27]. Specifically, the combination of steel fibers and polypropylene fibers shows positive synergy effect on the post-peak behavior of concrete composites (figure 3) before and after exposure to high temperature [11]. However, as synthetic fibers have a low melting point and ignition point, only steel fibers provide the stability and enhanced mechanical behavior to a concrete composite after exposing to elevated temperatures [22]. Considering steel to synthetic fiber content ratio, a concrete composite containing 1% synthetic fibers and 1% steel fibers by volume seems to produce the best results, balancing performance at high temperature with consideration of initial mechanical properties [30]. As a consequence, using hybrid fiber reinforced concrete (steel and synthetic fibers) might provide necessary safe guarantee for the rescue work and structure repair during and after a fire disaster.

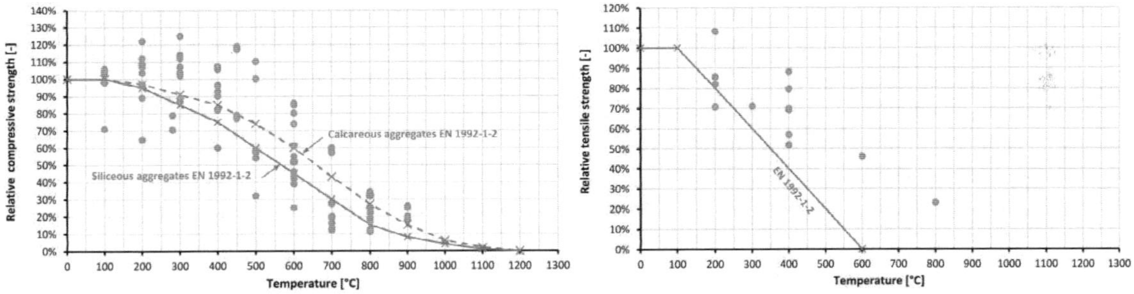

Figure 3. Residual compressive strength (left) and tensile strength (right) of hybrid (steel + synthetic) fiber reinforced concrete

3. Conclusion

The increasing utilization of FRC in the concrete industry is motivated by the physical and mechanical properties of the material which contribute to traditional concrete elements and structures various economical benefits. The extensive knowledge of mechanical properties of FRC exposed to elevated temperature seems to be decisive for a wider utilization of the material. Most of experimental investigations are conducted on test specimens at ambient temperature after high temperature exposure and only a few is performed on heated test specimens. The observed findings are as follows:

- Steel fibers in a concrete mix leads to an improvement in both mechanical properties and resistance to heating effects in comparison with unreinforced concrete. However, a higher content of steel fibers do not improve the compressive strength of concrete composites at elevated temperatures.
- The compressive strength of steel fiber reinforced reactive powder concrete and geopolymer concrete gradually increases when the material is heated up to 200–300 °C, but starts to decrease as temperature further increases.
- While the addition of steel fibers does not eliminate the spalling tendency of concrete, synthetic fibers enhance concrete spalling resistance.
- Polypropylene fibers have negative effect on the residual mechanical properties of PPFRC after high-temperature exposure as they significantly decrease the residual compressive strength, elastic modulus and tensile strength as well as they increase peak strain in comparison with unreinforced concrete.

- The combination of steel and synthetic fibers represents a promising alternative how to ensure good toughness of a concrete composite before heating and improve its residual mechanical behavior and spalling resistance as well as the ductility after heating.

Acknowledgments
This work was supported by the grant Models of steel and fiber composite columns exposed to fire, no. GACR 15-19073S, from Czech Science Foundation (GACR).

References
[1] Ma, Q, Rongxin G, Zhiman Z, Zhiwei L and Kecheng H. Mechanical properties of concrete at high temperature—A review 14 June 2015 *Construction and Building Materials* **93** 371–383

[2] Poon, C S, Shui Z H and Lam L . Compressive behavior of fiber reinforced high-performance concrete subjected to elevated temperatures 13 February 2004 *Cement and Concrete Research* **34** 2215–2222

[3] Drchalová J, Mňahončáková E, Vejmelka R, Kolísko J, Bayer P and Černý R. Hydric and mechanical properties of carbon fiber reinforced cement composites subjected to thermal load 6 April 2004 *Construction and Building Materials* **18** 567–578

[4] Noumowe A. Mechanical properties and microstructure of high strength concrete containing polypropylene fibres exposed to temperatures up to 200 °C 15 March 2005 *Cement and Concrete Research* **35**, 2192-2198

[5] Lau A. and Anson M.. Effect of high temperatures on high performance steel fibre reinforced concrete 23 March 2006 *Cement and Concrete Research* **36** 1698-1707

[6] Peng G, Wen-Wu Y, Zhao J, Ye-Feng L, Song-Hua B and Zhao L Explosive spalling and residual mechanical properties of fiber-toughened high-performance concrete subjected to high temperatures 14 December 2006 *Cement and Concrete Research* **36** 723-727

[7] Harun T Statistical analysis for mechanical properties of polypropylene fiber reinforced lightweight concrete containing silica fume exposed to high temperature 20 January 2009 *Materials and Design* **30** 3252-3258

[8] Sideris K K, Manita P and Chaniotakis E Performance of thermally damaged fibre reinforced concretes 27 September 2009 *Construction and Building Materials* **23** 1232-1239

[9] Sukontasukkul P, Pomchiengpin W and Songpiriyakij S Post-crack (or post-peak) flexural response and toughness of fiber reinforced concrete after exposure to high temperature 24 April 2010 *Construction and Building Materials* **24** 1967-1974

[10] Tai Y, Pan H and Kung Y Mechanical properties of steel fiber reinforced reactive powder concrete following exposure to high temperature reaching 800 °C 1 April 2011 *Nuclear Engineering and Design* **241** 2416-2424

[11] Ding Y, Azevedo C, Aguiar J B and Jalali S Study on residual behaviour and flexural toughness of fibre cocktail reinforced self compacting high performance concrete after exposure to high temperature 2 July 2011 *Construction and Building Materials* **26** 21-31

[12] Çavdar A A study on the effects of high temperature on mechanical properties of fiber reinforced cementitious composites 19 October 2012 *Composites: Part B* **43** 2452-2463

[13] Zheng W, Li H and Wang Y Compressive behaviour of hybrid fiber-reinforced reactive powder concrete after high temperature 21 May 2012 *Materials and Design* **41** 403-409

[14] Zheng W, Li H and Wang Y. Compressive stress–strain relationship of steel fiber-reinforced reactive powder concrete after exposure to elevated temperatures 15 June 2012 *Construction and Building Materials* **35** 931-940

[15] S ideris K K and Manita P. Residual mechanical characteristics and spalling resistance of fiber reinforced self-compacting concretes exposed to elevated temperatures 11 January 2013

Construction and Building Materials **41** 296-302

[16] Chen, G M, He Y H, Yang H, Chen J F and Guo Y C. Compressive behavior of steel fiber reinforced recycled aggregate concrete after exposure to elevated temperatures 7 September 2014 *Construction and Building Materials* **71** 1-15

[17] Qadi A, Al N S and Sleiman M A Effect of fibre content and specimen shape on residual strength of polypropylene fibre self-compacting concrete exposed to elevated temperatures 10 December 2014 *Journal of King Saud University – Engineering Sciences* **26** 33-39

[18] Bhat P, Chang S V and Li M Effect of elevated temperature on strain-hardening engineered cementitious composites 15 August 2014 *Construction and Building Materials* **69** 370-380

[19] Jameran A, Izni S I, Yazan H S S and Rahim S N A A. Mechanical properties of steel-polypropylene fibre reinforced concrete under elevated temperature 2015 *Procedia Engineering - The 5th International Conference of Euro Asia Civil Engineering Forum (EACEF-5)* **125** 818-824

[20] Kim J, Lee G and Moon D Y Evaluation of mechanical properties of steel-fibre-reinforced concrete exposed to high temperatures by double-punch test 23 January 2015*Construction and Building Materials* **79** 182-191

[21] Shaikh F U A and Hosan A Mechanical properties of steel fibre reinforced geopolymer concretes at elevated temperatures 31 March 2016 *Construction and Building Materials* **114** 15-28

[22] Choumanidis D, Badogiannis E, Nomikos P and Sofianos A The effect of different fibres on the flexural behaviour of concrete exposed to normal and elevated temperatures 3 November 2016 *Construction and Building Materials* **129** 266-277

[23] Serrano R, Cobo A, Prieto M I and González M D N Analysis of fire resistance of concrete with polypropylene or steel fibers 9 July 2016 *Construction and Building Materials* **122** 302-309.

[24] Zheng W, Luo B and Wang Y Stress–strain relationship of steel-fibre reinforced reactive powder concrete at elevated temperatures 18 April 2015 *Materials and Structures* **48** 2299-2314

[25] Aslani F and Samali B High Strength Polypropylene Fibre Reinforcement Concrete at High Temperature 4 March 2014 *Fire Technology* **50** 1229-1247

[26] Aslani F and Samali B. Constitutive Relationships for Steel Fibre Reinforced Concrete at Elevated Temperatures 4 September 2014 *Fire Technology* **50** 1249-1268

[27] Xiangjun D, Yining D and Tianfeng W Spalling and Mechanical Properties of Fiber Reinforced High-performance Concrete Subjected to Fire October 2008 *Journal of Wuhan University of Technology-Mater. Sci. Ed* 743-749

[28] Giaccio G M and Zerbino R L Mechanical behaviour of thermally damaged high-strength steel fibre reinforced concrete April 2005 *Materials and Structures* **38** 335-342

[29] Khaliq W and Kodur V Thermal and mechanical properties of fiber reinforced high performance self-consolidating concrete at elevated temperatures 24 June 2011 *Cement and Concrete Research* **41** 1112-1122

[30] Sanchayan S and Foster S J. High temperature behaviour of hybrid steel–PVA fibre reinforced reactive powder concrete 30 January 2016 *Materials and Structures* **46** 769-782.

[31] Kim J, Lee G and Moon D Y Evaluation of mechanical properties of steel-fibre-reinforced concrete exposed to high temperatures by double-punch test 23 January 2015 *Construction and Building Materials* **79** 182-191

[32] Bamonte P and Gambarova P G Thermal and Mechanical Properties at High Temperature of a Very High-Strength Durable Concrete June 2010 *Journal of Material in Civil Engineering* 545-555

[33] Bamonte P and P G Gambarova. A study on the mechanical properties of self-compacting concrete at high temperature and after cooling 16 February 2012 *Materials and Structures* **45** 1375–1387

[34] Xing Z, Beaucour A, Hebert R,Noumowe A and Béatrice Ledesert Influence of the nature of aggregates on the behaviour of concrete subjected to elevated temperature 11 January 2011 *Cement and Concrete Research* **41** 392-402

IOP Conf. Series: Materials Science and Engineering **246** (2017) 012046 doi:10.1088/1757-899X/246/1/012046

Two innovative solutions based on fibre concrete blocks designed for building substructure

J Pazderka[1] and P Hájek[1]

[1]Department of Building Structures, Faculty of Civil Engineering, Czech Technical University in Prague, Thákurova 7, 166 29 Praha 6, Czechia

E-mail: jiri.pazderka@fsv.cvut.cz

Abstract. Using of fibers in a high-strength concrete allows reduction of the dimensions of small precast concrete elements, which opens up new ways of solution for traditional construction details in buildings. The paper presents two innovative technical solutions for building substructure: The special shaped plinth block from fibre concrete and the fibre concrete elements for new technical solution of ventilated floor. The main advantages of plinth block from fibre concrete blocks (compared with standard plinth solutions) is: easier and faster assembly, higher durability and thanks to the air cavity between the vertical part of the block, the building substructure reduced moisture level of structures under the waterproofing layer and a comprehensive solution to the final surface of building plinth as well as the surface of adjacent terrain. The ventilated floor based on fibre concrete precast blocks is an attractive structural alternative for tackling the problem of increased moisture in masonry in older buildings, lacking a functional waterproof layer in the substructure.

1. Introduction

Today, thanks to the development of fiber and high-strength concrete, it is possible to start thinking about the new applications of these materials in building construction. Using of fibers in high-strength concrete allows reduction of dimensions of small precast concrete elements. This opens up new ways of solution for traditional construction details in buildings. The following text presents two new technical solutions for building substructure.

2. Special shaped plinth block from fibre concrete

2.1. The building plinth – short state of the art

The "plinth area" of buildings consists of two parts: the vertical part (surface of the peripheral wall of the building) and the horizontal part (adjacent terrain surface). The contemporary building plinth (vertical part) is solved by several ways: plaster (mostly as a surface layer of ETICS), ceramic or stone facing or grid+plates system (with many types of surface). The disadvantages of these traditional solutions of the building plinth (vertical part) consist mainly in: low durability (plasters), low vapor permeability (ceramic or stone facing), low resistance to mechanical damage (plasters, grid+boards systems)), high laboriousness (ceramic or stone facing, grid+boards systems) or wet process (plasters). In addition to the plinth surface, the adjacent terrain must be adjusted as well (usually concrete pavers or stone aggregate).

2.2. Special shaped plinth block from fibre concrete

The aforementioned disadvantages of existing solutions in the plinth area of buildings are significantly corrected with plinth based on precast blocks designed from high-strength concrete reinforced by fiber. This technical solution was registered by authors as a patent no. CZ305253(B6) [1]. The proposed plinth consists of a set of special shaped blocks form fibre concrete which solves the final surface of building plinth (vertical part) as well as the surface of adjacent terrain (horizontal part). The blocks have a rectangular shape, where one of the vertical "walls" is exceeded over the other thus forming a plinth surface on the peripheral wall of the building. The block comprises an internal space (open at the top) which serves to store stone aggregate which creates a surface of adjacent terrain. The side "walls" of the block (perpendicular to the wall surface of the building) are in their upper part designed in the shape of a reverse trapezoid. The reason is that they to be invisible after the deposit of stone aggregate thus ensuring the "continuous look" of surface treatment of an adjacent terrain.

The bottom of the inner space of the block (where the stone aggregate is placed) is designed in the slope, so that leaking rainwater drained towards the outer edge of the block where the drain openings are located. The purpose of the stone aggregate inside the block is: stabilization of precast block in the place, prevent splashing of falling rainwater, making nice appearance of terrain and prevent the growth of higher plants.

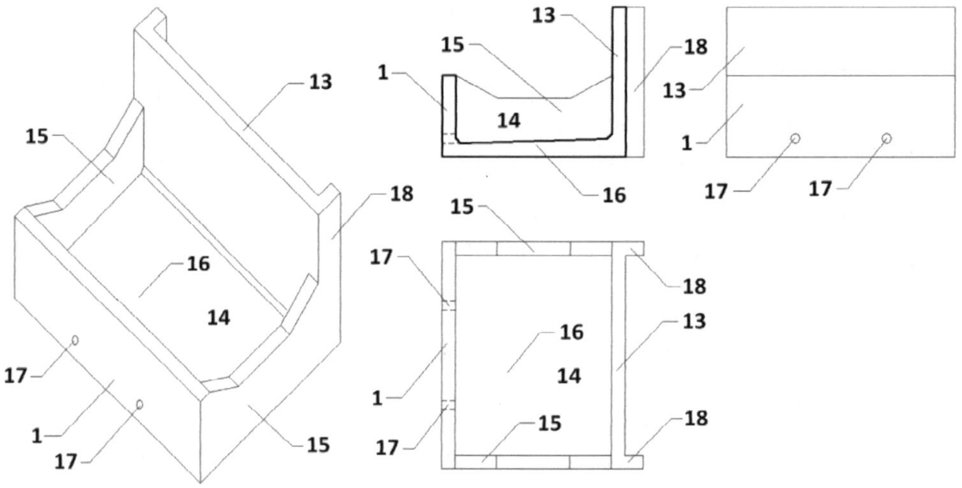

Figure 1. Special shaped plinth block from fibre concrete - description: 1 - plinth block from fibre concrete, 13 -vertical part which created a surface of building plinth, 14 - internal space (open at the top) which serves to store stone aggregate, 15 - the side "walls" in the shape of a reverse trapezoid, 16 - the bottom in the slope, 17 – the drain openings, 18 - the spacer strips to create a ventilated cavity between the block and the wall of the building

The rigidity of the thick "walls" of the block is ensured by the use of fibres in concrete in combination with high-strength concrete. There is placed no other reinforcement in the block. The high-strength concrete ensures the block durability in the moist environment as well because it has a closed structure that does not allow the seepage of water in the liquid phase. A significant advantage of the use of high-strength concrete in combination with fibres is a subtlety of block elements. This allows the manual handling of the block without the use of the lifting devices. The dimensions of precast element were considered as follows (can be adapted to each particular building): wall thickness: 30mm, height of the plinth part: 330 mm, width: 400mm, length: 500 mm. The weight of the block with these dimensions is only 44kg.

The proposed technical solution also reflects the problems of moisture accumulation under the waterproof layer in substructure (applies to buildings without basement). A banked vertical wall of the block (adjacent to the peripheral wall of the building) is equipped at its ends with vertical spacer strips. This adjustment creates an air cavity between the block and the wall of the building. This cavity serves for an intensification of diffusion of water vapor from the building substructure. To increase water vapor diffusion efficiency, it is possible to complement the solution by profiled foil (installed at the foundation structure and connected to cavity). For proper function of the cavity crated by the block, it is necessary to realize the gap (20-30 mm) between the top of the block and the adjacent ETICS (the gap must be fitted with insect screen) - Figure 2.

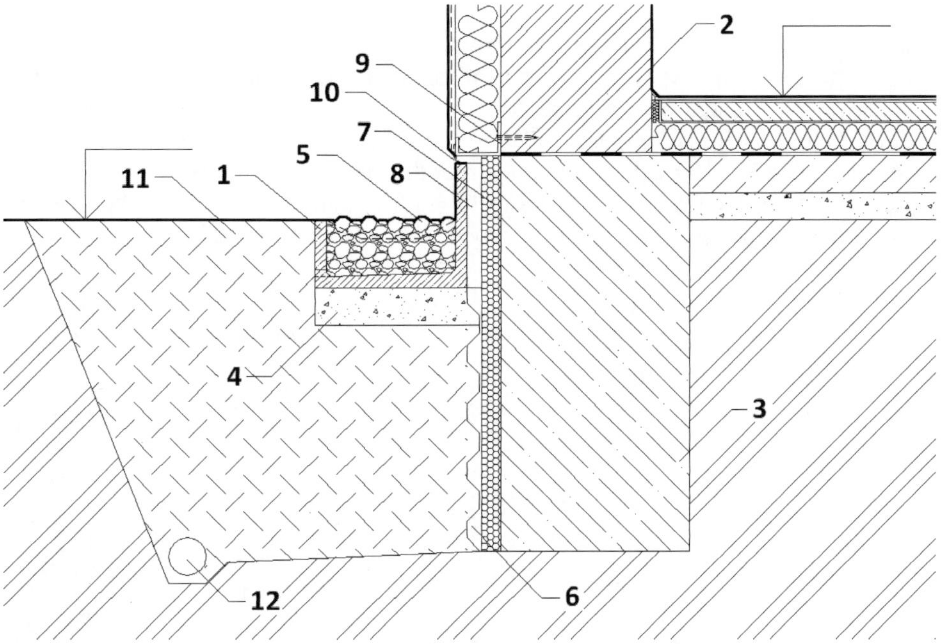

Figure 2. Example of block installation at the building – description: 1 - plinth block from fibre concrete, 2 - peripheral wall of the building, 3 - foundation structure, 4 –fine stone filler, 5 - stone aggregate inside the block, 6 – profiled foil, 7 - diffusion-open heat-insulating board, 8 – air cavity, 9 - ledge for ETICS, 10 - ventilation gap (with insect screen) , 11 – compacted filler, 12 – drainage.

2.3. Discussion

The main advantage of the above presented technical solution is that it solves the final surface of building plinth as well as the surface of adjacent terrain. Compared with standard plinth solutions, there are the following advantages: easier and faster assembly, higher durability and thanks to the air cavity between the vertical part of the block and the building substructure reduced moisture level of structures under the waterproofing layer. Special shaped plinth block can be used for refurbishment of older buildings as well as for new buildings. The disadvantage of technical solution is only a higher price of block, given by the use of high-strength concrete.

3. Ventilated floor consisting of blocks from fibre concrete

3.1. The ventilated floors – short state of the art

Ventilated floors (as a remediation measure) are currently implemented as either as the "zero" ceiling structure or in the form of plastic shaped blocks with a concrete load-bearing layer, or based on a system of ventilated piping installed below ground level [2]. The aim of these remediation measures is

IOP Conf. Series: Materials Science and Engineering **246** (2017) 012046 doi:10.1088/1757-899X/246/1/012046

to reduce the amount of rising water which increases moisture level in building walls (in case of malfunctioning or missing waterproofing layer) [3,4,5]. Today, the most often used type is the ventilated floor based on plastic shaped blocks covered with a layer of concrete. This is a system of plastic blocks (shaped like dome vault), placed side by side on the site of the future floor. Once the shaped blocks are placed, they have to be covered with a layer of concrete with a reinforcement mesh installed. This helps reinforce the gaps between the "legs" of the blocks with concrete, and they then serve as supports for the whole floor. The disadvantages of this systems is: adding a monolithic concrete slab to the floor structure (wet process), a certain risk of damage to the plastic blocks during installation, part of the concrete mixture can leaks between the shaped blocks (while the concrete slab is made).

3.2. Ventilated floor consisting of blocks from fibre concrete

The aforementioned disadvantages of floor from plastic shaped blocks with a concrete load-bearing layer are significantly corrected by a precast "fibre concrete floor" registered by the author as utility model no. CZ27462(U1) [6]. The essence of the precast fibre concrete floor is that it consists of a system of special shaped blocks made from high-strength concrete reinforced by fibers. The blocks are laid out so that they form a continuous ventilated air cavity underneath the floor surface. The system consists of two types of fibre concrete blocks: "cubes" and "slabs" (Figure 3).

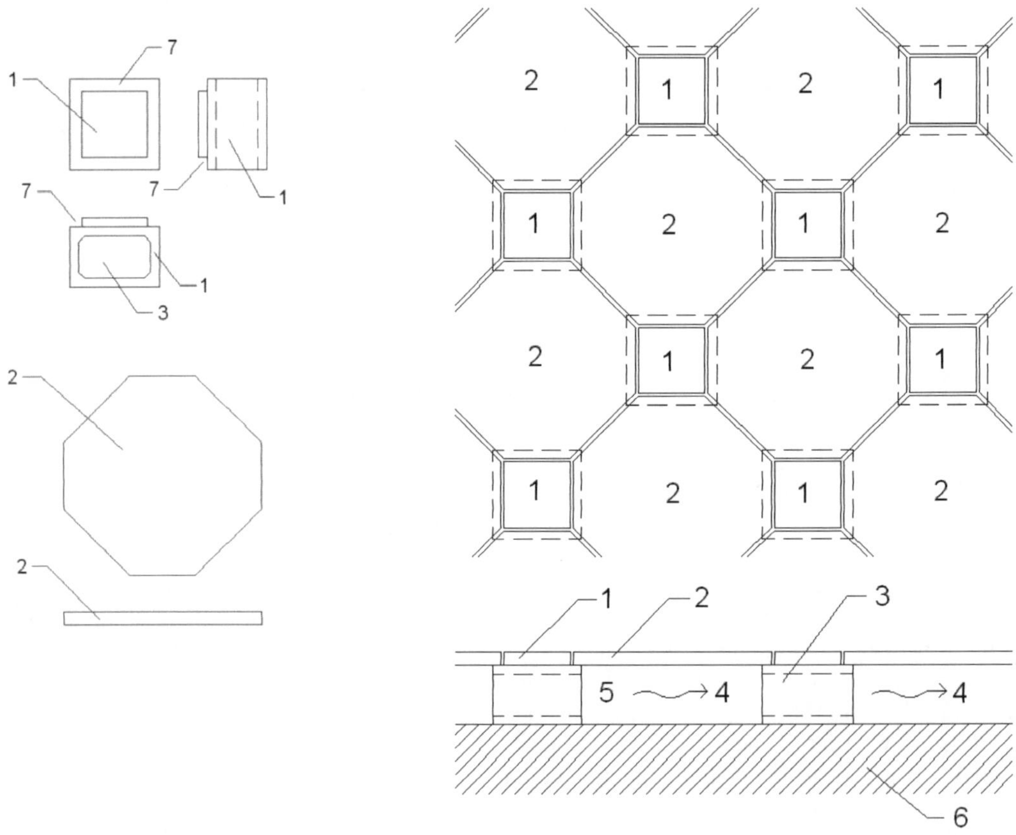

Figure 3. The ventilated floor from fibre concrete blocks: Basic elements „cube" and „slab" (left) and its arrangement in the floor (right) - description: 1-element „cube", 2-element „slab", 3-cavity in the „cube",4-air flow, 5-ventilated cavity, 6-terrain, 7- the recess for mounted of element „slab".

IOP Conf. Series: Materials Science and Engineering **246** (2017) 012046 doi:10.1088/1757-899X/246/1/012046

The "cubes" are cuboid-shaped elements lightened at the center by an extensive air duct. They have a raised part at the top, forming a recess along the edges for mounting the "slab" element. The "cube" has a square footprint. "Slabs" are mounted on the recess in the "cube" on all four sides; their footprint is an octagon. This shape of the "slab" makes it possible to space the "cubes" at a certain distance so as not to break the continuous air duct that the floor shapes in both directions. In one direction, the air flow through the cavity formed by the double floor is additionally facilitated by the cavity contained in the "cube" elements. Therefore, it is advisable to orient the "cube" blocks in the floor so that the cavity aligns with the direction of the expected dominant air flow.

Both the "cube" and "slab" elements are made of high-strength concrete, reinforced with non-metallic fibres (fibre-reinforced concrete). The use of this material permits delicate dimensions of the elements (40 mm thick), reducing the weight of both the "cube" and the "slab" to a level that permits manual handling of the precast elements without having the use lifting devices. Another advantage, particularly compared with plastic shaped blocks with a cover of concrete, is the low thickness of the horizontal load-bearing structure separating the ventilated cavity from the floor structure. The horizontal structure consists only of a delicate slab (40 mm), the supporting "cube" elements are located inside the duct and thus do not increase the height of the load-bearing structure. Moreover, thanks to its enclosed porous texture, the high-strength concrete gives the elements excellent properties in terms of durability in long-term contact with water and moisture.

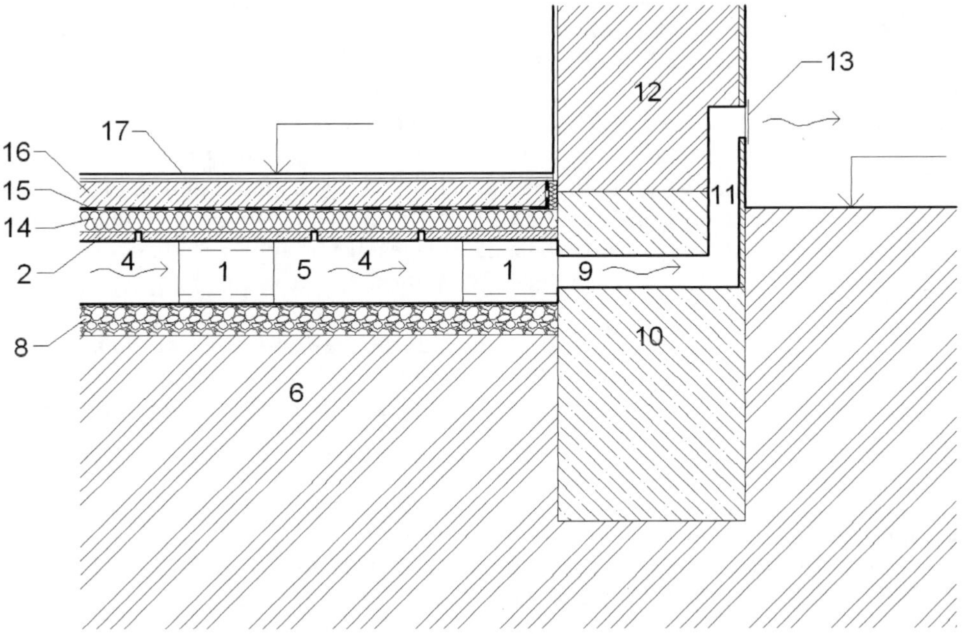

Figure 4. Example of "fibre concrete ventilated floor" installation in the building - description: 1 - element „cube", 2 - element „slab", 4 - air flow, 5 - ventilated cavity, 6 - terrain, 8 - gravel bedding, 9 - horizontal vent hole, 10 - foundation, 11 - vertical vent hole, 12 - masonry, 13 - cover grid, 14 - thermal insulation in the floor, 15 - separating layer, 16 - load distribution layer (eg. concrete screed or drywall etc.), 17 - the top floor (ceramic floor tiles).

Remediation of an existing diffusion-resistant floor by replacing it with the aforesaid technical solution can be made as follows: First the existing unsatisfactory floor is demolished (down to the ground). Then, the ground is deepened to the required level (depending on the height difference between the original and new floor). A layer of fine stone aggregate is applied on the ground,

IOP Conf. Series: Materials Science and Engineering **246** (2017) 012046 doi:10.1088/1757-899X/246/1/012046

thoroughly compacted in low layers. The floor system "cubes" are installed on this layer, 600 mm apart (both ways). Then, the octagonal "slabs" are mounted on the "cubes". Afterwards, the slabs are covered with a layer of thermal insulation (extruded polystyrene is recommended due to installation in a humid environment) of the desired thickness (in CZ depending on requirements of ČSN 730540-2 [7]). The thermal insulation layer is covered with a separating plastic sheet, on which a load-distributing layer is applied (either concrete screed with a mesh reinforcement, or a system of drywall boards laid in a criss-cross pattern, etc.). The walking floor layer is then rendered on the top of the load-distributing layer.

In protected heritage buildings, where intake holes cannot be made in the facade (negative effect on building appearance), the presented technical solution can be combined with a prefabricated ventilated tunnel as described in [8] or another type of ventilated tunnel. However, the efficiency of the whole system will then depend on the air flow speed in the tunnel duct (forced ventilation is recommended).

3.3. Discussion
A ventilated floor based on fibre concrete precast blocks is an attractive structural alternative for tackling the problem of increased moisture in masonry in older buildings, lacking a functional waterproof layer in the substructure. The chief advantage of this technical solution compared to existing double floors based on plastic shaped blocks with a concrete layer is the absence of a monolithic reinforced concrete slab in the load-bearing structure of the floor (shorter construction time, less demanding process). Another advantage is the very low thickness of the horizontal load-bearing structure (positive effect on structural floor height, less earth moving).

4. Conclusion
The special shaped plinth block from fibre concrete and the fibre concrete elements for new technical solution of ventilated floor demonstrated that using of fibers in high-strength concrete precast elements opens up new ways of solution for traditional construction details in buildings. In the future, a further specialized precast systems based on fibre concrete elements which will replace some traditional structures in buildings can be expected. The advantages of this approach are evident: easier and faster assembly, higher durability and new additional functions.

Currently, preparation for the production of the first prototype of the fibre concrete plinth block is performed. The abovementioned patent CZ305253(B6) and utility model CZ27462(U1) which contain the technical solutions presented in the paper, are valid only for Czech Republic, in other countries are free to use.

Acknowledgement
This work was supported by the research project granted by Technology Agency of the Czech Republic grant TH02020512 and by the CTU grant SGS17/117/OHK1/2T/11.

References
[1] Pazderka J and Hájek P 2015 *Plinth block - patent no. CZ305253(B6)* (Prague: Czech Industrial Property Office)
[2] Hájková E and Pazderka J 2015 Efficiency of air ducts in historical buildings *Building Engineer* **90**/9 22-25
[3] Torres I M and de Freitas V P 2017 Treatment of rising damp in historical buildings: wall base ventilation *Building and Environment* **42**/1 424-435
[4] Piaia J C Z, Cheriaf M, Rocha J C and Mustelier N L 2013 Measurements of water penetration and leakage in masonry wall: Experimental results and numerical simulation *Building and*

IOP Conf. Series: Materials Science and Engineering **246** (2017) 012046 doi:10.1088/1757-899X/246/1/012046

Environment **61** 18-26
[5] Grytli E, Kværness L, Rokseth L S and Ygre K F 2012 The impact of energy improvement measures on heritage buildings *Journal of Architectural Conservation* **18**/3 89-106
[6] Pazderka J 2014 *Ventilated floor - utility model no. CZ27462(U1)* (Prague: Czech Industrial Property Office)
[7] Tywoniak J, Svoboda Z and Matuška T 2011 *ČSN 73 0540 – 2 Thermal protection of buildings – Part 2: Requirements* (Prague: Czech office for standards, metrology and testing)
[8] Pazderka J 2016 Concrete with crystalline admixture for ventilated tunnel against moisture *Key Engineering Materials* **667**/1 108-113

IOP Conf. Series: Materials Science and Engineering **246** (2017) 012047 doi:10.1088/1757-899X/246/1/012047

Composite Material from By-products and Its Properties

K Šeps[1], I Broukalová[1] and J Vodička[1]

[1]Faculty of Civil Engineering, Czech Technical University in Prague, Thákurova 7, 166 29 Praha 6, Czech Republic

E-mail: karel.seps@fsv.cvut.cz

Abstract. The paper shows an example of utilization of specific textile admixture – fluffs of torn textiles from waste cars in production of composite with aggregate consisting entirely of unsorted recycled concrete. The admixture in the mixture of recycled concrete and cement binder fills the pores and voids in composite. The elaborated composite has working title STEREDconcrete. In the article, basic mechanical-physical properties of the composite are presented also the fire resistance of STEREDconcrete, which was determined in tests.

1. Introduction

The demand for increase of recycling creates new challenges for research institutions. Our team is engaged in investigations of utilization of crushed concrete for manufacturing of a new cement based composite.

The application of crushed concrete as aggregate for manufacturing of new concrete is quite common. Properties of concrete with recycled aggregate (RAC – recycled aggregate concrete) depend on ratio of recycled aggregate and natural aggregate used for manufacturing of the RAC. The density of RAC would not differ from normal concrete if up to 25 % of recycled aggregate is used; for higher content of recycled aggregate the RAC density used to be lower.

The compressive strength is mostly lower (4 – 45 %) [1, 2]. Other researches ascertain that RAC with recycled aggregate up to 25% reaches even higher compressive strength [3], the reason is probably in quality of the used recycled concrete aggregate crushed from high-class concrete. The tensile splitting strength is indicated lower up to 33% if more than 25% of natural aggregate is substituted [1, 4]; in contrary some studies introduced increase of by 20% [5].

We can summarize that RAC properties fluctuate, they are uncertain and thus their structural applications are economically disadvantageous. Thus an alternative approach was proposed, to design a new composite from waste for non-structural use. The intention was to consume a big amount of waste concrete fine and coarse aggregate without sorting to fractions. The composite has high void content – ca 30%. The high void content pointed to idea of filling the voids by additional admixture that could enhance the composite properties.

Properties of concrete composites are enhanced by application of steel or polymeric fibres, which affect mechanical and physical properties [6 - 8]. In a similar way fibres can be used for concrete with recycled aggregate. With respect to porous structure of RAC, synthetic the fibres are chosen.

Synthetic fibres randomly distributed in the structure of the compacted composite were the first tested admixture. Significant change of the response of the composite was detected – from brittle failure of the original composite to tough and resilient behaviour. The ability of carrying of tensile stresses after cracking is an essential feature of the composite with synthetic fibres.

Since 2012 admixture STERED has been tested. STERED is waste torn textile from car industry. The test results of mechanical properties have shown potential for practical use [9]. The desired properties of the composite for particular application can be engineered by dosage of STERED, related to amount of pores in the composite. The new composite was called STEREDconcrete and the development of the material was finalized by patent application n. 306029).

STERED is made from non-flammable upholstery textile. As a non-flammable material it should tribute to higher fire resistance of the composite. Therefore an application of the STEREDconcrete in insulating tilling boards was proposed. Thus further investigations focused on properties related to application in boards.

In the paper tests of mechanical properties are presented and tests focused in determination of fire resistance of the composite with textile shred.

2. Composition and tests of STEREDconcrete

2.1. Mixture

The composite mixture design is based on the demand of maximum waste consumption. The aggregate was recycled concrete with maximum grain size 16 mm (it was given by the specimen dimensions), binder was cement CEM I 42,5 R in dosage 400 kg/m^3 (the cement dosage was determined with respect to manufacturing of relatively thin board), water in dosage 100 kg/m^3 to reach fresh concrete consistency suitable for rolling and tamping. Dosage of the textile shred 29.5 kg/m^3 was determined to fill the volume of approximately 50% of voids in the composite.

In the first manufacturing step the recycled concrete aggregate is mixed with STERED, afterwards the cement is added and the last component added to fresh concrete (Fig. 1) is water.
The manufactured test specimens were prisms 100x100x400 mm and board with dimensions 500x1500 mm and thickness 80 mm. The prisms were compacted by tamping; the boards were compacted by combination of rolling and tamping. (Fig. 2, 3).

Figure 1. Fresh concrete composite with textile shred.

Figure 2. Compacting of the board by rolling.

Figure 3. Compacting of the board by compactor.

2.2. *Results of mechanical properties tests*

The prism specimens (100x100x400 mm) were examined according to ČSN EN 12390-5 standard [10] in four-point bending test (results are presented in the table 1 and Fig. 4) and then the fragments were used for determination of the compressive strength according to ČSN EN 12390-3 [11] (see table 2).

Table 1. Results of the four-point bending tests.

Specimen	Dimensions [mm]	Weight [g]	Bulk density [kg/m^3]	Flexural strength [MPa]
1.1	400x99x100.21	6350	1600	0.99
1.2	400x98.16x99.9	6585	1680	0.92
1.3	400x103.8x101.51	6495	1540	1.06
1.4	400x99.09x100.77	6510	1630	0.66
	Mean value		**1615**	**0.91**

IOP Conf. Series: Materials Science and Engineering **246** (2017) 012047 doi:10.1088/1757-899X/246/1/012047

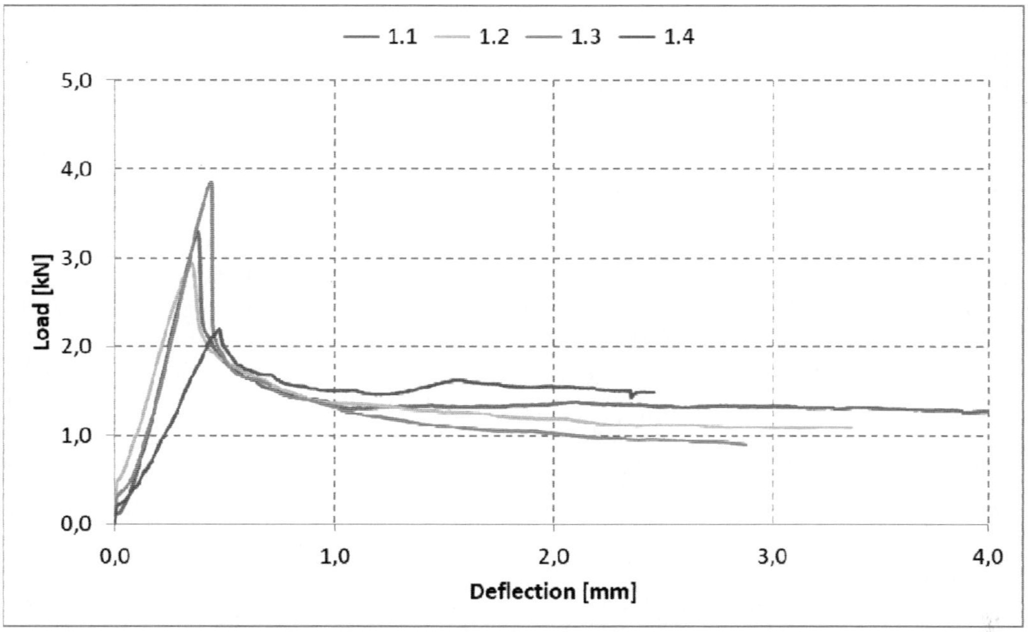

Figure 4. Load-deflection diagrams from bending test.

Table 2. Results of compressive tests on prism fragments.

Specimen	Dimensions [mm]	Force [kN]	Flexural strength [MPa]
1.1	100x99x100	57.02	5.76
1.2	100x98.16x99.9	40.96	4.18
1.3	100x103.8x100	35.93	3.46
1.4	100x99.09x100	42.8	4.32
	Mean value		**4.43**

The average value of compressive strength is 4.5 MPa, which is similar to aerated concrete. Flexural strength reaches ca 0.9 MPa; the value is relatively high compared to common concretes where the ratio of compressive and tensile strength is 1/10. The density of the composite STEREDconcrete varies from 1500 to 1700 kg/m^3 so that it can be classified light-weight concrete.

2.3. The fire test

Testing of the fire behaviour was performed with three boards.

Five temperature sensors were fixed on the board. The board was fixed in the test bench and loaded by fire. The flames heated the centre of the board. The sensors measured the temperature on the opposite side than flames were applied. At the same time the temperature of the flame and ambient temperature were measured. The gradients of temperature across the thickness of the board and between the centre and edges of the board were followed. Results of the test in relation temperature and time are plotted in the Fig. 5 and listed in the table 3.

IOP Conf. Series: Materials Science and Engineering **246** (2017) 012047 doi:10.1088/1757-899X/246/1/012047

Table 3. Measured values (in °C) from fire test.

Edge 1	Edge 2	Edge 3	Edge 4	Centre	Flame
21.80	16.80	21.30	27.70	21.20	22.7
19.20	18.60	21.70	28.40	22.40	63.8
28.50	20.10	23.70	30.20	24.60	158.4
24.00	20.90	25.70	29.10	24.30	237.9
26.00	21.60	29.00	30.20	25.10	409.9
20.40	20.20	26.70	29.50	23.90	523.0
32.60	26.00	36.60	37.30	31.60	632.4
30.20	22.40	33.70	39.90	31.00	700.8
27.50	26.50	41.40	39.30	31.90	785.2
34.30	45.60	57.70	48.40	49.30	849.8
33.60	55.00	65.30	53.30	56.90	710.1
34.30	57.40	70.90	58.70	57.40	622.4
36.80	58.70	67.60	62.40	59.20	576.9

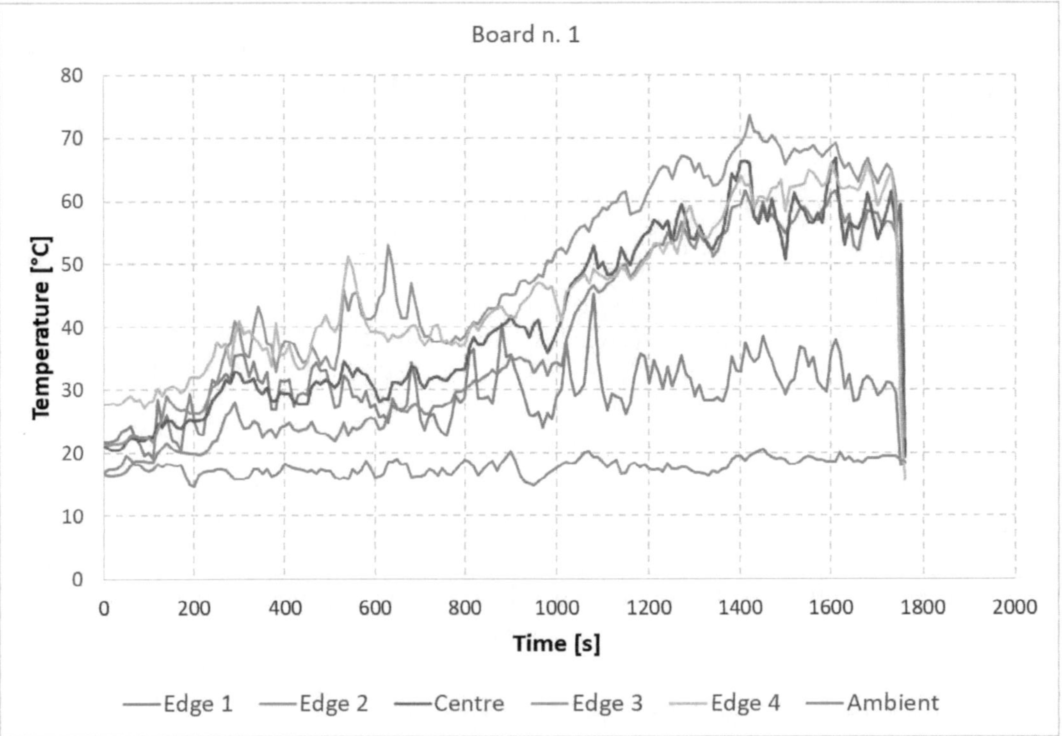

Figure 5. Results of temperature measuring in the test of board 1.

3. Conclusions

The proposed composite made from recycled concrete aggregate and material STERED showed its potential for practical applications. STEREDconcrete is a low strength, light-weight composite with beneficial behaviour in fire and satisfactory thermal insulating properties.

STEREDconcrete can be used for insulation lining, in low-floor structures as a base and as layer of line structures.

The research of composite made from waste resources will tribute to profitable use of construction and industrial waste materials.

Acknowledgements
The experimental work was supported by project TH02030649 Environmentally Efficient Construction and Demolition Waste for Structures and by project SGS17/123/OHK1/2T/11.

References
[1] Mas B, Cladera A, del Olmo T, Pitarch F 2012 Influence of the amount of mixed recycled aggregates on the properties of concrete for non-structural use *Construction and Building Materials* **27** 612–622.
[2] Leiva C, Solís-Guzmán J, Marrero M, Arenas C G 2013 Recycled blocks with improved sound and fire insulation containing construction and demoliion waste *Waste Management* **33** pp. 663-671.
[3] Hansen T C 1992 Recycled aggregates and recycled aggregate concrete, recycling of demolished concrete and masonry RILEM, London (RILEM Report No. 6).
[4] Puthussery J V, Kumar R, Garg A 2016 Evaluation of recycled concrete aggregates for their suitability in construction activities: An experimental study *Waste Management* **60** pp 270-276.
[5] McNeil K, Kang T H-K 2013 Recycled concrete aggregates: a review, *International Journal of Concrete Structures and Materials* **7** (1) pp. 61-69.
[6] Bílý P, Fládr J, Kohoutková A 2017 Behavior of anchorage areas in the steel-concrete composite structure loaded by longitudinal shear forces *Procedia Engineering* **172** pp. 104-110.
[7] Tesárek P, Holub P, Havrda J, Fládr J, Prošek Z, Trejbal J, Plachý T 2017 Mechanical Properties of Cement Composites Reinforced by Carbon Microfibers: Compressive and Bending Strength *Key Engineering Materials* **722** pp. 351-356.
[8] Štemberk P, Vaitová M, Hanzlová H 2016 Prediction of Early Age Tensile Creep of Concrete Using Fuzzy Logic *Mechanika 2016 Proceedings of the 21st International Scientific Conference* pp. 253-256.
[9] Šeps K and Vodička J 2012 Fibre Reinforced Concrete with Recycled Concrete and Insulation Material STERED. Advanced Materials Research **1054** pp 262-266.
[10] ČSN EN 12930-5 Testing hardened concrete – Part 5: Flexural strength of test specimens
[11] ČSN EN 12390-3 Testing hardened concrete – Part 3: Compressive strength of test specimens

IOP Conf. Series: Materials Science and Engineering **246** (2017) 012048 doi:10.1088/1757-899X/246/1/012048

Research of UHPC properties prepared with industrial mixer

E Šerelis[1], V Vaitkevičius[1] and V Kerševičius[1]

[1] Dept. of Building Materials, Kaunas University of Technology, Studentų 48, 51367 Kaunas, Lithuania

E-mail: evaldas.serelis@ktu.lt

Abstract. Ultra-high performance concrete (UHPC) mixture with advanced mechanical and durability properties was created using decent Zyklos ZZ50HE mixer. Zyklos ZZ50HE rotating pan mixer is similar to mixer which has common concrete plants. In experiment UHPC was prepared with Zyklos ZZ50HE mixer and thereafter best composition was selected and prepared with industrial HPGM 1125 mixer. Experiment results revealed that UHPC with W/C=0.29 and advanced mechanical and durability properties can be prepared. In experiment tremendous amount of micro steel fibres (up to 147 kg/m^3) were incorporated in UHPC. Concrete with excellent salt scaling resistance and great mechanical properties was obtained. Compressive strength was increased about 30 % from 116 MPa to 150 MPa and flexural strength was increased about 5 times from 6.7 to 36.2 MPa. Salt-scaling resistance at 40 cycles in 3 % NaCl solution varied from 0.006 kg/m^2 to 0.197 kg/m^2. There were a few attempts to create UHPC and UHPFRC with decent technology, however, unsuccessfully till now. In the world practice this new material is currently used in the construction of bridges and viaducts.

1. Introduction

Over a past few decades significant advances in research on cementitious materials have been made which led to develop a new class composition material with improved strength and durability properties. Probably the best durability and compressive strength properties could be attributed to ultra-high performance concrete (UHPC). In Europe UHPC is known as concrete which has compressive strength over 100 MPa. However by adding proper amount of necessary materials compressive strength could be increased up to 200 MPa [1-3]. Chemical admixtures are very important additives which helps reduce water to cement ratio below 0.30 and pozzolanic additives (fly ash, silica fume, ground-granulated blast-furnace slag and etc.) is very beneficial to gain additional mechanical strength during further hydration process. The most common pozzolanic material used in UHPC is silica fume [4-5]. Interaction of silica fume between reactive materials is very complicated and greatly affects workability and hydration process of cementitious materials. However silica fume is very expensive additive. Resent research showed, that silica fume better works with other pozzolanic materials such as glass powder. However deeper research is required to proper understand interaction of silica fume and glass powder on properties of ultra-high performance concrete.

Glass is a solid material which generally consists of non-crystalline silica, sodium oxide, calcium oxide and other components [6-8]. The chemical composition mainly depends on the raw materials used and differs slightly for each glass type. The most common soda-lime glasses consist of ≥70 % amorphous SiO_2, ≥12 % Na_2O and ≥5 % CaO [9-11]. Being amorphous and containing relatively large quantities of silicon, glass should be an excellent pozzolanic material for concrete industry. Thus when glass is finely grounded to powder, in theory, it could be used as cement replacement. However glass powder has very high amounts of Na_2O which could initiate alkali-silica reaction. Therefore glass powder can be characterized by a few controversial facts: it has enough amorphous SiO_2 to be considered as pozzolanic material; also one of the main chemical components is lime so it can be considered as cementitious material and it contains a very high amount of Na_2O which is a source for alkali silica reaction. These truths are highly debated between various scientists; however it is not entirely clear how glass can affect certain properties of concrete. In experiment was revealed some controversial truth about glass powder utilisation in UHPC.

The main aim of this research is to create UHPC composition which could be prepared with industrial mixer and without usage of advanced technology. New composition was prepared with different amount of micro steel fibres. Main properties of UHPC were determined by Chapelle test, mercury intrusion porosimetry (MIP), XRD analysis, compressive strength, flexural strength and salt-scaling. Proposed composition could be used as innovative material for various elements made of concrete which could be prepared with industrial mixer.

2. Used materials

Cement. Portland cement CEM I 52.5 R was used in experiment. Main properties: paste of normal consistency– 28.5 %; specific surface (by Blaine) – 4840 cm^2/kg; soundness (by Le Chatielier) – 1.0 mm; setting time (initial/final) – 110/210 min; compressive strength (after 2/28 days) – 32.3/63.1 MPa. Mineral composition: C_3S – 68.70; C_2S – 8.70; C_3A – 0.20; C_4AF – 15.90. Particle size distribution of cement, silica fume, quarts powder and quartz sand are shown in Fig.1.

Table 1. Chemical compositions of Portland cement, silica fume and glass powder

Components	Quantity, %		
	CEM I 52.5 R	Glass powder	Silica fume
SiO_2	20.61	72.76	92.08
TiO_2	-	0.04	-
Al_2O_3	5.45	1.67	1.16
Fe_2O_3	3.36	0.79	1.24
MnO	-	0.02	-
MgO	3.84	2.09	0.80
CaO	63.42	9.74	1.07
SO_3	0.80	0.10	1.27
Na_2O	0.20	12.56	1.13
K_2O	1.00	0.76	0.67
P_2O_5	-	0.02	-
$Na_2O_{eq.}$	0.86	13.06	1.57
Loss of ignition	1.00	1.00	-

Silica fume. Silica fume, also known as micro silica (MS) or condensed silica fume is a by-product of the production of silicon metal or ferrosilicon alloys. Main properties: density – 2532 kg/m^3; bulk density – 400 kg/m^3; pH – 5.3.

Quartz powder. Quartz powder was used in the experiment. Main properties: density 2671 kg/m^3; bulk density – 900 kg/m^3; average particle size – 18.12 μm; specific surface (by Blaine) – 3423 cm^2/g.

Quartz sand. Quartz sand was used in the experiment. Main properties: fraction: 0/0.5; density 2650 kg/m^3; specific surface (by Blaine) – 91 cm^2/g.

Glass powder. Glass powder was used in the experiment. Main properties: density 2528 kg/m^3; average particle size – 25.80 μm; specific surface (by Blaine) – 3350 cm^2/g. Particle size distribution is shown in Fig. 2.

Micro steel fibres. In experiment were used micro steel fibres with hooked ends. Main properties: length – 30 mm, diameter – 0.30 mm; tensile strength – 1000 MPa.

Chemical admixture. Superplasticizer (SP), based on polycarbocylisc ether (PCE) polymers, was used in the experiment. Main properties: appearance: dark brown liquid, specific gravity (20 °C) – 1.08±0.02 g/cm^3; pH – 7.0±1; viscosity – 128±30 m·Pa; alkali content ≤ 5.0 %, chloride content ≤ 0.1 %.

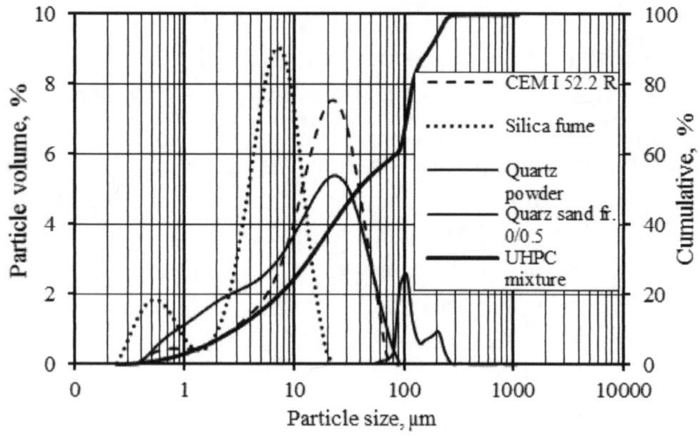

Figure 1. Particle size distribution of Portland cement, silica fume, quartz powder and 0/0.5 mm fr. quarts sand

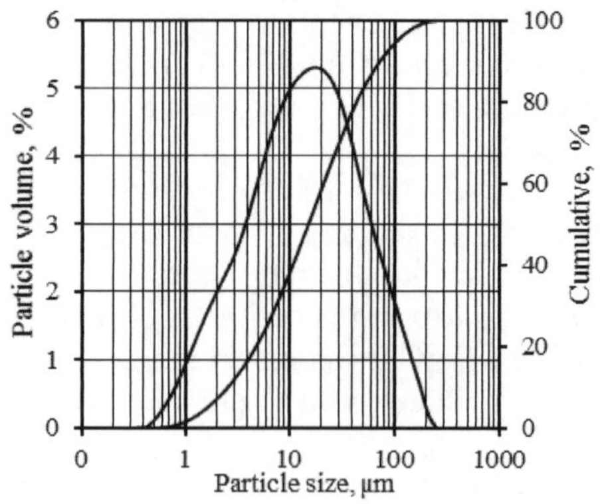

Figure 2. Particle size distribution of glass powder

3. Methods

Specific surface and particle size distribution. Specific surface was measured with Blaine instrument according to EN 196-6:2010 standard [12]. Particle size distribution was measured with "Mastersize 2000" instrument produced by Malvern Instruments Ltd.

Table 2. Compositions of ultra-high performance concrete

Composition	Water, l	Cement, kg/m^3	W/C	Micro filler, kg/m^3			Quarz sand, kg/m^3	SP, l	Micro fibres, kg/m^3
				Silica fume	Quarz powder	Glass powder			
QP/GP0	186	735	0,25	99	412	-	962	36.76	-
QP/GP100					-	412			
QP/GP100-F0	221	735	0,29	99	-	412	962	36.76	0
QP/GP100-F40									40
QP/GP100-F60									60
QP/GP100-F73.5									73.5
QP/GP100-F110.3									110.3
QP/GP100-F147.0									147.0

Mixing, sample preparation and curing. Some fresh concrete mixes (Table 2) were prepared with Zyklos ZZ50HE mixer and some with HPGM 1125.Main properties of Zyklos ZZ50HE were: capacity – 50 litres; high-speed whirler 1500 rpm (1.5/4 kW); power mixing star – 2.2 kW. Main properties of HPGM 1125 mixer were: the volume is 1125 litres; the effective mixing volume is 750 litres, the engine power is 30 kW and the mass of the mixer is 3800 kg. It took about 12 min of time to prepare UHPC mixture with both mixers and thereafter 1 extra min for proper fiber distribution. Mixtures were prepared from dry aggregates. Cement, aggregates and chemical admixtures were dosed by weight. Quartz powder and silica fume to glass powder were substituted by volume. In experiment various sizes beams, cubes (10x10x10 cm) and prisms (10x10x40 cm) were created. Homogeneous mixes were cast in moulds (without compaction) and kept for 28 days at 20 °C/95 RH.

Mercury Intrusion Porosimetry (MIP) test. At age of 28 days cylinder (d=50 mm, h=50 mm) of each composition was broken into small fragments ant then placed in isopropanol. Later these fragments were dried in the oven at 40 °C to remove all free water. All dried fragments were stored in sealed containers for mercury porosimetry tests. The pressure was applied from zero to 450 MPa. A constant contact angle of 140° and a constant surface tension of mercury of 480 mN/m were assumed for the pore size calculation.

X-ray diffraction (XRD) analysis. Hardened cement pastes were used for XRD analysis. The XRD measurements were performed with a XRD 3003 TT diffractometer of GE Sensing & Inspection Technologies GmbH with θ-θ configuration und CuKα radiation (λ=1.54 Å). The angular range was from 5 to 70 ° 2 Theta with a step width of 0.02 ° and a measuring time of 6 sec / step. For XRD quantitative phase analysis the samples were mixed with 20 wt. % ZnO (a standard material widely used in XRD analysis) as an internal standard and stored in argon atmosphere until measurement using the Rietveld refinement. This permits the estimation of the amount of non-crystalline phases by the Rietveld fitting procedure.

Figure 3. UHPC mixture preparation with Zyklos ZZ50HE (2-a) and with HPGM 1125 (2-c) mixer and the moulds filled with mixture (respectively 2-b and 2-d)

The pozzolanic activity of glass powder. Pozzolanic activity is assessed by Chapelle test. One gram of glass powder is mixed with 1g of CaO and 200 ml of water. The suspension is boiled for 16 h and the free $Ca(OH)_2$ is determined by means of sucrose extraction and titration with a HCl solution.

Compressive and flexural strength. Compressive and flexural strength were performed after 28 days according to EN 12390-4 and EN 12390-5 standard respectively [13, 14]. Compressive strength was obtained from 3 cubes (10x10x10 cm) and flexural strength was obtained from 3 prisms (10x10x40 cm) as an average value.

Salt-scaling. Salt-scaling of concrete was performed according to CEN/TS 12390-9:2006 standard [15]. 3 prisms (7x7x21 cm) were used for experiment.

4. Results and discussion

The main aim of the research is to create UHPC composition under factory conditions. In experiment two different types of mixers were used (Zyklos ZZ50HE and HPGM 1125). Mixer Zyklos ZZ50HE is rotating pan mixer which has similar mixing efficiency as larger mixers used under factory conditions. In the research were created various UHPC compositions with different amount of glass powder by using Zyklos ZZ50HE mixer. Thereafter best composition was selected and prepared with various amount of micro steel fiber by using industrial HPGM 1125 mixer. The HPGM 1125 is planetary counter current concrete mixer. Main results of the research will be discussed further.

IOP Conf. Series: Materials Science and Engineering **246** (2017) 012048 doi:10.1088/1757-899X/246/1/012048

4.1. Optimal fineness of glass powder

Optimal fineness of glass powder was determined by Chapelle test. According to Chapelle test (Fig. 3), the pozzolanic reactivity is evaluated on the basis of the amount of $Ca(OH)_2$ consumed during the pozzolanic reaction. Experiment results indicate, that pozzolanicity of glass powder after it oversteps fineness of 5240 cm^2/g tends to increase insignificantly. When fineness of glass powder varies from 2119 cm^2/g to 7399 cm^2/g consumed amount of $Ca(OH)_2$ varies from 465 mg to 652 mg respectively. Optimal fineness of glass powder according to the consumed amount of $Ca(OH)_2$ is 5240 cm^2/g. The same amount of silica fume consumes 1536 mg of $Ca(OH)_2$ and it is almost 3 times more reactive than glass powder. In Fig.4 specific surface of silica fume is not presented, because Blaine test method shows inaccurate results. Specific surface of silica fume usually varies from 15000 m^2/kg to 30000 m^2/kg by BET test method. Thereafter hardened cement pastes by hand were created and UHPC samples by Zyklos ZZ50HE mixer.

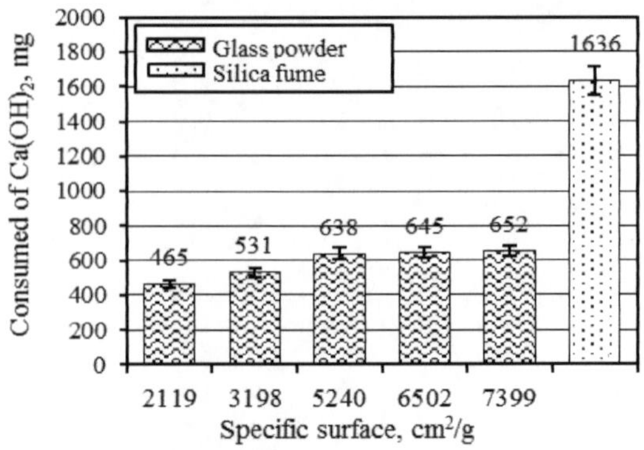

Figure 4. Results of Chapelle test for glass powder and silica fume

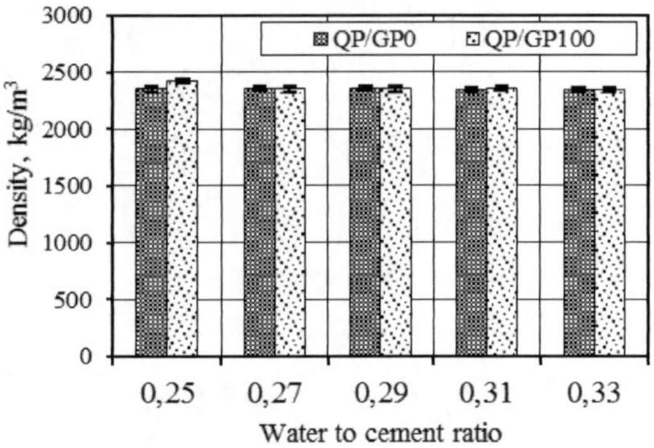

Figure 5. Relationship between water to cement ratio and density when heat treatment was not applied

4.2. XRD analysis

Fig. 6 illustrates the XRD patterns of two hardened cement pastes with different amount of glass powder. CH phase was found at 17.9°, 33.9°, 46.8°, 50.2° and 53.8°. Evidently the crystalline phase of CH was decreased with an increase of glass powder. C_2S and C_3S phases were found at 29.3°, 29.7°, 31.9°, 38.3°, 41.3°, 45.3°, 49.4°, 51.2° and 59.6°. It was noticed, what with increased amount of glass powder also decreased intensities of C_2S and C_3S. Decreased C_2S and C_3S peaks probably related with better solubility of clinker phase. Decreased CH peaks probably related with the consumption by pozzolanic reaction of silica fume and glass powder.

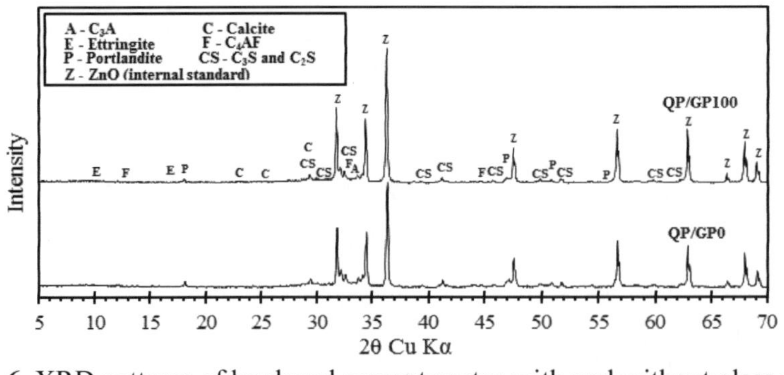

Figure 6. XRD patterns of hardened cement pastes with and without glass powder

4.3. Density of hardened UHPC
As it was expected density in all compositions was very similar and it was about 2400 kg/m^3 (Fig.5). It was noticed that density intends more decrease due to W/C ratio than due to used pozzolanic materials. Slightly decrease of density probably could be attributed due to higher amount of evaporable water which tends to increase capillary porosity of concrete and thus decrease overall density. It seems that glass powders do not influence overall density of UHPC or the influence is insignificant.

4.4. Pore size distribution of hardened UHPC
Experiment results revealed, that glass powder has positive effect on microstructure improvement of UHPC (Fig. 6 and Fig. 7). Significantly lowered porosity was obtained by composition (QP/GP100) with glass powder. Composition without glass powder (QP/GP0) had continuous pore distribution up to 700 μm, while pore distribution of composition with glass powder (QP/GP100) was smaller than 70 μm. Reduced macro porosity significantly affects mechanical properties of UHPC. Another interesting fact was observed, all compositions had higher concentration of porosity at Nano scale ≤0.1 μm. These types of pores do not influence mechanical properties, they mostly affect the shrinkage and creep of concrete. XRD analysis was applied in order to find out why composition with glass powder showed lowered pore size distribution.

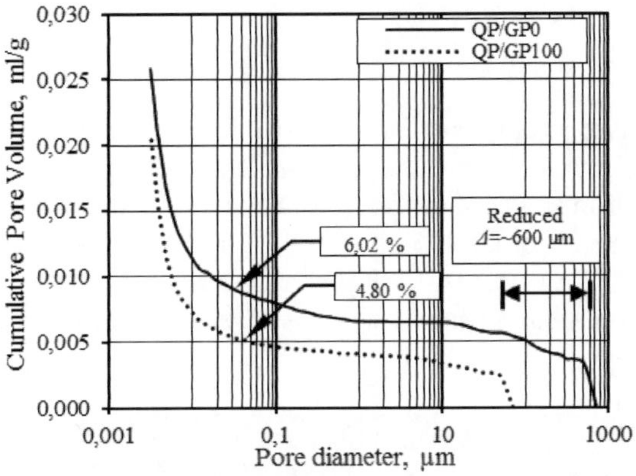

Figure 7. MIP pore diameter versus cumulative pore volume (W/C=0.25)

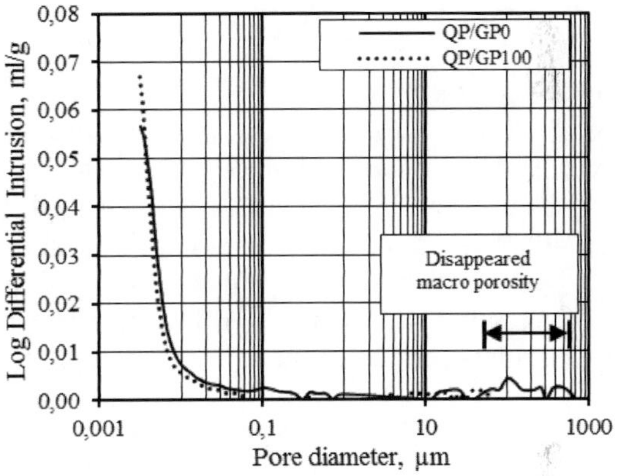

Figure 8. MIP pore diameter versus log differential intrusion (W/C=0.25)

4.5. Compressive and flexural strength of hardened UHPC
Interesting fact was noticed, when glass powder was incorporated in UHPC composition. UHPC compositions were prepared with Zyklos ZZ50HE mixer. With substitution up to 100 % of quartz powder by glass powder compressive strength increased about 40 MPa from 182 MPa (GP/GP0) to 221 MPa (GP/GP100), however, flexure strength remained almost the same - 20 MPa (Fig. 9). Such enormous compressive and flexure strength could be obtained in every concrete plant with advanced and sophisticated technology, however, most producers cannot afford such equipment. Water to cement ratio was increased up to 0.29. Modified mixture was prepared with various amount of micro steel fibres by using HPGM 1125 mixer (Fig.10).

Interesting fact was noticed (Fig. 10), that with increase up to 147 kg/m^3 of micro steel fibres, compressive strength increased about 30 % from 116 MPa (QP/GP100-F0) to 149 MPa (QP/GP100-F147), flexural strength increased more than 5 times from 6.7 MPa (QP/GP100-F0) to 36.2 MPa (QP/GP100-F147). According to the experiment results, micro steel fibres has positive effect on compressive ant flexural strength on UHPC. Despite the fact that optimal amount of micro steel fibres is about 40 kg/m^3, further increase up to 73.5 kg/m^3 has negligible effect on both flexural and compressive strength. Even further increase of micro steel fibres (up to 147 kg/m^3) has negligible effect on compressive strength and has positive effect on flexural strength. However, with extreme amount of steel fibres could be drastically reduced workability and durability of concrete. In order to denote this postulate another test method was applied.

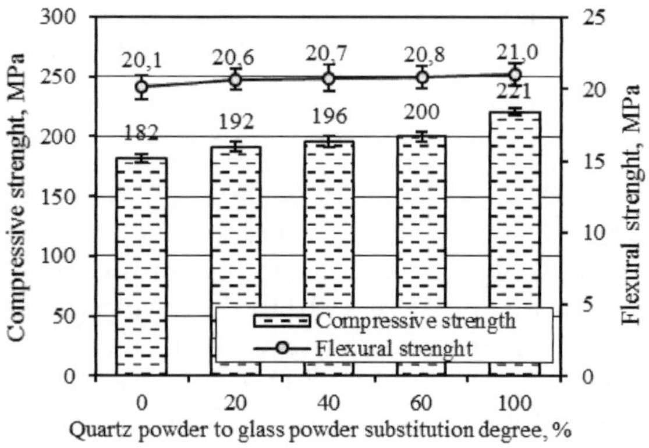

Figure 9. Compressive and flexural strength of UHPC with different amount of glass powder (W/C=0.25)

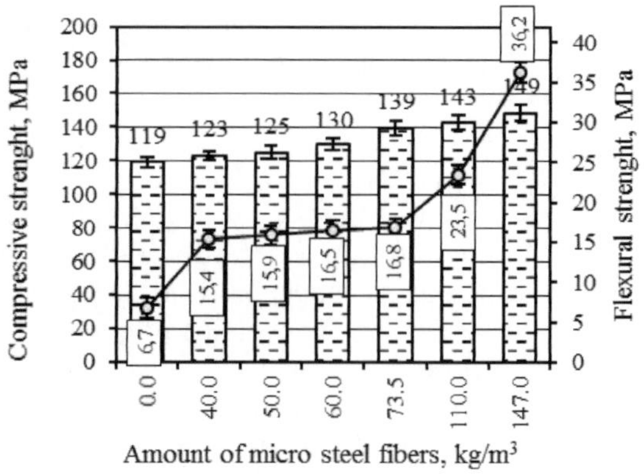

Figure 10. Compressive and flexural strength of UHPC with different amount of micro steel fibres (W/C=0.29)

4.6. *Salt-scaling*

In cold climate zones the main durability problem is insufficient resistance to frost damage. In order to find out how different amount of micro steel fibres affects resistance to frost damage, salt-scaling test methods was applied (Fig. 11). It was noticed that with increase up to 147 kg/m^3 of micro steel fibres after 40 cycles in 3 % NaCl solution salt-scaling increased from 0.006 kg/m^2 (QP/GP100-F0) to 0.197 kg/m^2 (QP/GP100-F147). Critical losses of mass at 28 cycles are 1 kg/m^2 according to CEN/TS 12390-9:2006 standard [15]. Although micro steel fibres has negative effect on salt-scaling, UHPC still has advanced resistance to frost damage.

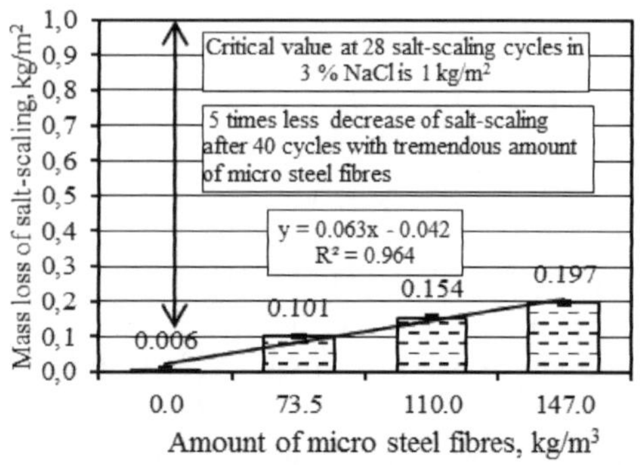

Figure 11. Salt-scaling of UHPC with different amount of micro steel fibres at 40 cycles (W/C=0.29)

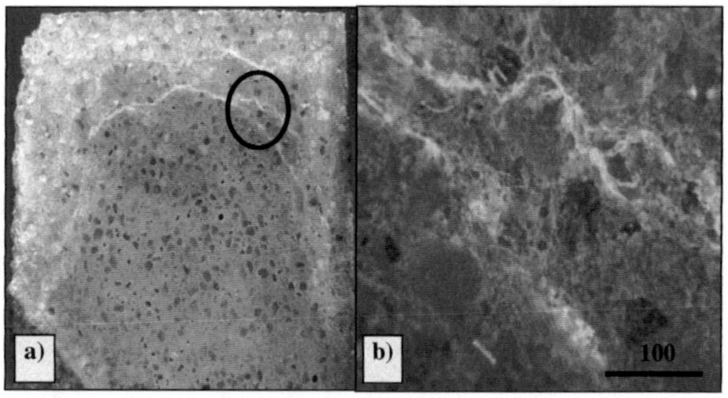

Figure 12. Defects of structure identified by fluorescence test method: a) without magnification and b) with 50x magnification

Visual inspection by fluorescence test method (Fig. 12) revealed that deterioration by surface scaling is not homogeneous. The greatest damage was observed on the frozen surface and gradually decreased in deeper layers. Surface scaling probably affected all cross section; however deterioration degree in each layer was different. Although concrete by nature is inhomogeneous material and in different circumstances applied methods should be perfect for structure analysis, however after cyclic

IOP Conf. Series: Materials Science and Engineering **246** (2017) 012048 doi:10.1088/1757-899X/246/1/012048

surface salt-scaling concrete should be considered as composite material, which has several layers with its own properties.

5. Conclusions

- Macro porosity of UHPC were eliminated. Largest porosity was observed at micro-scale (≤70 μm). All compositions of UHPC had highest concentration of pore at Nano-scale (≤0.1 μm).
- During experiment, it was founded that advanced mechanical properties of UHPC with industrial mixer can be obtained. Compressive strength can be increased by up to 30 % - from 116 MPa (without micro steel fibres) up to 149 MPa (with 147 kg/m^3 micro steel fibres). Flexural strength can be increased more than 5 times - from 6.7 MPa (without micro steel fibres) up to 36.2 MPa (with 147 kg/m^3).
- After 40 cycles, mass losses of researched compositions of salt-scaling were between 0.006 kg/m^2 to 0.197 kg/m^2. Highest mass losses were noticed in composition with 147 kg/m^3 of micro steel fibres.

References

[1] Singh L P, Karade S R, Bhattacharyya S K, Yousuf MM, Ahalawat S. 2013. *Beneficial role of nanosilica in cement based materials – A review*. Construction and Building Materials 47:1069–1077. http://dx.doi.org/10.1016/j.conbuildmat.2013.05.052

[2] Dils J, Boel V, De Schutter G. 2013. *Influence of cement type and mixing pressure on air content, rheology and mechanical properties of UHPC*. Construction and Building Materials 41:455–463.http://dx.doi.org/10.1016/j.conbuildmat.2012.12.050

[3] Yu R, Spiesz P. Brouwers, H J H. 2014. *Mix design and properties assessment of Ultra-High Performance Fibre Reinforced Concrete (UHPFRC)*. Cement and Concrete Research 56:29–39. http://dx.doi.org/10.1016/j.cemconres.2013.11.002

[4] Arroudj K, Zenati A, Oudjit M N, Bali A, Hamou A T. 2011. Reactivity of Fine Quartz in Presence of Silica Fume and Slag, Engineering 3:569-576. http://dx.doi.org/10.4236/eng.2011.36067

[5] Maroliya M K. 2012. *Micro Structure Analysis of Reactive Powder Concrete*. International Journal of Engineering Research and Development, 4(2):68-77. http://dx.doi.org/10.1007/s11595-009-3506-1

[6] Khatib J M, Negim E M, Sohl H S, Chileshe N. 2012. *Glass Powder Utilisation in Concrete Production*. European Journal of Applied Sciences, 4(4):173-176. http://dx.doi.org/10.5829/idosi.ejas.2012.4.4.1102

[7] Mirzahosseini M, Riding K A. 2014. *Effect of curing temperature and glass type on the pozzolanic reactivity of glass powder*. Cement and Concrete Research, Volume 58, Pages 103–111. http://dx.doi.org/10.1016/j.cemconres.2014.01.015

[8] Zanotto E D. 1991. *Surface crystallization kinetics in soda-lime-silica glases*. Journal of Non-Crystalline Solids, 129 183-190. DOI: no

[9] Pigeonneau F, Muller S. 2013. *The impact of iron content in oxidation front in soda-lime silicate glasses: An experimental and comparative study*. Journal of Non-Crystalline Solids, 380:86–94. http://dx.doi.org/10.1016/j.jnoncrysol.2013.09.003

[10] Venkatanarayanan H K, Rangaraju P R. 2013. *Decoupling the effects of chemical composition and fineness of fly ash in mitigating alkali-silica reaction*. Cement & Concrete Composites, 43:54–68. http://dx.doi.org/10.1016/j.cemconcomp.2013.06.009

[11] Pignatelli R, Comi C, Monteiro P J M. 2013. *A coupled mechanical and chemical damage model for concrete affected by alkali–silica reaction*. Cement and Concrete Research, 53:196–210. http://dx.doi.org/10.1016/j.cemconres.2013.06.011

[12] LST EN 196-6:2010. Methods of testing cement - Part 6: Determination of fineness.

[13] EN 12390-4:2000. Testing hardened concrete - Part 4: Compressive strength - Specification for testing machines.

[14] LST EN 12390-5:2009/P:2011. Testing hardened concrete – Part 5: Flexural strength of test specimens.
[15] CEN/TS 12390-9:2016. Testing hardened concrete - Part 9: Freeze-thaw resistance with de-icing salts – Scaling.

IOP Conf. Series: Materials Science and Engineering **246** (2017) 012049 doi:10.1088/1757-899X/246/1/012049

Realization of FRC interior and exterior furniture

Š Šonka[1], M Frantová[1], P Štemberk[1], J Havrda[1] and P Janouch[1]

[1] Department of Concrete and Masonry Structures, Faculty of Civil Engineering, CTU in Prague, Thákurova 7, 166 29 Prague 6, Czech Republic

E-mail: michaela.frantova@fsv.cvut.cz

Abstract. This article deals with the implementation of fibre reinforced concrete for interior and exterior furniture. The use of fibre reinforced concrete for non-traditional and small structures brings some specifics in design and realization. These are, in particular, the design of a suitable mixture, the choice of the shape of the structure in relation to the technological possibilities of realization, the static effects and finally the actual production of the element.

1. Introduction

It is well known that concrete is no longer just a load-bearing material hidden under cover of the final surface. Concrete is becoming a design complement of nowadays households, both interior and exterior. This trend emerged after spreading of fibre reinforced concrete and high-strength concrete, due to the properties of the composite formed by filler, binder and fibres, which are able to transfer tensile stresses in the structure. This enhancement enabled fundamental reduction in weight of the resulting product, creating subtle elements and complete omission of the classic reinforcement.

It should be mentioned that it is not only furniture, what is made from concrete, but also jewellery and house equipment such as bowls, lamps (Figure 1), etc. Many practical applications of fibre reinforced concrete can be found in [1].

Figure 1. PP fibre reinforced HPC lamp

2. Production process

The production of FRC tables, chairs, pots, lamps and other similar equipment brings specific requirements. We always require sufficient load-bearing capacity combined with the minimum thickness of the structure aiming to ease a manipulation with the product. Colour and surface quality requirements such as completely bubble-free surface or surface without visible cracks are common in this area of production. Fulfilling these requirements is essential for using concrete as design material because the product is observed by a user from far shorter distance than typical concrete surfaces are. Different scale of these structures compared to traditional concrete structures requires different approaches to most steps of the production process (e.g. load bearing test – Figure 2). This paper is discussing the specificity of mixture design, dyeing, formwork choice and casting relevant for small fibre reinforced concrete structures.

Figure 2. Load test of Nuselsky bridge (left, archive of DPP) and load test of FRC lamp (right)

2.1. Mixture design

For the mixture which is used for the furniture and house equipment smoothing of the grain curve, (especially the very small fractions of the aggregate) is very important, due to small size and visual requirements. As result very high cement content is needed.

Both meta-kaolin and micro silica fume were tested in the mixture. Better strength characteristics were achieved using meta-kaolin. Experimentally optimised mixture design is displayed in Table 1.

Table 1. Mixture design for FRC furniture

Component	Amount (kg.m^{-3})
Cement CEM I 42.5	700.0
Meta-kaolin	140.0
Water	168.0
Aggregate	1350.0
Plastificator	7.0
Stabilizator	3.5
PP fibres	4.0

Not only the mixture design, but also additional parameters of the homogenization procedure (e.g. type of the mixer, mixing speed, air pressure in the mixer, sequence of addition of the compounds, mixing times or temperature), are specific to fibre reinforced concrete. Many of these parameters are discussed in detail in [2-4].

2.2. Dyeing of concrete

Dyeing of concrete products is a very complex discipline. The fact, that concrete is alkaline, limits the usage of pigments only to inorganic dyes. Dyes are commonly available on the market in the form of a powder or a liquid. Manufacturers also supply colour samplers along with their dyes, but they are only informative because the resulting colour depends on many factors.

The main factor that affects colouring is especially the dose of pigment. Percentage of dosing refers typically to the amount of cement in kilograms. (When dispensing with a liquid suspension, the amount of water in the suspension has to be considered.) For pigment dose of more than 10% is necessary to verify the properties of the resulting concrete (compressive strength etc.). However, the authors of this paper have concluded experimentally that control samples should be rather casted for any amount of pigment dose, as the samplers from suppliers are very often inaccurate. Experimentally prepared control samples of grey-scale toning polypropylene fibre reinforced concrete flower pots are displayed in Figure 3.

Additional experimental observations relate to the amount, type and composition of cement used - it turned out that dark black colour of the product is only possible to make with white cement used in the mixture rather than with usual grey cement. Furthermore, it was observed that using the same type of cement from two different cement plants can lead to a different shade of the resulting product.

Additionally, several other factors impact the final colour, e.g. the amount of water in the mixture, the colour of the aggregate and the length of time the product is moulded and cured.

Figure 3. Grey-scale colour toning pots specimens made of FRC

2.3. Formwork

With such small structures as furniture, there are highly different requirements for the precision of production than for the load-bearing structures of buildings. Dimensional tolerances are routinely in the order of one millimetre, larger deviations are already visible. Many works about influence formwork on concrete surfaces were publicised, for example [5-7].

For classical plywood formwork, 15 high quality turnarounds were achieved for FRC flower pots. The first 4 pieces were of perfect quality and their surface did not have to be additionally grinded. The formwork must be thoroughly cleaned with formwork-oil. The cleaning highly affects the final surface of the concrete, especially amount of bubbles created.

Silicone formworks are used industrially for casting surfaces looking similar to a natural stone in tiles. When casting three-dimensional furniture parts, we encounter the problem of silicone shrinkage and loss of dimensional accuracy. Therefore, we recommend to test the shrinkage properties of the used silicone (properties are not same for all silicones) first, then adjust the shape to the expected deformation or change the silicone supplier. It is also recommended in FRC furniture production not to use the cheapest available silicones on the market due to their far worse properties. Properties also affects the surface quality of the final product.

2.4. *Casting Furniture*

Concrete mixture with scattered reinforcement was poured into the formwork and vibrated. The piece was removed from formwork after about 15 hours and put into the heat chamber where the increase in strength was accelerated. The resulting product had higher strength due to heating in the heat chamber and had higher suitability for potential grinding. When double taped a piece after removing from the heat chamber a ringing sound similar to the sound in clinker bricks was clearly hearable. With hydration acceleration, we were able to achieve a nominal pressure of up to 150 MPa and a tensile strength of 20 MPa in in the order of dozens of hours.

When jewellery was produced in small quantities (Figure 4), the mixing water was pumped through the syringe in order to achieve millilitre precision.

Figure 4. FRC necklace (left), FRC earing (right)

3. Conclusion

Fibre reinforced concrete home accessories will probably soon be widely spread, especially thanks to the ability to produce subtle elements with very high strength parameters. The industry of concrete furniture and jewellery is not likely to be the privilege of only large specialized companies, but it will be a chance for even small companies.

Authors of this paper studied the specificity of mixture design, dyeing, formwork choice and casting relevant for small fibre reinforced concrete structures. The findings provide useful guidance for manufactures and researchers of small FRC castings, in particular experimentally verified mixture design, approach to dyeing of products, formwork choice and design and casting process.

Acknowledgement

This work was financially supported by the Czech Technical University in Prague, project SGS17/048/OHK1/1T/11, which is gratefully acknowledged.

References

[1] Vodička J, Veselý V, Kolář K, Krátký J 2007 Practical Application of Fibre Concrete, In: Fibre Concrete 2007. Praha: Vydavatelství ČVUT, pp. 181-184.

[2] Dils J, De Schutter G, Boel V 2012 Influence of mixing procedure and mixer type on freshand hardened properties of concrete: a review *Materials and Structures* **45** pp 1673-1683.

[3] Mazanec O, Lowke D, Schiessl P 2009 Mixing of high performance concrete: effect of concrete composition and mixing intensity on mixing time. *Materials and Structures* **43** pp 357-365.

[4] Topçu I B, Uğurlu A 2003 Effect of the use of mineral filler on the properties of concrete. *Cement and Concrete Research* **33** pp 1071-1075.

IOP Conf. Series: Materials Science and Engineering **246** (2017) 012049 doi:10.1088/1757-899X/246/1/012049

[5] Courard L, Goffinet C, Migeotte N, Martin M, Pierard J and Polet V 2012 Influence of the reuse of OSB and marine plywood formworks on concrete surface aesthetics *Materials and Structures* **45** pp 1331–1343.

[6] Almusallam A A, Khan F M, Dulaijan S U, Al-Amoudi O S B 2003 Effectiveness of surface coatings in improving concrete durability *Cement and Concrite Composites* **25** pp 473–481.

[7] Goldammer K R 2012 DBV-guide to good practice for concrete cosmetics on exposed concrete surfaces *Beton- Stahlbetonbau* **107** 490–494.

IOP Conf. Series: Materials Science and Engineering **246** (2017) 012050 doi:10.1088/1757-899X/246/1/012050

Heat transfer in hybrid fibre reinforced concrete-steel composite column exposed to a gas-fired radiant heater

R Štefan[1], J Procházka[1], J Novák[1], J Fládr[1], F Wald[2], A Kohoutková[1], L Scheinherrová[3] and M Čáchová[3]

[1] Czech Technical University in Prague, Faculty of Civil Engineering, Department of Concrete and Masonry Structures, Thákurova 7, 166 29 Prague, Czech Republic

[2] Czech Technical University in Prague, Faculty of Civil Engineering, Department of Steel and Timber Structures, Thákurova 7, 166 29 Prague, Czech Republic

[3] Czech Technical University in Prague, Faculty of Civil Engineering, Department of Materials Engineering and Chemistry, Thákurova 7, 166 29 Prague, Czech Republic

E-mail: `radek.stefan@fsv.cvut.cz`

Abstract. In the paper, a gas-fired radiant heater system for testing of structural elements and materials at elevated temperatures is described. The applicability of the system is illustrated on an example of the heat transfer experiment on a hybrid fibre reinforced concrete-steel composite column specimen. The results obtained during the test are closely analysed by common data visualization techniques. The experiment is simulated by a mathematical model of heat transfer, assuming the material data of the concrete determined by in-house measurements. The measured and calculated data are compared and discussed.

1. Introduction

High temperature experiments and fire tests have been widely utilized for the determination of material properties (mechanical, thermal or other physical properties) of heated materials as well as for studying the behaviour of heated structures or their parts (structural collapse, spalling of concrete, ignition and burning of timber elements etc.) [1–4].

An experimental investigation of structural behaviour in fire is usually performed in fire furnaces, mainly for separate structural members or parts of a structure, e.g. [1, 2], or, less commonly, by the full-scale fire tests of entire structures, e.g. [5]. However, these complex experiments are highly cost-consuming and hence, many researches have focused on the development of alternative types of experimental set-up which would be cheaper, easy to use, while maintaining the accuracy and validity [4]. For illustration, the works in which a heat source is formed by a direct flame gas burner, e.g. [6], by ceramic pads heaters, e.g. [7, 8], or by radiant heaters – both electric or gas, e.g. [4, 9, 10], can be mentioned.

The ceramic pads heaters have been used at the Faculty of Civil Engineering (FCE), Czech technical university in Prague for many years, mainly for the investigation of high-temperature behaviour of steel structural members, e.g. [7], or for the determination of mechanical properties of heated concrete, e.g. [8].

Since 2014, a gas-fired radiant heater system has been developed at the FCE for helping to study the heat transfer in fibre reinforced concrete-steel composite members, within the scope of the research project *GA15-19073S – Models of steel and fibre concrete composite columns exposed to fire*.

The results obtained by this system are presented in this paper. These results were utilised for the design of a subsequent complex heat transfer experiment in a fire furnace, which was performed in PAVUS Fire Testing Laboratory at the end of 2016, see [11, 12].

2. The gas-fired radiant heater system

The experimental set-up of the gas-fired radiant heater system, see figure 1, consists of: (i) a radiant heater, (ii) a gas cylinder with a hose and regulator, (iii) a steel stand, (iv) a datalogger connected to a computer, (v) a test specimen with thermocouples, cf. [4].

Figure 1. Experimental set-up, cf. [4, Figs. 4, 5, 9].

The radiant heater used in the presented system is a commercially available device made by a Czech manufacturing company. Namely, the KASPO K30-STL type has been used, with the following parameters from the user manual [13]: nominal heat output of 30 kW, burner gas pressure of 70–100 kPa (for propane/butane mixture), gas consumption of 2.3 kg per hour (for propane/butane mixture), and the dimensions as shown in figure 2. For the heater, it is possible to use a propane/butane mixture (as in this case) or natural gas.

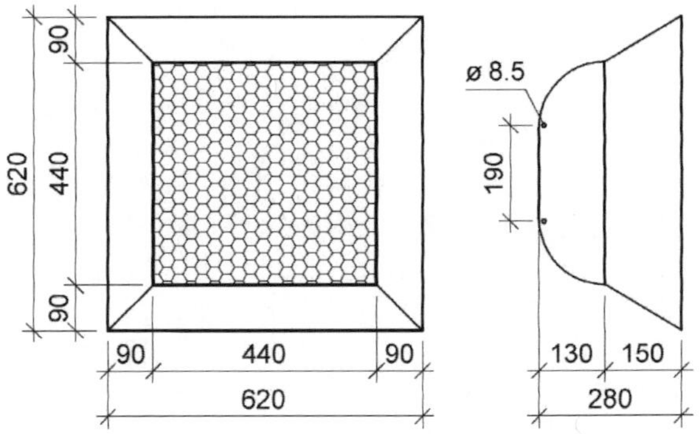

Figure 2. Dimensions of the radiant heater (in mm): front view (left), side view (right), according to [13].

As mentioned above, the propane/butane mixture is used in the system. It is stored in 10 kg gas cylinders. This amount of gas serves for about 4 hours of continual heating. The gas regulator is set to a constant value of 100 kPa.

The steel stand is an in-house product made of steel L-profiles. It enables to adjust the position of the heater in vertical direction. The position of the heater remains constant during the test, cf. [4].

The datalogger is a custom-made device made by a Czech company. It contains 24 thermocouple ports (NST 1200 type) in total and one USB-B port for connecting to a computer. For recording the measured data in the computer, a custom-made software (provided by the datalogger producer) is used. The data are stored in a *.dat* file format. The datalogger is destined for use in connection with K type thermocouples – both normal and sheathed.

A test specimen can be made of any material. During the development and application of the heating system at the FCE, concrete, steel-concrete composite, steel as well as timber elements have been tested. One of these experiments is described below.

3. Experiment

In the following part of the paper, the heat transfer experiment on one test specimen of a composite circular column is presented. The specimen was heated by the gas-fired radiant heater described above.

3.1. Test specimen

The test specimen represents a hybrid fibre reinforced concrete-steel composite column. A steel tube of the length of 600 mm, outer diameter of 273 mm and the wall thickness of 6.3 mm was filled by hybrid (steel and polypropylene) fibre reinforced concrete. Its composition is given in table 1.

Table 1. Concrete mixture composition.

Component	Content (kg/m^3)
Cement CEM I 42.5 R	490
Water	158
Aggregate 0/4 mm	890
Aggregate 4/8 mm	100
Aggregate 8/16 mm	745
Plasticizer	4.9
Steel fibres Dramix 3D 80/60 BN	40
Polypropylene fibres Forta Ferro 54 mm	3

During the concrete casting, thermocouples (TCs) and the steel support ensuring their position were inserted into the specimen, see figure 3. Nineteen K-type TCs ($2 \times \phi$ 0.5 mm) in total were used for measuring the internal temperature (17 TCs within the test specimen) as well as the outer temperature near the surface (2 TCs on the steel shell). In the vertical direction, the thermocouples were placed in the middle of the specimen height. The positions of the thermocouples in the cross-section of the specimen are illustrated in figure 4. The specimen was cast in December 2015 (the set of 3 specimens was prepared, one of them was employed for the presented experiment), the heat transfer test was performed in May 2016.

Figure 3. Thermocouples (TCs) in the steel tube before casting of concrete.

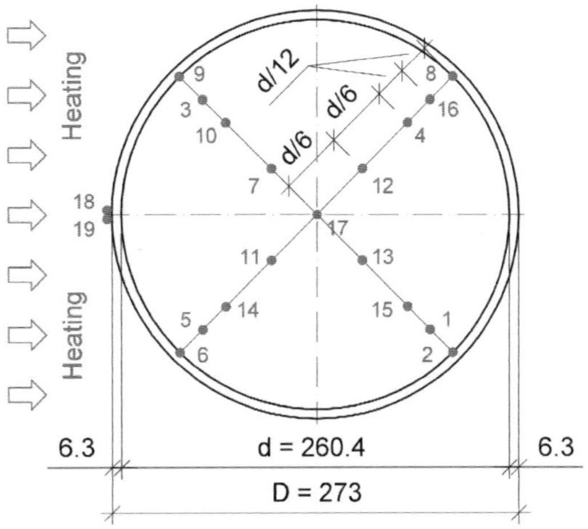

Figure 4. Scheme of the cross-section with the positions and numbers of TCs.

3.2. Test procedure

The test specimen was placed right in front of the heater, see figure 1, in the position illustrated in figure 4. The system was assembled as described above. The outer thermocouples (TC18, TC19) were fixed to the steel shell using a steel wire. The contact between the surface thermocouples and the shell was not perfect – the distance between the TCs and the shell was about 2–4 mm, which will be discussed in the next part of the paper. The heating lasted for 3 hours, thereafter, the test was terminated. The gas regulator was set to a constant value of 100 kPa. At the end of the test, the gas pressure slightly decreased, see figure 5, due to the small amount of gas in the cylinder.

3.3. Results analysis

The data obtained during the experiment have the form of a *.dat* file containing the measured temperatures for each thermocouple in defined time steps (10 s in this case). The basic method for depicting the results is to draw a graph with a temperature-time curve for the given thermocouple. In this case, an in-house MATLAB [14] code is used for data conversion and visualization. The time evolution of the outer temperatures (TC18, TC19) is illustrated in figure 5. The temperature-time curves for the selected inner TCs (9, 10, 17, 2; see figure 4) are shown in figure 6.

In order to better illustrate the temperature distribution within the specimen during the test, the in-house MATLAB [14] code was written. Based on the linear interpolation in polar coordinates, the code enables to draw the temperature field for a given time of heating, cf. [15].

In figure 7, the measured temperatures for the selected times of heating (90 min, 180 min) are displayed in the appropriate positions within the cross-section. These data serve as input parameters for the interpolation. In figure 8, the resulting temperature fields for the time of heating of 90 min are depicted.

From the results described above and illustrated in figures 5–8, the following findings can be drawn.

- The gas-fired radiant heater system is able to generate relatively high temperatures in the test specimen – above 400 °C in 3 hours of heating; which makes it highly promising for this type of experiments.

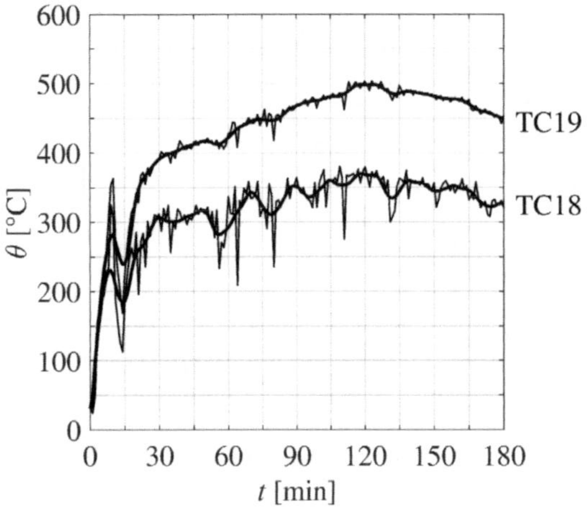

Figure 5. Temperature-time curves for outer TCs (18, 19); thin lines: measured; thick lines: smoothed graph for modelling.

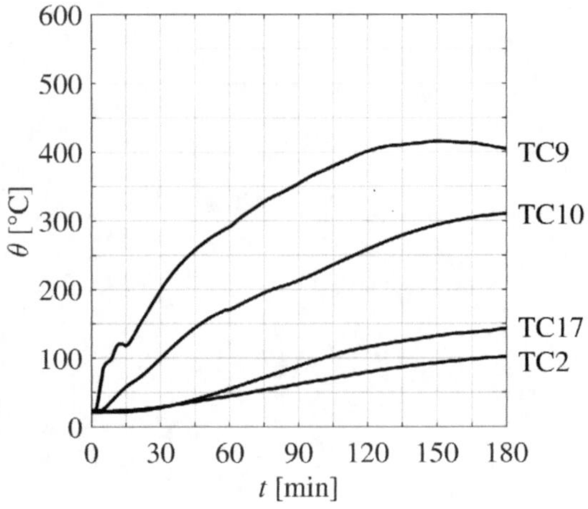

Figure 6. Measured temperature-time curves for selected inner TCs (9, 10, 17, 2) within the specimen, see figure 4.

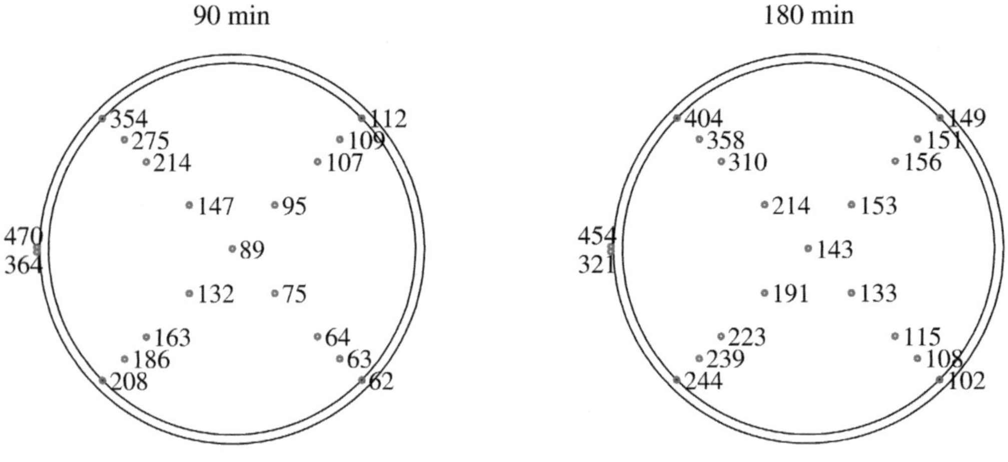

Figure 7. Measured temperatures in TCs for the time of heating of 90 min (left) and 180 min (right).

- The outer temperatures measured by TC18 and TC19 differ markedly – the difference reached about 100 °C. This is because the distances between these two thermocouples from the steel shell were not the same. It is obvious that it would be better to use a different type of thermocouples for measuring the boundary conditions in this case, e.g. sheathed thermocouples or plate thermometers instead of standard wire type thermocouples, see [15]. This is also discussed in the next part of the paper.
- The direction of heating was not parallel to the horizontal axis of the specimen's cross-section, as it was supposed, see figure 4. Firstly, it was believed that this was caused by the non-uniform air flow between the heater and the specimen due to wind – the experiment was performed outdoors. For that reason, a control experiment was done on the same specimen

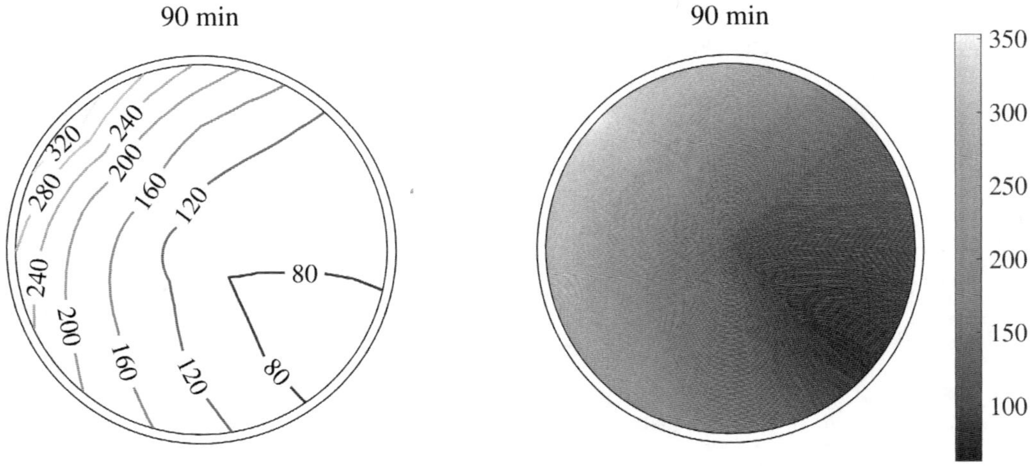

Figure 8. Temperature fields (in [°C]) obtained by linear interpolation in polar coordinates for the time of heating of 90 min: isolines (left) and isoareas (right), cf. [15].

several months later – in December 2016, within a student project [15]. For this control experiment, lateral barriers were added to the heating system in order to avoid the air flow. However, the results obtained from the control experiment were almost the same as from the original one. Hence, the probable cause of the discrepancy between the supposed heating direction and the obtained results is a small rotation of the the steel support with thermocouples during the casting of concrete. By performing a simple data analysis, based on the linear interpolation of the measured data, the angle of inclination of the heating direction was determined to be about 30° clockwise. This is taken into account in the modelling part.

4. Mathematical modelling

Mathematical modelling of heat transfer is applied in order to illustrate that the gas fired radiant heater experiment can be simulated by an appropriate model. It should be also verify that the model can be used for the hybrid fibre reinforced concrete, which is of particular interest for the subsequent complex heat transfer experiment in a fire furnace in PAVUS Fire Testing Laboratory in 2016 [12].

For the sake of simplicity, a common heat transfer mode is used, without any coupling with mass balance (the effect of moisture is neglected). A more comprehensive model will be employed in [12], see also [16, 17] and references therein.

It is necessary to determine the boundary conditions, which can be used in the model and which would represent the character of heating during the test. Generally, the Dirichlet or the Neumann boundary conditions can be used [2, 18]. For the first one, the temperature on the boundary is specified, for the second one, the heat flux on the boundary is assumed – as a function of temperature or as a prescribed function [2, 18].

For the presented gas-fired radiant heater system, it seems to be suitable to determine the boundary conditions based on the heat flux generated by the heater, see [4]. This is however not possible in this case since this variable was not measured during the test.

The Dirichlet boundary condition cannot be used since the surface temperature was not monitored – the contact between the outer TCs (18, 19) with the steel shell was not perfect.

Hence, it seems to be suitable to determine the boundary condition based on the heat flux, convective and radiative, which is also in accordance with the standard fire design approach, see [19]. The heat flux can be calculated based on the gas temperature measured by the outer TCs (TC18 and/or TC19, see below).

It should be noted that in literature, more sophisticated methods to determine the boundary condition during experiment can be found. Let us mention the approach based on the adiabatic surface temperature, see e.g. [20–22], which, however, cannot be employed for the presented experiment since the plate thermometers were not used.

4.1. Model

The model with the unknown $\theta(\boldsymbol{x},t)$, where $\boldsymbol{x} \in \Omega$, can be written in the form (taken from [2,18], see also [16,17]):

$$\text{for } t \in (0,T): \qquad \rho(\theta)c_p(\theta)\frac{\partial\theta}{\partial t} = \nabla\cdot(\lambda(\theta)\nabla\theta) \qquad \text{in } \Omega, \tag{1}$$

$$-(\lambda(\theta)\nabla\theta)\cdot\boldsymbol{n} = \alpha_c(\theta-\theta_\infty) + \Phi\epsilon\sigma(\theta^4-\theta_\infty^4) \qquad \text{on } \Gamma_E, \tag{2}$$

$$\text{for } t = 0: \qquad \theta = \theta_0 \qquad \text{in } \Omega, \tag{3}$$

where Ω is the domain, Γ_E is the boundary exposed to the heat flux prescribed by equation (2), θ [°C] is the temperature, ρ [kg m^{-3}], c_p [J kg^{-1} K^{-1}] and λ [W m^{-1} K^{-1}] are the density, the specific heat capacity, and the thermal conductivity of a material, respectively, α_c [W m^{-2} K^{-1}] is the convective heat transfer coefficient, Φ is the configuration factor, ϵ [−] is the surface emissivity, σ [W m^{-2} K^{-4}] is the Stefan-Boltzmann constant, θ_∞ [°C] is the ambient temperature, θ_0 [°C] is the initial temperature, t [s] is the time of heating and T [s] is the total time of exposure [2, 16–18].

4.2. Material properties

The material properties of the studied hybrid fibre reinforced concrete were determined by in-house measurements, see figures 9–11. The measurement procedures are described, e.g., in [23] and will be discussed in detail in [12].

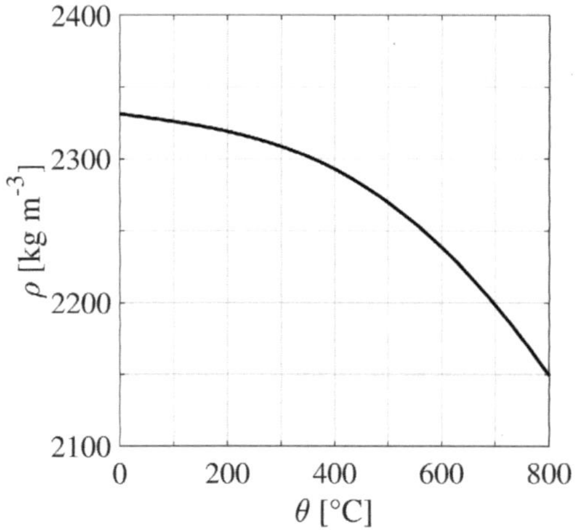

Figure 9. Density of the studied hybrid fibre reinforced concrete.

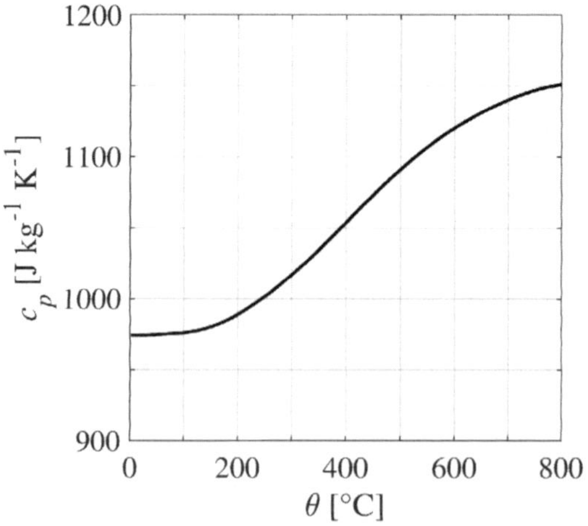

Figure 10. Specific heat capacity of the studied hybrid fibre reinforced concrete.

For steel, the material properties (ρ, $c_p(\theta)$ and $\lambda(\theta)$) are assumed according to [24].

IOP Conf. Series: Materials Science and Engineering **246** (2017) 012050 doi:10.1088/1757-899X/246/1/012050

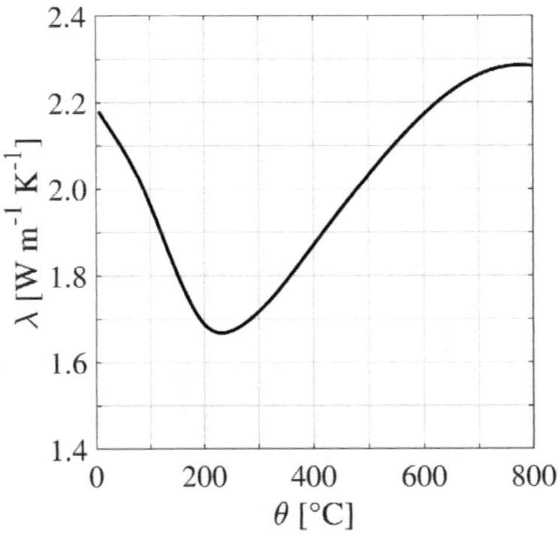

Figure 11. Thermal conductivity of the studied hybrid fibre reinforced concrete.

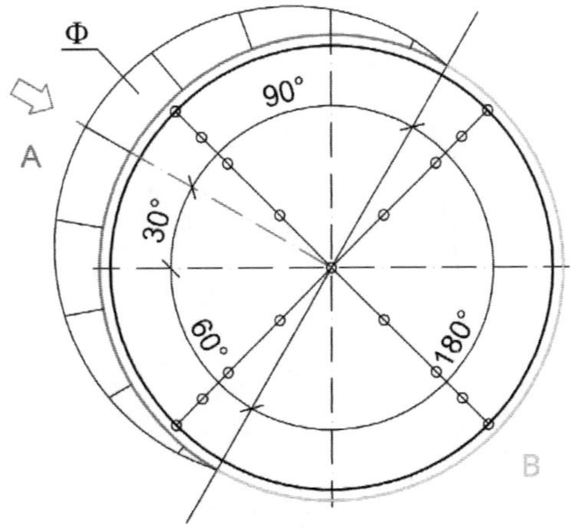

Figure 12. Boundary conditions assumed for the simulation.

4.3. Boundary and initial conditions

The boundary conditions are assumed according to equation (2). By performing a set of calculations, it was observed that more appropriate results can be obtained when assuming the outer temperature as measured by TC19 (hence, TC18 is omitted hereafter).

Based on the direction of heating, the boundary of the specimen cross-section is divided in two parts, see figure 12. For these parts, the parameters of equation (2) are set, according to [19, 24], as follows:

A: $\theta_\infty(t) = \theta_{TC19}(t)$, $\alpha_c = 25$ W m^{-2} K^{-1}, $\epsilon = 0.7$, $\Phi(\boldsymbol{x}) \in (0, 1)$ (see figure 12);

B: $\theta_\infty = 25$ °C (approximate air temperature during the test), $\alpha_c = 9$ W m^{-2} K^{-1}, $\epsilon = 0$;

where $\theta_{TC19}(t)$ is the time evolution of the outer temperature measured by TC19.

For the initial condition, see equation (3), the uniform distribution of temperature across the section is assumed, with $\theta_0 = 23.4$ °C, which is the mean value (with the standard deviation of 0.67 °C) of the inner TCs (TC1–TC17) temperatures measured before the start of the test.

4.4. Numerical simulation

System described by equations (1)–(3) was solved by the finite element method, implemented into the in-house MATLAB [14] code. For the spatial discretization, 3244 linear isoparametric triangular elements were used. Discretization in time was performed by the semi-implicit difference scheme, with the time step of 20 s [16, 17].

4.5. Results and discussion

In figures 13–14, the results obtained from the numerical simulation are compared with the data measured during the test. It is obvious that the gas fired radiant heater experiment can be simulated by an appropriate model, while achieving high levels of accuracy. The model, despite its simplicity, works quite well also for the hybrid fibre concrete used in this case. The small discrepancy between the measured and calculated results is probably caused by (i) neglecting the effect of moisture transfer and phase changes and employing a simple heat transfer model, (ii) assuming a simplified approach to determine the boundary conditions used in the model. The first point will be discussed in detail in [12]. The second one will be fixed in the next experiment by using the plate thermometers or the heat flux gauge, cf. [4].

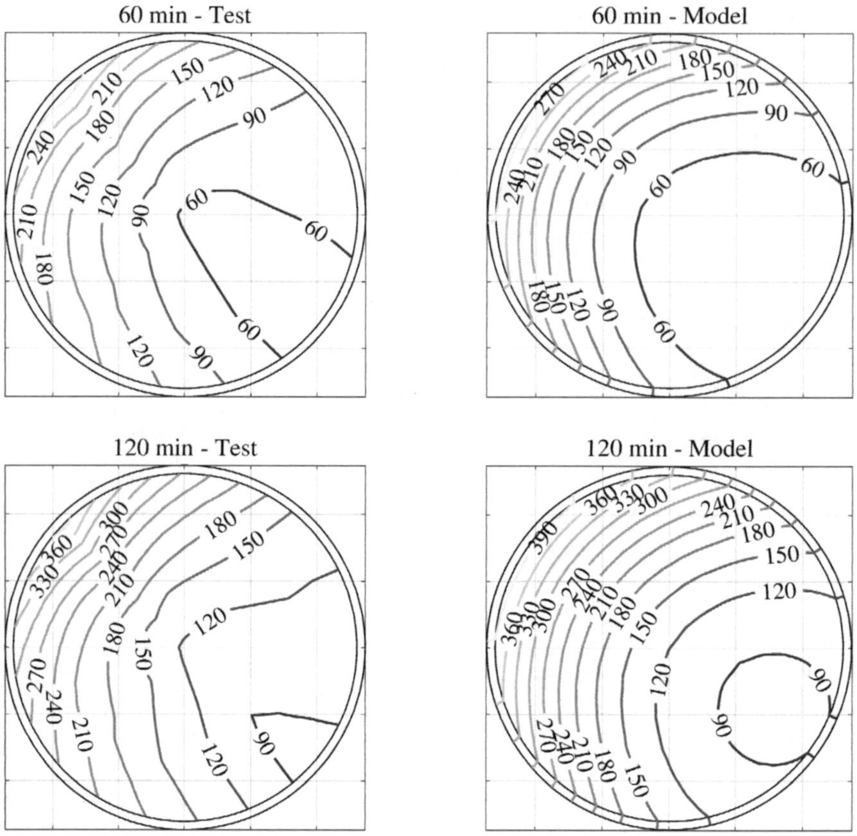

Figure 13. Comparison of temperature fields.

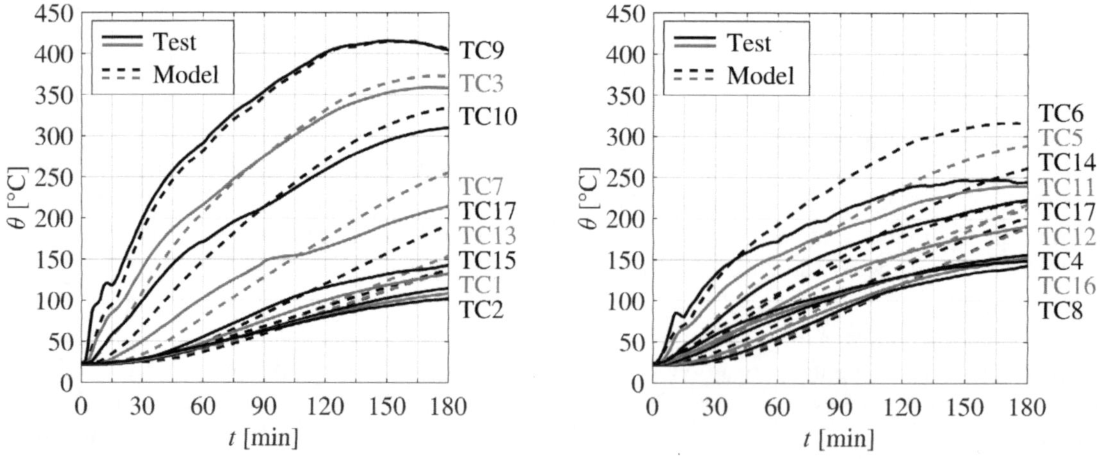

Figure 14. Comparison of temperatures for the thermocouples on the top-left to bottom-right diagonal (left) and on the bottom-left to top-right diagonal (right), see figure 4.

5. Conclusions

In the paper, the newly designed gas-fired radiant heater system for testing of structural elements and materials at elevated temperatures was described. The system has been developed at the CTU in Prague since 2014. The applicability of the system was illustrated on an example of the heat transfer experiment on the hybrid fibre reinforced concrete-steel composite column specimen. It was demonstrated that: (i) the gas fired radiant heater system is able to generate relatively high temperatures in the test specimen, which makes it highly promising for this type of experiments; (ii) the crucial aspect for modelling of such experiment is to determine the boundary conditions properly (e.g. based on the plate thermometers measurement); (iii) the gas fired radiant heater experiment can be simulated by an appropriate model, while achieving high levels of accuracy; (iv) the model works quite well also for the hybrid fibre concrete – if its material properties are known.

The results obtained by the system were utilised for the design of a subsequent complex heat transfer experiment performed in PAVUS Fire Testing Lab. furnace at the end of 2016 [11,12].

The next step in the development of the system will be focused on the improvement of boundary conditions measurements and simulation.

Acknowledgments

This work has been supported by the Czech Science Foundation, project No. GA15-19073S, which is gratefully acknowledged.

References

[1] Buchanan A H 2002 *Structural Design for Fire Safety* (Wiley)
[2] Purkiss J A 2007 *Fire safety engineering, Design of structures* 2nd ed (Elsevier. Butterworth-Heinemann)
[3] Guo Z and Shi X 2011 *Experiment and Calculation of Reinforced Concrete at Eleveted Temperatures* (Elsevier)
[4] Maluk C, Bisby L, Krajcovic M and Torero J L 2016 *Fire Safety J.* – URL `http://doi.org/10.1016/j.firesaf.2016.05.001`
[5] Wald F, Simões da Silva L, Moore D B, Lennon T, Chladná M, Santiago A, Beneš M and Borges L 2006 *Fire Safety J.* 509–522
[6] Bodnarova L, Valek J and Novosad P 2015 *The Scientific World Journal* 371913
[7] Strejček M, Řezníček J, Tan K H and Wald F 2011 *J. Constr. Steel Res.* **67** 1890–1899
[8] Novák J and Kohoutková A 2017 *Procedia Engineering* **172** 784–790
[9] Maluk C, Bisby L and Terrasi G P 2017 *Eng. Struct.* **141** 584–595
[10] Kalifa P, Menneteau F D and Quenard D 2000 *Cement Concrete Res.* **30** 1915–1927
[11] Kuča M 2016 *Experimentální analýza transportu tepla v betonových a ocelobetonových prvcích* Master thesis CTU in Prague *(in Czech)*
[12] Štefan R et al 2017 *Experimental and numerical analysis of heat transfer in hybrid fibre reinforced concrete and steel composite structural members exposed to fire* (Under Preperation)
[13] Světlé infrazářiče kaspo k7-k50 http://www.kaspo.cz/download/svetly/podrobne-informace-svetle-iz.pdf *(in Czech)*
[14] MATLAB 2015 Version 8.6.0 (R2015b) The MathWorks, Inc., Natick, Massachusetts, United States
[15] Lokvenc M, Douša M, Hášová T and Jiříček Z 2017 *Měření průběhu teplot v ocelobetonovém průřezu* Student project CTU in Prague *(in Czech)*
[16] Štefan R and Procházka J 2016 *Solid State Phenomena* **249** 246–252
[17] Štefan R 2015 *Transport Processes in Concrete at High Temperatures. Mathematical Modelling and Engineering Applications with Focus on Concrete Spalling* PhD thesis CTU in Prague
[18] Bergheau J M and Fortunier R 2008 *Finite Element Simulation of Heat Transfer* (Wiley)
[19] EN 1991-1-2 2002 *Eurocode 1: Actions on structures – Part 1-2: General actions – Actions on structures exposed to fire* (CEN)
[20] Wickström U 2008 *Fire Safety Science – Proceedings of the Ninth International Symposium* **9** 1227–1238
[21] Zhang C, Li G Q and Wang R 2013 *Int. J. Steel. Struct.* **13** 547–556
[22] Andreozzi A, Bianco N, Musto M and Rotondo G 2014 *Appl. Therm. Eng.* **65** 422–432
[23] Černý R and Vejmelková E 2011 *Measurement* **44** 1220–1228
[24] EN 1994-1-2 2005 *Eurocode 4: Design of composite steel and concrete structures – Part 1-2: General rules – Structural fire design* (CEN)

IOP Conf. Series: Materials Science and Engineering **246** (2017) 012051 doi:10.1088/1757-899X/246/1/012051

The Overall Research Results of Prestressed I-beams Made of Ultra-high Performance Concrete

P Tej[1], J Kolísko[1], P Kněž[1] and J Čech[1]

[1] Klokner Institute CTU in Prague, Czech Republic

E-mail: petr.tej@cvut.cz

Abstract. The design process of short-term and long-term loading of prestressed I-beams made of ultra-high performance concrete (UHPC) and the overall research results are presented in this article. The prestressed I-beams are intended and designed to replace steel HEB beams mainly in the construction of railway bridges with fully concreted height of the beams. These types of structures have the advantage of a low construction height. The prestressed I-beams were made of UHPC with dispersed steel fibres and are reinforced by prestressing cables in the bottom flange. Two specimens of 9 m span, three specimens of 7 m span and two specimens of 12 m span were made for the short-term loading. For the purpose of the long-term loading, two specimens of 12 m span were made and subsequently loaded for 450 days. All specimens were tested in four-point bending tests in the laboratory. The article presents also comparison of results of the experiments with computer simulations.

1. Introduction

UHPC has a number of unique properties, which are superior to conventional concrete. The main outstanding features include the ultra-high strength of the material under pressures exceeding 150 MPa and the high tensile bending strength which is about 15 MPa [1-5]. High levels of strength are provided by the ability to transmit tensions even after cracking using steel fibers. Fibers absorb energy and control cracks growth until pullout failure [6-11]. High water impermeability is provided by a dense cement matrix and a very low level of porosity with unconnected pores. It is caused by a very low water coefficient, close packing of fine grains of solid particles and by reaction of a very fine reactive material admixtures [2]. From high water impermeability are derived high resistance to frost and high durability [3, 12]. The surface of the material has negligible carbonation and so a high capability to protect the reinforcement is obtain [3-5]. The consistency of the concrete along with fine grain aggregates allows casting of complex thin-walled elements [4]. Due to demand for these high-quality properties some elements can be determined as a typical usage of UHPC.

In the case of horizontal structures, an example would be a thin-walled beam of high aspect ratio and a large span with a high preload. The advantage is the low weight that applies during the assembly of the components. In many instances the thin beams of UHPC could be a good substitute for steel beams, since they have the same weight while still maintaining the required bearing capacity, and a considerably longer service life with minimum need of maintenance [13, 14]. It is advisable to design the beams as T-shaped, Π, I or as a box girder [15, 16]. The advantage of this is an automatic camber of the structure during prestressing, while the steel construction needs to be elevated during production.

I-shaped beams are standard representatives of the basic range of classic reinforced concrete beams. As previously mentioned, the superior material properties of UHPC may be conveniently applied for use in prestressed elements. For test purposes, we designed the shape of a classic "I" element with prestressing with bond [17]. This type of beam is designed to be a replacement for steel beams of the type HEB, which is a very popular element in the construction of all types of structures due to its favorable ratio of weight and load bearing capacity. Steel beams of the HEB type are typically used as self-acting elements or composite elements complemented by a concrete slab only above the level of the upper flange. In this way they are usually used in civil engineering construction works and bridge construction for road bridges. They may also be used fully concreted, in which case the monolithic part would include the full structural height of the beam, except for the bottom flange, with the upper flange concreting usually being of up to 200 mm. Such a design is the most common type of construction for railway bridges with a span of up to 15 m.

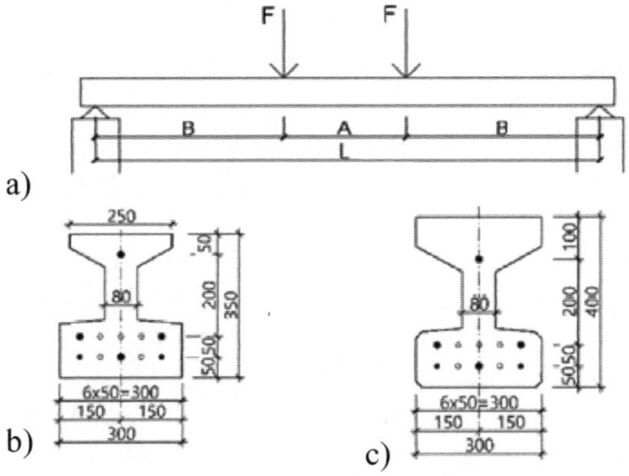

Figure 1.
a) Scheme of the short-term loading tests,
b) cross section of 9 m span beam,
c) cross section of 7 and 12m span beams

Table 1. Dimensions of the elements

Beam	L [m]	A [m]	B [m]
7 m	7	1.3	2.85
9 m	9	1.3	3.85
12 m	12	1.6	5.2

2. Material

The proposed prestressed I-beams are made of UHPC of class C110/130 with 120 kg/m3 of dispersed steel fibers. Although adequate strength for the inclusion of this material under the name of UHPC is not achieved, other material characteristics, such as a fine grained structure, resistance to frost and de-icing agents, allows the label to be used.

The material was composed from combinations of the following components:

Crushed stone (PR30, PR33), fibers MasterFiber 482, microsilica 940U, cements CEM II/A-S 42,5 R, CEM II/B-S 32,5 R, CEM II/A-LL 52,5 R, CEM I 52,5 R, slag and fly ash Dětmarovice, ground quartz MT8, MT12 resp. MT300 and additive Glenium ACE 300.

The use of silica fume Elkem MS 971 on comparative cylinder specimens that reduces the w/add to 0,19 leads to an increase in the compressive strength to a value of 175 MPa after 28 days. The mixture has a low water ratio w = 0,28 ((water + additive)/cement), and a very low binder coefficient b = 0,22 ((water + additive)/(cement + microsilica + slag)). The material has a very high resistance to de-icing agents according to ČSN 73 1326 method C peeled parts 25 g/m2 after 250 cycles. The specimens of UHPC produced in August 2010 and exposed to weather conditions, in parallel with those inside the hall, were tested on the depth of carbonation using phenolphthalein after 3 years. The thickness of the carbonated layer was max. 1.5 mm.

IOP Conf. Series: Materials Science and Engineering **246** (2017) 012051 doi:10.1088/1757-899X/246/1/012051

3. Detailed course of design and testing

3.1. 9m span beams

In the first instance the beams with a 9 m span were proposed and tested with a cross section of total height of 350 mm. Then after testing some changes, as described below, were carried out, and a 7 m and a final 12 m span beam were tested. The 7 and 12 m span beams had the same cross-section (final shape) with a total height of 400 mm. The upper flange had a width of 300 mm with an average thickness of 100 mm. The lower flange was designed with a width of 300 mm and had an average height of 135 mm. Ten straight cables in the bottom flange were used for the implementation of the stress. Five of them were placed at a distance of 1.3 m from the supports and the protectors were used in order to prevent an excess of stress at the ends of the beam. One straight cable leads in the top flange. All cables had the strength of 1570/1770 MPa and the diameter of 15.7 mm (see Figure 1 and Table 1). They were prestressed by a force of 206.25 kN, i.e. the stress in each cable 1375 MPa.

The stages of loading for 9 meter span beams regarding the operation and distribution of internal forces in the cross-section area and their boundaries were assumed to be as follows:

1. Achieving decompression of the lower strands cross-section, with fully flexible element behavior, the actual deformation corresponding to calculation according to Hook's law. The load for this limit is set at F = 2 x 52 kN, tension in the lower strands in concrete σx, c = 0 MPa, stress in the upper strands in concrete σy, c = - 55 MPa with an anticipated deflection of 42 mm.
2. The achievement of tension in the lower fibers at the limit of the tensile strength of UHPC. At this stage the formation of cracks is assumed, and the actual deformation will be slightly higher than deflections assuming fully elastic behavior, while there is a partial stress redistribution over the cross section. The load for this limit is set at F = 2 x 75 kN, tension in the lower strands of concrete σx, c = + 15 MPa, stress in the upper strands of concrete σy, c = - 76 MPa with an anticipated deflection of 66 mm.
3. Achieving tension near the compressive strength of the material in the upper strands and significantly exceeding the tensile strength of the material in the lower strands, a significant increase of deflections, plasticization under the compression zone and the eventual collapse of the cross section by press violation of the concrete. The load for this limit was set at F = 2 x 98 kN, tension in the upper strands of concrete σy, c = - 100 MPa with an expected deflection of 120 mm.
4. The test of the produced beams took place in April 2013. The test result was significantly different from the theoretical assumptions. Already in stage No.1 there were significant deflections of beams, leading to the formation of tensile cracks in the concrete, without the creation of any tensile stress. The collapse of the elements was due to a torn upper flange transverse tensile and shear separation of the upper and lower flanges with a crack in the wall.

The tests needed to be carefully analyzed in order to reveal the causes of major discrepancies between the theoretical assumptions and the test results. As possible causes, substantial diversion of behavior was considered:

- Production technology ensuring homogeneous material effect produced the specimens [18]
- Greater than expected losses due to stress from prestressed cables
- Variations in measurements of forces at a test facility in sync presses
- Subtle upper flange with imperfections affecting the production and the yaw

Based on these assumptions, it was decided to implement measures to eliminate or quantify major discrepancies. These would mainly be by change in the technological process of mold-filling, from a process of layering to the method of filling the full depth of the section.

To actually detect the introduced forces on the prestressing cables, the cables were equipped with sensors that made it possible to deduct the appropriate value of the tension in each cable. Only then was it possible to establish the correct value of the prestressed tension in the calculations.

IOP Conf. Series: Materials Science and Engineering **246** (2017) 012051 doi:10.1088/1757-899X/246/1/012051

The static scheme was changed for the tests which followed. Instead of using two independent presses synchronized by the introduction of force, only force action moving the press into two locations on the test element at a distance of 1.3 m will ensure spreading strap anchors.

The problem of the subtle top flange is a difficult one to tackle. In practice, the beam is first embedded in concrete in a monolithic plate, the load of the slabs in the construction state is not a significant load. A board ensures the stability of the flange when it is subsequently loaded. To ensure similar conditions for the functioning of the upper flange of the beam during the test itself is difficult.

For the above reasons, it was decided to further test the change of dimensions of the proposed element. The element had its shape and performance to provide compensation for HEB steel beams used in conventional designs.

Figure 2. and Figure 3. Setup of the experiment, deflection, bending and shear failure of 9m span beams

3.2. 7m span beams

The stages of loading for 7 meter span beams regarding the operation and distribution of internal forces in the cross-section area and their boundaries were assumed to be as follows:

1. Achieving decompression of the lower strands cross-section, with fully flexible element behavior, the actual deformation corresponding to calculation according to Hook's law. The load for this limit is set at F = 2 x 105 kN, tension in the lower strands in concrete σx, c = 0 MPa, stress in the upper strands in concrete σy, c = - 41 MPa with an anticipated deflection of 22 mm.
2. The achievement of tension in the lower fibers at the limit of the tensile strength of UHPC. At this stage the formation of cracks is assumed, and the actual deformation will be slightly higher than deflections assuming fully elastic behavior, while there is a partial stress redistribution over the cross section. The load for this limit is set at F = 2 x 150 kN, tension in the lower strands of concrete σx, c = + 14 MPa, stress in the upper strands of concrete σy, c = - 58 MPa with an anticipated deflection of 34 mm.
3. Achieving tension near the compressive strength of the material in the upper strands and significantly exceeding the tensile strength of the material in the lower strands, a significant increase of deflections, plasticization under the compression zone and the eventual collapse of the cross section by press violation of the concrete. The load for this limit was set at F = 2 x 210 kN, tension in the upper strands of concrete σy, c = - 90 MPa with an expected deflection of 85 mm.

The test of 3 produced beams took place in August 2013. The results were already published in 2014 [19]. The course of examination of specimens corresponded to the theoretical assumptions, while lower values were obtained. For all specimens, however, the desired torque carrying capacity in bending wasn't achieved. The elements collapsed by shear violation - separating the upper and lower flanges by cracking in the wall. In one case the reinforced concrete specimen failed during the experiment in bending, the top flange was simultaneously torn by lateral movement.

The tests were analyzed and the following conclusions were made:

The derogation in behavior of identical specimen sizes and the inserted preload can be caused by inhomogeneous properties of UHPC. Consequently, it is a matter of production technologies. Specimens with an identical load exhibited different deformations, while the prestressing was practically the same, and was verified by mounted sensors.

Technical reasons can be considered as a cause of premature failure of the sample. During the preparation of test specimens for further testing of 7 m span beams, the shear reinforcement would be used, which increases the resistance of the test specimen in shear. This specimens were, however, used as a basis for the testing of 12 meter span beams.

Figure 4. and Figure 5. The development of shear cracks during the loading tests of 7 m span beam

3.3. 12m span beams

The stages of loading for 12 meter span beams regarding the operation and distribution of internal forces in the cross-section area and their boundaries were assumed to be as follows:

1. Start of testing - the entire section is fully pressed, with completely elastic behavior of the element, the shape of the beam is determined by flexible camber caused by prestressing and creep of concrete. The beam is free from external load, the tension in the lower strands in the concrete is σy, c = - 40 MPa, stress in the upper strand in the concrete is σy, c = - 2 MPa.
2. The achievement of decompression in the lower strands of the cross-section, with fully flexible behavior, the actual deformation corresponding to Hook's law. The load for this limit was set at F = 2 x 55 kN, tension in the lower strands of the concrete σy, c = 0 MPa, stress in the upper strands in the concrete σy, c = - 42 MPa with an anticipated deflection of 55 mm.
3. Achieving stress in the lower strands before UHPC tensile strength limits, with elastic behavior of the element, the actual deformation corresponding to the number according to Hook's law. The load for this limit is set at F = 2 x 80 kN, tension in the lower strands of

IOP Conf. Series: Materials Science and Engineering **246** (2017) 012051 doi:10.1088/1757-899X/246/1/012051

the concrete σy, c = + 14 MPa, stress in the upper strands in the concrete σy, c = - 59 MPa with an expected deflection of 80 mm.

4. Achieving stress in the lower strands exceeding UHPC tensile strength limits. At this stage, a rise in the number and the moderate development of cracks is expected. The strain will be slightly increased compared with deflections determined assuming fully elastic behavior, while there will be a partial redistribution of stress over the cross section. The load for this limit was set at F = 2 x 110 kN, tension in the lower strands in the concrete σy, c = + 30 MPa, stress in the upper strands in the concrete σy, c = - 81 MPa with an anticipated deflection of 125 mm.
5. Achieving stress near compressive strength limits of the material in the upper strands, and the emphatic exceeding of the tensile strength of the material in the lower strands. Tensile stress is transmitted by prestressing reinforcement, a significant increase of deflections, plasticization in part of compression zone and eventual collapse of the cross section by the compressive violation of the concrete. The load for this limit was set at F = 2 x 170 kN, tension in the upper strands in the concrete σy, c = - 130 MPa with an expected deflection of 320 mm. Establishment of standard oblique shear cracks in the middle of the formation with vertical bending cracks of limited width.

The testing of two beams took place in 2014. The result of the test showed compliance with the assumptions of the calculation.

Figure 6. and Figure 7. Setup of the experiment, deflection and bending failure of 12 m span beams

4. Computer analysis

A computer model of an UHPC prestressed I-beam was created in a 3D environment using GiD 11 software, whereby the calculation was prepared. The calculation itself was carried out with ATENA Win Statics software [19, 20, 21]. The computer model of the specimen was modelled as a half of a symmetrical structure by means of one macro-element. The prestressing was modelled using line elements with assigned profiles of 15.7 mm according to the actual structure. The surface of the symmetry of the model is prevented from rotating about a horizontal axis (the axis perpendicular to the span) and the horizontal displacement. Steel spreading elements were added for transferring the load to the model [19]. For the modeling of steel fiber reinforced concrete the reduction coefficient of compressive strength was modified due to cracks growth control of fibers [18, 22, 23]. The loading was performed by displacement per 0.1 mm. The displacement and the corresponding force were monitored. In addition to this monitoring, the volume of concrete was monitored for checking the crack width. The model was meshed by hexahedra elements of a size of 0.05 m [19]. For the purpose of creep simulation, the B3 model [23] was chosen and the results were calculated using ATENA Win Creep software. The simulation was calculated in parallel to the 9 and 7 meter span beams. Adjusted parameters in the calculation were used in the calculation of the 12 m span beam before their testing. The setting of the model was extrapolated for the calculation of another span. In such a way a good congruence of the tests and the predicted calculation were reached.

Figure 8. Computer model - Isosurfaces of crack width

(Load bearing capacity to a value of the force 2 x 71 kN, deflection 119 mm)

Figure 9. Computer model - Isosurfaces of crack width

(Load bearing capacity to a value of the force 2 x 173 kN, deflection 57 mm)

Figure 10. Computer model - Isosurfaces of crack width

(Load bearing capacity to the value of the force 2 x 155 kN, deflection 268 mm)

5. Conclusions

5.1. Short-term loading

The prestressed I-beams of a span of 9 m failed during the experiment in bending and shear at value of the force 2 x 56,6/75,5 kN with a deflection of 81/122 mm in the middle of the span (refer with: Fig. 2 and 3). The numerical analysis showed a very similar load bearing capacity to a value of the force 2 x 71 kN with a 119 mm deflection. Isosurfaces of the cracks width at the maximum force are shown in the following figure (refer with: Fig. 8). Figure 11 shows a comparison of the L-D curves of the experiment and the FEM analysis.

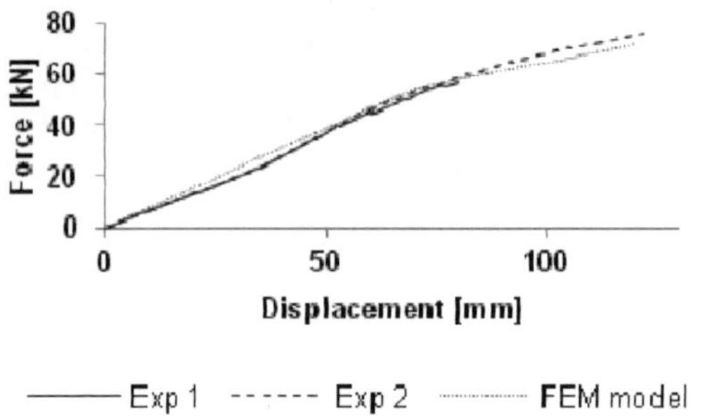

Figure 11. L-D for 9 m span beams

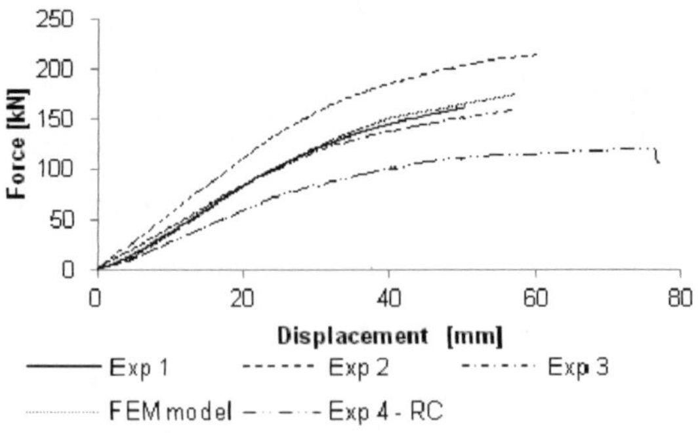

Figure 12. L-D for 7 m span beams

The prestressed I-beams of a span of 7 m failed during the experiment in shear at value of the force 2 x 161/212/158 kN with a deflection of 50/60/56 mm in the middle of the span (refer with: Fig. 4 and 5). The numerical analysis showed a very similar load bearing capacity to a value of the force 2 x 173 kN with a 57 mm deflection. Isosurfaces of the cracks width at the maximum force are shown in the following figure (refer to: Fig. 9). Figure 12 shows a comparison of the L-D curves of the experiment and the FEM analysis. The graph shows also comparison with reinforced concrete beam with the same cross section.

The final prestressed I-beams of a span of 12 m failed during the experiment in bending at a value of the force 2 x 170/171 kN with a deflection of 286/260 mm in the middle of the span (refer to: Fig. 6 and 7). The numerical analysis showed a very similar load bearing capacity to the value of the force 2 x 155 kN with a 268 mm deflection. Isosurfaces of the cracks width at the maximum force are shown in the following figure (refer to: Fig. 10). Figure 13 shows a comparison of the L-D curves of the experiment and the FEM analysis.

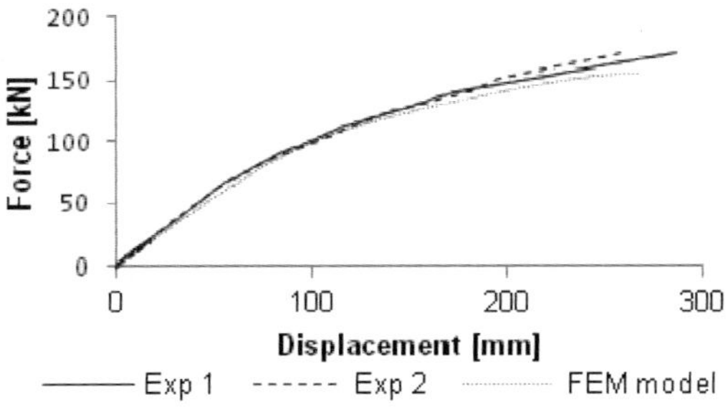

Figure 13. L-D for 12 m span beams

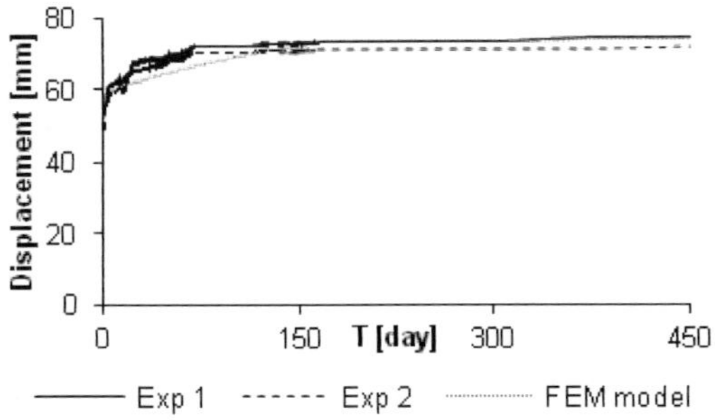

Figure 14. Creep tests of 12 m span beam

5.2. Long-term loading

For the purpose of creep testing, two specimens of a span of 12 m were loaded by a constant load of 22.6 t, which means a force of 111.8 kN for each beam. The prestressed I-beams of a span of 12 m were manufactured with initial deflection (camber) from prestressing of 75 mm in the middle of the span. Instant deflection just after the application of the load were measured as 49 / 45 mm, with a deflection after 450 days of 72 / 75 mm (refer with: Fig. 14). FEM analysis showed an instant deflection of 55 mm and a deflection in 450 days of 73 mm.

The aim of the numerical and experimental research of elements of the UHPC was to design a beam with I-shaped cross section. I-beams are intended to be fully concreted in the bridge slab, in which the monolithic part would include the full structural height of the beam, except for the bottom flange. Such a design is the suitable solution for construction of railway bridges with a span up to 15 m.

Extensive numerical and experimental research was carried out to design a UHPC I-beams for a 12 m span. First, several test specimens were made for a 9 and 7 m span. These specimens served to test beam production and were subjected to static short-term load tests. The resulting set of 12 m beams was subjected to short-term and long-term load tests. From the results of numerical and experimental analyses it can be stated that the beams thus designed can be used for the construction of railway bridges.

Acknowledgements

This project is supported by the grant GAČR17-22796S.

References

[1] Graybeal, B.: "Ultra-High Performance Concrete," FHWA Report HRT-11-038, 2011.
[2] Proceesings of the 2rd International symposium on UHPC, Kassel, March 5-7, 2008.
[3] Proceesings from HIPERMAT 3rd International symposium on UHPC and Nanotechnology for High Performance Material, Kassel, March 7-9, 2012.
[4] AFGC/SETRA; "Bétons fibrés à ultra-hautes performances. Recommandations." Documents scientifiques et techniques. Association Française de Génie Civil, Setra, Bagneux Codex, 2013.
[5] JSCE-USC; "Recommendations for Design and Construction of Ultra-High Strength Fiber-Reinforced Concrete Structures – Draft."

IOP Conf. Series: Materials Science and Engineering **246** (2017) 012051 doi:10.1088/1757-899X/246/1/012051

[6] R. Yu, P. Spiesz, H.J.H. Brouwers, "Mix design and properties assessment of Ultra-High Performance Fibre Reinforced Concrete (UHPFRC)," Cem. Concr. Res. 56, p. 29–39, 2014.

[7] SETRA/AFGC: Ultra High Performance Fibre-Reinforced Concretes: Interim Recommendations, 01/2002

[8] AFGC/SETRA: Ultra High Performance Fibre-Reinforced Comcretes, Recommandations (06/2013). Paris CEDEX.

[9] JSCE, Recommendations for Design and Construction of High Performance Fiber Reinforced Cement Composites with Multiple Fine Cracks, Rokugo K., (ed.) Concrete Eng. Series, 82, 2008.

[10] fib Model Code for Concrete Structures 2010, ©2013 fib, Ernst & Sohn.

IOP Conf. Series: Materials Science and Engineering **246** (2017) 012052 doi:10.1088/1757-899X/246/1/012052

Tensile properties of structural fibre reinforced concrete

M Tipka[1] and J Vašková[1]

[1] Department of Concrete and Masonry Structures, Faculty of Civil Engineering, CTU in Prague, Thákurova 7, 166 29 Prague 6, Czech Republic

E-mail: martin.tipka@fsv.cvut.cz

Abstract. The paper deals with the comparison of several loading tests, which are using for determination of tensile strength of cementitious composites. The paper describes several test methods, their advantages, disadvantages and possible outputs. In the experimental program several recipes of concrete and fibre reinforced concrete were tested in splitting test, 3-point and 4-point bending tests and in 2 variants of axial tension test. Tension strength ratios and conversion factors between loading tests were determined for each recipe, based on test results.

1. Introduction

Tensile strength is one of the fundamental material parameters, which is ordinarily used for structure design. It is usually determined experimentally based on destructive tests. There are several methods for testing the tensile strength of cementitious composites and their behaviour in tension. The test methods vary in their technical difficulty, accuracy and variance of results. Although all tests provide information on the strength in a particular strain, these strength values are different in different tests. A possibility of the results comparison is very important for the structural design in practice.

2. Test methods for determining the tensile strength

2.1. Splitting test

Splitting test is technically less difficult test for determining the tensile strength of cementitious composite. It can be realized on cube-shaped or cylindrical specimens (figure 1). The specimen is loaded by compressive force in constant loading race (increase of tensile splitting strength) 0.04 - 0.06 MPa/s [1]. This test provides the value of maximal material tensile splitting strength and residual tensile splitting strength depending on the transverse strain in case there are extensometers on the specimen surface (figure 2).

2.2. Bending test

Bending test is the most commonly used test for determining and verifying the tensile properties of fibre reinforced concrete and its behaviour after macrocracking. Here are a lot of arrangements of concrete and fibre reinforced concrete bending tests. These cases vary in the shape and dimensions of specimen, in supporting and load arrangement. Two modifications of bending test are standardized: 3-point bending test and 4-point bending test.

IOP Conf. Series: Materials Science and Engineering **246** (2017) 012052 doi:10.1088/1757-899X/246/1/012052

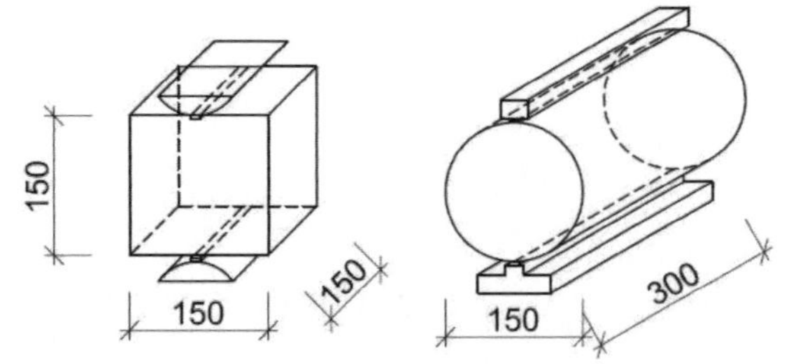

Figure 1. Shape of specimens for splitting test of concrete and fibre reinforced concrete.

Figure 2. Extensometers on the specimen surface in splitting test [2]

2.2.1. 3-point bending test. Beam specimen with standardly cross-section 150×150 mm is simply supported by span 500 mm and force loaded in the middle of span (figure 3). Un-notched or notched beams with the notch in the middle of span can be used. The width of potential notch is 5 mm a depth 25 mm [3]. This notch defines the cross-section, where the macrocrack is formed. The loading rate can be phased or it is constant throughout the test - increase of middle deflection 0.2 mm/min. Dependence load - central deflection (F_R - δ_l) is recorded during the test. In the case of notched beam it is possible to install the extensometer near the notch for notch opening displacement measuring [9]. The output is than dependence load - crack mouth opening displacement (F_R - *CMOD*). Disadvantages of this arrangement are direct definition of failure cross-section, where isn't generally the worst material properties and complicated stress near the notch. The results are higher values of measured strengths.

2.2.2. 4-point bending test. Only un-notched beams are used for 4-point bending test. The standard cross-section is 150×150 mm a length 700 mm [4]. Beams are simply supported by span 600 mm and loaded with pair of forces in thirds of span (figure 4). The loading rate can again be phased into 3 laps or it is constant throughout the test - increase of middle deflection 0.2 mm/min [8]. Dependence load - central deflection (F_R - δ_l) is recorded during the test. Advantage of this arrangement is the fact that failure occurs in the middle third of span, where shear force isn't and critical cross-section is loaded only with bending moment. Disadvantage is the fact that the failure point is unknown and it isn't possible to install the extensometers for crack mouth opening displacement measurement.

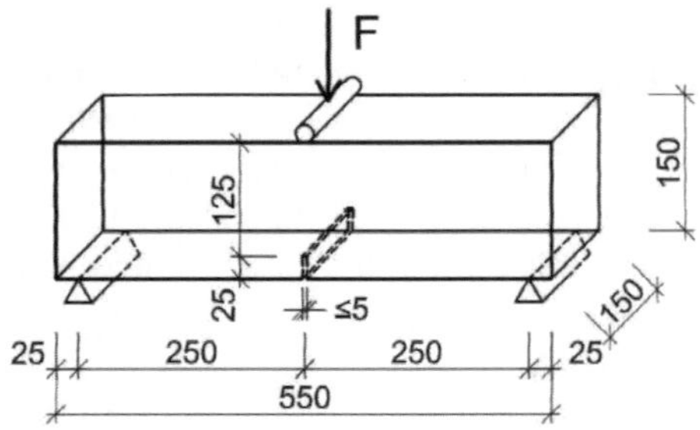

Figure 3. 3-point bending test

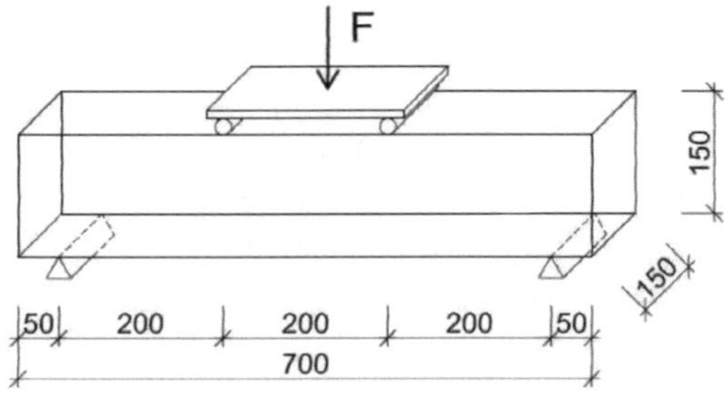

Figure 4. 4-point bending test

IOP Conf. Series: Materials Science and Engineering **246** (2017) 012052 doi:10.1088/1757-899X/246/1/012052

2.3. Axial tension test

Axial tension test isn't commonly used, because it is very technically difficult. At present there are few rules and recommendations that clearly define the test arrangement of fibre reinforced concrete in axial tension.

Recommendation RILEM [5] prescribe a notched cylindrical specimen with diameter of 150 mm and min. height of 150 mm. The notch should be performed over the whole periphery with a constant depth of 15 mm and a constant width max. 5.0 mm (figure 5). The specimen is fixed via glue joint on the ends of cylinder. In the notch there are installed three LVDT sensor, which measure notch mouth opening displacement and also control loading. Loading is divided to 2 phases with loading rate of 0.005 mm/min and 0.1 mm/min. Dependence F_R - *CMOD* is recorded. Disadvantages of this arrangement are permanent connection of specimen, which complicates quick test repeatability and again direct definition of critical cross-section. It is also suitable only for material with softening behaviour, because only one macrocrack can form.

They are also known other arrangements of axial tension test, which eliminate these disadvantages. One of them has been developed at Department of Concrete and Masonry Structures, Faculty of Civil Engineering, CTU in Prague. It use dogbone specimen with rectangular cross-section and its two-level reduction (figure 6). Longitudinal deformations of central part of specimens are measured with two sensors, which are situated on the opposite surfaces of specimen. The specimen is mechanically clamped, it allow easy test repeatability.

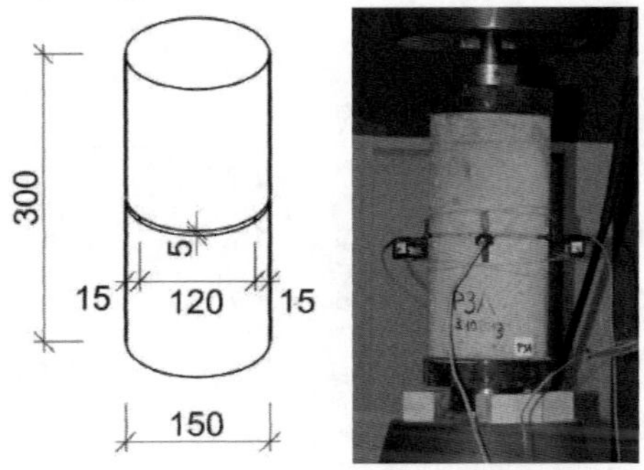

Figure 5. Axial tension test - notched cylinder

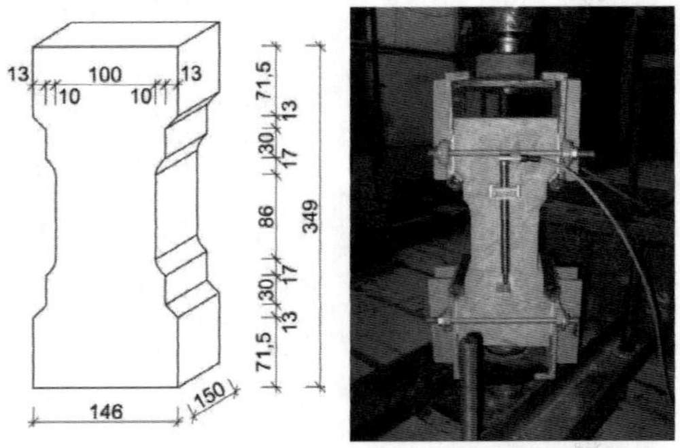

Figure 6. Axial tension test - dogbone

3. Experimental comparison of loading tests

The aim of experiments was to compare the values of tensile strength of various composites, which were determined in different loading test and quantify the conversion factors between loading tests for each tested recipe.

Six recipes of concrete and fibre reinforced concrete were tested in splitting test, 3-point and 4-point bending tests and two variants of axial tension test. All recipes had the same concrete matrix and various type and dose of fibres. Portland cement CEM I 42.5 R, granite aggregate in fractions 0/4 - 4/8 - 8/16 and superplasticizer Sika Visco Crete - 1035 CZ were used. The fibres were TriTreg 50/1,05mm, Dramix RC 80/60 BN, Fibrex A1 and Dramix 5D 65/60 BG in dose of 0 to 3.0 % by volume. Cubed specimens with edge 150 mm were used for splitting tests, the other test were arranged as shown in chapter 2 (figures 3 to 6). Only maximal loading force was recorded in splitting test. Dependence load force - central deflection (F_R - δ_t) was recorded in bending tests and dependence load force - deformation of monitored area (F_R - Δ_t) was recorded in axial tension tests.

A detailed description of the recipes and realized tests is given in [6].

3.1. General experiment results

The tests have showed clearly positive effect of fibres for specimen resistance both at and after macrocracking. The values of tensile strength and strain at macrocracking increase with increasing dose of fibres. Ductility of material increase also with dose of fibres, but type and length of fibres have significant influence.

Brittle failure at macrocracking is typical for ordinary concrete in all tests variants, there isn't possible to record the residual part of stress-strain diagram. The material with lower quality fibres or with lower dose of fibres has exhibited the softening behaviour after macrocracking in bending tests, but there was brittle failure in axial tension test. The materials with higher dose of fibres have exhibited hardening behaviour in bending test and softening behaviour in axial tension tests.

3.2. Comparison of test results, material properties

The values and ratios of tensile strengths at macrocracking for tested material are shown in table 1, the values and ratios of stain at macrocracking are shown in table 2. The values and ratios of residual tensile strengths are shown in table 3.

Table 1. Mean values of tensile strength at the macrocracking $f_{t,cr}$ [MPa].

Recipe		**splitting test** cube 150×150×150mm		**3-point bending** notched beam 150×150×600mm		**4-point bending** un-notched beam 150×150×700mm		**axial tension** notched cylinder 150×300mm		**axial tension** dogbone 146×150×349mm
		$f_{t,sp}$ [MPa]	$f_{t,sp}$ / $f_{t,ax}$	$f_{t,3P}$ [MPa]	$f_{t,3P}$ / $f_{t,ax}$	$f_{t,4P}$ [MPa]	$f_{t,4P}$ / $f_{t,ax}$	$f_{t,ax,cyl}$ [MPa]	$f_{t,ax,cyl}$ / $f_{t,ax}$	$f_{t,ax}$ [MPa]
1	reference concrete	3.60	***1.32***	4.87	***1.79***	4.84	***1.78***	3.37	***1.24***	2.72
2	0.5% TT	4.45	***1.29***	5.18	***1.51***	4.82	***1.40***	3.83	***1.11***	3.44
3	1.0% TT	4.70	***1.33***	5.93	***1.68***	5.55	***1.57***	4.06	***1.15***	3.53
4	1.0% Dr 80/60	6.20	***1.53***	-	-	6.30	***1.55***	-	-	4.06
5	3.0% TT+Fib	8.80	***1.78***	-	-	6.48	***1.44***	-	-	4.50
6	0.64% Dr 65/60	-	-	-	-	5.17	***1.58***	-	-	3.28
Standard:			***1.18***		***1.67***		***1.45***		***1.00***	

Table 2. Mean values of tensile strain at the macrocracking $\varepsilon_{t,cr}$ [‰].

Recipe		**3-point bending** notched beam 150×150×600mm		**4-point bending** un-notched beam 150×150×700mm		**axial tension** notched cylinder 150×300mm		**axial tension** dogbone 146×150×349mm
		$\varepsilon_{t,3P}$ [‰]	$\varepsilon_{t,3P}$ / $\varepsilon_{t,ax}$	$\varepsilon_{t,4P}$ [‰]	$\varepsilon_{t,4P}$ / $\varepsilon_{t,ax}$	$\varepsilon_{t,ax,cyl}$ [‰]	$\varepsilon_{t,ax,cyl}$ / $\varepsilon_{t,ax}$	$\varepsilon_{t,ax}$ [‰]
1	reference concrete	0.30	***1.43***	0.22	***1.05***	0.18	***0.86***	0.21
2	0.5% TT	0.34	***1.89***	0.29	***1.61***	0.21	***1.17***	0.18
3	1.0% TT	0.45	***1.73***	0.33	***1.27***	0.22	***0.85***	0.26
4	1.0% Dr 80/60	-	-	0.33	***1.14***	-	-	0.29
5	3.0% TT+Fib	-	-	0.27	***0.69***	-	-	0.39
6	0.64% Dr 65/60	-	-	0.22	***1.38***	-	-	0.16

The strength ratios between splitting test $f_{t,cr,sp}$ and axial tension test $f_{t,ax}$ are in the range, which is higher than the values in document [7]. It is also evident, that the ratio increases with the strength of material and with the dose of fibres.

The 3-point bending test shows higher values of tensile strength at macrocracking than 4-point bending test in all recipes. The strength ratios between bending tests ($f_{t,cr,3P}$ and $f_{t,cr,4P}$) and axial tension test $f_{t,cr,ax}$ are in ranges of values specified in documents [3] and [4], wherein the clear dose-dependency of the used fibres has not been demonstrated.

The axial tension test with notched cylindrical specimens shows higher values by 11 to 24 percent of strength at macrocracking than the test with dogbone specimens. This is because the notch defines the failure point, which may not be the place with the worst material properties.

The values of material tension strain at macrocracking $\varepsilon_{t,cr}$ are strongly dependent on the type of used test. The difference in strain $\varepsilon_{t,cr}$ between 4-point bending and axial tension is relatively small, while 3-point bending shows significant increase of this value.

Table 3. Mean values of residual tensile strength $f_{t.res}$ [MPa].

$\varepsilon_{t.res}$	Recipe		**3-point bending** notched beam 150×150×600mm			**4-point bending** un-notched beam 150×150×700mm		**axial tension** notched cylinder 150×300mm		**axial tension** dogbone 146×150×349mm
			$f_{t,3P}$ [MPa]	$f_{t,3P}$ / $f_{t,ax}$	$f_{t,3P}$ / $f_{t,4P}$	$f_{t,4P}$ [MPa]	$f_{t,4P}$ / $f_{t,ax}$	$f_{t,ax,cyl}$ [MPa]	$f_{t,ax,cyl}$ / $f_{t,ax}$	$f_{t,ax}$ [MPa]
2.0 ‰	2	0.5% TT	1.62	-	***1.11***	1.46	-	0	-	0
	3	1.0% TT	2.86	***2.42***	***1.26***	2.27	***1.92***	1.20	***1.02***	1.18
	4	1.0% Dr 80/60	-	-	-	3.28	***2.34***	-	-	1.40
	5	3.0% TT+Fib	-	-	-	3.43	***1.54***	-	-	2.23
5.0 ‰	2	0.5% TT	1.44	-	***1.12***	1.29	-	0	-	0
	3	1.0% TT	2.64	***2.34***	***1.10***	2.41	***2.13***	1.18	***1.04***	1.13
	4	1.0% Dr 80/60	-	-	-	3.67	***2.22***	-	-	1.65
	5	3.0% TT+Fib	-	-	-	3.17	***1.46***	-	-	2.17
10.0 ‰	2	0.5% TT	1.31	-	***1.17***	1.12	-	0	-	0
	3	1.0% TT	2.39	***2.28***	***1.06***	2.26	***2.15***	1.13	***1.08***	1.05
	4	1.0% Dr 80/60	-	-	-	3.50	***2.35***	-	-	1.49
	5	3.0% TT+Fib	-	-	-	2.69	***1.43***	-	-	1.88
20.0 ‰	2	0.5% TT	1.11	-	***1.13***	0.98	-	0	-	0
	3	1.0% TT	1.98	***2.08***	***1.08***	1.83	***1.93***	1.02	***1.07***	0.95
	4	1.0% Dr 80/60	-	-	-	2.98	***2.29***	-	-	1.30
	5	3.0% TT+Fib	-	-	-	1.84	***1.67***	-	-	1.10
Standard:				***1.67***			***1.45***		***1.00***	

Higher tensile strength in 3-point bending test than in 4-point bending test is also evident after macrocracking (figure 7). Comparison of test results in bending tests and axial tension test is limited to recipe 3 in case 3-point bending test (figure 7) and to recipes 3, 4 and 5 in case 4-point bending test (figure 8). Here was a brittle failure at the material with lower dose of fibres in axial tension test, so it was not possible to record residual part of F_R-Δ_t diagrams.

The differences between residual strengths in bending tests and axial tension test are significantly higher than at macrocracking (table 5). The ratios of values significantly exceed the values, which are defined in documents [3] and [4].

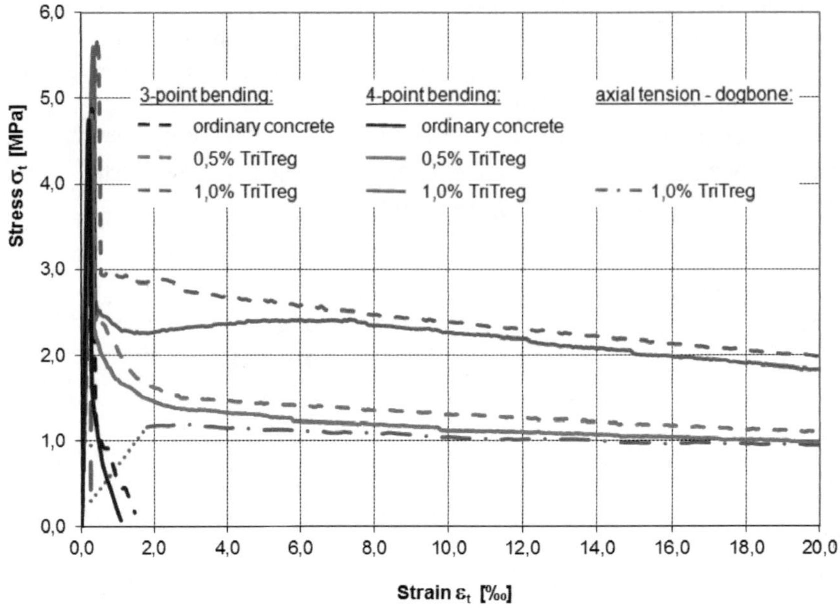

Figure 7. Comparison of stress-strain diagram in 3-point bending test, 4-point bending test and axial tension test - recipes 1, 2, 3

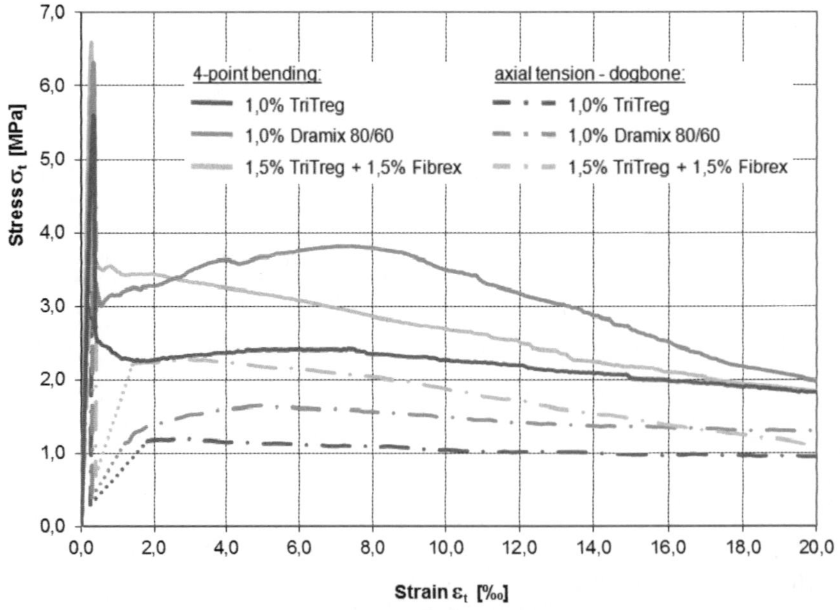

Figure 8. Comparison of stress-strain diagram in 4-point bending test and axial tension test - recipes 3, 4, 5

It was unable to derive parametric conversion relationships for tensile strength in different tests, where the variable parameter would be the value of tensile strain $\varepsilon_{t,i}$. An influence of type and dose of fibres isn't also clear and conversion relationships need to be calibrated for each specific composition of cementitious composite separately.

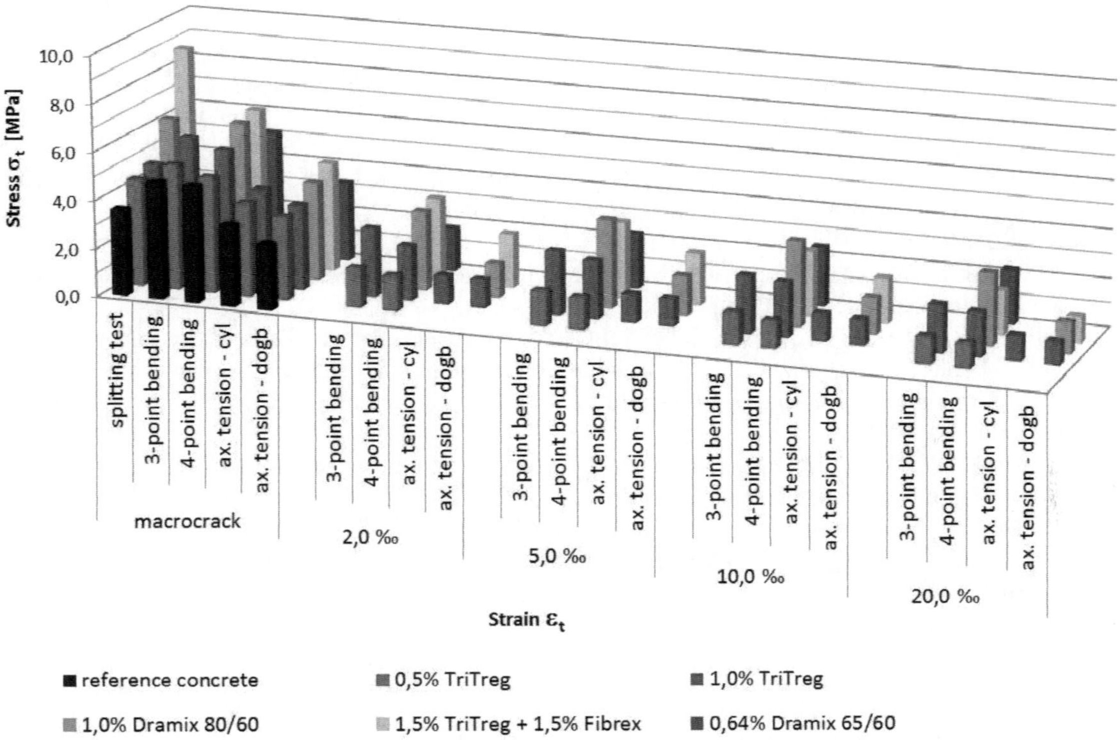

Figure 9. Evaluation of experiments - tensile strength of fibre reinforced concrete

4. Conclusion

The tensile strength of fibre reinforced composites with different recipe in different types of tests was experimentally compared. The outputs are the conversion factors between strengths for specific tested composites. It has been shown that values of conversion factors are different for each recipe and also depends on the value of deformation (both in and after macrocracking). It can be clearly recommended to determine these factors for each specific composition, if the tensile strengths are not tested in axial tension test.

Acknowledgments

This paper was prepared in support of the projects GA16-08937S „ State of stress and strain of fiber reinforced composites in interaction with the soil environment" and SGS17/049/OHK1/1T/11 "Fibre Reinforced Concrete with Specific Structures - Behavioural Analysis and Applications".

IOP Conf. Series: Materials Science and Engineering **246** (2017) 012052 doi:10.1088/1757-899X/246/1/012052

References

[1] ČSN EN 12390-6 (73 1302) Testing hardened concrete - Part 6: Tensile splitting strength of test specimens (in Czech). ÚNMZ, Prague 2009

[2] Abrishambaf A, Barros J A O, Cunha V M C F 2015 Tensile stress–crack width law for steel fibre reinforced self-compacting *Cement and Concrete Composites* **57** p 153-165 ISSN 0958-9465

[3] RILEM TC 162-TDF 2002 Test and design methods for steel fibre reinforced concrete: Bending test. Final recommendation *Materials and Structures* **35** p 579-582 ISSN 1359-5997

[4] ČSN P 73 2452: Fibre-reinforced concrete - Testing of hardened fibre-reinforced concrete. ÚNMZ, Prague 2015

[5] RILEM TC 162-TDF 2001 Test and design methods for steel fibre reinforced concrete: Uni-axial tension test for steel fibre reinforced concrete *Material and Structures* **34** p 3-6 ISSN 1359-5997

[6] Tipka M 2017 Tensile properties of structural fibre reinforced concrete. *Ph.D. Thesis*, CTU in Prague, Faculty of Civil Engineering

[7] TP FC 1-1: Technical rules 1: Fibre-reinforced concrete - Part 1: Testing of fibre-reinforced concrete - Evaluation of destructive tests and characteristic stress-strain diagram determination for desing of fibre-reinforced concrete structures. CTU in Prague and Betotech s.r.o. Prague 2009

[8] Tipka M, Vašková J, Drahorád M 2011 Effect of loading rate on the behaviour of fibre reinforced concrete elements *6th International Conference Fibre concrete 2011 - Technology, Design, Application*, CTU in Prague p. 427-433 ISBN 978-80-01-04917-4

[9] Tipka M, Vašková J 2015 Dependence of crack opening on deflection during bending test of fibre reinforced concrete *8th International Conference Fibre concrete 2015 - Technology, Design, Application* CTU in Prague p 1-8 ISBN 978-80-01-05683-7

IOP Conf. Series: Materials Science and Engineering **246** (2017) 012053 doi:10.1088/1757-899X/246/1/012053

Influence of Basalt Mesh Induced Increase of Heterogeneity of Cement Composites with Dispersed Fibers on Its Resistance under Near-Field Blast

J Zíma[1] and M Foglar[1]

[1] Department of Concrete and Masonry Structures, Czech Technical University in Prague, Faculty of Civil Engineering, Thákurova 7/2077, Prague, Czech Republic

E-mail: jakub.zima@fsv.cvut.cz

Abstract. This paper describes the influence of multiple basalt meshes in the cement composite specimens on the damage induced by near-field blast. Experimental measurements performed in the Boletice military area in 2014 and 2015 are evaluated by numerical simulations. The evaluation of the results is mainly focused on the stress propagation in the cement composite with dispersed fibers, the propagation of the overpressure caused by the blast and velocity of the ejected parts from the specimen. The influence of the presence and position of the basalt meshes in the specimen on its damage induced by delamination is also examined.

1. Experimental program

The experiments were performed in the Boletice military training area, in cooperation with the Armed Forces of the Czech Republic, in fall 2014 and 2015. The location and the layout of the experiment was very similar to the authors' previous experiments [1, 2, 3], though some improvements were gradually made to the layout. Two specimens in total were tested; consecutively named no. 17 and no. 18 with respect to the history of the experimental program. Specimens no. 17 and 18 were concrete slabs 6 m in length, 1.5 m in width and 0.3 m in thickness with the same amount of steel reinforcement, i.e. 11 pcs ∅ 16 mm reinforcing bars every 140 mm on both surfaces, ∅ 10 mm every 150 mm as an outer transverse reinforcement, and shear reinforcement was provided by ∅ 8 mm links (9 pcs/m^2). Concrete cover of 50 mm was taken to the surface of the transverse reinforcement. Reinforcement arrangement was essentially the same as in [1, 2, 3]. The scheme of the experiment is shown in Fig. 1. The ground beneath the slabs was excavated to a depth of 2 m to avoid the results being influenced by the shock wave bouncing off the ground. 25 kg TNT charges were placed on steel chairs in the middle of each slab. The chairs provided 450 mm standoff from the slab. Information on the instrumentation used during the experiment can be found in Kovar et al. [4], Foglar et al. [5] and Hajek et al. [6].

IOP Conf. Series: Materials Science and Engineering **246** (2017) 012053 doi:10.1088/1757-899X/246/1/012053

Figure 1. Specimen P19

Specimen No. 17 and No. 18 were made of ultra-high performance fiber-reinforced concrete (UHPFRC). Two types of high-strength steel (HSS) fibers were used: 35 mm in length and 13 mm in length. The yield strength of the fiber material is guaranteed by the manufacturer to be higher than 2200 MPa.

Specimens contained basalt fiber meshes in multiple layers along the depth of the specimen in distances about 50 mm (Fig. 2). The melting point of the basalt fibers is 1350 °C. The tensile strength of the basalt fabric was about 4200 MPa with a tensile modulus of elasticity of about 85 GPa. The unit weight of the mesh is 250 g/m^2 with weight density of 2.67 g/cm^3. Each string of the mesh consists of multiple fibers so the diameter of each string cannot be accurately defined. From parameters presented above, an effective area of each string equal to 1.4 mm^2 was calculated instead.

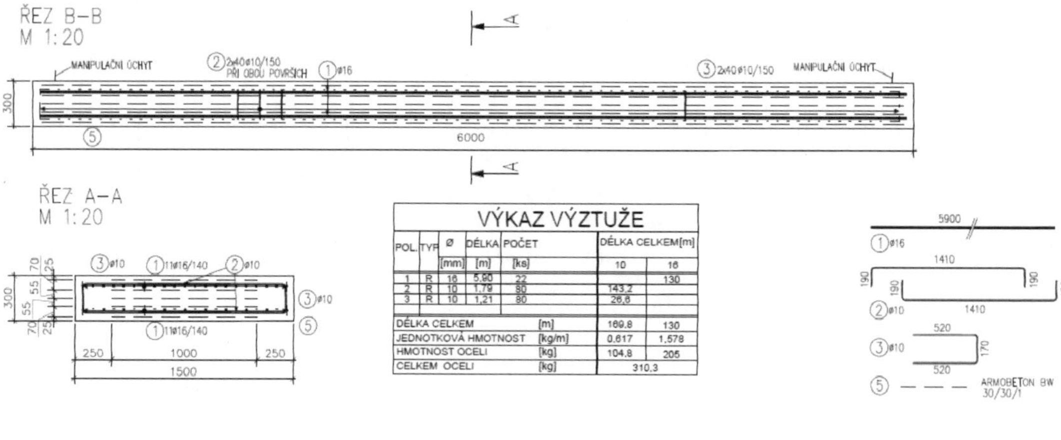

POL.	TYP	Ø [mm]	DÉLKA [m]	POČET [ks]	DÉLKA CELKEM [m] 10	DÉLKA CELKEM [m] 16
1	R	16	5,90	22		130
2	R	10	1,79	80	143,2	
3	R	10	1,21	80	26,6	
DÉLKA CELKEM				[m]	169,8	130
JEDNOTKOVÁ HMOTNOST				[kg/m]	0,617	1,578
HMOTNOST OCELI				[kg]	104,8	205
CELKEM OCELI				[kg]	310,3	

Figure 2. Concrete details drawing Specimen No. 18

IOP Conf. Series: Materials Science and Engineering **246** (2017) 012053 doi:10.1088/1757-899X/246/1/012053

2. Numerical simulations

Numerical simulations were performed in ANSYS LS-DYNA software. Near-field blast was simulated with the build-in function Load Blast Enhanced which converts the blast load into pressure impulse. Using axis symmetry of the experiment, one quarter of the specimen is considered. Parameters of the used material correspond to the parameters of concrete with cylindrical compressive strength of 80 MPa. Individual parameters of concrete were introduced into the program using the cubic strength and fracture energy of the material model. High-performance concretes with a strength up to 120 MPa were used in experimental measurements. However, the material model used allows the setting of cylindrical compressive strength only 80 MPa [7, 8]. Nevertheless, this fact does not influence the outcomes of the numerical study, since the purpose is to study the effect of the basalt mesh layers, not the effect of the compressive strength of concrete.

Three types of numerical models were created:

- Model with reinforced concrete and dispersed fibers
- Model with reinforced concrete, dispersed fibers and one layer of basalt mesh at the soffit of the specimen in the half of the reinforcement cover
- Model with reinforced concrete, dispersed fibers and five layers of basalt meshes as seen in Figure 2

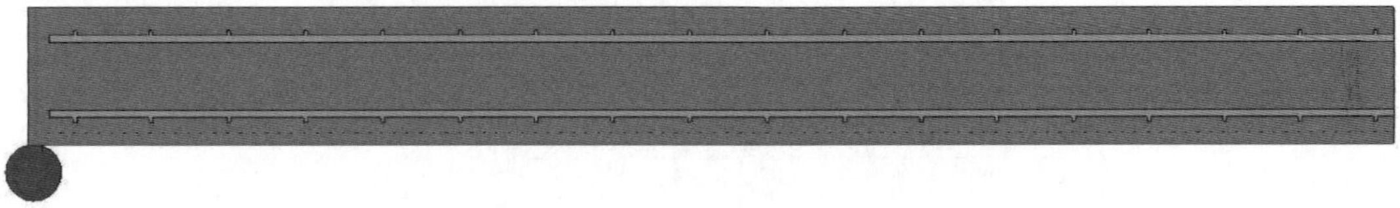

Figure 3. Layout of basalt meshes, 1 layer at the soffit of the specimen in the half of the reinforcement cover

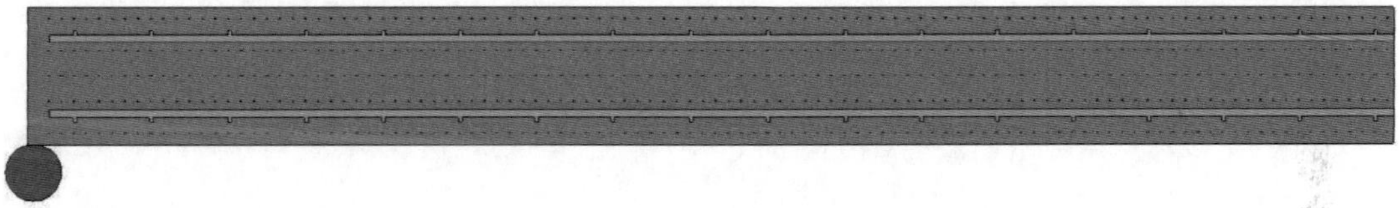

Figure 4. Layout of basalt meshes, 5 layers, as presented in Figure 2

For velocity evaluation of the ejected parts of the specimen and comparison with the experimental measurement, points were determined on the lower surface of the specimen. During the the experimental measurement, the points were measured in the middle of the specimen span, 150, 300 and 450 millimetres from the center. A specimen of five layers of basalt mesh was used for comparison of the experimental measurement and the numerical model. Numerical model with one layer of basalt mesh could not be compared with experimental measurement performed by the PDV device (see below), only macroscopic damage could be performed.

As can be seen in Figure 7, the velocity of the ejected parts in the middle of the span, 150 and 300 millimetres from the center do not correspond to the numerical models. This difference in velocity is mainly caused by the presence of the steel chair (on which the charge is placed). The chair behaves as a projectile, which is why the velocity of the scabbed parts is higher than in the places where there is no effect of the chair. For this reason, the numerical model could not be exactly compared with one layer of mesh in the points where the effect of the steel chair is predominant. The velocity of the flying debris only in the middle of the span was measured during the experimental measurement.

IOP Conf. Series: Materials Science and Engineering **246** (2017) 012053 doi:10.1088/1757-899X/246/1/012053

- Channel position CH1, Node 332, Coordinate [3000; 0; 0]
- Channel position CH2, Node 317, Coordinate [2850; 0; 0]
- Channel position CH3, Node 302, Coordinate [2700; 0; 0]
- Channel position CH4, Node 289, Coordinate [2550; 0; 0]

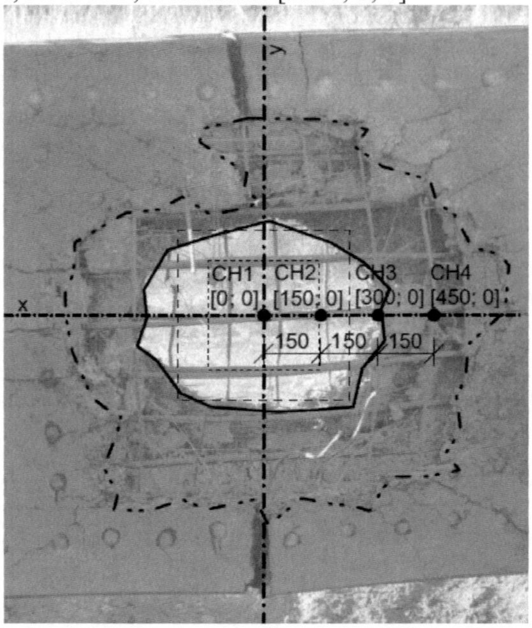

Figure 5. Experimental results for the specimen no. 18 – Damaged specimen; channel position scheme from the measurement

Figure 6. Position of the measurement points, view from the soffit of the specimen; measurement point are located at the centre-line of symmetry

In the following graphs, the influence of the basalt mesh layers in the specimen can be observed with respect to the velocity of the ejected parts and then the comparison with experimental measurements is performed.

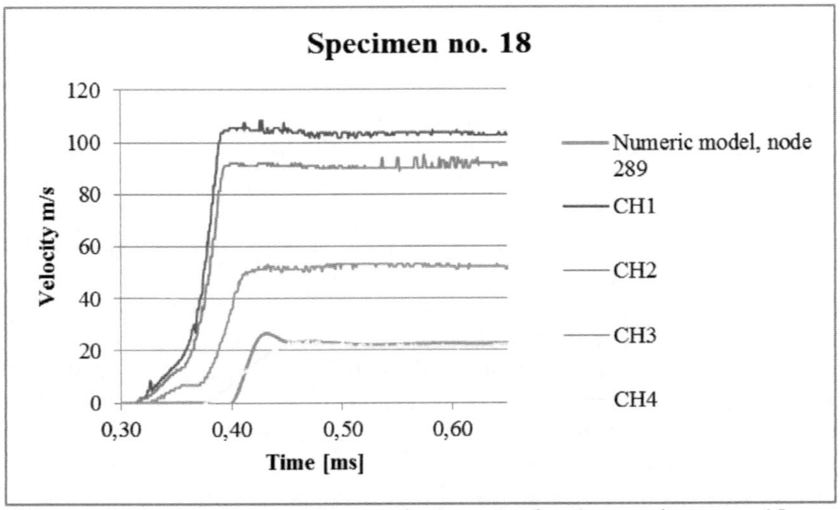

Figure 7. Experimental and numerical results for the specimen no. 18

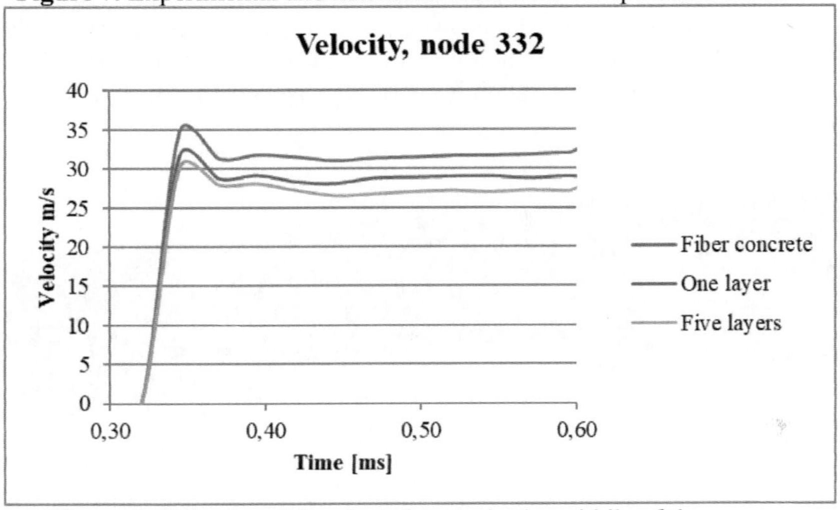

Figure 8. Velocity of the ejected parts, in the middle of the span

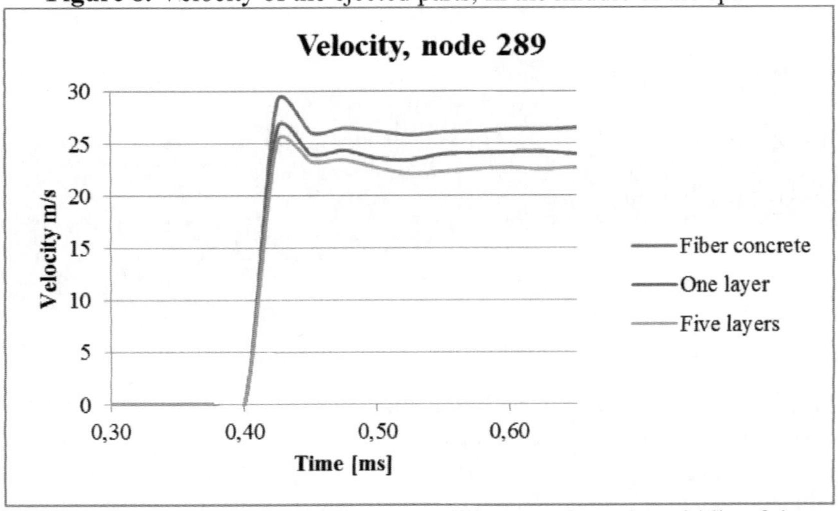

Figure 9. Velocity of the ejected parts, 450 millimetres from the middle of the span

The graphs indicate the effect of basalt meshes on the velocity of the ejected parts from the specimen. In the middle of the span and 150 millimetres from the center of the specimen, the velocity of the scabbed parts is higher. This effect is most probably caused by the influence of a flying steel chair. The half of the steel chair has size 250 millimetres. The measurement point 300 millimetres from the middle of the specimen is affected by the load distribution. At the measurement point where the velocity of the ejected parts is half compared to the measured points under the impact of the chair. The compliance with the experimental measurement at the measurement point was reached at a distance of 450 millimeters from the midpoint of the specimen span, because there is no direct influence of the steel chair on the velocity of the ejected parts.

Damage to the soffit has increased with the number of nets, the scabbed parts are smaller in size. This can increase the effectiveness in protection of persons.

In Figures 10 to 13, the initialization of the delamination of the specimen without, with one and with five basalt meshes can be observed. The difference of the material properties of the concrete and basalt causes internal reflection of the overpressure wave. The delamination of the specimen in numerical model is in good agreement with the results of experimental measurement. In the specimen without basalt mesh, the section is delaminated only at the steel reinforcement. In the specimen with one layer basalt mesh, delamination occurs in the area of the concrete reinforcement at the top surface and at the basalt mesh. At the specimen with five layers of basalt meshes, there is a clear delamination near all the basalt meshes. The delamination of the specimen with basalt meshes does not occur only around the plan center of the specimen, but it is distributed over a larger specimen area from the centre point of the blast loading.

Figure 10. Specimen without basalt meshes – delamination at steel reinforcement

Figure 11. Specimen with basalt mesh in the middle of the concrete cover, delamination of the specimen due to the placement of basalt meshed and at steel reinforcement

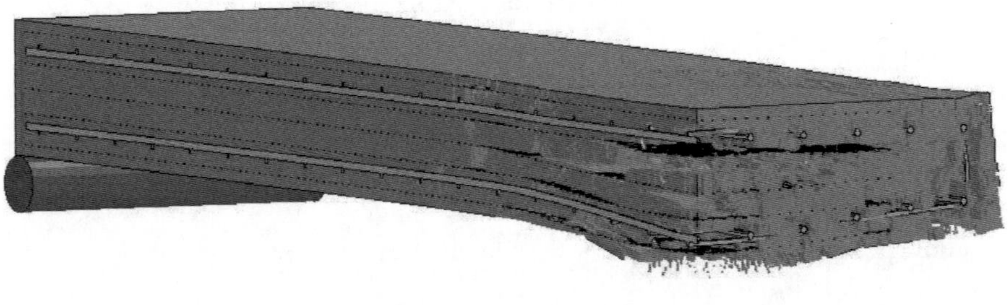

Figure 12. Specimen with five basalt meshes, delamination of the specimen due to the basalt meshes and steel reinforcement

Figure 13. Delamination of the specimen due to the presence of basalt meshes, experimental finding

Evaluation of the stress development in the specimen during the blast showed no influence of basalt meshes on the decreasing of compression on the upper face of the specimen. Stress values are practically identical in all measured points from blast initialization to 0.5 ms. From this time, the stress in evaluated numerical models starts to change. Course of the stress damping is similar for each numerical model but differs in the scale. Pressure in the measured points reaches the initial phase value of 80 MPa which is considered as a strength limit for the material.

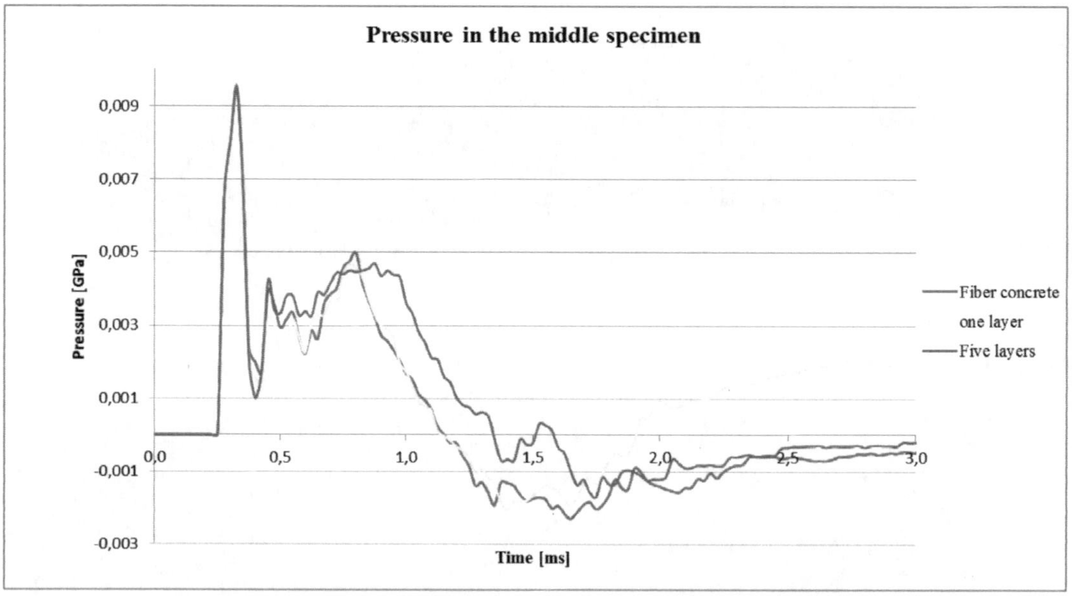

Figure 14. Pressure in the specimen, in the middle of the span

3. Conclusion

The increase of heterogeneity by addition of the basalt meshes into concrete can decrease the volume of concrete debris in the case of the blast. Placement of basalt reinforcements can reduce the velocity of the ejected parts. When comparing the velocity of the scabbed parts with experimental measurements, match can only be achieved at the measurement points where the steel chair does not interact with the specimen. The steel chair behaves as a projectile and the velocity of the scabbed parts at the impact site is several times higher. The basalt meshes increase heterogeneity of the material, increase the number of internal overpressure propagation rebounds and therefore helps to reduce the size of the ejected parts. The pressure in the specimen is not significantly changed by the basalt meshes. A further important step for precise modelling is the incorporation of the steel chair in the phenomenon, which significantly affects the extent of damage and the rate of the scabbed parts of the specimen at the point of its impact.

Acknowledgements

Financial support from Grant Agency of the Czech Republic project no. 17-23067S is gratefully acknowledged.

References

[1] Foglar, M., Kovar, M. Conclusions from experimental testing of blast resistance of FRC and RC bridge decks (2013), International Journal of Impact Engineering, 59, pp. 18-28. DOI: 10.1016/j.ijimpeng.2013.03.008

[2] Foglar, M., Hajek, R., Kovar, M., Štoller, J. Blast performance of RC panels with waste steel fibers (2015) *Construction and Building Materials*, 94, art. no. 6897, pp. 536-546. DOI: 10.1016/j.conbuildmat.2015.07.082

[3] Foglar, M., Hajek, R., Fladr, J., Pachman, J., Stoller, J. Full-scale experimental testing of the blast resistance of HPFRC and UHPFRC bridge decks (2017). Construction and Building Materials.

[4] Kovar, M., Foglar, M., Hajek, R., Blast performance of real-scale reinforced concrete specimens with varying fiber type and content. In: *WIT Transactions on The Built Environment*, 2014, 141, p. 159-169.

[5] Foglar, M., Pachman, J., Pelikan, V., Kunzel, M., Kovar, M., The structural response of a reinforced concrete specimen subjected to adjacent blast loading, In: *WIT Transactions on the Built Environment*. 2014, 141, p. 171-179.
[6] Hajek, R., Kovar, M., Foglar, M., Pachman, J., Stoller, J., Field testing of concrete members subjected to contact and adjacent blast. In: *Advanced Materials Research, 2015*, 1106, p. 164-167.
[7] LS-DYNA (2006). Theory Manual. Livermore Software Technology Corp. (LSTC).
[8] LS-DYNA (2007). Keyword User's Manual, Version 971. LSTC.

IOP Conf. Series: Materials Science and Engineering **246** (2017) 012054 doi:10.1088/1757-899X/246/1/012054

Production of a textile reinforced concrete protective layers with non-woven polypropylene fabric

J Žák[1], P Štemberk[1] and J Vodička[1]

[1] Czech Technical University, Thákurova 7, 166 29 Prague 6, Czech Republic

E-mail: jakub.zak@fsv.cvut.cz

Abstract. Textile concrete with nonwoven polypropylene fabric can be used for protective layers of reinforced concrete structures, reducing the thickness of the cover layer or reducing the water penetration rate into the structure. The material consists of cement matrix with fine-grained aggregate and nonwoven textile reinforcement. The maximum grain size of the mixture suitable for the nonwoven textile infiltration is 0.25 mm. The interlayer contains larger aggregates and short fibers. Tensile loading causes a large amount of microcracks in the material. The material can withstand strain over 25% without collapsing. Increased quality and water-cement ratio reduction was achieved using the plasticizers and distribution of the mixture into a fabric using a vibrating trowel. It is possible to make flat plates and even curved structures from this material. Larger curvatures of structures should be solved by cutting and overlapping the fabric. Small curvatures can be solved within the deformability of the fabric. Proper infiltration of the cement mixture into the fabric is the most important task in producing this material.

1. Introduction

Textile reinforced concrete is currently used as a load bearing material [1], a part of sandwich panels [2, 3], or to reinforcing an existing structure [4]. The large advantage of textile reinforced concrete is increased tensile strength, high deformability and crack width reduction.

Textile reinforced concrete is thin-walled composite material which consists of cement matrix with fine-grained aggregate and textile reinforcement. Nonmetallic reinforcement allows small thickness of the structure. Carbon, alkaline-resistant glass or polypropylene fabric are the most commonly used. Fiber cords of reinforcing fabrics are oriented orthogonally. Effectiveness of the reinforcement is reduced by the inadequate activation of the internal fibers inside the cords [5]. This problem is solvable by nonwoven textile.

2. Production of a material with a nonwoven polypropylene fabric

Textile reinforced concrete with oriented fibers is produced in two ways. A layer of cement mortar is applied on the formwork and a reinforcing fabric is pressed into it. Or, a 3D reinforcement insert is prepared in advance and it is filling by a suitable cement mixture. Textile reinforced concrete with nonwoven fabric can be produced only as a layered structure.

Oriented fiber textiles have defined fiber distances such as 4.3 x 4.3 mm. Therefore, the maximum grain size of the mixture can be estimated. Nonwoven fabric is consisted of randomly oriented fibers with very small distances. Polypropylene nonwoven fabric was used for the experiment (figure 1). The

IOP Conf. Series: Materials Science and Engineering **246** (2017) 012054 doi:10.1088/1757-899X/246/1/012054

cement mixture infiltration into the nonwoven is influenced by grain size, water quantity, chemical admixtures and method of compaction.

Figure 1. Polypropylene nonwoven fabric.

2.1. Infiltration of aggregate into the fabric

The maximum grain size of the mixture was determined experimentally because surface tension reduces water permeability (figure 2). Mixtures with a maximum grain size of 1 mm, 0.5 mm and 0.25 mm were prepared. Aggregates larger than 0.25 mm were not able to infiltrate the fabric (figure 3). The ideal size of the grain for the tested fabric was determined to 0-0.1 mm after further tests. A mixture with this size of aggregate is not suitable for the entire volume of material. Therefore, two different mixtures have to be used for the production of nonwoven multi-layered textile reinforced concrete.

Figure 2. A water drop on the surface of a nonwoven polypropylene fabric.

Figure 3. Infiltrated nonwoven textile surfaces.

2.2. Water-cement ratio and chemical admixtures

The amount of water and chemical admixtures influences significantly the process of mixture infiltration into the fabric. A larger amount of water (w/c ratio ≥ 0.5) facilitates the penetration of the mixture. After the saturation of the fabric, the excess water is accumulating on the surface of the textile-reinforced layer (figure 4) and degrades the material properties. Chemical admixtures can reduce the water-cement ratio to about 0.35. Plasticizing admixtures usually have retarding effects that slow down the production process. Using large amounts of plasticizing admixtures is common in textile reinforced concrete production [6].

Figure 4. Excess water on the fully saturated fabric.

2.3. Compaction

The mixture is distributed within the fabric by vibration. Initially, the use of the vibratory table was tested. The mixture was laid on the fabric during vibration. Water infiltrated the fabric, but the aggregate began to move horizontally across the fabric surface. The next method involved pressing the mixture into the fabric with a vibrating trowel. But, the fabric was compressed and the mixture could not penetrate it. Therefore, a suitable method is to use a vibrating trowel to roll the mixture across the fabric. This method ensures infiltration of all mixture components (figure 5). This compaction method can reduce the w/c ratio even below 0.35.

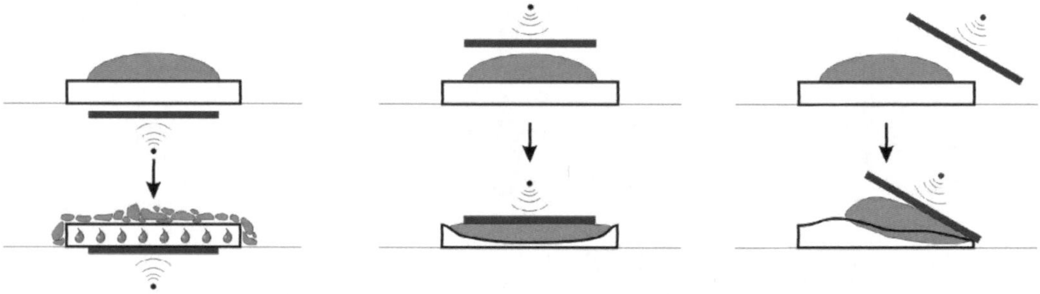

Figure 5. Vibration methods and material response.

The quality of the mixture infiltration into the fabric determines total material properties. This quality can be assessed on the material cut surface. Inadequate infiltrated fibers stand out of the cut surface. Well infiltrated fibers are not visible at first glance of the cut surface (figure 6).

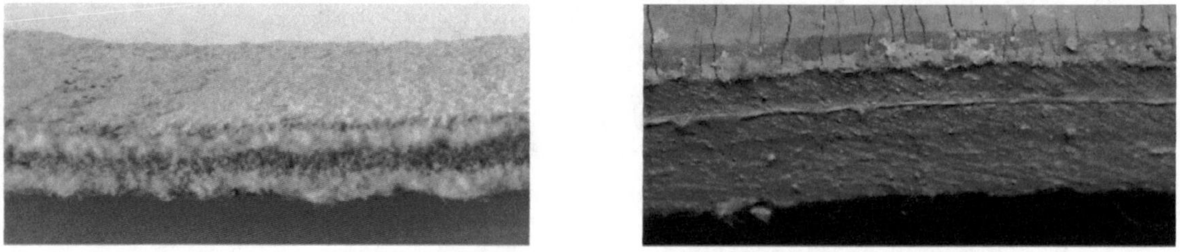

Figure 6. Material cut surfaces:. left – inadequate infiltration; right – suitable infiltration.

IOP Conf. Series: Materials Science and Engineering **246** (2017) 012054 doi:10.1088/1757-899X/246/1/012054

2.4. Production process

The material is produced by layering the infiltrated textiles. It is also possible to make an interlayer from the mixture with a larger aggregate. Shear cohesion should be improved by using very short filaments (e.g. 6 mm long).

Flat prefabricated plate can be produced by infiltration of mixtures into the fabric directly in formwork. It is better to create space-curved elements by layering already infiltrated fabrics in the formwork, when the fabric must be pre-shaped into curved shape. Large curvatures can be solved by cutting and overlapping the fabric (figure 7). A small curvature can be solved within the deformability of the polypropylene textile (figure 8). The maximum usable stretching in this case is 20%, which can be determined from the uniaxial tensile test (figure 9).

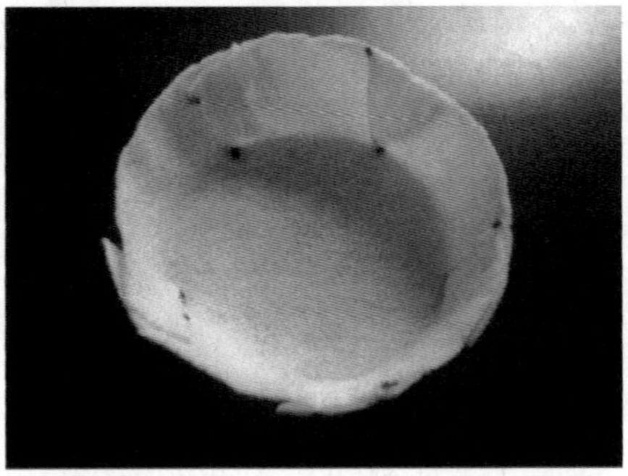

Figure 7. Cut and overlapping fabric.

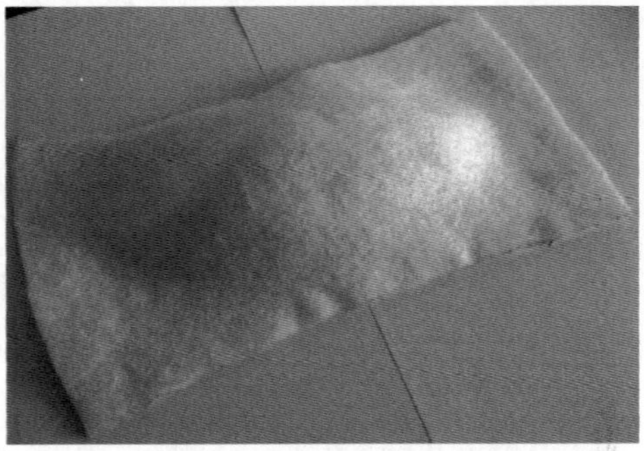

Figure 8. Textile deformability.

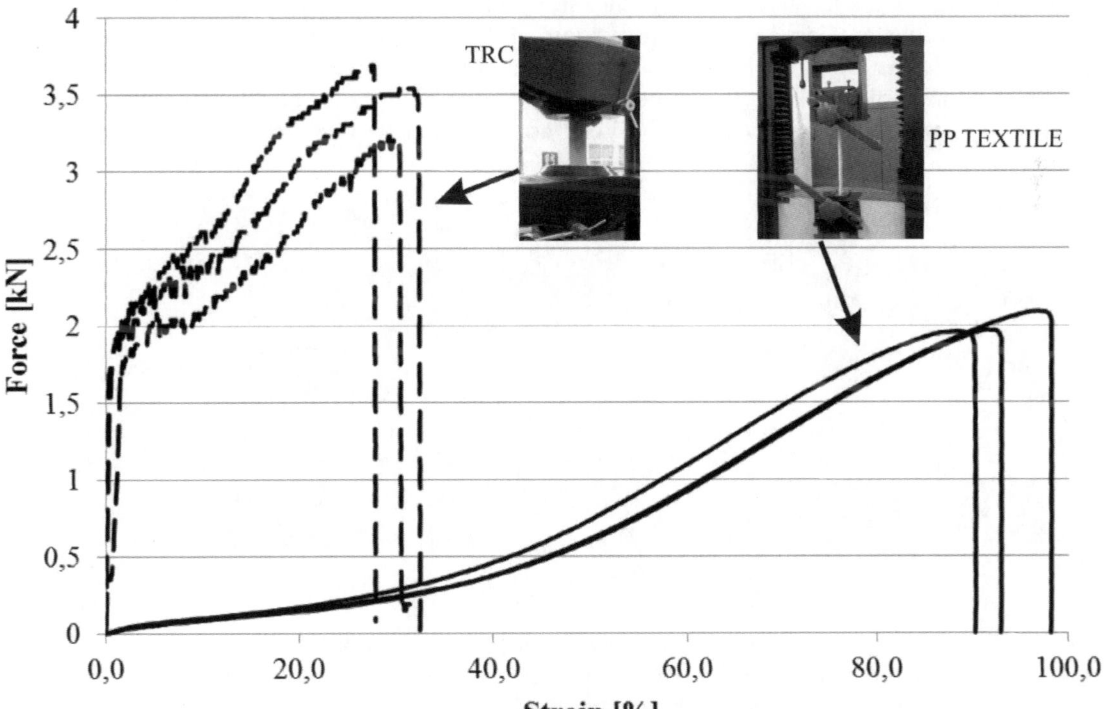

Figure 9. Results of uniaxial tensile tests.

3. Possible application

Textile concrete with polypropylene nonwoven reinforcement layers is suitable for protective layers of reinforced concrete structures as only microcracks are generated within the textile concrete layer during large deformations (figure 10). The macrocrack in reinforced concrete is then branched to more small cracks in the protective layer of textile concrete (figure 11) [7]. Reinforcing layers with glass and carbon fabric may cause tearing of the original cover layer due to their higher stiffness [8]. However, the protective layer with the polypropylene textile is less stiff than the protected structure and thus this risk is reduced.

Figure 10. Microcracks.

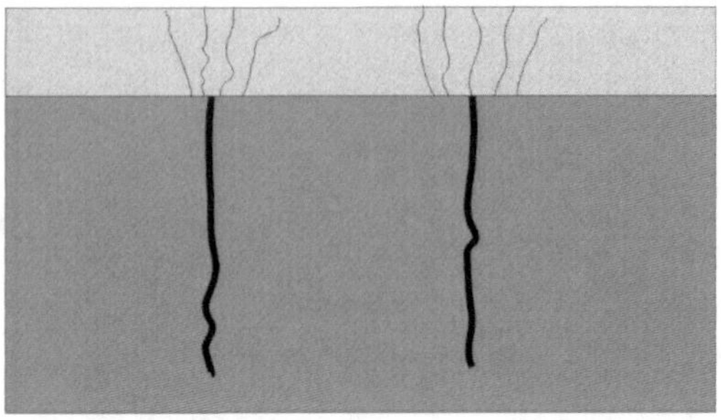

Figure 11. Cracks development.

One possible application is reducing the thickness of the cover layer. For example, the above-ground parts of noise barriers according to the Technical Quality Conditions of the Czech Railways have a minimum covering thickness of 45 mm due to environmental conditions (XF4 + XD3), 50 years of lifetime and other resistance requirements. Non-metallic reinforcement in textile reinforced concrete can reduce the cover layer or even replace completely the metallic reinforcement and thereby reduce significantly the overall thickness and transport weight of the wall panel.

Another possible application is reconstruction of bridge pillars with broken cover layer and corroded reinforcement (figure 12). Rehabilitation of pillars normally involves removal of surface concrete, welding of new reinforcement and concreting new sufficient layer of great quality concrete. Applying a protective layer of textile reinforced concrete can prevent the repeated damage of the covering layer.

Figure 12. Pillar with damaged cover layer.

The surface layer of textile reinforced concrete reduces the water absorption into structure [9]. This can be used for making white tanks. Textile reinforced concrete can be used as a permanent formwork and at the same time it can reduce water penetration into the structure. Water penetration management is an important factor for the correct functionality of the crystallization additives.

IOP Conf. Series: Materials Science and Engineering **246** (2017) 012054 doi:10.1088/1757-899X/246/1/012054

4. Conclusions

In this paper, a method of production and application of textile reinforced concrete with polypropylene nonwoven fabric has been presented. The material is produced by layering the reinforcing and filler layers. The maximum grain size for infiltrating the nonwoven fabric was set at 0.25 mm. The mixture for the filler layer may include larger aggregates and short fibers to improve shear cohesion. The best method for textile infiltration is rolling the mixture over the fabric with a vibrating trowel. Sufficient mixture infiltration into the fabric was achieved with 0-0.1 mm aggregate, plasticizing additives and a water-cement ratio of 0.35.

Textile reinforced concrete with polypropylene fabric has the density between 1200-1500 kg/m^3. The material is suitable as a protective layer for reinforced concrete structures as many microcracks are created within the protective layer. The tested material was able to deform over 25%. A possible application is reducing the cover layer thickness for outdoor structures, rehabilitation of surface layers of structures or reduction of the rate of water penetration into waterproof reinforced concrete structures.

Acknowledgments

This work was financially supported by the Czech Technical University in Prague, project SGS17/048/OHK1/1T/11, Czech Science Foundation, project 15-11753S and 16-08937S, which are gratefully acknowledged.

References

[1] Hegger J, Will N, Bruckermann O, Voss S, *Load–bearing behaviour and simulation of textile reinforced concrete*, Materials and Structures 39 (2006) 765-776

[2] Shams A, Horstmann M, Hegger J, *Experimental investigations on Textile-Reinforced Concrete (TRC) sandwich sections*, Composite Structures 118 (2014) 643-653

[3] Shams A, Hegger J, Horstmann M, *An analytical model for sandwich panels made of textile-reinforced concrete*, Construction and Building Materials 64 (2014) 451–459

[4] Brückner A, Ortlepp R, Curbach M, *Textile reinforced concrete for strengthening in bending and shear*, Materials and Structures 39 (2006) 741–748

[5] Si Labri A, Agbossou A, Hamelin P, *Experimental and numerical investigations about textile-reinforced concrete and hybrid solutions for repairing and/or strengthening reinforced concrete beams*, Composite Structures 99 (2013) 152–162

[6] Colombo I G, Magri A, Zani G, Colombo M, di Prisco M, *Textile Reinforced Concrete: experimental investigation on design parameters*, Materials and Structures 46 (2013) 1933–1951

[7] Mechtcherine V, Lieboldt M, *Permeation of water and gases through cracked textile reinforced concrete*, Cement & Concrete Composites 33 (2011) 725–734

[8] Brückner A, Ortlepp R, Curbach M, *Anchoring of shear strengthening for T-beams made of textile reinforced concrete (TRC)*, Materials and Structures 41 (2008) 407–418

[9] Lieboldt M, Mechtcherine V, *Capillary transport of water through textile-reinforced concrete applied in repairing and/or strengthening cracked RC structures*, Cement and Concrete Research 52 (2013) 53–62

Institute of Physics
Dirac House, Temple Back
Bristol BS1 6BE UK

ISSN: 1757-8981
ISBN 978-1-5108-4994-5

9 781510 849945

CURRAN ASSOCIATES INC.
proceedings.com

International Conference on Mechanical Engineering, Energy, and Advanced Materials (ICMEEAM 2018)

IOP Conference Series: Materials Science and Engineering
Volume 606

Kota Kinabalu, Malaysia
29 - 30 November 2018

IOP
Institute of Physics